Kern's
Process Heat Transfer
Second Edition

Scrivener Publishing
100 Cummings Center, Suite 541J
Beverly, MA 01915-6106

Publishers at Scrivener
Martin Scrivener (martin@scrivenerpublishing.com)
Phillip Carmical (pcarmical@scrivenerpublishing.com)

Kern's
Process Heat Transfer
Second Edition

Ann Marie Flynn
Toshihiro Akashige
Louis Theodore

Scrivener
Publishing

WILEY

This edition first published 2019 by John Wiley & Sons, Inc., 111 River Street, Hoboken, NJ 07030, USA and Scrivener Publishing LLC, 100 Cummings Center, Suite 541J, Beverly, MA 01915, USA
© 2019 Scrivener Publishing LLC . First edition © Geoffrey L. Kern.
For more information about Scrivener publications please visit www.scrivenerpublishing.com.

1st edition (1950), 2nd edition (2019)

Wiley Global Headquarters
111 River Street, Hoboken, NJ 07030, USA

For details of our global editorial offices, customer services, and more information about Wiley products visit us at www.wiley.com.

Limit of Liability/Disclaimer of Warranty
While the publisher and authors have used their best efforts in preparing this work, they make no representations or warranties with respect to the accuracy or completeness of the contents of this work and specifically disclaim all warranties, including without limitation any implied warranties of merchantability or fitness for a particular purpose. No warranty may be created or extended by sales representatives, written sales materials, or promotional statements for this work. The fact that an organization, website, or product is referred to in this work as a citation and/or potential source of further information does not mean that the publisher and authors endorse the information or services the organization, website, or product may provide or recommendations it may make. This work is sold with the understanding that the publisher is not engaged in rendering professional services. The advice and strategies contained herein may not be suitable for your situation. You should consult with a specialist where appropriate. Neither the publisher nor authors shall be liable for any loss of profit or any other commercial damages, including but not limited to special, incidental, consequential, or other damages. Further, readers should be aware that websites listed in this work may have changed or disappeared between when this work was written and when it is read.

Library of Congress Cataloging-in-Publication Data

Names: Flynn, Ann Marie, author. | Kern, Donald Quentin, 1914- author. |
  Akashige, Toshihiro, author. | Theodore, Louis, author.
Title: Kern's process heat transfer / Ann Marie Flynn, Toshihiro Akashige,
  Louis Theodore.
Description: Second edition. | Hoboken, New Jersey : John Wiley & Sons, Inc.
  ; Salem, Massachusetts : Scrivener Publishing, [2018] | Includes
  bibliographical references and index. |
Identifiers: LCCN 2018041527 (print) | LCCN 2018044072 (ebook) | ISBN
  9781119364177 (ePub) | ISBN 9781119364832 (Adobe PDF) | ISBN 9781119363644
  (hardcover)
Subjects: LCSH: Heat--Transmission | Chemical processes. | Thermodynamics. |
  Heating
Classification: LCC TP363 (ebook) | LCC TP363 .F59 2018 (print) | DDC
  621.402/2--dc23
LC record available at https://lccn.loc.gov/2018041527

Cover images: Russell Richardson
Cover design by: Russell Richardson

Set in size of 11pt and Minion Pro by Exeter Premedia Services Private Ltd., Chennai, India

10 9 8 7 6 5 4 3 2 1

Donald Q. Kern
1914–1971

Donald Quentin Kern was born in New York City in 1914. He studied at the Massachusetts Institute of Technology and received his Bachelor's, Master's and Ph.D. in Chemical Engineering from the Polytechnic Institute of Brooklyn (now New York University Tandon School of Engineering) in 1942. He was employed by Foster Wheeler from 1940 to 1947, and became Director of the Process Engineering Division at the Patterson Foundry & Machine Company in 1948.

In 1950, he published what is now considered the landmark text, *Process Heat Transfer*. As the first applied heat transfer book, engineers worldwide have come to know *Process Heat Transfer* as the definitive applied heat transfer reference. Eventually, the term "process heat transfer" became a recognized specialty within heat transfer, particularly for chemical engineers.

In 1953, Dr. Kern moved to Cleveland where he became the Director of Engineering for the Chemical and Process division at Colonial Iron Works Company. He formed his own firm in Cleveland in 1954, D.Q. Kern and Associates. The company specialized in thermal process technology and served clients in the chemical, petroleum, nuclear, and assorted equipment industries. Kern also consulted for the Atomic Energy Commission and the Department of the Interior, and taught graduate courses at the Polytechnic Institute of Brooklyn and Case Western Reserve University.

Donald Kern's fame would have been secured by *Process Heat Transfer* alone, but he also co-authored *Extended Surface Heat Transfer* (1972), published 60 papers and articles in heat transfer design and economics, and lectured widely to engineering groups. Kern was also active professionally as a member of AIChE and ASME. He was a founder and Chair of the Heat Transfer and Energy Conversion Division of AIChE and Chair of the National Heat Transfer Conference Coordinating Committee. The legacy and contribution of Donald Q. Kern was ensured in 1973 when the AIChE Heat Transfer and Energy Conversion Division (now Transport and Energy Processes Division) commemorated their most prestigious, annual award in his honor.

With thanks to A.E. Bergles and W.J. Warner.

Acknowledgement

Kern's *Process Heat Transfer, Second Edition* was nearly a decade in the making, and a dream for Dr. Ann Marie Flynn. Introduced to the original text as both an undergraduate and graduate student, she subsequently adopted the book during her term as a professor of chemical engineering at Manhattan College. Nearly 70 years after its original publication, Dr. Kern's *Process Heat Transfer* was still an extraordinary teaching tool.

It was through her efforts — the search to contact the Kern family, securing the copyright to the first edition, convincing Scrivener Publishing of the need for the book, and assembling the team to accomplish the job — that made her dream become a reality.

Dr. Flynn was able to locate the first member of the Kern family – Dr. Kern's nephew, the son of his sister, Helen. His nephew recounted stories with Dr. Flynn: the day he served as ring bearer at his uncle's wedding to the former Natalie Weiss; how he watched as the nearly 900 pages of the original manuscript were typed on a manual typewriter in the middle of Kern's living room in New York city.

Dr. Kern's son was located in Chicago (Kern's wife and daughter had since passed). Kern's son shared that he was very young when his father died in 1971, so he knew very little about his father's work. He also shared that as an adult, while he and his wife were hiking in Canada, they came across an engineering firm and decided to stop in. By chance, he mentioned that his father had once written a book – maybe they had heard of it? Dr. Kern's son was amazed to discover that the engineers described his father's book as the "Bible."

Dr. Kern – son, husband, father, and brother passed away on March 2, 1971 at his home in Shaker Heights, Ohio. Services were held on March 4, 1971 at Riverside Chapel in New York city. In the 21 years since its first printing, *Process Heat Transfer* had been translated into Russian, Japanese, and Spanish by the time Donald Quentin Kern was laid to rest. He was 56 years old.

On behalf of all students who have already, or will benefit from Dr. Kern's extraordinary work, the authors will be forever grateful for the blessing and consent provided by the Kern family to go forward with this project.

Contents

(First Edition)

PREFACE vii

INDEX TO THE PRINCIPAL APPARATUS CALCULATIONS xi

CHAPTER
1. Process Heat Transfer 1
2. Conduction 6
3. Convection 25
4. Radiation 62
5. Temperature 85
6. Counterflow: Double-pipe Exchangers 102
7. 1-2 Parallel-counterflow: Shell-and-Tube
    Exchangers 127
8. Flow Arrangements for Increased Heat
    Recovery 175
9. Gases 190
10. Streamline Flow and Free Convection 201
11. Calculations for Process Conditions 221
12. Condensation of Single Vapors 252
13. Condensation of Mixed Vapors 313
14. Evaporation 375
15. Vaporizers, Evaporators, and Reboilers 453
16. Extended Surfaces 512
17. Direct-contact Transfer: Cooling Towers 563
18. Batch and Unsteady State Processes 624
19. Furnace Calculations 674
20. Additional Applications 716
21. The Control of Temperature and Related
    Process Variables 765

Appendix of Calculation Data... 791

Author Index.. 847

Subject Index 851

Preface
(First Edition)

It is the object of this text to provide fundamental instruction in heat transfer while employing the methods and language of industry. This treatment of the subject has evolved from a course given at the Polytechnic Institute of Brooklyn over a period of years. The possibilities of collegiate instruction patterned after the requirements of the practicing process engineer were suggested and encouraged by Dr. Donald F. Othmer, Head of the Department of Chemical Engineering. The inclusion of the practical aspects of the subject as an integral part of the pedagogy was intended to serve as a supplement rather than a substitute for a strong foundation in engineering fundamentals. These points of view have been retained throughout the writing of the book.

To provide the rounded group of heat-transfer tools required in process engineering it has been necessary to present a number of empirical calculation methods which have not previously appeared in the engineering literature. Considerable thought has been given to these methods, and the author has discussed them with numerous engineers before accepting and including them in the text. It has been a further desire that all the calculations appearing in the text shall have been performed by an experienced engineer in a conventional manner. On several occasions the author has enlisted the aid of experienced colleagues, and their assistance is acknowledged in the text. In presenting several of the methods some degree of accuracy has been sacrificed to permit the broader application of fewer methods, and it is hoped that these simplifications will cause neither inconvenience nor criticism.

It became apparent in the early stages of writing this book that it could readily become too large for convenient use, and this has affected the plan of the book in several important respects.

A portion of the material which is included in conventional texts is rarely applied in the solution of run-of-the-mill engineering problems. Such material, as familiar and accepted as it may be, has been omitted unless it qualified as important fundamental information. Secondly, it was not possible to allocate space for making bibliographic comparisons and evaluations and at the same time present industrial practice. Where no mention has been made of a recent contribution to the literature no slight was intended. Most of the literature references cited cover methods on which the author has obtained additional information from industrial application.

The author has been influenced in his own professional development by the excellent books of Prof. W. H. McAdams, Dr. Alfred Schack, and others, and it is felt that their influence should be acknowledged separately in addition to their incidence in the text as bibliography.

For assistance with manuscript indebtedness is expressed to Thomas H. Miley, John Blizard, and John A. Jost, former associates at the Foster Wheeler Corporation. For checking the numerical calculations credit is due to Krishnabhai Desai and Narendra R. Bhow, graduate students at the Polytechnic Institute. For suggestions which led to the inclusion or exclusion of certain material thanks are due Norman E. Anderson, Charles Bliss, Dr. John F. Middleton, Edward L. Pfeiffer, Oliver N. Prescott, Everett N. Sieder, Dr. George E. Tait, and to Joseph Meisler for assistance with proof. The Tubular Exchanger Manufacturers Association has been most generous in granting permission for the reproduction of a number of the graphs contained in its Standard. Thanks are also extended to Richard L. Cawood, President, and Arthur E. Kempler, Vice-President, for their personal assistance and for the cooperation of the Patterson Foundry & Machine Company.

Donald Q. Kern
New York City, N.Y.
April 1950

To:

Donald Q. Kern,

Without whom, all of this would not have been possible.

ANN MARIE FLYNN

To:

My beloved parents, Hidenori and Mieko Akashige,
My dearest friends, MD Azim, Corine Laplanche,
Christopher Cacciavillani, Kleant Daci,
Michael Pryor, and Anet Kashoa,
My brother, Tetsuya
My peers from the classes of 2017 and 2018 of Manhattan College,
Coauthor and colleague, Dr. Louis Theodore,
and
My extraordinary mentor, Dr. Ann Marie Flynn.

TOSHIHIRO AKASHIGE

To:

Ann Marie Flynn,

A very special person, dedicated to education,
who has somehow managed to survive Manhattan College,
and for inviting me to contribute to this unique undertaking.

LOUIS THEODORE

Contents to the Second Edition

Acknowledgement		vii
Contents (First Edition)		ix
Preface (First Edition)		xi
Dedication		xiii
Contents to the Second Edition		xv
Preface to the Second Edition		xxi
Part I	**Fundamentals and Principles**	**1**
1	**Introduction to Process Heat Transfer**	**3**
	Introduction	3
	1.1 Units and Dimensional Analysis	4
	1.2 Key Physical Properties	10
	1.3 Key Process Variables and Concepts	14
	1.4 Laws of Thermodynamics	22
	1.5 Heat-Related Theories and Transfer Mechanisms	26
	1.6 Fluid Flow and Pressure Drop Considerations	28
	1.7 Environmental Considerations	35
	1.8 Process Heat Transfer	39
	References	40
2	**Steady-State and Unsteady-State Heat Conduction**	**43**
	Introduction	43
	2.1 Flow of Heat through a Plane Wall	46
	2.2 Flow of Heat through a Composite Plane Wall: Resistances in Series	50
	2.3 Flow of Heat through a Pipe Wall	54

2.4 Flow of Heat through a Composite Pipe Wall:
Resistances in Series   57
2.5 Steady-State Conduction: Microscopic Approach   63
2.6 Unsteady-State Heat Conduction   68
2.7 Unsteady-State Conduction: Microscopic Approach   71
References   77

3 Forced and Free Convection   **79**
Introduction   79
3.1 Forced Convection Principles   82
3.2 Convective Resistances   87
3.3 Heat Transfer Coefficients: Quantitative Information   89
3.4 Convection Heat Transfer: Microscopic Approach   105
3.5 Free Convection Principles and Applications   108
3.6 Environmental Applications   120
References   127

4 Radiation   **129**
Introduction   129
4.1 The Origin of Radiant Energy   132
4.2 The Distribution of Radiant Energy   133
4.3 Radiant Exchange Principles   138
4.4 Kirchoff's Law   139
4.5 Emissivity Factors and Energy Interchange   145
4.6 View Factors   153
References   157

Part II   Heat Exchangers   **159**

5 The Heat Transfer Equation   **161**
Introduction   161
5.1 Heat Exchanger Equipment Classification   162
5.2 Energy Relationships   163
5.3 Log Mean Temperature Difference (LMTD)
Driving Force   166
5.4 The Overall Heat Transfer Coefficient (U)   183
5.5 The Heat Transfer Equation   208
References   216

6 Double Pipe Heat Exchangers   **217**
Introduction   217
6.1 Equipment Description and Details   218

6.2	Key Describing Equations	225
6.3	Calculation of Exit Temperatures	244
6.4	Pressure Drop in Pipes and Pipe Annuli	251
6.5	Open-Ended Problems	254
6.6	Kern's Design Methodology	262
6.7	Practice Problems from Kern's First Edition	285
	References	286

7   Shell-and-Tube Heat Exchangers — **289**

	Introduction	289
7.1	Equipment Description and Details	290
7.2	Key Describing Equations	307
7.3	Open-Ended Problems	333
7.4	Kern's Design Methodology	339
7.5	Other Design Procedures and Applications	350
7.6	Computer Aided Heat Exchanger Design	372
7.7	Practice Problems from Kern's First Edition	376
	References	379

8   Extended Surface/Finned Heat Exchangers — **381**

	Introduction	381
8.1	Fin Details	382
8.2	Equipment Description	388
8.3	Key Describing Equations	390
8.4	Fin Effectiveness and Performance	398
8.5	Kern's Design Methodology	418
8.6	Other Fin Considerations	433
8.7	Practice Problems from Kern's First Edition	434
	References	435

9   Other Heat Exchangers — **437**

	Introduction	437
9.1	Condensers	439
9.2	Evaporators	450
9.3	Boilers and Furnaces	470
9.4	Waste Heat Boilers	480
9.5	Cogeneration/Combined Heat and Power (CHP)	488
9.6	Quenchers	492
9.7	Cooling Towers	498
9.8	Heat Pipes	508
	References	510

Part III   Peripheral Topics   513

10 Other Heat Transfer Considerations   515
Introduction   515
10.1 Insulation and Refractory   516
10.2 Refrigeration and Cryogenics   533
10.3 Instrumentation and Controls   546
10.4 Batch and Unsteady-State Processes   555
10.5 Operation, Maintenance, and Inspection (OM&I)   561
10.6 Economics and Finance   569
References   585

11 Entropy Considerations and Analysis   589
Introduction   589
11.1 Qualitative Review of the Second Law   590
11.2 Describing Equations   591
11.3 The Heat Exchanger Dilemma   595
11.4 Application to a Heat Exchanger Network   603
References   606

12 Health and Safety Concerns   607
Introduction   607
12.1 Definitions   611
12.2 Legislation   620
12.3 Material Safety Data Sheets   623
12.4 Health Risk versus Hazard Risk   628
12.5 Health Risk Assessment   629
12.6 Hazard Risk Assessment   640
References   650

Appendix   653

Tables
AT.1 Conversion Constants   654
AT.2 Thermodynamic Properties of Steam/Steam Tables   658
AT.3 Properties of Water (Saturated Liquid)   667
AT.4 Properties of Air at 1 atm   669
AT.5 Properties of Selected Liquids at 1 atm and 20 °C (68 °F)   670
AT.6 Properties of Selected Gases at 1 atm and 20 °C (68 °F)   670
AT.7   Dimensions, Capacities, and Weights of Standard
          Steel Pipes   674
AT.8   Dimensions of Heat Exchanger Tubes   676

AT.9 Tube-Sheet Layouts (Tube Counts) on a Square Pitch   678
AT.10 Tube-Sheet Layouts (Tube Counts) on
          a Triangular Pitch   680
AT.11 Approximate Design Overall Heat Transfer
          Coefficients (Btu/hr·ft^2·°F)   683
AT.12 Approximate Design Fouling Coefficient
          Factors (hr·ft^2·°F/Btu)   684

Figures

AF.1 Fanning Friction Factor (*f*) vs. Reynolds Number (*Re*) Plot 688
AF.2 Psychometric Chart: Low Temperatures:
          Barometric Pressure, 29.92 in. Hg.   689
AF.3 Psychometric Chart: High Temperatures:
          Barometric Pressure, 29.92 in. Hg.   690

Index   **691**

Preface to the Second Edition

A second edition? After 65 plus years? Is it reasonable? Does it make sense from a technical and publication perspective? The answer is ordinarily "*No.*" But for Donald Q. Kern's classic heat transfer book, *Process Heat Transfer*, the answer is definitely "*Yes.*"

The first edition sold approximately 65,000 copies over its lifetime. And for good reason. It stands alongside the powerhouse classics in the chemical engineering literature: Treybal's *Mass Transfer Operations*, McCabe and Smith's *Unit Operations of Chemical Engineering*, Bird, Stewart, and Lightfoot's *Transport Phenomena* (with the second edition arriving on the scene after a half century), etc.

As Kern put it in his Preface: "the object of this text is to promote instruction in heat transfer while employing the methods and language of industry by providing the heat transfer tools required in process engineering." Unbelievably, this book still achieves many of its original objectives as many practicing (chemical) engineers involved with heat transfer design include this book as part of their library – as evidenced by an email from a former Manhattan College student who is now working as a Process Engineer:

Subject Line: Can't Escape Kern
Date: 5/27/14

I had some calcs to do;
Director handed me these off his desk.

Regards,
Will Gallei

Even though the relentless passage of time has brought about many changes to the past, there have been relatively few truly innovative changes to the heat transfer equipment employed by industry since 1950 – and that may or may not be good. Still, there are changes that need to be addressed if Kern's original work is to continue to remain relevant in the 21st century process engineering literature. Topics that are part of the current engineers' vocabulary but need to be addressed include (but are not limited to) energy conservation and the associated topic of quality energy, nanomaterials, environmental considerations, health and safety (and the accompanying topic of risk), packaged calculation programs, the move from engineering units to the International System of Units (SI), etc.

Some of the above factors convinced the authors of the need for an update to Kern's classic work so that students would find an easy transition from classroom examples to industrial applications. From an educational perspective, the lead author of this second edition (Flynn) has employed Kern's *Process Heat Transfer* as the primary text in the junior-year chemical engineering course at Manhattan College for 18 years. One of the co-authors (Theodore) attempted to model his recent 2013 heat transfer book, *Heat Transfer Applications for the Practicing Engineering* (John Wiley & Sons) after Kern's book.

Kern's second edition is divided into three Parts: Fundamentals and Principles, Heat Exchangers, and Peripheral Topics. The first Part provides a series of chapters concerned with introductory topics that are required when solving heat transfer problems. This part of the book deals with heat transfer principles; topics that receive treatment include steady-state heat conduction, unsteady-state heat conduction, forced convection, free convection, and radiation. Part II is considered by the authors to be the "meat" of the book – addressing heat transfer equipment design procedures and applications. In addition to providing a more meaningful treatment of the various types of heat exchangers, this Part also examines the roles of computers on predicting the performance and the design of heat transfer equipment. It also should be noted that Kern's original practice problems were included in this part. The concluding Part of the book examines other related topics of interest including insulation and refractory, refrigeration and cryogenics, boilers, cooling towers, quenchers, heat pipes, and batch and unsteady-state processes health and hazard risk, and entropy considerations.

A comment on units and notations. The original units and notations employed by Kern were essentially retained. A short write-up on the International System of Units (SI) is provided in the Appendix to accommodate the clamor for metric units; unit conversion tables are also included. This accommodation was included despite industry's continual use of British Engineering units. Finally, and for obvious reasons, Kern adopted chemical engineering notation; fortunately, they have – for the most part – been retained by industry.

The changes to the present edition have evolved from a host of sources, including: course notes, homework assignments, and exam problems prepared by Ann Marie Flynn for a core, three-credit, undergraduate course, "Chemical Engineering Principles II: Heat Transfer," offered by Manhattan College; I. Farag and J. Reynolds, *Heat Transfer*, A Theodore Tutorial, East Williston, N.Y., 1994; and J. Reynolds, J. Jeris, and L. Theodore, *Handbook of Chemical and Environmental Engineering Calculations*, John Wiley & Sons, Hoboken, NJ, 2004. Although the bulk of the new material is original and/or taken from sources that the authors have been directly involved with, every effort has been made to acknowledge material drawn from other sources.

Our sincere thanks are extended to Kleant Daci, Michael Pryor, and Anet Kashoa as contributing authors for Chapter 9, Chapter 10, and Chapter 11, respectively. We also appreciate the extraordinary insight and guidance provided by Francesco Ricci and Paul Farber during the writing of this book.

Ann Marie Flynn
Toshihiro Akashige
Louis Theodore
Floral Park, N.Y.
March, 2019

Note: The authors are in the process of developing a useful resource in the form of a website which will contain over 150 additional problems and 15 hours of exams; *solutions* for these problems and exams will be available for those who *adopt* the book for training and/or academic purposes.

Part I

FUNDAMENTALS AND PRINCIPLES

1

Introduction to Process Heat Transfer

Introduction

The science of thermodynamics deals with the quantitative transitions and rearrangements of energy as heat in bodies of matter. *Heat transfer* is the science which deals with the rates of exchange of heat between hot and cold bodies called the *source* and *receiver*, respectively. The equipment employed to bring about this heat exchange is referred to as a *heat exchanger*.

Perhaps the simplest example of the various types of heat exchangers is the double pipe heat exchanger, a unit that will receive extensive treatment in Part II, Chapter 6. A simple diagram representing this exchanger is provided in Figure 1.1. This unit consists of two concentric pipes. Each of the two fluids — hot and cold — flow either through the inside of the inner pipe or through the annulus formed between the outside of the inner pipe and the inside of the outer pipe. Generally, it is more economical (from heat efficiency and design perspectives) for the hot fluid (the source) to flow-through the inner pipe and the cold fluid (the receiver)

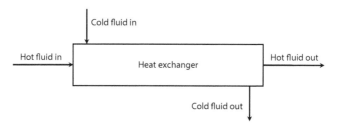

Figure 1.1 Line diagram of a cocurrent flow heat exchanger.

through the annulus, thereby reducing heat losses from the hot fluid to the surroundings.

Fundamentally, a temperature difference between the two bodies in close proximity (or between two parts of the same body) results in heat flow from higher to lower temperatures. There are three different (and classic) mechanisms by which this heat transfer can occur: conduction, convection, and radiation. When the heat transfer is the result of molecular motion (e.g., the vibrational energy of molecules in a solid being passed along from molecule to molecule), the mechanism of transfer is *conduction*. When heat transfer results from macroscopic motion, such as currents in a fluid, the mechanism is *convection*. When heat is transferred by electromagnetic waves, *radiation* is the mechanism. In most industrial applications, multiple mechanisms are usually involved in the transmission of heat. However, since each mechanism is governed by its own set of physical laws, it is beneficial to discuss them independently of each other (see Chapters 2–4).

This introductory chapter consists of the following eight sections:

1.1   Units and Dimensional Analysis
1.2   Key Physical Properties
1.3   Key Process Variables and Concepts
1.4   Laws of Thermodynamics
1.5   Heat-related Theories and Transfer Mechanisms
1.6   Fluid Flow and Pressure Drop Considerations
1.7   Environmental Considerations
1.8   Process Heat Transfer

1.1   Units and Dimensional Analysis

This section is primarily concerned with units. The units used in the text are consistent with those adopted by the engineering profession in the United States [1–3]. One usually refers to them as English or engineering units. Since engineers are often concerned with units and conversion of units, both the English and SI (International System of Units) units are used throughout the book. All quantities of the physical and chemical properties to be discussed are expressed using either or both of these two systems. Although the English (engineering) system of units is primarily employed in this text (Kern originally used engineering units), a discussion on the metric and SI system of units is warranted. Background histories are provided in the next two subsections.

The Metric System [1, 2]. The need for a single worldwide coordinated measurement system was recognized nearly 350 years ago. In 1670, Gabriel Mouton, Viscar of St. Paul's church in Lyon, proposed a comprehensive decimal measurement system based on the length of one minute of arc of a great circle of the Earth. In 1671, Jean Picard, a French astronomer, proposed that the period of an instrument called a "seconds pendulum" be precisely two seconds - thereby denoting that one unit of time, a second, be one swing of the seconds pendulum. (Such a pendulum would have been fairly easy to reproduce, thus facilitating the widespread distribution of uniform standards.) Other proposals were made, but over a century elapsed before any action was taken.

In 1790, in the midst of the French Revolution, the National Assembly of France requested the French Academy of Sciences to "deduce an invariable standard for all the measures and weights." The Commission appointed by the Academy created a system that was, at once, simple and scientific. The unit of length was to be a portion of the Earth's circumference. Measures for capacity (volume) and mass (weight) were to be derived from the unit of length, thus relating the basic units of the system to each other and to nature. Furthermore, the larger and smaller versions of each unit were to be created by multiplying or dividing the basic units by 10 and its multiples. This feature provided a great convenience to users of the system by eliminating the need for such calculations as dividing by 16 (to convert ounces to pounds) or by 12 (to convert inches to feet). Similar calculations in the metric system could be performed simply by shifting the decimal point. Thus, the metric system is a *base-10* or *decimal* system.

The Commission assigned the name *metre* (which is now spelled meter) to the unit of length. This name was derived from the Greek word *metron* meaning "a measure." The physical standard representing the meter was to be constructed so that it would equal one ten-millionth of the distance from the north pole to the equator along the meridian of the Earth running near Dunkirk in France and Barcelona in Spain.

The metric unit of mass, called the *gram*, was defined as the mass of one cubic centimeter (a cube that is 1/100 of a meter on each side) of water at its temperature of maximum density. The cubic decimeter (a cube 1/10 of a meter on each side) was chosen as the unit of fluid capacity. This measure was given the name *liter.*

Although the metric system was not accepted with enthusiasm at first, adoption by other nations occurred steadily after France made its use compulsory in 1840. The standardized character and decimal features of the metric system made it well suited to scientific and engineering work. Consequently, it is not surprising that the rapid spread of the system coincided with an age of rapid technological development. In the United States, by Act of Congress in 1866, it was made "lawful throughout the United States of America to employ the weights and measures of the metric system in all contracts, dealings, or court proceedings."

By the late 1860s, even better metric standards were needed to keep pace with scientific advances. In 1875, an international treaty, the "Treaty of the Meter," set up well-defined metric standards for length and mass. The treaty also established permanent machinery to recommend and adopt further refinements in the metric system. This treaty, known as the *Metric Convention*, was signed by 17 countries, including the United States. As a result of the treaty, metric standards were constructed and distributed to each nation that ratified the Convention. Since 1893, the internationally agreed to metric standards have served as the fundamental weights and measures standards of the United States.

By 1900, a total of 35 nations — including the major nations of continental Europe and most of South America — had officially accepted the metric system. Today, with the exception of the United States and a few small countries, the entire world is predominantly using the metric system or is committed to such use. In 1971, the Secretary of Commerce, in transmitting to Congress the results of a 3-year study authorized by the Metric Study Act of 1968 recommended that the U.S. change to the use of the metric system through a coordinated national program.

The International Bureau of Weights and Measures located at Sevres, France, serves as a permanent secretariat for the Metric Convention, coordinating the exchange of information about the use and refinement of the metric system. As measurement science develops more precise and easily reproducible ways of defining the measurement units, the General Conference of Weights and Measures — the diplomatic organization made up of adherents to the Convention — meets periodically to ratify improvements in the system and the standards.

The SI System [1, 2]. In 1960, the General Conference adopted an extensive revision and simplification of the system. The name *Le Systeme*

International d'Unites (International System of Units), with the international abbreviation SI, was adopted for this modernized metric system. Further improvements in and additions to SI were made by the General Conferences in 1964, 1968, and 1971.

The basic units in the SI system are the *kilogram* (mass), *meter* (length), *second* (time), *Kelvin* (temperature), *ampere* (electric current), *candela* (the unit of luminous intensity), and *radian* (angular measure). All are commonly used by the practicing engineer and scientist. The Celsius scale of temperature (0 °C = 273.15 K) is commonly used with the absolute Kelvin scale. The important derived units are the *newton* (SI unit of force), the *joule* (SI unit of energy), the *watt* (SI unit of power), the *pascal* (SI unit of pressure), and the *hertz* (SI unit of frequency). There are a number of electrical units: *coulomb* (charge), *farad* (capacitance), *henry* (inductance), *volt* (potential), and *weber* (magnetic flux). As noted above, one of the major advantages of the metric system is that larger and smaller numbers are given in powers of ten. In the SI system, a further simplification is introduced by recommending employing only those units with multipliers of 10^3. Thus, for lengths in engineering, the *micrometer* (previously referred to as a *micron*), *millimeter*, and *kilometer* are recommended, and the *centimeter* is generally avoided. A further simplification is that the decimal point may be replaced by a comma (as in France, Germany, and South Africa), while the other number, before and after the comma, is separated by spaces between groups of three, e.g., 123, 456, 789.

Equations are generally dimensional and involve several terms. For the equality to hold, each term in the equation must have the same dimensions (i.e., the equation must be dimensionally homogenous or consistent). This condition can be easily proved. Throughout the text, great care is exercised in maintaining the dimensional formulas of all terms and the dimensional consistency of each equation. The approach employed will often develop equations and terms in equations by first examining each in specific units (feet rather than length), primarily for the English system. Hopefully, this approach will aid the reader and will attach more significance to each term and equation.

A conversion constant (factor) is a term that is used to obtain units in a more convenient and/or useful form. All conversion constants have magnitude and units in the term, but can be shown to be equal to 1.0 (unity) with *no* units. Conversion constants are provided in Table AT.1 in the Appendix for both the English and SI system of units.

Terms in equations must also be consistent from a "magnitude" viewpoint. Differential terms cannot be equated with finite or integral terms. Care should also be exercised in solving differential equations. In order to solve differential equations to obtain a description of the pressure, temperature, composition, etc., of a system, it is necessary to specify boundary and/or initial conditions for the system. This information arises from a description of the problem or the physical situation. The number of boundary conditions (BC) that must be specified is the sum of the highest-order derivative for each independent differential term. A value of the solution on the boundary of the system is one type of boundary condition. The number of initial conditions (IC) that must be specified is the highest-order time derivative appearing in the differential equation. The value for the solution at time equal to zero constitutes an initial condition. For example, the following differential equation for temperature, T,

$$\frac{d^2 T}{dz^2} = 0 \qquad (1.1)$$

requires 2 BCs (in terms of the position coordinate z). The equation

$$\frac{dT}{dt} = 0; \quad t = \text{time} \qquad (1.2)$$

requires 1 IC. And finally, the equation

$$\frac{\partial T}{\partial t} = \alpha \frac{\partial^2 T}{\partial y^2}; \quad \alpha = \text{thermal diffusivity} \qquad (1.3)$$

requires 1 IC and 2 BCs (in terms of y).

Dimensionless Groups. Problems are frequently encountered in heat transfer and other engineering subjects that involve several variables. Engineers are generally interested in developing functional relationships (equations) between these variables. When these variables can be grouped together in such a manner that they can be used to predict the performance of similar pieces of equipment, independent of the scale or size of the operation, something very valuable has been accomplished.

Dimensional analysis is a powerful tool that is employed in planning experiments, presenting data compactly, and making practical predictions from models without detailed mathematical analysis. The first step in an analysis of this nature is to write down the units of each variable. The end result of a dimensional analysis is a list of pertinent *dimensionless groups of numbers* [1, 2]. Readers will soon come to realize that dimensionless numbers, as well as dimensionless groups, play an important role in engineering analysis of heat transfer phenomena. One of the useful aspects of dimensional analysis is its ability to provide a relationship among the variables when the information about a (heat transfer) phenomena is incomplete.

The Buckingham π (pi) theorem may be used to determine the number of independent dimensionless groups required to obtain a relation describing a physical phenomenon [2, 3]. The theorem requires that equations relating variables must be dimensionally homogenous (consistent). The approach has been applied to the problem of correlating experimental heat transfer data (see also Chapter 3).

There are several dimensionless numbers encountered in heat transfer applications. Some of the more important and more commonly used numbers are provided in alphabetical order in Table 1.1 with some of

Table 1.1 Key dimensionless groups in heat transfer.

Biot number (Bi)	$\dfrac{hL}{k}$
Graetz number (Gz)	$\dfrac{\dot{m}c_p}{kL}$
Grashof number (Gr)	$\dfrac{L^3 \rho^2 g \beta \Delta T}{\mu^2}$
Liquid Jacob number (Ja)	$\dfrac{c_p \Delta T_e}{h_{vap}}$
Nusselt number (Nu)	$\dfrac{hD}{k}$
Peclet number (Pe)	$\dfrac{D\bar{v}\rho c_p}{k}$
Prandtl number (Pr)	$\dfrac{c_p \mu}{k}$
Reynolds number (Re)	$\dfrac{D\bar{v}\rho}{\mu}, \quad \dfrac{DG}{\mu}$
Stanton number (St)	$\dfrac{h}{c_p \bar{v}\rho}, \quad \dfrac{h}{c_p G}$

the terms to be defined later. With reference to Table 1.1, also note that $Nu = (St)(Re)(Pr)$.

1.2 Key Physical Properties

Perhaps the most important physical properties to practicing (chemical) engineers are the properties of water [1, 2]. And the most important properties of water are those that appear in and have come to be defined as the "steam tables." These are provided in the Appendix in Table AT.2. However, there are other specific properties associated with other substances of importance to the practicing engineer involved with heat transfer. Some of the more important ones are provided in the subsections below.

Density. All matter has a density, measured as mass/unit volume. In general, density ↑ when pressure ↑. In contrast, density ↓ when temperature ↑.

Liquids that are not affected by changes in pressure are said to be *incompressible*. This text will assume that liquids are incompressible, and that the density of a liquid changes only if there are extreme changes in temperature. For the most part, this text will also assume that the density of a liquid is not a strong function of temperature, and is essentially constant.

In the case of gases, the density is appreciably impacted by both temperature and pressure. However, this text assumes that the density of gases can be assumed constant for small changes in temperature and pressure.

The *specific gravity* (SG) is the ratio of the density of a substance to the density of a reference substance at a specific condition:

$$SG = \frac{\rho}{\rho_{ref}} \tag{1.4}$$

(It should be noted that Kern employs the term S and s for specific gravity throughout his text). The reference most commonly used for *solids* and *liquids* is water at its maximum density, which occurs at 4 °C; this reference density is 1.000 g/cm³, 1000 kg/m³, or 62.43 lb/ft³. Note that, since the specific gravity is a ratio of two densities, it is *dimensionless*. Therefore, any set of units may be employed for the two densities as long as they are consistent. The specific gravity of *gases* is used only rarely; when it is, air at the same conditions of temperature and pressure as the gas is usually employed as the reference substance.

Another dimensionless quantity related to density is the API (American Petroleum Institute) gravity, which is often used to indicate densities of fuel oils. The relationship between the API scale and specific gravity is

$$\text{degrees API} = {}^{\circ}\text{API} = \frac{141.5}{SG(60/60^{\circ}F)} - 131.5 \tag{1.5}$$

where SG(60/60 °F) = specific gravity of the liquid at 60 °F using water at 60 °F as the reference. Note that petroleum refining is a major industry. Petroleum products serve as an important fuel for the power industry, and petroleum derivatives are the starting point for many syntheses in the chemical industry. Petroleum is also a mixture of a large number of chemical compounds.

Tabulated values of the density of water, air, selected liquids, and selected gases are provided in Tables AT.3, AT.4, AT.5, and AT.6, respectively.

Viscosity. Viscosity, μ, is an important fluid property that provides a measure of the resistance to flow. Although all real fluids possess viscosity, an ideal fluid is a hypothetical fluid that has a viscosity of zero and possesses no resistance to shear. The viscosity is frequently referred to as the *absolute* or *dynamic* viscosity. The principal reason for the difference in the flow characteristics of water and of molasses is that molasses has a much higher viscosity than water. Note also that the viscosity of a liquid ↓ with ↑ temperature, while the viscosity of a gas ↑ with ↑ temperature.

One set of units of viscosity in SI units is g/cm·s, which is defined as a poise (P). Since this numerical unit is somewhat high for many heat transfer applications, viscosities are frequently reported in centipoises (cP) where one poise is 100 centipoises. In English or engineering units, the dimensions of viscosity are in lb/ft·s. To convert from poises to this unit, one may multiply by (30.48 / 453.6) or (0.0672); to convert from centipoises, multiply by 6.72×10^{-4}. To convert centipoises to lb/ft·hr, multiply by 2.42. Kinematic viscosity, v, is the absolute viscosity divided by the density (μ / ρ) and has the dimensions of (volume) / (length · time). The corresponding unit to the poise is the Stoke, having the SI dimensions of cm²/s.

Viscosity is a fluid property listed in many heat transfer and engineering books, including *Perry's Chemical Engineers' Handbook* [4]. Tabulated values of the viscosities of water, air, selected liquids, and selected gases are provided in Tables AT.3, AT.4, AT.5, and AT.6, respectively.

Surface Tension. A liquid forms an interface with another fluid. At the surface, the molecules are more densely packed than those within the fluid. This results in surface tension effects and interfacial phenomena. The surface tension coefficient, σ, is the force per unit length of the circumference of the interface, or the energy per unit area of the interface area. Surface tension values for water are provided in *Perry's Chemical Engineers' Handbook* [4].

Heat Capacity. The heat capacity of a substance is defined as the quantity of heat required to raise the temperature of that substance by one degree on a unit mass (or mole) basis. The term *specific heat* is frequently used in place of *heat capacity*. This is not strictly correct, because specific heat has been traditionally defined as the ratio of the heat capacity of a substance to the heat capacity of water. However, since the heat capacity of water is approximately 1 cal/g·°C or 1 Btu/lb·°F, the term *specific heat* has come to imply heat capacity.

For gases, the addition of heat to cause a $1°$ temperature rise may be accomplished either at constant pressure or at constant volume. Since the amounts of heat necessary are different for the two cases, subscripts are used to identify which heat capacity is being used — c_p for constant pressure and c_v for constant volume. For liquids and solids, this distinction does not have to be made since there is little difference between the two values. Note, however, that the notation C (hot fluid) and c (cold fluid) is employed for heat capacity later in the book. Values of heat capacity are provided in the Appendix for water, air, selected liquids, and selected gases in Table AT.3, AT.4, AT.5, and AT.6, respectively.

Heat capacities are often employed on a *molar* basis instead of a *mass* basis, in which case the units become cal/gmol·°C or Btu/lbmol·°F. To distinguish between the two bases, uppercase letters (C_p, C_v) have been used to represent the molar-based heat capacities, and lowercase letters (c_p, c_v) for the mass-based heat capacities or specific heats [5, 6].

Heat capacities are functions of both temperature and pressure, although the effect of pressure is generally small and is neglected in almost all engineering calculations. It should be noted that heat capacities for solids are a weak function of both temperature and pressure. Liquids are less dependent on pressure than solids, but are slightly influenced by temperature. Gases exhibit a strong temperature dependence while the effect of pressure is small, except near the critical state where the pressure dependence diminishes with increasing temperature [5, 6].

The effect of temperature on C_p can be described by

$$C_p = \alpha + \beta T + \gamma T^2 \qquad (1.6)$$

or

$$C_p = a + bT + cT^{-2} \qquad (1.7)$$

Values for a, β, γ, and a, b, c, as well as the average heat capacity information are provided in tabular form by Theodore *et al.* [7]. *Average* or *mean* heat capacity data over specific temperature ranges are also available.

Thermal Conductivity [7]. Experience has shown that when a temperature difference exists across a solid body, energy in the form of heat will transfer from the high-temperature region to the low-temperature region until thermal equilibrium (same temperature) is reached. The mode of heat transfer, where vibrating molecules pass along kinetic energy through the solid, is called *conduction*. Liquids and gases may also transport heat in this fashion. The property of *thermal conductivity*, k, provides a measure of how fast (or how easily) heat flows through a substance. It is defined as the amount of heat that flows per unit time through a unit surface area of unit thickness as a result of a unit difference in temperature. It is also a property of the material. Typical units for conductivity are Btu·ft/hr·ft²·°F or Btu/hr·ft·°F or W/m·°F. Typical values of k are given in Table 1.2.

Thermal conductivities of solids are primarily dependent on temperature. In general, thermal conductivities for a pure metal decrease with temperature; alloying elements tend to reverse this trend. Liquids are for the most part temperature dependent but insensitive to pressure. The thermal conductivities of most liquids decrease with increasing temperature. It should also be noted that water has the highest thermal conductivity of all liquids except the so-called liquid metals. The thermal conductivity of a gas increases with increasing temperature; it is a weak function of pressure for pressures close to atmospheric conditions. However, the effect of pressure is significant for high pressures. Also note that the thermal conductivity of steam exhibits a strong pressure dependence.

Thermal conductivity values are provided for water, air, selected liquids, and selected gases in Tables AT.3, AT.4, AT.6, respectively. With regard to heat transfer applications, this particular property finds extensive application in designing heat exchangers.

Table 1.2 Typical thermal conductivity values.

Material	k, W/m·°C
Solid Metals	15 – 400
Liquids	0.1 – 10
Gases	0.01 – 0.2

Thermal Diffusivity. A useful combination of terms already considered is the *thermal diffusivity*, α; it is defined by

$$\alpha = \frac{k}{\rho c_p} \tag{1.8}$$

As with the kinematic viscosity, v, the units of α are ft²/hr or m²/s. As one might expect, thermal energy diffuses rapidly through substances with high α and slowly through those with low α. There is a strong dependence of α for gases for both pressure and temperature.

1.3   Key Process Variables and Concepts

Reynolds Number [8, 9]. The Reynolds number, *Re*, is a dimensionless quantity, and is a measure of the relative ratio of inertia to viscous forces in a fluid.

$$Re = \frac{L\bar{v}\rho}{\mu} = \frac{L\bar{v}}{v} \tag{1.9}$$

where,
   L = a characteristic length
   $\bar{v}$ = average velocity
   ρ = fluid density
   μ = dynamic (or absolute) viscosity
   v = kinematic viscosity

In flow through round pipes and tubes, L is replaced by D, the inside diameter. The Reynolds number provides information on flow behavior. It is particularly useful in scaling up bench-scale or pilot plant heat transfer data to full-scale applications.

Laminar flow is usually encountered at a Reynolds number, *Re*, below approximately 2100 in a circular duct, but it can persist up to higher Reynolds numbers. Under ordinary conditions of flow, the flow (in circular ducts) is turbulent at a Reynolds number above approximately 4000. Between 2100–4000, where the type of flow may be either laminar or turbulent (often referred to as the transition zone), the predictions are unreliable. The Reynolds number at which the fluid flow changes from laminar to transition or transition to turbulent are termed *critical* numbers. In the case of flow in circular ducts there are two critical Reynolds numbers, namely the aforementioned 2100 and 4000. Different *Re* criteria exist for geometries other than pipes. The Reynolds number will be revisited later in this chapter and in the individual heat transfer equipment chapters provided in Part II.

Kinetic Energy. A body of mass, m, may be acted upon by a force, F. If the mass is displaced a distance, dL, during a differential interval of time, dt, the kinetic energy (KE) expended is given by

$$dE_{KE} = m\left(\frac{a}{g_c}\right)dL \tag{1.10}$$

Since acceleration is defined as $a = d\bar{v} / dt$,

$$dE_{KE} = \left(\frac{m}{g_c}\right)\left(\frac{d\bar{v}}{dt}\right)dL = \left(\frac{m}{g_c}\right)\left(\frac{dL}{dt}\right)d\bar{v} \tag{1.11}$$

Noting that $\bar{v} = dL / dt$, the above expression becomes:

$$dE_{KE} = m\left(\frac{\bar{v}}{g_c}\right)d\bar{v} \tag{1.12}$$

If this equation is integrated from $\bar{v}_1$ to $\bar{v}_2$, the change in kinetic energy is

$$\Delta E_{KE} = \frac{m}{g_c}\int_{\bar{v}_1}^{\bar{v}_2}\bar{v}d\bar{v} = \frac{m}{g_c}\left(\frac{\bar{v}_2^2}{2} - \frac{\bar{v}_1^2}{2}\right) \tag{1.13}$$

or

$$\Delta E_{KE} = \left(\frac{m\bar{v}_2^2}{2g_c} - \frac{m\bar{v}_1^2}{2g_c}\right) = \Delta\left(\frac{m\bar{v}^2}{2g_c}\right) \tag{1.14}$$

The term above is defined as the change in kinetic energy.

The reader should note that for flow through heat exchanger pipes and tubes, the above kinetic energy term can be retained as written if the velocity profile is uniform, i.e., the local velocities at all points in the cross-section are the same. Ordinarily, there is a velocity gradient across the passage; this introduces an error in the above calculation, the magnitude of which depends on the nature of the velocity profile and the shape of the cross section. For the usual case where the velocity is approximately uniform (e.g., turbulent flow), the error is not serious and, since the error tends to cancel because of the appearance of kinetic terms on each side of any energy balance equation, it is customary to ignore the effect of velocity gradients. When the error cannot be ignored, the introduction of a correction factor is needed [8, 9].

Potential Energy. A body of mass, m, may be raised vertically from an initial position z_1 to z_2. For this condition, an upward force at least equal to the weight of the body must be exerted on it, and this force must move through the distance $z_2 - z_1$. Since the weight of the body is the force of gravity on it, the minimum force required is given by Newton's law:

$$F = \frac{ma}{g_c} = m\left(\frac{g}{g_c}\right); \quad \text{if } a = g \tag{1.15}$$

where g is the local acceleration due to gravity. The minimum work required to raise the body is the product of this force and the change in vertical displacement, i.e.,

$$\Delta E_{PE} = F\left(z_2 - z_1\right) = m\left(\frac{g}{g_c}\right)\left(z_2 - z_1\right) = \Delta\left(m\frac{g}{g_c}z\right) \tag{1.16}$$

The term above is defined as the change in the potential energy of the mass.

Ideal Gas Law. Observations based on physical experimentation can often be synthesized into simple mathematical equations called laws. These laws are never perfect and hence are only an approximate representation of reality. The *ideal gas law* (IGL) was derived from experiments in which the effects of pressure and temperature on gaseous volumes were measured over moderate temperature and pressure ranges. This law works well in the pressure and temperature ranges that were used in taking the data; extrapolations outside of the ranges have been found to work well in some cases and poorly in others. As a general rule, this law works best when the molecules of the gas are far apart, i.e., when the pressure is low and the temperature is high. Under these conditions, the gas is said to behave *ideally*, i.e., its behavior is a close approximation to the so-called *perfect* or *ideal gas*, a hypothetical entity that obeys the ideal gas law perfectly. For engineering calculations, and specifically for most heat exchanger applications, ideal gas conditions are assumed (unless an abnormal condition is noted that would make this assumption erroneous).

The two precursors of the ideal gas law were *Boyle's* and *Charles'* laws. Boyle found that the volume of a given mass of gas is inversely proportional to the *absolute* pressure if the temperature is kept constant:

$$P_1 V_1 = P_2 V_2 \tag{1.17}$$

where,

V_1 = volume of a gas at absolute pressure P_1 and temperature T
V_2 = volume of a gas at absolute pressure P_2 and temperature T

Charles found that the volume of a given mass of gas varies directly with the *absolute* temperature at constant pressure:

$$\frac{V_1}{T_1} = \frac{V_2}{T_2} \tag{1.18}$$

where,

V_1 = volume of a gas at absolute pressure P and temperature T_1
V_2 = volume of a gas at absolute pressure P and temperature T_2

Boyle's and Charles' laws may be combined into a single equation in which neither temperature nor pressure need be held constant:

$$\frac{P_1 V_1}{T_1} = \frac{P_2 V_2}{T_2} \tag{1.19}$$

Moreover, experiments with different gases showed that Equation (1.19) could be expressed in a far more generalized form. If the number of moles (n) is introduced, the above equation may be rewritten as:

$$\frac{PV}{nT} = R \tag{1.20}$$

where,
R = universal gas constant

Equation (1.20) is referred to as the ideal gas law. Numerically, the value of R depends on the units used for P, V, T, and n [6]. Other useful forms of the ideal gas law are shown in Equations (1.21) and (1.22) below. Equation (1.21) applies to gas flow rather than to a gas confined in a container.

$$Pq = \dot{n}RT \tag{1.21}$$

where,

   q = gas volumetric flow rate (ft³/hr)
   P = absolute pressure (psia)
   $\dot{n}$ = molar flow rate (lbmol/hr)
   T = absolute temperature (°R)
   R = 10.73 psia· ft³/lbmol·°R

Equation (1.20) may also be expressed in terms of the density:

$$P(MW) = \rho RT \tag{1.22}$$

where,

   MW = molecular weight of gas (lb/lbmol)
   ρ = density of gas (lb/ft³)

Volumetric flow rates are often not specified at the actual conditions of pressure and temperature but at arbitrarily chosen standard conditions (STP, standard temperature and pressure). To distinguish between flow rates based on the two conditions, the letters "a" and "s" are often used as part of the unit. The units *acfm* and *scfm* stand for actual cubic feet per minute and standard cubic feet per minute, respectively. The ideal gas law can be used to convert from *standard* to *actual* conditions, but since there are many standard conditions in use, the STP being used *must* be known. The reader is cautioned on the incorrect use of *acfm* and/or *scfm*. The use of standard conditions is a convenience; when predicting the performance of or designing heat transfer equipment, the *actual* conditions *must* be employed. Designs based on standard conditions can lead to disastrous results because the equipment is usually underdesigned. Equation (1.23), which is a form of Charles' law, can be used to correct flow rates from standard to actual temperature conditions:

$$q_a = q_s \left(T_a / T_s \right) \tag{1.23}$$

where,

   q_a = volumetric flow rate at actual conditions (ft³/hr)
   q_s = volumetric flow rate at standard conditions (ft³/hr)
   T_a = actual absolute temperature (°R)
   T_s = standard absolute temperature (°R)

The reader is again reminded that *absolute* temperatures and pressures must be employed in all ideal gas law calculations.

As noted above, an *ideal* gas is a hypothetical entity that obeys the ideal gas law perfectly. But, in heat transfer applications involving gases, one almost always deals with *real* gases. Although most heat exchanger applications involving gases occur at conditions approaching ideal gas behavior, there are rare occasions when the deviation from ideality is significant. Detailed calculation procedures are available [4–6] to account for these deviations. Although no real gas obeys the ideal gas law exactly, the "lighter" gases (hydrogen, oxygen, air, and so on) at ambient conditions approach ideal gas law behavior. The "heavier" gases such as sulfur dioxide and hydrocarbons, particularly at high pressures and low temperatures, can deviate considerably from the ideal gas law. Despite these deviations, the ideal gas law is routinely used in not only heat transfer but also other engineering calculations.

Numerous attempts have been made to develop an all-purpose gas law. Although it is beyond the scope of this book to review these theories in any great detail, a brief outline of one (and perhaps the most popular) approach to account for deviations from ideality is to include a "correction factor," Z, which is defined as the *compressibility coefficient* or *factor*,

$$PV = ZnRT \tag{1.24}$$

Note that Z approaches 1.0 as P approaches a vacuum. For an ideal gas, Z is exactly unity. This equation may also be written as

$$PV' = ZRT \tag{1.25}$$

where V' is now the *specific* volume (not the total volume or velocity) with units of volume/mole. Details on calculating Z are available in the literature [2, 5, 6].

Partial Pressure and Partial Volume [5, 6]. Mixtures of gases are more often encountered than single or pure gases in engineering practice. The ideal gas law is based on the *number* of moles present in the gas volume; the *kind* of molecules is not a significant factor, only the number. This law applies equally well to mixtures and pure gases alike. Dalton and Amagat both applied the ideal gas law to mixtures of gases.

Dalton defined the partial pressure of a component as the pressure that would be exerted if the same mass of the component gas occupied the same total volume *alone* at the same temperature as the mixture. The sum of these partial pressures would then equal the total pressure:

$$P = P_A + P_B + P_C + \cdots + P_n = \sum_{i=A}^{n} P_i \tag{1.26}$$

where,
P = total pressure
n = number of components
P_i = partial pressure of component i

Equation (1.26) is known as *Dalton's law*. Applying the ideal gas law to one component (A) only,

$$P_A V = n_A RT \tag{1.27}$$

where,
n_A = number of moles of component A

Eliminating R, T, and V between Equation (1.26) and (1.27) yields

$$\frac{P_A}{P} = \frac{n_A}{n} = y_A \quad \text{or} \quad P_A = y_A P \tag{1.28}$$

where,
y_A = mole fraction of component A

Amagat's law is similar to Dalton's. Instead of considering the total pressure to be made up of partial pressures, where each component occupies the total container volume, Amagat considered the total volume to be made up of partial volumes in which each component is at the total pressure. The definition of the *partial volume* is therefore the volume occupied by a component gas alone at the same temperature and pressure as the mixture. For this case:

$$V = V_A + V_B + V_C + \cdots + V_n = \sum_{i=A}^{n} V_i \tag{1.29}$$

Applying Equation (1.20), as before,

$$\frac{V_A}{V} = \frac{n_A}{n} = y_A \tag{1.30}$$

where,
V_A = partial volume of component A

It is common in heat exchanger applications to describe low concentrations of corrosive components in gaseous mixtures in parts per million (ppm) by volume. Since partial volumes are proportional to mole fractions, it is only necessary to multiply the mole fraction of the component by 1 million (10^6) to obtain the concentration in parts per million (6). For liquids and solids, parts per million (ppm) is also used to express concentration, although it is usually on a *mass* basis rather than a *volume* basis. The terms ppmv and ppmw are sometimes used to distinguish between volume and mass bases, respectively.

The Conservation Laws [2, 6]. There are three conservation laws of interest to the practicing engineer involved with heat transfer and heat transfer applications. The three laws are concerned with mass, energy, and momentum. A brief introduction to each law is provided below.

The *conservation law* for mass can be applied to any process, equipment, or system. The general form of this law is given by Equations (1.31) and (1.32).

$$
\begin{Bmatrix} Mass \\ In \end{Bmatrix} - \begin{Bmatrix} Mass \\ Out \end{Bmatrix} + \begin{Bmatrix} Mass \\ Generated \end{Bmatrix} = \begin{Bmatrix} Mass \\ Accumulated \end{Bmatrix} \quad (1.31)
$$

$$
I \quad - \quad O \quad + \quad G \quad = \quad A
$$

or on a time rate basis by

$$
\begin{Bmatrix} Rate\,of \\ Mass \\ In \end{Bmatrix} - \begin{Bmatrix} Rate\,of \\ Mass \\ Out \end{Bmatrix} + \begin{Bmatrix} Rate\,of \\ Mass \\ Generated \end{Bmatrix} = \begin{Bmatrix} Rate\,of \\ Mass \\ Accumulated \end{Bmatrix} \quad (1.32)
$$

$$
\dot{I} \quad - \quad \dot{O} \quad + \quad \dot{G} \quad = \quad \dot{A}
$$

The *law of conservation of mass* states that mass can neither be created nor destroyed. Nuclear reactions, in which interchanges between mass and energy are known to occur, provide a notable exception to this law. Even in chemical reactions, a certain amount of mass-energy interchange takes place. However, in normal heat transfer applications, nuclear reactions do not occur and the mass-energy change in chemical reactions is so miniscule that it is not worth taking into account.

The *law of conservation of energy*, which like the law of conservation of mass, applies for all processes that do not involve nuclear reactions, states that energy can neither be created nor destroyed. As a result, the energy level of the system can change only when energy crosses the system boundary, i.e.,

$$
\Delta\left(Energy\ Level\ of\ System\right) = Energy\ Crossing\ Boundary \quad (1.33)
$$

(Note: Here, the symbol "Δ" means "change in".) This law is also referred to as the first law of thermodynamics and is treated in more detail in the next section.

The *law of conservation of momentum* is treated in the last section of this chapter. An offshoot of this law is pressure drop and a host of fluid flow equations — two topics of interest to the (chemical) engineers involved with heat transfer applications.

1.4   Laws of Thermodynamics

Prior to undertaking the writing of this text, one of the authors earlier co-authored a text entitled *Thermodynamics for the Practicing Engineer* [6]. It soon became apparent that some overlap existed between thermodynamics and heat transfer (the subject of this text). Even though the former topic is broadly viewed as a science, heat transfer is one of the unit operations and can justifiably be classified as an engineering subject. But what are the similarities and what are the differences?

The similarities that exist between thermodynamics and heat transfer are grounded in the three aforementioned conservation laws: mass, energy, and momentum. Both topics are primarily concerned with energy-related subject matter and both, in a very real sense, supplement each other. However, thermodynamics deals with the transfer of energy and the conversion of energy into other forms of energy, with consideration generally limited to systems in equilibrium. As noted earlier, the topic of heat transfer deals with the transfer of energy in the form of heat; the applications almost exclusively occur with heat exchangers that are employed in the chemical, petrochemical, petroleum (refinery), and engineering industries.

In terms of introduction to the laws of thermodynamics, there are three laws of thermodynamics. The first law is primarily concerned with energy conservation. There are many definitions of the second law, but it would serve the objectives of this text to state that it is used to define the term entropy, noted as S. The third law is almost an afterthought, but it does serve to specify entropy's values of pure substances at *absolute* zero temperature as zero. Interestingly, thermodynamics was once defined by Webster as "the science which deals with the intertransformation of heat and work." The fundamental principles of thermodynamics are contained in the aforementioned first, second, and third laws of thermodynamics. These principles have been defined as "pure" thermodynamics. These laws were developed and extensively tested in the latter half of the nineteenth century and are essentially based on experience. (The third law was developed during the twentieth century.) Details on each of these first two laws follows.

The *law of conservation of energy* (the first law), which, like the law of conservation of mass, holds for all processes that do not involve nuclear reactions, states that energy can be neither created nor destroyed. Equation (1.33) can be rewritten with respect to rate by stating that the rate of change of energy in the system can change only when there is a change in the rate of energy crossing the system boundary:

Rate of Change of Energy of the System = Rate of Change of Energy
Crossing System Boundary

$$(1.34)$$

This law is usually written as:

$$\Delta U = Q + W \tag{1.35}$$

where U refers to internal energy, and where kinetic energy and potential energy effects or changes are neglected as is usually the case with most heat transfer applications. When dealing with open or flow systems, it is more convenient to write the first law as:

$$\Delta H = Q + W_s \tag{1.36}$$

where H and W_s are the *enthalpy* and mechanical (or shaft) *work*, respectively. For most heat exchanger systems, the work term can be neglected so that the above equation reduces to:

$$\Delta H = Q \tag{1.37}$$

The Enthalpy Term. Many types of enthalpy are included when applying Equation (1.37), including the following:

1. Sensible (temperature)
2. Latent (phase)
3. Dilution (with water), e.g., HCL with H_2O
4. Solution (nonaqueous), e.g., HCL with a solvent other than H_2O
5. Reaction (chemical)

What follows is concerned with effects (1) and (2).

Sensible Enthalpy Calculations. Sensible enthalpy effects calculations are associated with temperature. This subsection provides methods that can be employed to calculate these changes. These methods include the use of

1. Enthalpy values
2. Average heat capacity values
3. Heat capacity as a function of temperature

Detailed calculations on effects 1–3 above are provided by Theodore *et al.* [6]. Two expressions for heat capacity are considered in topic (3) employing *a*, *b*, *c* constants and *a*, *β*, *γ* constants.

If enthalpy values are available, the enthalpy change is given by

$$\Delta h = h_2 - h_1 \quad (\text{mass basis}) \tag{1.38}$$

$$\Delta H = H_2 - H_1 \quad (\text{mole basis}) \tag{1.39}$$

If *average* molar heat capacity data are available, Equation (1.37) below may be written as

$$\Delta H = \overline{C}_p \Delta T \tag{1.40}$$

Average molar heat capacity data are provided in the literature where $\overline{C}_p$ is the average molar value of C_p in the temperature range ΔT. [4–6]

A more rigorous approach to enthalpy calculations can be provided if heat capacity variation with temperature is available. If the heat capacity is a function of temperature, the enthalpy change is written in differential form:

$$dH = C_p dT \tag{1.41}$$

If the temperature variation of the heat capacity is given by

$$C_p = \alpha + \beta T + \gamma T^2 \tag{1.42}$$

then Equation (1.41) may be combined with (1.42) and integrated between an initial temperature (T_1) and a final temperature (T_2):

$$\Delta H = H_2 - H_1$$

$$\Delta H = a(T_2 - T_1) + \frac{\beta}{2}(T_2^2 - T_1^2) + \frac{\gamma}{3}(T_2^3 - T_1^3) \tag{1.43}$$

Equation (1.41) may also be integrated if the heat capacity is a function of temperature of the form

$$C_p = a + bT + cT^{-2} \tag{1.44}$$

The enthalpy change is then given by

$$\Delta H = a\left(T_1 - T_0\right) + \frac{b}{2}\left(T_1^2 - T_0^2\right) + c\left(T_1^{-1} - T_0^{-1}\right) \tag{1.45}$$

Tabulated values of a, β, γ and a, b, c for a host of compounds, including chlorinated organics, are available in the literature [5, 6].

Latent Enthalpy Calculations. The term *phase*, when used to describe a pure substance, refers to a state of matter that is a gas, liquid, or solid. *Latent* enthalpy effects are associated with *phase* changes. These phase changes involve no change in temperature, but there is a transfer of energy to and from the substance. There are three possible latent effects: (1) vapor-liquid, (2) liquid-solid, and (3) vapor-solid.

Vapor-liquid changes are referred to as *condensation* when the vapor is condensing and *vaporization* when liquid is vaporizing. *Liquid-solid* changes are referred to as *melting* when the solid melts to liquid and *freezing* when a liquid solidifies. *Vapor-solid* changes are referred to as *sublimation*. Also, there are enthalpy effects associated with a phase change of a solid to a solid of another form; however, this enthalpy effect finds no application in heat exchanger applications.

Finally, the brief discussion of energy conservation above leads to an important *second-law* consideration: that energy has *quality* as well as quantity. Because work is 100 percent convertible to heat whereas the reverse situation is not true, work is a *more valuable* form of energy than heat. Although it is not as obvious, it can also be shown through second-law arguments that work has a quality-related value which varies according to the temperature at which heat is discharged from a system. The higher the temperature at which heat transfer occurs, the greater the potential for energy transformation into work. Thus, thermal energy stored at higher temperatures is generally more useful to society than that available at lower temperatures. While there is an immense quantity of energy stored in the oceans, for example, its present *availability* to society for performing useful tasks is essentially nonexistent [5, 6]. Calculations related to the second law primarily key on entropy, a topic reviewed in Part III, Chapter 11.

1.5 Heat-Related Theories and Transfer Mechanisms

The study of heat transfer would be greatly enhanced by a sound understanding of the nature of heat. However, it must be noted that the nature of heat cannot be fully understood from a single perspective. Because so many manifestations of heat have been discovered and studied, no simple theory covers them all. Laws which apply to mass transitions may not be applicable to molecular or atomic transitions, and those which are applicable at low temperatures may not apply at high temperatures. For the purposes of engineering analysis, it is necessary to undertake the study with basic information on a broad range of heat-related phenomena.

The phases of a single substance: solid, liquid, or gas, are related to its energy content. In the solid phase, the molecules or atoms are close together, giving it rigidity. In the liquid phase, sufficient thermal energy is present to increase the distance between molecules so that rigidity is lost. In the gas phase, the presence of additional thermal energy results in a complete separation of atoms or molecules so that they may wander anywhere in a confined space. It is also recognized that whenever a change of phase occurs outside the critical region [6, 7], a large amount of energy is involved in the phase change.

The various thermal properties have different orders of magnitude for the same substance in its different phases. As an example, the heat capacity per unit mass is very low for solids, high for liquids, and usually intermediate for gases. Similarly, in any body absorbing or losing heat, special consideration must be given whether the change is one of sensible or latent heat, or both. Still further, it is also known that a hot source is capable of such great subatomic excitement that it emits energy without any direct contact with the receiver, and this is the underlying principle of radiation. In effect, each type of change exhibits its own peculiarities.

There are three distinct mechanisms in which heat may pass from a source to a receiver, although some engineering applications are combinations of two or three. These three mechanisms are *conduction*, *convection*, and *radiation*. Each are briefly described in the following text.

Conduction is the transfer of heat through a fixed material such as the stationary wall shown in Figure 1.2. The direction of heat flow is at right angles to the wall if the wall surfaces are isothermal and the body homogeneous and isotropic. Assume that the source of heat exists on the left face of the wall and a receiver of heat exists on the right face. It has been known and later it will be confirmed by derivation that the flow of heat per unit of time is proportional to the change in temperature through the wall and the area of the wall A. If T is the temperature at any point in the wall and x is

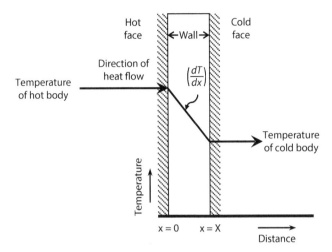

Figure 1.2 Heat flow through a wall.

the thickness of the wall in the direction of heat flow, the rate of heat flow Q is given by

$$Q = kA\left(-\frac{dT}{dx}\right); \quad \text{Btu}/\text{hr} \tag{1.46}$$

The term $-dT/dx$ is referred to as the *temperature gradient* and has a negative sign if the temperature has been assumed higher at the face of the wall where $x = 0$ and lower at the face where $x = X$. In other words, the instantaneous quantity of heat transfer is proportional to the area and temperature difference dT, which drives the heat through the wall of thickness dx. The proportionality constant k is peculiar to conductive heat transfer and is known as the aforementioned *thermal conductivity*. It is normally evaluated experimentally and is basically defined by Equation (1.46). The thermal conductivities of solids have a wide range of numerical values depending upon whether the solid is a relatively good conductor of heat such as a metal or a poor conductor such as asbestos. The latter serve as *insulators*. Although heat conduction is usually associated with heat transfer through solids, it is also applicable with limitations to liquids and gases.

Convection is the transfer of heat between relatively hot and cold portions of a fluid by mixing. Suppose a can of liquid were placed over a hot flame. The liquid at the bottom of the can becomes heated and less dense than before owing to its thermal expansion. The liquid adjacent to the bottom is also less dense than the cold upper portion and rises through it, transferring its heat by mixing as it rises. The transfer of heat from the hot liquid at the bottom of the can to the remainder is *natural* or *free convection*. If any other agitation occurs, such as that produced by a stirrer, it is *forced convection*; this type of heat transfer may be described in an equation which imitates the form of the conduction equation and is given by

$$dQ = hAdT \qquad (1.47)$$

The proportionality constant h is a term which is influenced by the nature of the fluid and the nature of the agitation and must be evaluated experimentally. It is referred to as the *heat-transfer coefficient*. When Equation (1.47) is written in integrated form, $Q = hA\Delta T$, it is referred to by some as Newton's law of cooling.

Radiation involves the transfer of radiant energy from a source to a receiver. When radiation issues from a source to a receiver, part of the energy is absorbed by the receiver and part reflected by it. Based on the second law of thermodynamics Boltzmann established that the rate at which a source gives off heat is

$$dQ = \sigma \epsilon \, dA T^4 \qquad (1.48)$$

This is known as the fourth-power law in which T is the *absolute* temperature. The term σ, is a dimensional constant, but ϵ is a factor peculiar to radiation and is called the *emissivity*. The emissivity, like the thermal conductivity k or the heat-transfer coefficient h, is normally determined experimentally.

Details on each of the three heat transfer mechanisms receives treatment in the next three chapters.

1.6   Fluid Flow and Pressure Drop Considerations

Historically, the design of a heat exchanger was primarily based on physical requirements — mainly the heat transfer area. Theodore [7] and Farag [8] introduced potential constraints associated with pressure drop and fluid flow concerns, while Theodore [6] addressed the issues of energy conservation from both an entropy and energy perspective. The latter issue was briefly introduced in an earlier section (Laws of Thermodynamics) and will be expanded upon in earnest in Part III, Chapter 11. The issue of fluid flow and pressure drop considerations is addressed in this section and its application to real world applications will become apparent in several illustrative example prepared by Kern [9] — based on his design methodology — that appear in Part II, Chapters 6–8.

Conduits. Fluids are usually transported in tubes or pipes; both serve as the "arteries" and "veins" that supply the fluids enabling energy transfer to occur in heat exchangers. Generally speaking, pipes are heavy-walled and have a relatively large diameter. Tubes are thin-walled and often come in coils.

Pipes are specified in terms of their diameter and wall thickness. The nominal diameters range from 1/8 to 30 inches for steel pipes. Standard dimensions of steel pipe are available in the literature and Table AT.7 in the Appendix, and are known as IPS (iron pipe size) or NPS (nominal pipe size). The pipe wall thickness is indicated by the schedule number where tube sizes are indicated by the outside diameter. The pipe wall thickness is related to the outside diameter. The tube wall thickness is usually given by the Birmingham wire gauge (BWG) number; the smaller the BWG, the heavier the tube.

The wall thickness of a pipe, indicated by the schedule number, can be approximated from 1000 (P/S) where P is the maximum internal service pressure (psi) and S is the allowable bursting stress in the pipe material (psi). The S value varies by material, grade of material and temperature; allowable S values may be found in Piping Handbooks.

Tube sizes are labeled by the outside diameter. The wall thickness is usually given a BWG number. Table AT.8 in the Appendix lists the sizes and wall thicknesses of condenser and heat exchanger tubes. For example, a ¾-inch 16 BWG tube has an outside diameter (OD) of 0.75 in, an inside diameter (ID) of 0.62 in, a wall thickness of 0.065 in, and a weight of 0.476 lb/ft [8, 10].

Additional details on tubes and pipes are provided in the Appendix in Tables AT.7 and AT.8, respectively.

Mechanical Energy Equation – Modified Form. Abulencia and Theodore [10] provide the following equation:

$$\frac{\Delta P}{\rho} + \frac{\Delta \bar{v}^2}{2g_c} + \frac{g}{g_c}\Delta z_h - \eta W_s + \sum F = 0; \quad F = \text{frictional losses} \qquad (1.49)$$

where,

ΔP = a change in pressure
$\Delta \bar{v}$ = a change in velocity
Δz_h = a change in height
η = efficiency
W_s = work done on the system

They referred to this as the *mechanical energy equation*. Equation (1.49) can be rewritten without the pump work and friction terms.

$$\frac{\Delta P}{\rho} + \frac{\Delta \bar{v}^2}{2g_c} + \frac{g}{g_c}\Delta z_h = 0 \tag{1.50}$$

This equation was defined as the basic form of the *Bernoulli equation*. Equation (1.49) can also be written as

$$\left(\frac{P_1}{\rho}\right)\left(\frac{g_c}{g}\right) + \frac{\bar{v}_1^2}{2g} + z_{h,1} = \left(\frac{P_2}{\rho}\right)\left(\frac{g_c}{g}\right) + \frac{\bar{v}_2^2}{2g} + z_{h,2} + h_s\left(\frac{g_c}{g}\right) + h_f\left(\frac{g_c}{g}\right) \tag{1.51}$$

note that h_s and h_f have replaced ηW_s and ΣF, respectively.

The two h terms were included above to represent the loss of energy due to friction in the system. Frictional loss can take several forms. An important chemical engineering problem is the calculation of these losses. It was noted (earlier) that the fluid can flow in either of two modes — laminar or turbulent. For laminar flow, an equation is available from basic theory [11] to calculate friction loss in a pipe. In practice, however, fluids (particularly gases) are rarely moving in laminar flow.

Laminar Flow Through a Circular Tube. Fluid flow in circular tubes (or pipes) is encountered in many applications, and is always accompanied by friction. Consequently, there is energy loss associated with a pressure drop in the direction of flow. As noted above, one can theoretically derive the h_f term for laminar flow [10, 11]. The equation can be shown to take the form

$$h_s = \frac{32\mu\bar{v}L}{\rho g_c D^2} \tag{1.52}$$

for a fluid flowing through a straight cylinder of diameter D and length L.

A friction factor, f, that is dimensionless, may now be defined (for laminar flow):

$$f = \frac{16}{\text{Re}} \tag{1.53}$$

so that Equation (1.52) takes the form

$$h_f = \frac{4fL\bar{v}^2}{2g_c D} \tag{1.54}$$

Although this equation describes friction loss across a conduit of length, L, it can also be used to predict the pressure drop due to friction per unit length of conduit, i.e., $\Delta P/L$, by simply dividing the above equation by L. It should also be noted that another friction factor term exists which differs from that presented in Equation (1.53). In this other case, f_D is defined as

$$f_D = \frac{64}{Re} \qquad (1.55)$$

The f_D term is used to distinguish the difference of Equation (1.53) from that of Equation (1.55). In essence:

$$f_D = 4f \qquad (1.56)$$

The term f is defined as the Fanning friction factor while f_D is defined as the Darcy or Moody friction factor [12]. Care should be taken as to which of the friction factors are being used in calculations. This will become more apparent shortly. In general, chemical engineers employ the Fanning friction factor; other engineers prefer the Darcy (or Moody) factor. This text primarily employs the Fanning friction factor, although Kern chose to employ both in his original work [9].

Employing Equation (1.56), Equation (1.51) may be extended in the absence of pump work and rewritten as

$$\frac{\Delta P}{\rho} + \frac{\Delta \bar{v}^2}{2g_c} + \Delta z_h \frac{g}{g_c} + \frac{4fLv^2}{2g_c D} = 0 \qquad (1.57)$$

The symbols Σh_c and Σh_e, representing the sum of the contraction and expansion losses, respectively, may also be added to this equation as provided below in Equation (1.58). These effects are discussed in more detail by Abulencia and Theodore [10].

$$\frac{\Delta P}{\rho} + \frac{\Delta \bar{v}^2}{2g_c} + \Delta z_h \frac{g}{g_c} + \frac{4fL\bar{v}^2}{2g_c D} + \Sigma h_c + \Sigma h_e = 0 \qquad (1.58)$$

Note that the second and third terms in the left-hand side of this equation representing kinetic and potential energy effects, respectively, can be safely neglected in nearly all heat exchanger calculations.

Turbulent Flow Through a Circular Conduit. It is important to note that almost all the key fluid flow equations presented for laminar flow apply as well to turbulent flow, provided the appropriate friction factor is employed.

The effect of the Reynolds number of the Fanning friction factor is provided in Figure AF.1 in the Appendix. Note that f from Equation (1.53) appears on the far left-hand side of Figure AF.1

In the turbulent regime, the "roughness" of the pipe becomes a consideration. In his original work on the friction factor, Moody [12] defined the term k (or ε), as the roughness and the ratio, k/D, as the relative roughness. Thus, for rough pipes/tubes in turbulent flow:

$$f = f\left(Re, k/D\right) \tag{1.59}$$

This equation reads that the friction factor is a function of *both* the Re and k/D. However, as can be seen in Figure AF.1, the dependency on the Reynolds number is a weak one in turbulent flow.

Two Phase Flow [2, 8, 10]. The simultaneous flow of two phases in pipes (as well as other conduits) is complicated by the fact that the action of gravity tends to cause settling and "slip" of the heavier phase; the result is that the lighter phase flows at a different velocity in the pipe than does the heavier phase. The results of this phenomena are different depending on the classification of the two phases, the flow regime, and the inclination of the pipe (conduit). As one might suppose, the major industrial application in this area is gas (G) – liquid (L) flow in pipes.

Prime Movers [8, 10]. Three devices convert electrical energy into the mechanical energy that is applied to various streams in heat transfer applications. These devices are: fans, which move low-pressure gases; pumps, which move liquids and liquid-solid mixtures such as slurries, suspensions, and sludge; and compressors, which move (compress) high-pressure gases.

There are three general process classifications of prime movers — centrifugal, rotary, and reciprocating — that can be selected. Except for special applications, centrifugal units are normally employed in industry. These units are usually rated in terms of four characteristics listed below:

1. Capacity: the quantity of fluid discharge per unit time (the mass flow rate)
2. Increase in pressure, often reported for pumps as *head*: head can be expressed as the energy supplied to the fluid per unit mass and is obtained by dividing the increase in pressure (the pressure change) by the fluid density
3. Power: the energy consumed by the mover per unit time
4. Efficiency: the energy supplied to the fluid divided by the energy supplied to the unit

Finally, the net effect of almost all prime movers is to increase the pressure of the fluid.

Valves and Fittings [8, 10]. Pipes and tubing (and other conduits) are used for the transportation of gases, liquids, and slurries. These ducts are often connected and may also contain a variety of valves and fittings, including expansion and contraction units. Types of connecting conduits include:

1. Threaded
2. Bell-and-spigot
3. Flanged
4. Welded

As indicated above, pipes and tubes, as well as other conduits, are used for the transportation of gases, liquids, and slurries. These ducts are often connected and may also contain a variety of valves and fittings, including expansion and contraction joints. The two major types of connections in conduits include:

1. Threaded
2. Welded

Extensive information on these two classes of connections is available in the literature [8, 9].

Because of the diversity of the types of systems, fluids, and environments in which valves [9] must operate, a vast array of valve types employed in heat exchangers have been developed. Examples of the common types are the globe valve, gate valve, ball valve, plug valve, pinch valve, butterfly valve, and check valve. Each type of valve has been designed to meet specific needs and are almost always located external to the exchanger.

A fitting is a piece of equipment that has for its function one or more of the following:

1. The joining of two pieces of straight pipe or tube (e.g., couplings and unions)
2. The changing of flow direction (e.g., elbows and T's)
3. The changing of pipe or tube diameter (e.g., reducers and bushings)
4. The terminating of a pipe or tube (e.g., plugs and caps)
5. The joining of two streams (e.g., T's and Y's)

Note that fittings find application primarily with pipes as opposed to tubes.

Other Considerations. Another important concept is that referred to as a "calming," "entrance," or "transition" length. This is the length of conduit required for a velocity profile to become fully developed following some form of disturbance in the conduit. This disturbance can arise because of a valve, a bend in the line, an expansion in the line, etc. This is often an important concern. An estimate of this "calming" length, L_c, for laminar flow is

$$\frac{L_c}{D} = 0.05 \, Re \tag{1.60}$$

For turbulent flow, one may employ

$$L_c = 50 D \tag{1.61}$$

Flow in conduits that are not cylindrical (e.g., a rectangular duct) are treated as if the flow occurs in a pipe. For this situation, a hydraulic radius, r_h, is defined as:

$$r_h = \frac{\text{Cross-sectional area perpendicular to flow}}{\text{Wetted perimeter}} \tag{1.62}$$

For flow in a circular tube

$$r_h = \frac{\left(\pi D^2 / 4\right)}{\pi D} = \frac{D}{4} \quad \text{and} \quad D = 4 r_h \tag{1.63}$$

One may extend this concept to any cross-section such that

$$D_{eq} = 4 r_h \tag{1.64}$$

For flow in the annular space between two concentric pipes of diameter D_1 and D_2,

$$D_{eq} = \frac{4\pi \left(D_1^2 - D_2^2\right)}{4\pi \left(D_1 + D_2\right)} = D_1 - D_2 \tag{1.65}$$

Although this approach strictly applies to turbulent flow, it may be employed for laminar flow situations if no other approaches are available.

Finally, flow through a number of pipes or conduits often arise in heat exchanger applications. If flow originates from the same source and exits at the same location, the pressure drop across each conduit *must be the same*. Thus, for flow through conduits 1, 2, and 3, one may write:

$$\Delta P_1 = \Delta P_2 = \Delta P_3 \qquad (1.66)$$

1.7   Environmental Considerations

Environmental factors, both technical and social, have become an integral part of not only heat exchanger design and analysis but also the general subject of process heat transfer. The greatest impact has been in the field of pollution prevention, as it applies to the conservation of energy. Unfortunately, the term pollution prevention has come to mean different things to different people. The term pollution prevention, in this text, is defined as that process or operation that attempts to reduce or eliminate waste of quality energy and/or any accompanying pollutants that are emitted into the environment. However, irrespective of the definition employed, the main focus of pollution prevention today and in the future will be to conserve both energy and resources [13].

Applying pollution prevention strategies — the environmental management option of the future — will not conserve all energy and resources from all production processes. Rather, pollution prevention strategies offer a cost-effective means of minimizing the waste of energy and resources while facilitating compliance with local, state, and federal environmental health and safety regulations. Industrial ecology, green science and engineering, and sustainability are logical extensions to the concepts of pollution prevention that focus on improved energy utilization [13].

Many believe that the key to solving environmental management problems centers on pollution prevention and the related topics of industrial ecology, green science and engineering, and sustainability. The three aforementioned terms — industrial ecology, green science and engineering, and sustainability — are relevant to current and future practices as energy-related industries balance costs and benefits when evaluating process improvement opportunities. Each of these topics in its own way attempts to provide systematic design procedures that can assist in the development, design, operation, and uses of heat exchangers plus energy conservation and processes costs. Analysis of the benefits of process changes can be expanded beyond simple savings in energy and operating expenditures to more broadly include avoided environmental control costs, avoided liability, and in some cases, improve relations with local communities. Each of the above topics are briefly discussed below from a generic perspective.

Industrial Ecology. The term *industrial* in industrial ecology refers to all human activities occurring within modern technological society. Tourism, housing, medical services, transportation, agriculture, etc., are all considered part of the industrial systems in addition to what is traditionally considered industrial activities that produce energy, goods, and products [13, 14]. The term *ecology* refers to relationships between organisms and their past, present, and future environments. This includes physiological responses of individuals, structures, and the dynamics of populations, interactions among species, organization of biological communities, and the processing of (in particular) energy and matter in ecosystems [13–16].

Industrial ecology is a developing conceptual framework that attempts to expand knowledge and understanding of an industrial system's environmental impact and to identify and implement strategies that reduce negative impacts at the local and global scale, with the goal of creating sustainable industries. Industrial ecology encompasses the study of physical, chemical, biological, and energy interactions and interrelationships within and between industrial and ecological systems. Industrial ecology often focuses on reducing the environmental impacts of goods, services, and energy uses using a systems-based approach and on innovations that can significantly improve environmental performance [13–16].

Industrial ecology may also be viewed as encompassing all components of the industrial economy as a whole and its interaction and relationship to the biosphere. It is a systematic, comprehensive view that involves identifying the complex patterns of energy and material flow within and outside the industrial system, in contrast to traditional approaches that mostly consider the economy in abstract monetary terms. Thus, understanding energy and materials flows and their impact on the environmental systems, along with redesigning products, services, and production methods, can reduce negative environmental impacts. In conclusion, industrial ecology applications lead to more efficient use of energy and materials and involves pollution prevention, energy conservation, and total quality environmental management practices in order to develop and design sustainable industrial systems.

Green Chemistry and Green Engineering. Activities in the field of green chemistry and green engineering are increasing at a near exponential rate. This subsection aims to familiarize the reader with both green chemistry and green engineering by defining and providing principles to each.

Before beginning, it is important that the term *green* should not be considered a new method or type of chemistry or engineering. Rather, it should be incorporated into the way scientists and engineers design for categories that include energy, the environment, manufacturability, disassembly, recycle, serviceability, and compliance. Today, the major green element is to search for technology to reduce and/or eliminate waste from operations and processes (the pollution prevention approach), with an important priority of not wasting valuable energy and not creating waste in the first place.

Green chemistry [13, 17], also called "clean chemistry," refers to that field of chemistry dealing with the synthesis, processing, and use of both energy and materials that reduce risks to humans and the environment. It has also been defined as the invention, design, and application of chemical products and processes to reduce or to eliminate the generation of pollutants. Green engineering is similar to green chemistry in many respects, as witnessed by the underlying urgency of attention to the environment seen in both sets of the principles. The U.S. Environmental Protection Agency [18] offered the following comments:

> "Green engineering is the design, commercialization, and use of processes and products in a way that reduces pollution, promotes sustainability, and minimizes risk to human health and the environment without sacrificing economic viability and efficiency. Green engineering embraces the concept that decisions to protect human health and the environment can have the greatest impact and cost-effectiveness when applied early in the design and development phase of a process or product."

In effect, green engineering primarily involves the development of equipment for factoring the environment into the design of processes and products. Therefore, green engineering also supports incremental improvements in materials, machine efficiencies, and energy use and conservation

What is the difference between green engineering and green chemistry? From the definitions given previously, one would conclude that green engineering is concerned with the design, commercialization, and use of all types of processes and products, whereas green chemistry covers just a very small subset of this, that is, the design of chemical processes and products. Although green chemistry may be viewed as a subset of green engineering, it is, in fact, a very broad field, encompassing everything from improving energy efficiency in manufacturing processes to developing plastics from renewable resources. One important aspect in this area is the development of mathematically based tools that aid in decision making when faced with alternatives. Another is the discovery and development of new technology that makes the design, commercialization, and use of processes that reduce or eliminate pollution.

Sustainability. The term *sustainability* has many different meanings to different people. To sustain is defined as to "support without collapse" [13, 17, 19]. Discussion of how sustainability should be defined was initiated by the Bruntland Commission. This group was assigned a mission to create a "global agenda for change" by the General Assembly of the United Nations (UN) in 1984. They defined *sustainable* very broadly: humanity has the ability to make development sustainable — to ensure that it meets the needs of the people present without compromising the ability of future generations to meet their own needs. In a very real sense, "sustainability" involves simultaneous programs in these major areas: human, economic, technological, energy, and environmental. Sustainability requires conservation of energy and resources, minimizing depletion of nonrenewable resources, and using sustainable practices for managing renewable resources. There can be no product development or economic activity of any kind without available resources. Except for solar energy, the supply of resources is finite. Depletion of nonrenewable resources and overuse of otherwise renewable resources limit their availability to future generations. As a side note, to add perspective to this issue, sustainability in many ways conflicts with inevitable changes in the socioeconomic landscape, politics, human evolution, and the basic principles of entropy [6] as well as other thermodynamic principles (see Part III, Chapter 11).

The "energy resource" is a sustainability topic within itself. Consider fossil fuels. One of the greatest challenges facing humanity during this century will surely be that of providing everyone on the planet access to safe, clean energy supplies. The use of energy has been central to the functioning and development of human societies throughout history. However, in recent years, fossil fuel energy usage and pricing have run amuck. World petroleum resources currently appear to be abundant with the recent accelerated expansion of oil and gas extraction from alternative oil supplies, i.e., tar sands, shale, and using new extraction technologies (e.g., fracking). Despite recent painfully high prices for petroleum (the price of crude oil exceeded $100 per barrel in 2014), the price of crude oil at the time of the preparation of this manuscript was below $40 per barrel due to the aforementioned expansion of natural gas and crude oil supplies. Furthermore, the International Energy Agency projected that more than 80% of the world energy demand will continue to be met by fossil fuels through 2030. Fossil fuels will be a major energy source into the near future, and therefore, there is an immediate need to increase the present efficiency of fossil fuel usage to reduce its impact on human health and the environment and make this fuel more sustainable.

Natural resources were initially abundant relative to needs. In the earlier years of the Industrial Revolution, production was limited by technology and labor. However, population is in surplus and technology has reduced the need for human labor. Increasingly, production is becoming limited by earth's natural environment that includes the availability of natural resources. The demand for most resources has increased at a near exponential rate. The emergence of newly developing economies, particularly those in the highly populated countries of China and India, has further increased the demand for energy resources. Humans need to realize that conservation of nonrenewable energy sources is essential to sustainability.

The reader should note that two later chapters of the book are devoted to environmental management topics as they relate to process heat transfer. Chapter 11 in Part III is concerned with "quality energy" via the application of entropy and second law thermodynamics principles [6] to heat exchanger design. Chapter 12 in Part III is concerned with accidents; it addresses hazard (particularly) and health risk assessment issues [20] as they apply to the general subject of heat transfer.

Note: The bulk of the material is this section was adapted from: R. Dupont, K. Ganesan and L. Theodore, *Pollution Prevention: Industrial Ecology, Green Science and Engineering, and Sustainability,* 2nd edition, CRC Press/Taylor & Francis Group, Boca Raton, FL, 2017 [13].

1.8   Process Heat Transfer

Heat transfer has been described as the study of the rates at which heat is exchanged between heat sources and receivers. *Process heat transfer* deals with the rates of heat exchange as they occur in the heat-transfer equipment of the engineering and chemical processes. The study of heat transfer, especially in Part II of this text, brings to better focus the importance of the temperature difference between the source and receiver, which is, after all, the driving force whereby the transfer of heat is accomplished. A typical problem of process heat transfer is concerned with the quantities of heats to be transferred, the rates at which they may be transferred because of the natures of the bodies, the driving potential, the extent and arrangement of the surface separating the source and receiver, and the amount of mechanical energy which may be expended to facilitate the transfer. Since heat transfer involves an *exchange* in a *system*, the loss of heat by one body will equal the heat absorbed by another *within the confines of the same system*.

The next three chapters which follow will review the three individual heat-transfer phenomena discussed earlier.

References

1. I. Farag and J. Reynolds, *Heat Transfer*, A Theodore Tutorial, Theodore Tutorials, East Williston, NY, originally published by USEPA/APTI, RTP, NC, 1996.
2. J. Reynolds, *Material and Energy Balances*, A Theodore Tutorial, Theodore Tutorials, East Williston, NY, originally published by USEPA/APTI, RTP, NC, 1992.
3. E. Buckingham, Phys. Rev., 4, 345–376, location unknown, 1914.
4. D. Green and R. Perry (editors), *Perry's Chemical Engineers' Handbook*, 8th edition, McGraw-Hill, New York City, NY, 2008.
5. L. Theodore and J. Reynolds, *Thermodynamics*, A Theodore Tutorial, Theodore Tutorials, East Williston, NY, originally published by USEPA/APTI, RTP, NC, 1991.
6. L. Theodore, F. Ricci, and T. VanVliet, *Thermodynamics for the Practicing Engineer*, John Wiley & Sons, Hoboken, NJ, 2009.
7. L. Theodore, *Heat Transfer Applications for the Practicing Engineer*, John Wiley & Sons, Hoboken, NJ, 2013.
8. I. Farag, *Fluid Flow*, A Theodore Tutorial, Theodore Tutorials, East Williston, NY, originally published by USEPA/APTI, RTP, NC, 1994.
9. D. Kern, *Process Heat Transfer*, McGraw-Hill, New York City, NY, 1950.
10. P. Abulencia and L. Theodore, *Fluid Flow for the Practicing Chemical Engineer*, John Wiley & Sons, Hoboken, NJ, 2009.
11. L. Theodore, *Transport Phenomena for Engineers*, Theodore Tutorials, East Williston, NY, originally published by International Textbook Co. Scranton, PA, 1971.
12. L. Moody, *Friction Factors for Dye Flow*, Trans. Am. Soc. Mech. Engrs., 66, 671–84, New York City, NY, 1944.
13. R. Dupont, K. Ganesan and L. Theodore, *Pollution Prevention: Industrial Ecology, Green Science and Engineering, and Sustainability*, 2nd edition, CRC Press/Taylor & Francis Group, Boca Raton, FL, 2017.

14. D. O'Rourke, L. Connelly, and C. Koshland. Industrial Ecology: A Critical Review. *International Journal of Environment and Pollution* 6(2/3):89–112, Washington, D.C., 1996.

15. A. Garner and G. Keoleian, *Industrial Ecology: An Introduction*. Ann Arbor, MI: National Pollution Prevention Center for Higher Education, University of Michigan. http://www.umich.edu/~nppcpub/resources/compendia/INDEpdfs/INDEintro.pdf, 1995.

16. V. Thomas et. at. Industrial Ecology: Policy Potential and Research Needs. *Environmental Engineering Science* 20(1):1–9, Washington, D.C., 2003.

17. K. Skipka and L. Theodore. *Energy Resources: Availability, Management, and Environmental Issues.* CRC Press/Taylor & Francis Group. Boca Raton, FL, 2014.

18. U.S. EPA, About Green Engineering? http://www2.epa.gov/green-engineering/about-green-engineering#definition, Washington, D.C., 2015.

19. M.K. Theodore and L. Theodore, *Introduction to Environmental Management*, 2nd edition, CRC Press/Taylor & Francis Group, Boca Raton, FL, 2010.

20. L. Theodore and R. Dupont, *Environmental Health and Hazard Risk Assessment: Principles and Calculations*, CRC Press/Taylor & Francis Group, Boca Raton, FL, 2012.

2

Steady-State and Unsteady-State Heat Conduction

Introduction

The laws of heat conduction were established nearly two centuries ago and are generally attributed to Joseph Fourier, a French mathematician and physicist. In numerous systems involving flow, e.g., heat flow, fluid flow, or flow of electricity, etc., it has been observed that the flow is directly proportional to a *driving force* or *potential* and inversely proportional to the *resistance*(s) impacting the system. This may be expressed in the following manner:

$$\text{Flow} \propto \frac{\text{driving force}}{\text{resistance}} \tag{2.1}$$

Fluid flows through a pipe or tube from regions of high pressure to regions of low pressure. The roughness of the pipe is one of the resistances to fluid flow. In a similar manner, heat transfer through a plane wall flows from regions of high temperature to regions of low temperature. The physical properties of the wall provides a measure of the resistance to heat flow.

In fluid flow, the pressure difference along the pipe, ΔP, is the driving force. The driving force in heat transfer is the temperature difference, ΔT, between the hot and cold side of the wall – more formally known as the *temperature difference driving force*.

As previously stated, the wall properties creates a resistance to heat flow. For example, a metal wall offers little resistance to heat transfer, while a wall of insulation offers a large resistance. Since *conductance* is the reciprocal of *resistance*, Equation (2.1) may be rewritten as:

$$\text{Flow} \propto \left(\text{conductance}\right)\left(\text{driving force}\right) \tag{2.2}$$

In order to make Equation (2.2) an equality, the proportionality must be replaced with a constant. The conductance is a measured property of the entire wall – although it has also been found experimentally that the flow of heat is influenced independently by the thickness and the area of the wall. The conductance of a plane wall is a function of the wall thickness, Δx [=] ft, the heat-flow area perpendicular/normal to the direction of energy flow, A [=] ft², and the thermal conductivity of the wall material, k [=] Btu/hr·ft² (see Chapter 1). The relationship is provided below:

$$\text{Conductance}_{\text{plane wall}} = \frac{kA}{\Delta x}\ [=]\text{Btu} / \text{hr} \cdot {}^{\circ}\text{F} \tag{2.3}$$

When the temperature difference from one side of the wall to the other is defined as ΔT [=] °F, Equation 2.2 becomes:

$$Q = k\left(\frac{A}{\Delta x}\right)\Delta T\ [=]\text{Btu} / \text{hr} \tag{2.4}$$

The term, Q, is the quantity of heat flow transferred from one side of the wall to the other.

Thermal conductivity, k, is literally the ability *to conduct therms* (a therm is a unit of heat = 100,000 Btu). The conductivity of solids are greater than those of liquids, which in turn are greater than those of gases. Therefore, it is easier to transfer heat through a solid than a liquid, and easier to transfer heat through a liquid than a gas.

Some solids, such as metals, have high thermal conductivities and are called *conductors*. Others have low conductivities, are poor conductors of heat, and are referred to as *insulators*. Thermal conductivities of three such insulators is provided in Table 2.1 in English units.

The thermal conductivities for materials at different states in SI units is provided in Table 2.2.

Table 2.1 Thermal conductivities of three common insulating materials.

	k (Btu/hr·ft·°F)
Asbestos – cement boards	0.430
Fiber, insulating board	0.028
Glass, 1.5 lb/ft³	0.022

Table 2.2 Thermal conductivities for materials of different states.

	k (W/m·°C)
Solid metals	15 – 400
Liquids	0.0 – 100
Gases	0.01 – 0.2
Note that 1 Btu/hr·ft· °F = 1.7307 J/sec·m· °C	

The conductivities of solids may either increase or decrease with temperature. The conductivities of most liquids decrease with increasing temperature (although water is a notable exception). For all common gases and vapors, conductivity increases with increasing temperature. The relationship between thermal conductivity, k, and temperature, T [=] is expressed by the simple linear equation:

$$k = a + bT \ [=] \text{Btu} / \text{hr} \cdot \text{ft} \cdot {}^{\circ}\text{F} \qquad (2.5)$$

where a is the conductivity of the material at 0 °F, and b is a constant denoting the change in the conductivity per degree change in temperature. However, for most practical problems in this text, the variation of the thermal conductivity with temperature is relatively small and is neglected.

The influence of pressure on the conductivities of solids and liquids appears to be negligible, and the reported data on gases are too inexact owing to the effects of free convection (see Chapter 3) and radiation (see Chapter 4) to permit any generalization. From the kinetic theory of gases it can be concluded that the influence of pressure should be small unless a very low vacuum is encountered.

Equations (2.1) to (2.4) provide an introduction to heat transfer via conduction. Equation (2.4) written in differential form is known as *Fourier's Law*, and has broadest applicability for conduction problems going forward:

$$Q = kA \frac{dT}{dx} \qquad (2.6)$$

Seven sections follow that complement the above introductory section:

    2.1   Flow of Heat through a Plane Wall
    2.2   Flow of Heat through a Composite Plane Wall:
         Resistances in Series
    2.3   Flow of Heat through a Pipe Wall
    2.4   Flow of Heat through a Composite Pipe Wall:
         Resistances in Series
    2.5   Steady-State Conduction: Microscopic Approach
    2.6   Unsteady-State Conduction
    2.7   Unsteady-State Conduction: Microscopic Approach

2.1   Flow of Heat through a Plane Wall

Equation (2.4) is used as a starting point to calculate the heat flow through a plane wall:

$$Q = k\left(\frac{A}{\Delta x}\right)\Delta T \tag{2.4}$$

The reader should note that:
Q'(single prime) = Q/L: heat transfer rate/ unit length.
Q''(double prime) = Q/A: heat transfer rate/unit area and defined as *heat flux*.
As previously stated, the *resistance* to heat transfer through a plane wall, R, is the reciprocal of conductance, $k(A/\Delta x)$, and is defined as follows:

$$R = \frac{\Delta x}{kA} \; [=] \, \text{hr} \cdot {}^{\circ}\text{F} / \text{Btu} \tag{2.7}$$

Equation (2.4) can then be rewritten as:

$$Q = \frac{\Delta T}{R} \tag{2.8}$$

Example 2.1 – Flow of Heat through a Wall.
The faces of a 6 in. thick wall measuring 12-ft by 16-ft are maintained at 1500 and 300 °F, respectively. The wall is made of kaolin insulating brick whose thermal conductivity at 932 °F is 0.15 Btu/(hr·ft³·°F/ft). How much heat will escape through the wall?
    SOLUTION: The average temperature of the wall is 900 °F. "Extrapolation" of k to 900 °F will not change this value appreciably. Apply Equation (2.4):

$$Q = k\left(\frac{A}{\Delta x}\right)\Delta T \tag{2.4}$$

where,

$$\Delta T = 1500 - 300 = 1200\ °F$$

$$A = (16)(12) = 192\ ft^2$$

$$\Delta x = \frac{6}{12} = 0.5\ ft$$

Substituting,

$$Q = 0.15\left(\frac{192}{0.5}\right)(1200) = 69,200\ Btu\ /\ hr$$

Example 2.2 – Heat Flow Calculation.

A new 1-ft thick insulating material was recently tested for heat resistant properties. The data recorded temperatures of 70 °F and 210 °F on the cold and hot sides, respectively. If the thermal conductivity of the insulating material is 0.026 Btu/ft·hr·°F, calculate the rate of the heat flux, Q'' or Q/A, through the wall in Btu/ft²·hr. Resolve the problem in SI units.

SOLUTION: The problem statement information is used to calculate the following:

$$\Delta T = 210 - 70 = 140\ °F$$

$$\Delta x = 1.0\ ft$$

Rearrangement of Equation (2.4) followed by substitution yields:

$$\frac{Q}{A} = k\left(\frac{\Delta T}{\Delta x}\right) = (0.026)\left(\frac{140}{1.0}\right) = 3.64\ Btu\ /\ ft^2 \cdot hr$$

The heat flux, Q'', in SI units is given by:

$$Q'' = \frac{Q}{A} = (3.64)\left(\frac{252\ cal\ /\ hr}{1\ Btu\ /\ hr}\right)\left(\frac{1\ ft^2}{0.093\ m^2}\right)$$

$$= 9863\ cal\ /\ m^2 \cdot hr = 9.863\ kcal\ /\ m^2 \cdot hr$$

Example 2.3 – Furnace Application.

The following information is provided. A rectangular furnace wall is made from fire clay (height, $H = 3$ m, width, $W = 1.2$ m, thickness, $\Delta x = 0.17$ m).

The temperature of the inside surface (area = $H \times W$), T_H is 1592 K, and of the outside surface, T_C is 1364 K. Determine the temperature gradient, the heat transfer rate, and the heat transfer flux.

SOLUTION: Determine the temperature at which the wall properties should be calculated. In general, thermophysical properties are calculated at the arithmetic average of the wall temperatures:

$$T_{avg} = (T_H + T_C)/2$$
$$= (1592 + 2364)/2$$
$$= 1478 \, K$$

Obtain the fire clay properties (see Tables in the Appendix). For T = 1478 K,

$$\rho = 2645 \, kg/m^3$$
$$k = 1.8 \, W/m \cdot K$$
$$c_p = 960 \, J/kg \cdot K$$

Calculate the thermal diffusivity, noting that:

$$1 ft^2/s = 0.0929 \, m^2/s$$

and,

$$a = \frac{k}{\rho c_p}$$

Substituting,

$$a = \frac{1.8}{(2645)(960)}$$
$$= 7.09 \times 10^{-7} \, m^2/s$$
$$= \frac{7.09 \times 10^{-7}}{0.0929} = 7.63 \times 10^{-6} \, ft^2/s$$

Since the temperature gradient may be approximated by:

$$\frac{dT}{dx} \approx \frac{T_H - T_C}{X_C - X_H} = \frac{T_H - T_C}{\Delta x}$$

Calculate the temperature gradient:

$$\frac{dT}{dx} = \frac{1364 - 1592}{0.17}$$
$$= -1341 \, K/m$$
$$= -1341°C/m$$

The heat transfer area is:

$$A = H \times W$$
$$= (3)(1.2)$$
$$= 3.6 \, m^2$$

Calculate the heat transfer rate using Equation (2.4):

$$Q = \frac{kA(T_H - T_C)}{\Delta x}$$

Substituting,

$$Q = \frac{(1.8)(3.6)(1592 - 1364)}{0.17}$$
$$= 8691 \, W$$
$$\cong 8.7 \, kW; \quad \text{note that } 1 \, W = 3.412 \, Btu/hr$$
$$= (8691)(3.412)$$
$$= 29,653 \, Btu/hr$$

Example 2.4 – Surface Wall Application.
A concrete wall has a surface area of 30 m² and is 0.30 m thick. It separates warm room air from cold ambient air. The inner surface of the wall is known to be at a temperature of 25 °C, while the outer surface is at −15 °C. The thermal conductivity of the concrete is 1.0 W/m·K.

1. Describe the conditions that must be satisfied in order for the temperature distribution in the wall to be linear.
2. What is the driving force for the transfer of heat?
3. What is the heat loss through the wall?

SOLUTION: Once again, assume:

1. steady-state heat transfer
2. homogenous isotropic medium (2)

The driving force for heat transfer is:

$$\Delta T = T_H - T_C = 25 - (-15) = 40\,^\circ C$$

The thermal resistance due to conduction is:

$$R = \frac{x}{kA}$$

Substituting,

$$R = \frac{0.3}{(1.0)(30)} = 0.01\,^\circ C\,/\,W = 0.0005275\,^\circ F \cdot hr\,/\,Btu$$

To calculate the heat loss, Q, through the wall, apply Equation (2.4) and substitute:

$$Q = \frac{\text{temperature driving force}}{\text{thermal resistance}}$$

$$= \frac{40\,^\circ C}{0.01\,^\circ C\,/\,W} = 4000\ \text{W} = 4\,\text{kW}$$

2.2   Flow of Heat through a Composite Plane Wall: Resistances in Series

Equation (2.8) is of interest when a wall consists of several wall materials in series, such as in the construction of a furnace or boiler firebox. Several types of refractory brick are usually employed since those capable of withstanding higher inside temperatures are more fragile and expensive than those required near the outer surface where the temperatures are considerably lower. Figure 2.1 is an example of a *composite plane wall*. Three different refractory materials, indicated by the subscripts a, b, and c, are layered in series.

When examining this composite plane wall:

1. The heat flow through material a must overcome the resistance, R_a. In a sequential/series fashion, the heat must then flow through material b by overcoming resistance, R_b, followed by material c and its associated resistance, R_c.

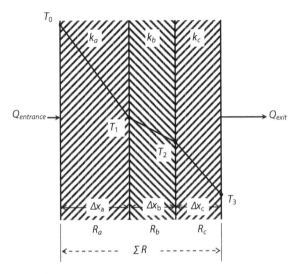

Figure 2.1 Heat flow through a composite wall.

2. Since steady state conditions apply, the heat flow entering the left face of the composite wall, $Q_{entrance}$, must be equal to the heat flow exiting the right face, Q_{exit}.
3. Similarly, the heat flow through material a, Q_a, must be equal to the heat flow through material b, Q_b, and equal to the heat flow through material c, Q_c, i.e., $Q_{entrance} = Q_a = Q_b = Q_c = Q_{exit}$.

Note: When first introduced to heat transfer, many undergraduate students make the following mistake: $Q_{exit} = Q_a + Q_b + Q_c$. THIS IS NOT CORRECT. The above statements are presented in mathematical form:

$$Q_a = \frac{\Delta T_a}{R_a} = \frac{(T_0 - T_1)}{R_a} \quad \text{where } R_a = \frac{\Delta x_a}{k_a A} \qquad (2.10)$$

$$Q_b = \frac{\Delta T_b}{R_b} = \frac{(T_1 - T_2)}{R_b} \quad \text{where } R_b = \frac{\Delta x_b}{k_b A} \qquad (2.11)$$

$$Q_c = \frac{\Delta T_c}{R_c} = \frac{(T_2 - T_3)}{R_c} \quad \text{where } R_c = \frac{\Delta x_c}{k_c A} \qquad (2.12)$$

A summation results in:

$$Q_a R_a + Q_b R_b + Q_c R_c = (T_0 - T_1) + (T_1 - T_2) + (T_2 - T_3) = T_0 - T_3 \quad (2.13)$$

Given that:

$$\text{Since } Q_a = Q_b = Q_c = Q$$
$$Q(R_a + R_b + R_c) = (T_0 - T_1) + (T_1 - T_2) + (T_2 - T_3) \quad (2.14)$$

$$Q = \frac{(T_0 - T_3)}{(R_a + R_b + R_c)} \quad (2.15)$$

$$Q = \frac{T_0 - T_3}{\dfrac{\Delta x_a}{k_a A} + \dfrac{\Delta x_b}{k_b A} + \dfrac{\Delta x_c}{k_c A}} \quad (2.16)$$

$$Q = \frac{\Delta T}{\Sigma R} \quad (2.17)$$

with,

$$\Sigma R = R_a + R_b + R_c = R \quad (2.18)$$

The terms ΣR and R are used interchangeably throughout this text.

Example 2.5 – Flow of Heat Through a Composite Wall.

The wall of an oven consists of three layers of brick. The inside is built from 8 in. of firebrick, k = 0.68 Btu/(hr·ft²·°F/ft), surrounded by 4 in. of insulating brick, k = 0.15 Btu/(hr·ft²·°F/ft), and an outside layer of 6 in. of building brick, k = 0.40 Btu/(hr·ft²·°F/ft). The oven operates at 1600 °F. It is anticipated that the outer side of the wall can be maintained at 125 °F by the circulation of air. How much heat will be lost per square foot of surface area and what are the temperatures at the interfaces of the layers?

SOLUTION:

$$\text{For the firebrick, } R_a = \frac{\Delta x_a}{k_a A} = \frac{8}{(12)(0.68)(1)} = 0.98 \, °F \cdot hr / Btu$$

$$\text{Insulating brick, } R_b = \frac{\Delta x_b}{k_b A} = \frac{4}{(12)(0.15)(1)} = 2.22 \, °F \cdot hr / Btu$$

$$\text{Building brick, } R_c = \frac{\Delta x_c}{k_c A} = \frac{6}{(12)(0.40)(1)} = 1.25 \, °F \cdot hr / Btu$$

In addition,

$$\Sigma R = R = 0.98 + 2.22 + 1.25 = 4.45 °F \cdot hr / Btu$$

The heat loss is given as:

$$Q = \frac{\Delta T}{R} = \frac{1600 - 125}{4.45} = 332 \text{ Btu / hr} \tag{2.8}$$

For the individual layers:

$$\Delta T = QR, \text{ and } \Delta T_a = QR_a, \text{ etc.}$$

Therefore,

$$\Delta T_a = (332)(0.98) = 325 °F$$
$$\Delta T_b = (332)(2.22) = 738 °F$$
$$\Delta T_c = (332)(1.25) = 415 °F$$

and,

$$T_1 = 1600 - 325 = 1275 °F$$
$$T_2 = 1275 - 738 = 537 °F$$

Example 2.6 – Interface Temperatures.
The inside and outside temperatures of a composite wall are T_1 = 27 °C and T_3 = 68.7 °C, respectively. Refer to Figure 2.1 for a visual representation of the wall. The walls consist of 6 inch concrete (c), 8 inch cork-board (b), and 1-inch wood (w) with corresponding thermal conductivities of 0.762, 0.0433, and 0.151 W/m·K, respectively. Calculate the heat transfer rate across the wall. Determine the temperature at the interface between the wood and cork-board.
SOLUTION: Based on the data provided:

$$T_C = T_1 = 27 °C$$
$$T_H = T_3 = 68.7 °C$$
$$\Delta x_c = (6)(0.0254) = 0.1524 \text{ m}$$
$$\Delta x_b = (8)(0.0254) = 0.2032 \text{ m}$$
$$\Delta x_w = (1)(0.0254) = 0.0254 \text{ m}$$

Perform the calculations on 1.0 m² of area, so that A = 1.0. From Equation (2.6):

$$R_c = \frac{\Delta x_c}{k_c} = \frac{0.1524}{0.762} = 0.200\,K/W$$

$$R_b = \frac{\Delta x_b}{k_b} = \frac{2.032}{0.0433} = 4.693\,K/W$$

$$R_w = \frac{\Delta x_w}{k_w} = \frac{0.0254}{0.151} = 0.1682\,K/W$$

Equation (2.9) is employed to calculate the heat transfer rate:

$$Q = \frac{T_1 - T_3}{R_c + R_b + R_w} = \frac{27 - 68.7}{0.200 + 4.693 + 0.1682} = -8.239\,W$$

Note: the solution to this problem assumed that heat flowed from T_1 to T_2. The negative sign indicates that that assumption was incorrect, and that heat flows from T_2 to T_1, i.e., from the hot to the cold surface. If the wall area is 4.0 m², the heat rate is 32.96W.

The interface temperature between the wood and the cork-board, , can now be calculated:

$$Q = \frac{T_1 - T_2}{R_w}$$

$$-8.239 = \frac{27 - T_2}{0.1682}$$

$$T_2 = 28.4\,°C = 301.4\,K$$

2.3   Flow of Heat through a Pipe Wall

As heat flows through the composite plane wall in Figure 2.1, the area for heat transfer does not change from the entrance to exit, i.e., the area is constant throughout the entire heat flow path. In contrast, referring to Figure 2.2 showing a length of pipe, L, the surface area for heat transfer is *not* constant as it flows through the pipe wall from $r_i \rightarrow r_o$.

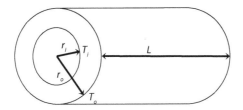

Figure 2.2 Heat Flow through a Pipe Wall.

Since the surface area, A, is defined as $2\pi r L$ it follows that, $2\pi r_i L < 2\pi r_o L$
Rather than use an arithmetic average of the two surface areas, A_i and A_o,
to calculate the heat transferred across a pipe wall, it is preferred to use a
logarithmic mean of the two areas – denoted as A_{LM}.
By definition, the log mean of any two *non-negative* numbers is calculated
as:

$$X_{LM} = \frac{X_1 - X_2}{\ln \dfrac{X_1}{X_2}} \tag{2.19}$$

Depending on the information provided in a problem statement, A_{LM} can
be calculated in a variety of ways (subscripts i and o refer to the inside and
outside surfaces of the pipe, respectively):

$$A_{LM} = \frac{A_o - A_i}{\ln \dfrac{A_o}{A_i}} \tag{2.20}$$

or,

$$A_{LM} = \frac{\pi L (D_o - D_i)}{\ln \dfrac{D_o}{D_i}} \tag{2.21}$$

or,

$$A_{LM} = 2\pi r_{LM} L \qquad r_{LM} = \left(\frac{r_o - r_i}{\ln \dfrac{r_o}{r_i}} \right) \tag{2.22}$$

If the heat flows from the inside of the pipe outwards, the temperature gradient is dT/dr. Equation (2.5) then becomes:

$$Q = kA\left(\frac{dT}{dr}\right)[=]\text{Btu}/\text{hr}\cdot\text{ft} \tag{2.23}$$

where,

$$A = 2\pi rL \tag{2.24}$$

and,

$$Q = k(2\pi rL)\left(\frac{dT}{dr}\right) \tag{2.25}$$

Equation (2.25) may be integrated using general boundary conditions:

BC(1): $T = T_i$ at $r = r_i$
BC(2): $T = T_o$ at $r = r_o$

$$Q = \frac{k(2\pi L)(T_o - T_i)}{\ln\dfrac{r_o}{r_i}} \tag{2.26}$$

The thickness of the pipe wall, Δx_w, is defined as:

$$\Delta x_w = r_o - r_i \tag{2.27}$$

The log mean average radius of the pipe, r_{LM} is defined as:

$$r_{LM} = \frac{r_o - r_i}{\ln\dfrac{r_o}{r_i}} \quad or, \quad \frac{r_{LM}}{r_o - r_i} = \frac{1}{\ln\dfrac{r_o}{r_i}} \tag{2.28}$$

Substitution of equation (2.28) into (2.26) yields:

$$Q = \frac{k(2\pi r_{LM}L)(T_o - T_i)}{r_o - r_i}, \quad \text{where } 2\pi r_{LM}L = A_{LM} \text{ and } r_o - r_i = \Delta x_w \tag{2.29}$$

Substitution produces:

$$Q = \frac{k(A_{LM})(T_o - T_i)}{\Delta x_w} \tag{2.30}$$

Finally, note the similarities between the expression for heat transfer through a *pipe wall*:

$$Q = k \left(\frac{A_{LM}}{\Delta x_w} \right) \Delta T \qquad (2.31)$$

and the expression for heat transfer through a *plane wall*:

$$Q = k \left(\frac{A}{\Delta x} \right) \Delta T \qquad (2.4)$$

2.4   Flow of Heat through a Composite Pipe Wall: Resistances in Series

The general expression that defines heat transfer through a composite *pipe* wall is the same as the expression that defines heat transfer through a composite *plane* wall. Specifically:

$$Q = \frac{\Delta T}{\Sigma R} \qquad (2.17)$$

A composite pipe wall with cylindrical resistances in series is shown schematically in Figure 2.3. Referring to Figure 2.3, there are two composite cylindrical resistances in series—one resistance, R_a, for material a and a second resistance, R_b, for material b. The resistance for material a is calculated as follows:

$$R_a = \frac{\Delta x_{wa}}{k_a A_{LMa}} = \frac{r_2 - r_1}{k_a (2 \pi r_{LMa} L)} \qquad (2.32)$$

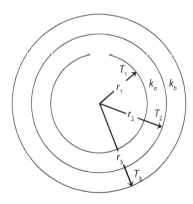

Figure 2.3 Cylindrical resistances in series.

where,

$$r_{LMa} = \frac{r_2 - r_1}{\ln \dfrac{r_2}{r_1}}$$

(2.33)

Similarly, R_b is calculated as:

$$R_b = \frac{\Delta x_{wb}}{k_b A_{LMb}} = \frac{r_3 - r_2}{k_b (2\pi r_{LMb} L)}$$

(2.34)

and,

$$r_{LMb} = \frac{r_3 - r_2}{\ln \dfrac{r_3}{r_2}}$$

(2.35)

Substitution into Equation (2.17) produces:

$$Q = \frac{T_1 - T_3}{\dfrac{\Delta x_{wa}}{k_a A_{LMa}} + \dfrac{\Delta x_{wb}}{k_b A_{LMb}}}$$

(2.36)

Consider a slightly more complicated configuration of a pipe in Figure 2.4 covered (lagged) with rock wool insulation and carrying steam at a temperature, T_s. The steam temperature is considerably higher than that of the atmosphere, T_a. The overall temperature difference driving force through the pipe is $T_s - T_a$.

The resistances to heat transfer, taken in order, from T_s to T_a are:

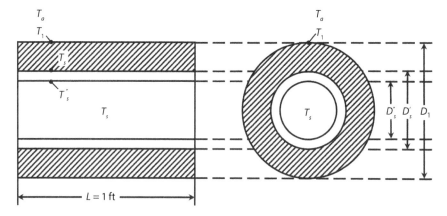

Figure 2.4 Heat loss from an insulated pipe.

1. R_s, the convective heat transfer resistance of the steam. This resistance has been found experimentally to be very small due to a large *convection heat transfer coefficient, h,* so that $T_s \approx T_s''$ (see Chapter 3 - Convection Heat Transfer where, $Q_{conv} = h_{conv} A \Delta T$ for additional details).
2. R_{pipe}, the resistance of the metal pipe wall. This is also assumed to be very small due to a large k_{pipe}, resulting in $T_s' \approx T_s''$. The exception to this assumption occurs for very thick-walled conduits).
3. R_{ins}, the resistance of the rock wool insulation
4. R_{air}, the resistance of the surrounding air

The last resistance is appreciable, although the removal of heat is also effected by the natural convection (also see Chapter 3 for additional details) of the ambient air in addition to the radiation caused by the temperature difference between the outer surface and colder air.

The flow of heat from steam inside an insulated pipe to ambient air is represented mathematically as follows:

Convection heat transfer from steam to the inner pipe wall:

$$Q_s = h_s \left(\pi D_s'' L \right)\left(T_s - T_s'' \right) \ and \ R_s = \frac{1}{h_s \pi D_s''} \qquad (2.37)$$

Conduction heat transfer from the inner pipe wall to the outer pipe wall:

$$Q_{pipe} = \frac{k_{pipe} \left(A_{LMpipe} L \right)}{\Delta x_{pipe}} \left(T_s'' - T_s' \right) \ and \ R_{pipe} = \frac{\Delta x_{pipe}}{k_{pipe} \left(\pi D_{LMpipe} L \right)} \qquad (2.38)$$

where,

$$\Delta x_{pipe} = \frac{D_s'' - D_s'}{2} \ and \ D_{LMpipe} = \frac{D_s'' - D_s'}{\ln \dfrac{D_s''}{D_s'}} \qquad (2.39)$$

Conduction heat transfer from the inner diameter of insulation to outer diameter of insulation:

$$Q_{ins} = \frac{k_{ins} \left(A_{LMins} L \right)}{\Delta x_{ins}} \left(T_s' - T_1 \right) \ and \ R_{ins} = \frac{\Delta x_{ins}}{k_{ins} \left(\pi D_{LMins} L \right)} \qquad (2.40)$$

where,

$$\Delta x_{ins} = \frac{D'_s - D_1}{2} \text{ and } D_{LMins} = \frac{D'_s - D_1}{\ln\dfrac{D'_s}{D_1}} \tag{2.41}$$

Convection heat transfer from the outside of insulation to air:

$$Q_{air} = h_{air}\left(\pi D_1 L\right)\left(T_1 - T_a\right) \text{ and } R_{air} = \frac{1}{h_{air}\left(\pi D_1 L\right)} \tag{2.42}$$

Combining the above expressions leads to:

$$Q = \frac{\Delta T}{\Sigma R} = \frac{T_s - T_a}{\dfrac{1}{h_s \pi D''_s} + \dfrac{\Delta x_{pipe}}{k_{pipe}\left(\pi D_{LMpipe} L\right)} + \dfrac{\Delta x_{ins}}{k_{ins}\left(\pi D_{LMins} L\right)} + \dfrac{1}{h_{air}\left(\pi D_1 L\right)}} \tag{2.43}$$

As previously stated, of the four resistances in the above calculation, the resistance due to steam, R_s, and the resistance of the pipe wall, R_{pipe}, are usually very small compared to the insulation and air resistance, R_{ins} and R_{air}, respectively, based on the following assumptions:

$$R_s = \frac{1}{h_s \pi D''_s} \quad h_s \text{ is very large } \uparrow, \text{ therefore } R_s \text{ is very small } \downarrow. \tag{2.44}$$

and,

$$R_{pipe} = \frac{\Delta x_{pipe}}{k_{pipe}\left(\pi D_{LMpipe} L\right)} \quad k_{pipe} \text{ is very large } \uparrow, \tag{2.38}$$

$$\text{therefore } R_{pipe} \text{ is very small } \downarrow.$$

Equation (2.43) then reduces to:

$$Q = \frac{T_s - T_a}{\dfrac{\Delta x_{ins}}{k_{ins}\left(\pi D_{LMins} L\right)} + \dfrac{1}{h_{air}\left(\pi D_1 L\right)}} \tag{2.45}$$

Example 2.7 – Heat Flow through a Pipe Wall.

A glass pipe has an outside diameter of 6.0 in. and an inside diameter of 5.0 in. It will be used to transport a fluid which maintains the inner surface at 200 °F. It is expected that the outside of the pipe will be maintained at 175 °F. The glass has a thermal conductivity of 0.63 Btu/hr·ft²·°F/ft. What heat flow will occur?

SOLUTION: Rearrange Equation (2.26):

$$Q' = \frac{Q}{L} = \frac{k(2\pi)(T_o - T_i)}{\ln\dfrac{r_o}{r_i}}$$

(2.46)

Substituting,

$$Q' = \frac{Q}{L} = \frac{2\pi(0.63)(200-175)}{\ln\left(\dfrac{6.0}{5.0}\right)} = 542 \text{ Btu / hr} \cdot \text{ft}$$

Example 2.8 – Pipe Heat Loss.

A 3-in. outside diameter steel pipe is covered with a 0.5-in. layer of asbestos (a), which in turn is covered with a 2-in. layer of glass wool (b). Determine the steady-state heat transfer per foot of pipe, L, if the pipe outer surface temperature is 500 °F and the glass wool outer temperature is 100 °F. Assume an asbestos-glass wool interfacial temperature of 300 °F, an average asbestos (a) temperature of 200 °F, and glass wool (b) temperature of 400 °F. Based on a literature review, asbestos and glass wool thermal conductivity values have been estimated to be:

$$k_{a(200°F)} = 0.120 \text{ Btu / hr} \cdot \text{ft} \cdot °F$$

$$k_{b(400°F)} = 0.0317 \text{ Btu / hr} \cdot \text{ft} \cdot °F$$

SOLUTION: Apply Equation (2.36) to the materials asbestos (a) and glass wool (b) :

$$Q = \frac{T_H - T_C}{\left[\dfrac{\Delta x_a}{k_a A_{a,LM}} + \dfrac{\Delta x_b}{k_b A_{b,LM}}\right]}$$

For the asbestos (a), apply Equation (2.21) while maintaining dimensional consistency:

$$A_{a,LM} = \frac{\pi L (D_2 - D_1)/12}{\ln(D_2 / D_1)}, \qquad L = \text{length of pipe, ft}$$

Substituting,

$$A_{a,LM} = \frac{\pi L (4.0 - 3.0)/12}{\ln(4.0/3.0)} = 0.910 L \, \text{ft}^2$$

Similarly, for glass wool (b):

$$A_{b,LM} = \frac{\pi L (8.0 - 4.0)/12}{\ln(8.0/4.0)} = 1.51 L \, \text{ft}^2$$

Substituting into the expression for Q above gives:

$$Q = \frac{500 - 100}{\left[\frac{0.5/12}{(0.120)(0.910L)} + \frac{2.0/12}{(0.0317)(1.51L)} \right]}$$

$$Q = \frac{400}{\dfrac{0.382}{L} + \dfrac{3.48}{L}}$$

Substitution and factoring out L produces:

$$Q' = \frac{Q}{L} = \frac{400}{3.864} = 103.5 \, \text{Btu}/\text{hr} \cdot \text{ft}$$

Example 2.9 – Interfacial Temperature Calculation.
Refer to the previous example. Calculate the outer asbestos temperature, T_i (i.e., the interfacial temperature between the asbestos and glass wool).

SOLUTION: Since the heat transfer rate is now known, a single-layer equation can be used to determine the interfacial temperature. For the glass wool layer, a rearranged form of Equation (2.4) is then applied:

$$\Delta T = \frac{LQ'}{kA}$$

$$500 - T_i = \frac{(2.0/12)(103.5)}{(0.0317)(1.51)}$$

$$500 - T_i = 360$$

$$T_i = 140\,°\text{F}$$

Alternatively, for the asbestos layer:

$$T_i - 100 = \frac{(103.5)(0.5/12)}{(0.120)(0.910)}$$

$$T_i = 100 + 39.5 = 139.5°F$$

The interfacial temperatures are in reasonable agreement.

2.5 Steady-State Conduction: Microscopic Approach

This *microscopic approach* section has been included to complement the earlier material in this chapter, particularly for those readers interested in a more theoretical approach to the treatment of conduction heat transfer. The analysis of Fourier's law was presented previously for a rather simple system. Generally, a fluid will possess three temperature gradients with corresponding heat-flux components in a given coordinates system. When this is the case, three heat-flux terms, Q_x'', Q_y'', and Q_z'', arise at every point in a system described by rectangular coordinates. The flux terms are given by:

$$Q_x'' = -k\frac{dT}{dx} \quad Q_y'' = -k\frac{dT}{dy} \quad Q_z'' = -k\frac{dT}{dz}$$

All components of the heat flux in a given coordinate system can be expressed in terms of the thermal conductivity of a material and the temperature gradient. These are presented in Table 2.3 for rectangular, cylindrical, and spherical coordinates.(2,3)

Table 2.3 Heat-flux components.

Component	Rectangular coordinates	Cylindrical coordinates	Spherical coordinates
Q_1''	$Q_x'' = -k\dfrac{\partial T}{\partial x}$	$Q_r'' = -k\dfrac{\partial T}{\partial r}$	$Q_r'' = -k\dfrac{\partial T}{\partial r}$
Q_2''	$Q_y'' = -k\dfrac{\partial T}{\partial y}$	$Q_\phi'' = -k\dfrac{1}{r}\dfrac{\partial T}{\partial \phi}$	$Q_\theta'' = -k\dfrac{1}{r}\dfrac{\partial T}{\partial \theta}$
Q_3''	$Q_z'' = -k\dfrac{\partial T}{\partial z}$	$Q_z'' = -k\dfrac{\partial T}{\partial z}$	$Q_\phi'' = -k\dfrac{1}{r\sin\theta}\dfrac{\partial T}{\partial \phi}$

The equation describing energy transfer in *stationary* solids serves as an excellent starting point for the microscopic presentation of the general equation for energy transfer. The energy-transfer equation for solids is developed by applying the conservation law for energy on a time-dependent basis to a fixed-volume element. The derivation is available in the literature.(2–4) The resulting equation describes the temperature variation in solids due to energy transfer, subject to the assumptions in the development. It applies to either steady-state or unsteady-state systems and is independent of any particular coordinate system. The describing equations can be expanded into rectangular, cylindrical, and spherical coordinates for *steady-state conduction* (the *unsteady-state* conduction equation can be found in the next two sections). The results are presented in Table 2.4 for a solid that behaves in accordance with Fourier's law.

Table 2.4 provides various differential equations that describe steady-state energy transfer. Integration constants must be specified before a complete solution to the differential equations can be obtained. These integration constants are usually obtained from information at specified locations at the boundary and/or knowledge of initial conditions. The most common conditions specified for these energy transfer problems are:

1. The temperature is specified at a solid boundary or surface.
2. The temperatures on either side of an interface are equal.

Table 2.4 Steady-state energy-transfer equation for stationary solids.

Rectangular coordinates:	
$$\alpha \left[\frac{\partial^2 T}{\partial x^2} + \frac{\partial^2 T}{\partial y^2} + \frac{\partial^2 T}{\partial z^2} \right] + \frac{A}{\rho C_p} = 0$$	(1)
Cylindrical coordinates:	
$$\alpha \left[\frac{1}{r} \frac{\partial}{\partial r} \left(r \frac{\partial T}{\partial r} \right) + \frac{1}{r^2} \frac{\partial^2 T}{\partial \phi^2} + \frac{\partial^2 T}{\partial z^2} \right] + \frac{A}{\rho C_p} = 0$$	(2)
Spherical coordinates:	
$$\alpha \left[\frac{1}{r^2} \frac{\partial}{\partial r} \left(r^2 \frac{\partial T}{\partial r} \right) + \frac{1}{r^2 \sin\theta} \frac{\partial}{\partial \theta} \left(\sin\theta \frac{\partial T}{\partial \theta} \right) + \frac{1}{r^2 \sin^2 \theta} \frac{\partial^2 T}{\partial \phi^2} \right] + \frac{A}{\rho C_p} = 0$$	(3)
Source term,	A = rate of heat generated in solid; energy / time · volume α = thermal diffusivity; $k / \rho C_p$

3. The heat flux is specified at a boundary.
4. The heat flux on either side of an interface are equal.
5. The gradients on either side of an interface may be specified.
6. The value of the temperature at time zero is given (for unsteady-state systems).

Example 2.10 – Temperature Profile in a Cylinder.

A long hollow *cylinder* has its inner and outer surfaces maintained at temperatures T_b and T_a, respectively. The inner and outer radii are b and a, respectively. Calculate the temperature profile in the solid section of the cylinder and determine the flux at both surfaces. Comment on the results. Assume steady-state conditions.

Solution: This system is represented in Figure 2.5 and solved using cylindrical coordinates. If end effects are neglected, T is not a function ϕ and z. Equation (2) in Table 2.4 reduces to:

$$\frac{1}{r}\frac{\partial}{\partial r}\left(r\frac{\partial T}{\partial r}\right) = 0$$

or,

$$\frac{d}{dr}\left(r\frac{\partial T}{\partial r}\right) = 0, \text{ since } T \text{ depends solely on } r$$

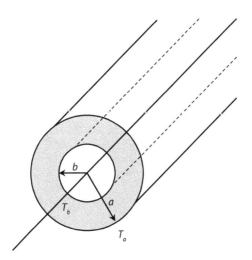

Figure 2.5 Heat conduction, hollow cylinder.

Integrating once yields:

$$r\frac{\partial T}{\partial r} = C$$

Integrating again gives:

$$T = C \ln r + B$$

The boundary conditions (BC) are:

$$BC(1): \ T = T_b \text{ at } r = b$$

$$BC(2): \ T = T_a \text{ at } r = a$$

Applying the BC's to Equation (1) in Table 2.4 yields:

$$T_b = C\ln b + B$$

$$T_a = C\ln a + B$$

Solving the above equations simultaneously gives:

$$C = \frac{T_a - T_b}{\ln(a/b)} \quad and \quad B = T_b - \left[\frac{T_a - T_b}{\ln(a/b)}\right]\ln b$$

Therefore,

$$T = T_b + \left[\frac{T_a - T_b}{\ln(a/b)}\right]\ln\left(\frac{r}{b}\right)$$

The heat flux is given by:

$$Q_r'' = -k\frac{dT}{dr}$$

But,

$$\frac{dT}{dr} = \frac{C}{r}$$

so that,

$$Q_r'' = -k\frac{C}{r} = -k\left[\frac{T_a - T_b}{\ln(a/b)}\right]\left(\frac{1}{r}\right)$$

Therefore,

$$Q_r''\Big|_{r=a} = -k\left[\frac{T_a - T_b}{\ln(a/b)}\right]\left(\frac{1}{a}\right) \quad and \quad Q_r''\Big|_{r=b} = -k\left[\frac{T_a - T_b}{\ln(a/b)}\right]\left(\frac{1}{b}\right)$$

Although the fluxes are not equal, one can easily show that the rate of energy transfer, Q, at both surfaces are equal. If the reader so desires, he/she is invited to review the literature [2, 3] in order to examine and better appreciate how the microscopic approach can be employed in generating the describing equations for various systems.

Example 2.11 – Energy Generation in a Solid.
A constant rate of energy/unit volume A is uniformly generated in a solid cylinder. The temperature at the outer surface of the cylinder ($r = a$) is maintained at T_a. Calculate the temperature profile and the heat flux at the outer wall under steady-state conditions.

SOLUTION: This problem is solved once again using cylindrical coordinates. Based on the problem statement and physical grounds, it is concluded that the temperature is solely a function of r. Equation (2) in Table 2.4 reduces to:

$$\frac{a}{r}\frac{d}{dr}\left(r\frac{dT}{dr}\right) + \frac{A}{\rho C_p} = 0$$

Integrating this equation once and rearranging the coefficients gives:

$$r\frac{dT}{dr} = -\frac{Ar^2}{4k} + B$$

Integrating again yields:

$$T = -\frac{Ar^2}{4k} + B\ln r + C$$

The BC's are now written as:

$$BC(1):\ T = T_a \text{ at } r = a$$

$$BC(2):\ T \neq -\infty \text{ at } r = 0$$

or,

$$\frac{dT}{dr} = 0 \text{ at } r = 0,\ \text{ due to symmetry}$$

Substituting the BC into the equation above gives:

$$B = 0 \quad and \quad C = T_a + \left(\frac{A}{4k}\right)a^2$$

so that,

$$T = T_a + \frac{A}{4k}\left(a^2 - r^2\right)$$

The heat flux at the wall is given by:

$$Q_r'' = -k\frac{dT}{dr}$$

However,

$$\frac{dT}{dr} = -\frac{Ar}{2k}$$

so that,

$$Q_r'' = \frac{Aa}{2}$$

The maximum temperature is obtained at $r = 0$, i.e.,

$$T_{max} = T_a + \frac{Aa^2}{4k}$$

2.6   Unsteady-State Heat Conduction

While engineers and scientists spend time working with systems that operate under steady-state conditions, there are some processes that are transient in nature, i.e., varying with *time, t*. These are defined as *unsteady-state* systems.

It is necessary to be able to predict how process variables change with time, as well as how these effects will impact the design and performance of systems. The prediction of the unsteady-state temperature distribution in solids is an example of one such process. It can be accomplished very effectively using conduction equations; the energy balance equations can be solved to calculate the spatial and time variation of the temperature within the solid.

In a very real since, this material is an extension of that presented in the previous section. The relationships developed in the preceding chapter applied only to the steady-state conditions in which the heat flow and spatial-temperature profile were constant with time. *Unsteady-state* processes are those in which the heat flow, or the temperature, or both, vary with time at a fixed point in space. Batch heat-transfer processes are typical unsteady-state processes. For example, heating reactants in a tank or the startup of a cold furnace are two unsteady-state applications. Other common examples include: the rate at which heat is conducted through a solid while the temperature of the heat source varies; the daily, periodic variations of the heat of the sun on various solids; the quenching of steel in an oil or cold-water bath; cleaning or regenerating processes, or, in general, any processes that can be classified as intermittent.

A good number of heat transfer conduction problems are time dependent. These *unsteady,* or *transient,* situations usually arise when the boundary conditions of a system are changed. For example, if the surface temperature of a solid is changed, the temperature at each point in the solid will also change. For some cases, the changes will continue to occur until a *steady-state* temperature distribution is reached.

Transient effects also occur in many industrial heating and cooling processes involving solids. Solids are generally separated into one of three physical categories:

1. Finite
2. Semi-infinite
3. Infinite

In order to treat common applications of batch and unsteady-state heat transfer, Kern [1] defined processes as either liquid (fluid) heating/cooling or solid heating/cooling. Some examples are outlined below:

1. Heating and cooling liquids
   a. Liquid batches
   b. Batch reactors
   c. Batch distillation
2. Heating and cooling solids
   a. Constant solid temperature
   b. Periodically varying temperature
   c. Regenerators
   d. Granular solids in stationary beds
   e. Granular solids in fluidized beds

The *physical* representation of several solid systems is provided below:

1. Finite wall (or slab or plate)
2. Semi-infinite solid
3. Semi-infinite flat wall
4. Infinite flat wall
5. Finite rectangular parallelepiped
6. Finite hollow rectangular parallelepiped
7. Semi-infinite rectangular parallelepiped
8. Infinite rectangular parallelepiped
9. Short finite cylinder
10. Long finite cylinder
11. Short finite hollow cylinder
12. Long finite hollow cylinder
13. Semi-infinite cylinder
14. Semi-infinite hollow cylinder
15. Infinite cylinder
16. Infinite hollow cylinder
17. Sphere
18. Hollow sphere

The reader should also note that most of these geometric systems are employed to describe not only conduction systems, but also forced convection (Chapter 3), free convection (Chapter 3), and radiation systems (Chapter 4). Although a comprehensive treatment of all of the above is beyond the scope of this text, the reader is referred to the classic work of Carslaw and Jaeger [4] for a truly all-encompassing treatment of nearly all of these systems.

The above categories can be further classified to include specific applications involving unsteady-state heat conduction in *solids.* These include:

1. Walls of furnaces
2. Structural supports
3. Mixing elements
4. Cylindrical catalysts
5. Spherical catalysts
6. Extended surfaces/ Fins (Chapter 8)
7. Insulating materials (Chapter 9)

Transient heat transfer in infinite plates, infinite cylinders, finite cylinders, spheres, bricks, and other composite shapes has been studied extensively in the past. Farag and Reynolds [5] provide an excellent review that is supplemented with numerous illustrative examples; a semi-theoretical approach to describing the time-positions variations in these systems has been simplified by use of a host of figures and tables.

2.7   Unsteady-State Conduction: Microscopic Approach

This *microscopic approach* section has been included to complement the qualitative material presented in the previous section. It should serve the needs of those readers interested in a more theoretical approach and treatment of unsteady-state conduction. The microscopic equations describing transient heat conduction can be found in Table 2.5 [2, 3]. The equations are provided for rectangular, cylindrical, and spherical coordinates, and are valid subject to the assumptions specified for Table 2.4 in the steady-state section. This section concludes with three illustrative examples adopted from the earlier work of Theodore [2] plus Carslaw and Jaeger [4].

Example 2.12 – Transient Solid Temperature Profile.
A constant rate of energy per unit volume is *uniformly* liberated in a solid of arbitrary shape. The solid is insulated. Obtain the temperature of the solid as a function of position and time if the initial temperature of the solid is zero everywhere.

Table 2.5 Unsteady-state energy-transfer equation for stationary solids.

Rectangular coordinates:	
$$\alpha\left[\frac{\partial^2 T}{\partial x^2}+\frac{\partial^2 T}{\partial y^2}+\frac{\partial^2 T}{\partial z^2}\right]+\frac{A}{\rho C_p}=\frac{\partial T}{\partial t}$$	(1)
Cylindrical coordinates:	
$$\alpha\left[\frac{1}{r}\frac{\partial}{\partial r}\left(r\frac{\partial T}{\partial r}\right)+\frac{1}{r^2}\frac{\partial^2 T}{\partial \phi^2}+\frac{\partial^2 T}{\partial z^2}\right]+\frac{A}{\rho C_p}=\frac{\partial T}{\partial t}$$	(2)
Spherical coordinates:	
$$\alpha\left[\frac{1}{r^2}\frac{\partial}{\partial r}\left(r^2\frac{\partial T}{\partial r}\right)+\frac{1}{r^2\sin\theta}\frac{\partial}{\partial \theta}\left(\sin\theta\frac{\partial T}{\partial \theta}\right)+\frac{1}{r^2\sin^2\theta}\frac{\partial^2 T}{\partial \phi^2}\right]+\frac{A}{\rho C_p}=\frac{\partial T}{\partial t}$$	(3)

SOLUTION: Based on the problem statement *and* physical grounds, it is concluded that the temperature of the solid is *not* a function of position. Therefore, the describing equation(s) are independent of the coordinate system. The equation(s) in Table 2.5 reduce to:

$$\frac{\partial T}{\partial t} = \frac{A}{\rho c_p}$$

or,

$$\frac{dT}{dt} = \frac{A}{\rho c_p}$$

The temperature, T, is solely a function of t. Integrating this equation gives:

$$T = \left(\frac{A}{\rho c_p}\right)t + B$$

The initial condition (IC) is:

$$T = 0 \text{ at } t = 0$$

Therefore,

$$B = 0$$

and,

$$T = \left(\frac{A}{\rho c_p}\right)t$$

The temperature of the solid will increase linearly with time.

Example 2.13 – Time-Position Temperature Variation in a Solid.
Consider the rectangular, insulated copper rod pictured in Figure 2.6. If the rod is initially ($t = 0$) at T_A and the ends of the rod are maintained at T_S at $t \geq 0$, provide an equation that describes the temperature (profile) in the rod as a function of both position and time.

SOLUTION: The geometry of the system is described by rectangular coordinates. The describing equation is Equation (1) from Table 2.5, and it takes the form:

$$\frac{\partial T}{\partial t} = a\frac{\partial^2 T}{\partial x^2}$$

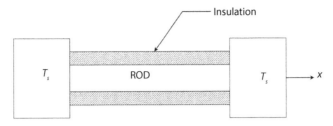

Figure 2.6 Transient rod temperature system; Example 2.15.

The IC and BC are given by:

$$\text{IC: } T = T_A \text{ at } t = 0$$

$$\text{BC (1): } T = T_S \text{ at } x = L$$

$$\text{BC (2): } T = T_S \text{ at } x = 0$$

The solution to this equation can be obtained via separation of variables, resulting in [2, 5]:

$$T = T_S + \left(T_A - T_S\right) \sum_{n=1}^{\infty} \left[\frac{(-1)^{n+1} + 1}{n\pi} \right] e^{-\left[\alpha\left(\frac{n\pi}{L}\right)^2 t\right]} \sin\left(\frac{n\pi x}{L}\right)$$

Example 2.14 – Time-Position Temperature Calculation.

Refer to Example 2.13. If T_S represents saturated steam at 15 psig, the initial temperature T_A is 71 °F, and the rod is 20 inches in length of stainless steel with $k = 9.1$ Btu/hr·ft·°F, $\rho = 0.29$ lb/in.³, and $c_p = 0.12$ Btu/lb·°F, calculate the temperature 0.875 inches from one of the ends after 30 minutes.

Solution: First calculate α:

$$\alpha = \frac{k}{\rho c_p}$$

$$= \frac{9.1}{(0.29)(17.28)(0.12)} = 0.151 \approx 0.15$$

From the steam tables in the Appendix, at $P = 15 + 14.7 = 29.7$ psig, $T_S = 249.7$ °F. Assuming only the first term (i.e., n = 1), in the infinite series in Equation (2) in Example 2.13 above contributes significantly to the solution, one obtains:

$$T = 249.7 + (71 - 249.7) \left[\left[\frac{(2)(-1)^2 + 1}{(1)\pi} \right] e^{-\left[0.15 \left(\frac{(1)\pi}{20/12} \right)^2 (0.5) \right]} \sin \left(\frac{(1)(\pi)(0.875)}{20} \right) \right]$$

$$= 232°F$$

The reader is left with the exercise of determining the effect on the calculation by including more terms in the infinite series. Note that the Monte Carlo method can be applied in the solution to some partial differential equations of this form [6–8].

Example 2.15 – Analytical Analysis.
Outline how to verify that the solution in Example 2.14 describes the system.
   SOLUTION: The solution may be verified if it satisfies the boundary and initial conditions *and* if the equation can be differentiated in order to satisfy Equation (1). For example, for BC(1),

$$T = T_S \text{ at } x = L$$

so that,

$$T = T_S + (T_A - T_S) \left[\sum_{n=1}^{\infty} \left[\frac{4}{n\pi} \right] e^{-\left[\alpha \left(\frac{n\pi}{L} \right)^2 t \right]} \sin \left(\frac{n\pi L}{L} \right) \right]$$

$$= T_S + (T_A - T_S)(0)$$
$$= T_S$$

For BC(2),

$$T = T_S \text{ at } x = 0$$

so that,

$$T = T_S + (T - T_S) \left[\sum_{n=1}^{\infty} \left[\frac{4}{n\pi} \right] e^{-\left[\alpha \left(\frac{n\pi}{L} \right)^2 t \right]} \sin(0) \right]$$

$$= T_S + (T - T_S)(0)$$
$$= T_S$$

Differentiating the describing equation to determine if it satisfies Equation (1) is left as an exercise for the reader.

Example 2.16 – Infinite Solid.
A plane membrane, impervious to the transfer of heat, separates an infinite solid into two equal parts. One half of the solid's temperature is initially at T_0 while the other half is at zero. At time $t = 0$, the membrane is removed and the solids are brought into direct contact with each other. Calculate the temperature in the solid as a function of position and time.

SOLUTION: The initial temperature profile in the system is shown in Figure 2.7.
The PDE describing this system can easily be shown to be:

$$\frac{\partial T}{\partial t} = a \frac{\partial^2 T}{\partial y^2} \text{ where } T = T(y,t)$$

The BC and IC are:

BC(1): T = finite at $y = \infty$, for all t and approaching $T_0/2$ as $t \to \infty$

BC(2): T = finite at $y = -\infty$, for all t and approaching $T_0/2$ as $t \to \infty$

IC: $T = T_0$ at $t = 0$, for $-\infty \le y \le 0$

$T = 0$ at $t = 0$, for $0 \le y \le +\infty$

The above equation can be solved by using Laplace transforms or Fourier integrals. Theodore provides the details of the solution [2].

$$T = \frac{T_0}{2}\left[1 - \mathrm{erf}\left(\frac{y}{\sqrt{4at}}\right)\right]$$

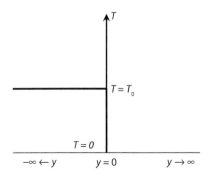

Figure 2.7 Infinite solid separated by a membrane: initial profile; Example 2.16.

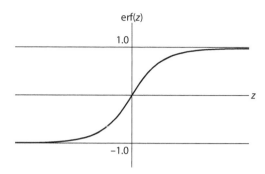

Figure 2.8 Error function.

The error function is tabulated in most advanced mathematics texts [6–8]. A plot of erf(z) against z is presented in Figure 2.8. The reader should note that the notation z used in the error function is a generic variable. Note that this function has the properties:

$$\text{erf}(\infty) = 1.0, \quad \text{erf}(0) = 0, \quad \text{erf}(-\infty) = -1.0$$

and,

$$\text{erf}(-z) = -\text{erf}(z) \tag{2.48}$$

It can also be evaluated numerically by expanding $e^{-\xi^2}$ in a power series:

$$e^{-\xi^2} = 1 - \xi^2 + \frac{1}{2}\xi^4 - \cdots$$

The RHS of this equation can then integrated term by term to give:

$$\text{erf}(z) = \frac{2}{\sqrt{\pi}} \sum_{n=0}^{\infty} \frac{(-1)^n z^{2n+1}}{(2n+1)!}$$

This power series converges for all z, although very slowly for large z. The reader can now verify that Equation (2) is a solution to Equation (1). A check of the BC and/or IC yields:

$$\text{BC:} \quad \text{at } t = \infty \text{ and any } y, \quad T = \frac{T_0}{2}$$

$$\text{IC:} \quad \text{at } t = 0 \text{ and } y > 0, \quad T = \frac{T_0}{2}\left[1 - \mathrm{erf}\left(\infty\right)\right]$$

$$= \frac{T_0}{2}\left[1 - 1\right]$$

$$= 0$$

$$\text{at } t = 0 \text{ and } y < 0, \quad T = \frac{T_0}{2}\left[1 - \mathrm{erf}\left(-\infty\right)\right]$$

$$= \frac{T_0}{2}\left[1 - \left(-1\right)\right]$$

$$= T_0$$

References

1. Donald Q. Kern, *Process Heat Transfer*, McGraw-Hill, New York City, NY, 1950.
2. L. Theodore, *Transport Phenomena for Engineers*, Theodore Tutorials, East Williston, NY, originally published by International Textbook Co., Scranton, PA, 1971.
3. R. Bird, W. Stewart, and E. Lightfoot, *Transport Phenomena*, 2nd edition, John Wiley & Sons, Hoboken, NJ, 2002.
4. H. Carslaw and J. Jaeger, *Conduction of Heat in Solids*, 2nd edition, Oxford University Press, London, 1959.
5. I. Farag and J. Reynolds, *Heat Transfer*, A Theodore Tutorial, Theodore Tutorials, East Williston, NY, originally published by the USEPA/APTI, RTP, NC, 1996.
6. S. Shaefer and L. Theodore, *Probability and Statistics Applications for Environmental Science*, CRC / Taylor & Francis Group, Boca Raton, FL, 2007.
7. L. Theodore and K. Behan, *Introduction to Optimization for Environmental and Chemical Engineers*, CRC Press/Taylor & Francis Group, Boca Raton, FL, 2018.
8. C. Prochaska and L. Theodore, *Introduction to Mathematical Methods for Environmental and Chemical Engineers*, Scrivener-Wiley, Salem, MA, 2018.

3

Forced and Free Convection

Introduction

When a pot of water is heated on a stove, the portion of water adjacent to the bottom of the pot is the first to experience an increase in temperature. Eventually, the water at the top will also become hotter. Although some of the heat transfer from the bottom to the top is explainable by conduction through the water, most of the heat transfer is due to a second mechanism of heat transfer – *convection.* As the water at the bottom is heated, its density becomes lower. This results in convection currents as gravity causes low density water to move upwards while being replaced by the higher density, cooler water from above. This macroscopic mixing is occasionally a far more effective mechanism for transferring heat in the form of energy through fluids compared to conduction. This convective effect is attributed to the buoyant forces that arise due to the aforementioned density differences within a system and is termed *natural* or *free* convection because no external forces, other than gravity, need be applied to transport the energy in the form of heat. It is treated analytically as another external force term in the momentum equation. The momentum (velocity) and energy (temperature) effects are therefore interdependent; consequently, both equations must be solved simultaneously in analytical applications. This treatment is beyond the scope of this text but is available in the literature.

In most industrial applications, however, it is more economical to speed up the mixing action by artificially generating a current by the use of a pump, agitator, fan, compressor, etc. This is referred to as *forced* convection and practicing engineers are primarily interested in this mode of heat transfer (i.e., most industrial applications involve heat transfer by convection) [1]. *Forced convection* is due to the bulk motion of the fluid caused by external forces of the mechanical devices mentioned above and is essentially independent of "thermal" effects. Also note that *both* forced and free convection may exist in some applications [1].

As noted above, heat transfer by either convective effect is due to fluid motion. Free or natural convection occurs when the fluid motion is not implemented by mechanical agitation. But when the fluid is mechanically agitated, the heat is transferred by forced convection. The mechanical agitation may be supplied by stirring, although, in most process applications it is induced by circulating the hot and cold fluids at rapid rates on the opposite sides of pipes or tubes. Forced- and free-convection heat transfer occur at very different speeds, the former being the more rapid and therefore the more common. Also note that the factors which promote high rates for forced convection do not necessarily have the same effect on free convection.

Kern provided the following introduction to this subject [2]. In the flow of heat from steam through a pipe to air it can be shown that the passage of heat into air is not accomplished solely by conduction. Instead, it occurs partly by radiation and partly by free convection. A temperature difference exists between the pipe surface and the *average* temperature of the air. Since the distance from the pipe surface to the region of average air temperature is indefinite, the resistance cannot be computed from $R_a = L_a/k_a A$, using k for air. Instead the resistance must be determined experimentally by appropriately measuring the surface temperature of the pipe, the temperature of the air, and the heat transferred from the pipe as evidenced by the quantity of steam condensed in it. The resistance for the entire surface is then computed from $R_a = \Delta T_a/Q \; (\text{hr} \cdot {}^\circ\text{F})/(\text{Btu})$. If desired, L_a can also be calculated from this value of R_a and would represent the length of a fictitious conduction film of air equivalent to the combined resistances of conduction, free convection, and radiation. The length of the film is of little significance, although the concept of the fictitious film finds numerous applications. Instead it is preferable to deal directly with the reciprocal of the unit resistance h, (defined in the previous chapter) which has an experimental origin. Because of the use of the unit resistance L/k is so much more common than the use of the total surface resistance L/kA, the letter R will now be used to designate L/k, $\left(\text{hr}\cdot\text{ft}^2\cdot{}^\circ\text{F}\right)/\text{Btu}$ and it will simply be called the resistance.

Not all effects other than conduction are necessarily combinations of two effects. Particularly in the case of free or forced convection to liquids and, in fact, to most gases at moderate temperatures and temperature differences, the influence of radiation (see Chapter 4) may be neglected and the experimental resistance corresponds to forced or free convection alone as the case may be.

Consider now either a flat plate or a pipe wall with forced convection of different magnitudes on both sides of the pipe as shown in Figure 3.1. On the inside, heat is transferred from a hot flowing liquid, and on the outside, heat is received by a cold flowing liquid. Either resistance can be measured independently by obtaining the temperature difference between the surface and the average temperature of the fluid. The heat transfer can be determined from the sensible-heat change in either fluid over the length of the plate or pipe in which the heat transfer occurs. Designating the resistance on the inside by R_i and on the outside R_o, the inside and outside solid wall temperatures by T_p and t_w, and applying an expression for steady state:

$$Q = \frac{A_i\left(T_i - T_p\right)}{R_i} = \frac{A_o\left(t_w - t_o\right)}{R_o}; \quad R = \text{unit resistance} \qquad (3.1)$$

where $\Delta T_i = T_i - T_p$ and $\Delta t_w = t_w - t_o$

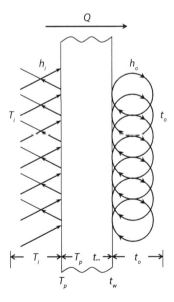

Figure 3.1 Two convection coefficients.

where T_i is the temperature of the hot fluid on the inside and t_o the temperature of the cold fluid on the outside. Replacing the resistances by their reciprocals h_i and h_o, respectively, has come to be defined by some as Newton's law of cooling.

$$Q = h_i A_i \Delta T_i = h_o A_o \Delta t_o \qquad (3.2)$$

The reciprocals of the heat-transfer resistances have the dimensions of Btu / (hr · ft² · °F of temperature difference) and are referred to simply as *film coefficients*. A slightly different approach is provided in the next section.

Inasmuch as the film coefficient is a measure of the heat flow for unit surface and unit temperature differences, it provides a measure of the rate or speed with which fluids having a variety of physical properties and under varying degrees of agitation transfer heat. Other factors influence the film coefficient such as the size of the solid surface and whether or not the fluid is considered to be on the inside or outside of the surface. With so many variables, each having its own degree of influence on the rate of heat transfer (film coefficient), it is fairly understandable why a rational derivation is not available for the direct calculation of the film coefficient. On the other hand, it is impractical to run an experiment to determine the coefficient each time heat is to be added or removed from a fluid. Instead, it is desirable to study some method of correlation whereby several basic experiments performed with a wide range of variables can produce a relationship which will hold for any other combinations of the variables. This is discussed later in the chapter.

The remainder of the chapter consists of six sections:

> 3.1. Forced Convection Principles
> 3.2. Convective Resistances
> 3.3. Heat Transfer Coefficients: Quantitative Information
> 3.4. Convection Heat Transfer: Microscopic Approach
> 3.5. Free Convection Principles and Applications
> 3.6. Environmental Applications

3.1   Forced Convection Principles

In order to circumvent the difficulties encountered in the analytical solution of microscopic heat-transfer problems, it is common practice in engineering to write the rate of heat transfer in terms of the aforementioned heat transfer coefficient h, a topic that will receive extensive treatment in this chapter [3, 4]. If a surface temperature is now designated as T_S, and T_M represents the temperature of the bulk fluid, one may write that

$$Q = hA\left(T_S - T_M\right) \tag{3.3}$$

Since h and T_S are usually functions of the area A, the above equation may be rewritten in differential form as

$$dQ = h\left(T_S - T_M\right)dA \tag{3.4}$$

Integrating for the area gives

$$\int_0^Q \frac{dQ}{h\left(T_S - T_M\right)} = \int_0^A dA \tag{3.5}$$

This concept of a heat-transfer coefficient is an important concept in heat transfer, and is often also referred to in the engineering literature as the *individual film* coefficient.

The expression in Equation (3.3) may be better understood by referring to the heat transfer to a fluid flowing in a conduit. For example, if the resistance to heat transfer is thought of as existing only in a laminar film [2] adjacent to the wall of the conduit, the coefficient h may then be viewed as equivalent to $k/\Delta x_e$, where Δx_e is the equivalent thickness of a stationary film that offers the same resistance corresponding to the observed value of h. This is represented pictorially in Figure 3.2. In effect, this concept simply replaces the real resistance with a hypothetical one.

The reader should note that for flow past a surface, the velocity at the surface (s) is *zero* (no slip) [2]. The only mechanism for heat transfer at the surface is therefore conduction. One may therefore write

$$Q = -kA \frac{dT}{dx}\bigg|_z \tag{3.6}$$

where z = any solid-surface location

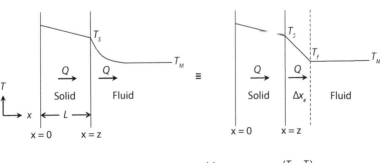

Figure 3.2 Convection temperature profile.

If the temperature profile in the fluid can be determined (i.e., $T = T(x)$), the gradient, dT / dx, can be evaluated at all points in the system including the surface, $dT / dx|_z$.

The heat transfer coefficient, h, was previously defined by

$$Q = hA(T_S - T_M) \tag{3.3}$$

Therefore, combining Equations (3.3) and (3.6) leads to

$$-kA\frac{dT}{dx}\bigg|_z = hA(T_S - T_M) \tag{3.7}$$

Since the thermal conductivity, k, of the fluid is usually known, information on h may be obtained. This information is provided later in this chapter.

To summarize, the transfer of energy by convection is governed by Equation (3.3) and is referred to by many as *Newton's Law of Cooling*,

$$Q = hA(T_S - T_M) \tag{3.3}$$

where Q is the convective heat transfer rate (Btu/hr); A, the surface area available for heat transfer (ft^2); T_S, the surface temperature (°F); T_M, the bulk fluid temperature (°F); and h is the aforementioned convection heat transfer coefficient, also termed *film coefficient* or *film conductance* (Btu / hr · ft^2 · °F or W/m^2 · K). Note that Btu / hr · ft^2 · °F = 5.6782 W / m^2 · K.

The magnitude of h depends on whether the transfer of heat between the surface and the fluid is by forced convection or by free convection, radiation, or boiling or condensation (to be discussed in later chapters). Typical values for h are given in Tables 3.1 and 3.2.

Table 3.1 Typical film coefficients.

	h	
Mode	Btu / hr · ft^2 · °F	W / m · K
Forced convection		
Gases	5–50	25–250
Liquids	10–4000	50–20,000
Free convection (see later section)		
Gases	1–5	5–25
Liquids	10–200	50–1000
Boiling/condensation	500–20,000	2500–100,000

Table 3.2 Film coefficients in pipes.

	h, inside pipes	h, outside pipes[b,c]
Gases	10–50	1–3 (n), 5 – 20 (f)
Water (liquid)	200–2000	20–200 (n), 100 – 1000 (f)
Boiling water[d]	500–5000	300–9000
Condensing steam[d]		1000–10,000
Nonviscous fluids	50–500	50–200 (f)
Boiling liquids[d]		200–2000
Condensing vapor[d]		200–400
Viscous fluids	10–100	20–50 (n), 10 – 100 (f)
Condensing vapor[d]		50–100

[a] h = Btu / hr · ft² · °F
[b] (n) = natural convection
[c] (f) = forced convection

Example 3.1 – Convective Flow Over a Flat Plate. Hot gas at 530 °F flows over a flat plate of dimensions 2 ft by 1.5 ft. The convection heat transfer coefficient between the plate and the gas is 48 Btu / ft² · hr · °F. Determine the heat transfer rate in Btu / hr and kW from the air to one side of the plate when the plate is maintained at 105 °F.

SOLUTION: Once again, assume steady-state conditions and constant properties. Write Newton's law of cooling to evaluate the heat transfer rate. Note that the gas is hotter than the plate. Therefore, Q will be transferred from the gas to the plate so that one may write

$$Q = hA(T_S - T_M) \qquad (3.3)$$

Substituting

$$Q = (48)(3)(530 - 105) = 61,200 \text{ Btu/hr}$$
$$= 61,200 / 3.4123 = 17,935 \text{ W}$$
$$= 17.94 \text{ kW}$$

Example 3.2 – Fluid Temperature Calculation. The glass window of area 3.0 m² shown in Figure 3.3 has a temperature at the outer surface of 10 °C. The conductivity of the glass is 1.4 W/m · K. The convection coefficient (heat transfer coefficient) of the air is 100 W/m² · K. The heat transfer is 3.0 kW. Calculate the bulk temperature of the fluid.

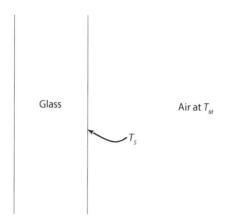

Figure 3.3 Convective glass.

SOLUTION: Write the equation for heat convection. Equation (3.3) is once again given by

$$Q = hA(T_S - T_M)$$ (3.3)

where T_S = surface temperature at the wall and T_M = temperature of the bulk fluid. Solving for the unknown, the air temperature T_M, and substituting values yields

$$T_M = T_S - \frac{Q}{hA}$$

$$T_M = (273 + 10) - \frac{3000}{(100)(3)} = 273\,\mathrm{K} = 0°\mathrm{C}$$

Example 3.3 – Cylindrical Reactor Calculation. Consider a closed cylindrical reactor vessel of diameter $D = 1$ ft, and $L = 1.5$ ft. The surface temperature of the vessel, T_1, and the surrounding temperature, T_2, are 390 °F and 50 °F, respectively. The convective film heat transfer coefficient, h, between the vessel wall and surrounding fluid is 4.0 Btu / ft · hr · °F. Calculate the thermal resistance in °F·hr/Btu.

SOLUTION: Write the heat transfer rate equation:

$$Q = hA(T_1 - T_2)$$ (3.3)

Since $D = 1$ ft and $L = 1.5$ ft, the total heat transfer area may be determined.

$$A = \pi D + \frac{2\pi D^2}{4}$$

Substituting

$$A = \pi (1)(1.5) + \frac{2\pi (1)^2}{4} = 6.28 \,\text{ft}^2$$

Calculate the rate of heat transfer.

$$Q = (4)(6.28)(390 - 50) = 8541 \,\text{Btu/hr}$$

Finally, calculate the thermal resistance associated with the film coefficient.

$$R = \frac{1}{hA} \tag{3.8}$$

Substituting

$$R = \frac{1}{(4.0)(6.28)} = 0.0398°\text{F} \cdot \text{hr} / \text{Btu}$$

Example 3.4 – Referring to the previous example, convert the resistance to K/W and °C/W.

SOLUTION: First note that

$$1\,\text{W} = 3.412 \,\text{Btu/hr}$$

and $1\,°\text{C}$ change corresponds to a $1.8\,°\text{F}$ change. Therefore,

$$R = (0.0398)(3.412)/1.8$$
$$= 0.075°\text{C} / \text{W}$$
$$= 0.075 \,\text{K} / \text{W}$$

3.2   Convective Resistances

Consider heat transfer across a flat plate, as pictured in Figure 3.4. The total resistance (R_t) may be divided into three contributions: the inside film (R_i), the wall (R_w), and the outside film (R_o),

$$R_t = \sum R = R_i + R_w + R_o$$

where the total resistance can be expressed as

$$R_t = \frac{1}{h_i A} + \frac{\Delta x}{kA} + \frac{1}{h_o A} \tag{3.9}$$

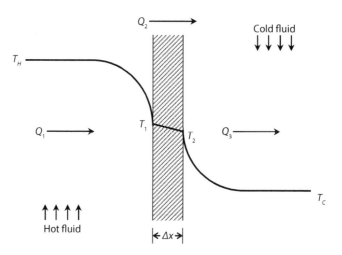

Figure 3.4 Heat flow through a flat plate.

The term h_i is the *inside* film coefficient (Btu / hr · ft² · °F); h_o, the *outside* film coefficient (Btu / hr · ft² · °F); A, the surface area (ft²); Δx, the thickness (ft); and, k, the thermal conductivity (Btu / hr · ft · °F) [3, 4].

Returning to Newton's law (see introduction section), the above development is now expanded and applied to the system pictured in Figure 3.4. Based on Newton's law of cooling,

$$Q_1 = h_i A \left(T_H - T_1 \right) \qquad (3.10)$$

$$Q_2 = \frac{kA}{\Delta x} \left(T_1 - T_2 \right) \qquad (3.11)$$

$$Q_3 = h_o A \left(T_2 - T_C \right) \qquad (3.12)$$

Assuming steady-state ($Q = Q_1 = Q_2 = Q_3$), and combining the above in a manner similar to the development provided earlier yields:

$$Q = \frac{T_H - T_C}{\left(\dfrac{1}{h_i A} + \dfrac{\Delta x}{kA} + \dfrac{1}{h_o A} \right)} \qquad (3.13)$$

which may also be written as

$$Q = \frac{T_H - T_C}{R_i + R_w + R_o} = \frac{T_H - T_C}{R_t} \qquad (3.14)$$

3.3   Heat Transfer Coefficients: Quantitative Information

Many heat transfer film coefficients have been determined experimentally. Typical ranges of film coefficients are provided in Tables 3.3–3.5. Explanatory details are provided in the subsections that follow. In addition to those provided in Tables 3.3–3.5 that follow, many empirical correlations can be found in the literature for a wide variety of fluids and flow geometries [1].

One of the more critical steps in solving a problem involving convection heat transfer is the estimation of the convective heat transfer coefficient. Cases considered in this section involve forced convection, while natural convection is considered later in the chapter. During forced convection heat transfer, the fluid flow may be either external to the surface (e.g., flow over a flat plate, flow across a cylindrical tube, or across a spherical object) or inside a closed surface (e.g., flow inside a circular pipe or a duct).

When a fluid flows over a flat plate maintained at a different temperature, heat transmission takes place by forced convection. The nature of the flow (laminar or turbulent) influences the forced convection heat transfer rate. The correlations used to calculate the heat transfer coefficient between the plate surface and the fluid are usually presented in terms of the average Nusselt number, $\overline{Nu}_x$, the average Stanton number, $\overline{St}$, (or, equivalently, the Colburn j_H factor) [2], the local Reynolds number, Re_x, and the Prandtl number, Pr. In some cases, the Peclet number, Pe, is also used (see also Table 1.1).

Correlation details as presented in Tables 3.3 and 3.4 apply in this section for:

1. Convection from a plane surface
2. Convection in circular pipes
   a. laminar flow
   b. turbulent flow
3. Convection in non-circular conduits
4. Convection normal to a cylinder
5. Convection normal to a number of circular tubes
6. Convection for spheres
7. Convection between a fluid and a packed bed

Topics (2) and (5) receive the bulk of the treatment that follows.

Flow in a Circular Tube. Many industrial applications involve the flow of a fluid in a conduit (e.g., fluid flow in a circular tube). When the Reynolds

Table 3.3 Summary of forced convection heat transfer correlations for external flow.

Geometry	Correlation			Conditions
Flat plate: laminar	Hydrodynamic boundary layer	δ/x	$= 5\,\mathrm{Re}_x^{-1/2}$	Laminar flow, properties at the fluid film temperature, T_f
	Local friction factor	f_x	$= 0.664\,\mathrm{Re}_x^{-1/2}$	Laminar, local, T_f
	Local Nusselt number	Nu_x	$= 0.332\,\mathrm{Re}_x^{1/2}\mathrm{Pr}^{1/3}$	Laminar, local, T_f, $0.6 \leq \mathrm{Pr} \leq 50$
	Thermal boundary layer thickness	δ_t	$= \delta\,\mathrm{Pr}^{-1/3}$	Laminar, T_f
	Average friction factor	$\bar{f}$	$= 1.328\,\mathrm{Re}_x^{-1/2}$	Laminar, average, T_f
	Average Nusselt number	$\overline{\mathrm{Nu}}_x$	$= 0.644\,\mathrm{Re}_x^{1/2}\mathrm{Pr}^{1/3}$	Laminar, average, T_f, $0.6 \leq \mathrm{Pr} \leq 50$
Flat plate: liquid metal	Liquid metal	Nu_x	$= 0.565\,\mathrm{Pr}_x^{1/2}$	Laminar, local, T_f, $\mathrm{Pr} \leq 0.05$
Flat plate: turbulent flow	Local friction factor	f_x	$= 0.592\,\mathrm{Re}_x^{-1/5}$	Turbulent, local, T_f, $\mathrm{Re}_x \leq 10^8$
	Hydrodynamic boundary layer	δ/x	$= 0.37\,\mathrm{Re}_x^{-1/5}$	Turbulent, local, T_f, $\mathrm{Re}_x \leq 10^8$
	Local Nusselt number	Nu_x	$= 0.0296\,\mathrm{Re}_x^{4/5}\mathrm{Pr}^{1/3}$	Turbulent, local, T_f, $\mathrm{Re}_x \leq 10^8$, $0.6 \leq \mathrm{Pr} \leq 60$
	Average Nusselt number	$\overline{\mathrm{Nu}}$	$= 0.037\,\mathrm{Re}_L^{4/5}\mathrm{Pr}^{1/3}$	Turbulent, average, T_f, $\mathrm{Re} \geq \mathrm{Re}_{x,c} = 5\times10^5$, $0.6 < \mathrm{Pr} < 60$, $L \gg x_c$
	Average friction factor	$\bar{f}$	$= 2\,j_H = 0.064\,\mathrm{Re}_x^{-1/5}$	Turbulent, average, T_f

Table 3.3 Cont.

Geometry	Correlation			Conditions
Flat plate: mixed flow	Average friction factor	$\bar{f}$	$= 0.074\,\mathrm{Re}_L^{-1/5} - 1472\,\mathrm{Re}_L^{-1}$	Mixed average, T_f, $\mathrm{Re}_{x,c} = 5\times10^5$, $\mathrm{Re}_L \leq 10^8$
	Average Nusselt number	$\overline{\mathrm{Nu}}$	$= \left(0.037\,\mathrm{Re}_L^{4/5} - 871\right)\mathrm{Pr}^{1/3}$	Mixed average, T_f, $\mathrm{Re}_{x,c} = 5\times10^5$, $\mathrm{Re}_L \leq 10^8$, $0.6 < \mathrm{Pr} < 60$
Cylinder	Kundsen and Katz equation of average Nusselt number (Table 3.4 provides C and m information)	$\overline{\mathrm{Nu}}_D$	$= C\mathrm{Re}_D^m \mathrm{Pr}^{1/3}$	Average, T_f, $0.4 < \mathrm{Re}_d < 4\times10^5$, $\mathrm{Pr} \gtrsim 0.7$
Sphere	Whitaker equation for the average Nusselt number	$\overline{\mathrm{Nu}}_D$	$= 2 + \left(0.4\mathrm{Re}_D^{1/2} + 0.06\mathrm{Re}_D^{2/3}\right)\mathrm{Pr}^{0.4}\left(\dfrac{\mu}{\mu_s}\right)^{1/4}$	Average, properties at T_∞, $3.5 < \mathrm{Re}_D < 7.6\times10^4$, $0.71 < \mathrm{Pr} < 380$, $1.0 < (\mu/\mu_s) < 3.2$
Falling drop		$\overline{\mathrm{Nu}}$	$= 2 + 0.6\mathrm{Re}_D^{1/2}\mathrm{Pr}^{1/3}\left[25\left(\dfrac{x}{D}\right)^{-0.7}\right]$	Average, T_∞
Packed bed of spheres		$\varepsilon\,\bar{j}_H$	$= \varepsilon\,\bar{j}_m = 2.06\mathrm{Re}_D^{-0.575}$	Average, $\bar{T}$, $90 < \mathrm{Re}_D < 4000$, $\mathrm{Pr} \approx 0.7$

Notes: These correlations assume isothermal surfaces. $\bar{j}_H$ = Colburn j_H factor = $\mathrm{St}\,\mathrm{Pr}^{2/3}$ = $\mathrm{Nu}\,\mathrm{Pr}^{-1/3}$. The term ε is the void fraction or porosity. For packed beds, properties are evaluated at the average fluid temperature, $\bar{T} = (T_i + T_o)/2$, or the average film temperature, $\bar{T}_f = (T_s + \bar{T})/2$.

number exceeds 2100, the flow is normally assumed to be turbulent. Two cases are considered below. For commercial (rough) pipes, the friction factor employed is a function of the Reynolds number and the relative roughness of the pipe, ε / D [5]. In the region of complete turbulence (high Re and/or large ε / D), the friction factor depends mainly on the relative roughness [5]. Typical values of the roughness, ε, for various kinds of new commercial piping are available in literature [5].

1. *Fully Developed Turbulent Flow in Smooth Pipes*
   When the temperature difference is moderate, the Dittus and Boelter [3] equation may be used

   $$Nu = 0.023\,Re^{0.8}Pr^{n} \tag{3.15}$$

   where Nu = Nusselt number = hD / k, Re = $D\bar{v} / v = D\bar{v}\rho / \mu$, and Pr = $c_p\mu / k$. The properties in this equation are evaluated at the average (or mean) fluid bulk temperature, T_m, and the exponent n is 0.4 for the heating fluid and 0.3 for the cooling fluid. The equation is valid for fluids with Prandtl numbers, Pr, ranging from approximately 0.6 to 100.

   If wide temperature differences are present in the flow, there may be an appreciable change in the fluid properties between the wall of the tube and the central flow. To take into account the property variations, the Sieder and Tate [3, 4] equation should be used:

   $$Nu = 0.027\,Re^{0.8}Pr^{1/3}\left(\mu/\mu_s\right)^{0.14} \tag{3.16}$$

   All properties are evaluated at bulk temperature conditions, expect μ_s which is evaluated at the wall surface temperature, T_s.

2. *Turbulent Flow in Rough Pipes*
   The recommended equation by Kern is the Chilton-Colburn j_H analogy between heat transfer and fluid flow. It relates the j_H factor [2] $(= St\,Pr^{2/3})$ to the *Darcy* friction factor, f:

   $$j_H = St\,Pr^{2/3} = Nu\,Re^{-1}\,Pr^{-1/3} = f/8 \tag{3.17}$$

   or

$$Nu = (f/8) Re Pr^{1/3} \tag{3.18}$$

where f is obtained from the Moody chart [5]. The average Stanton number (St) is based on the fluid bulk temperature, T_m, while Pr and f are evaluated at the film temperature, i.e., $(T_s + T_{av})/2$ where T sub s is the temperature of the fluid at the surface and T sub ave is the average bulk temperature of the fluid.

Internal flow film coefficients are provided in Table 3.5. Special conditions that apply are provided in the note at the bottom of the Table.

Convection Across Cylinders. Forced air coolers and heaters, forced air condensers, and cross-flow heat exchangers (to be discussed in Part III) are examples of equipment that transfer heat primarily by the forced convection of a fluid flowing across a cylinder. For engineering calculations, the average heat transfer coefficient between a cylinder (at temperature T_s) and a fluid flowing across the cylinder (at a temperature T_∞) is calculated from the Knudsen and Katz equation (see Table 3.4)

$$\overline{Nu}_D = \frac{\bar{h}D}{k} = C Re^m Pr^{1/3} \tag{3.19}$$

All the fluid properties are evaluated at the *mean film temperature* (the arithmetic average temperature of the cylinder surface temperature and the bulk fluid temperature). The constants C and m depend on the Reynolds number of the flow and are given in Table 3.4.

The Biot Number. An important dimensionless number that arises in some heat transfer studies and calculations is the Biot number, Bi. It plays a role in conduction calculations that also involve convection effects, and

Table 3.4 Constants for the Knudsen and Katz equation for heat transfer to cylinders in cross flow.

Re	C	m
0.04–4	0.989	0.330
4–40	0.911	0.385
40–4000	0.683	0.466
4000–40,000	0.193	0.618
40,000–400,000	0.027	0.805

Table 3.5 Summary of forced convection correlations for internal flow.

	Correlation	Conditions
	Darcy friction factor: [2]	
(1)	$f = \dfrac{64}{Re}$	Laminar, fully developed
(2)	$Nu = 4.364$	Laminar, fully developed, constant Q', UHF, $Pr \geq 0.6$
(3)	$Nu_D = 3.658$	Laminar, fully developed, constant T_S, UWT, $Pr \geq 0.6$
	Seider and Tate equation:	
(4)	$\overline{Nu} = 1.86\left(Re_D Pr\dfrac{D}{L}\right)^{1/3}\left(\dfrac{\mu}{\mu_s}\right)^{0.14}$	Laminar, combined entry length, properties at mean bulk temperature of the fluid, $\left[\left(Re_D Pr\dfrac{D}{L}\right)^{1/3}\left(\dfrac{\mu}{\mu_s}\right)^{0.14}\right] \geq 2$, Constant wall temperature, T_S, $0.48 < Pr < 16,700$, $0.0044 < (\mu/\mu_s) < 9.75$
	$= 1.86 Gz^{1/3}\left(\dfrac{\mu}{\mu_s}\right)^{0.14}$	
(5)	$f = 0.316\,Re_D^{-0.25}$	Turbulent, fully developed, smooth tubes, $Re \leq 2\times10^4$
(6)	$f = 0.184\,Re_D^{-0.2}$	Turbulent, fully developed, smooth tubes, $Re \geq 2\times10^4$
	Dittus-Boelter equation:	
(7)	$Nu = 0.023\,Re_D^{0.8}Pr^n$	Turbulent, fully developed, $0.6 \leq Pr \leq 160$, $Re \geq 10,000$, $L/D \geq 10$, $n = 0.4$ for $T_S > T_m$ and $n = 0.3$ for $T_S < T_m$

Table 3.5 (continued)

(8)	Seider and Tate turbulent flow equation: $$Nu = 0.027 Re_D^{0.8} Pr^{1/3} \left(\frac{\mu}{\mu_s}\right)^{0.14}$$	Turbulent, fully developed, $0.7 \leq Pr \leq 16,700$, $Re_D \geq 10,000$, $L/D \geq 10$, large change in fluid properties between the wall of the tube and the bulk flow
(9)	Nusselt turbulent entrance region equation: $$Nu_D = 0.036 Re_D^{0.8} Pr^{1/3} \left(\frac{D}{L}\right)^{0.055} \left(\frac{\mu}{\mu_s}\right)^{0.14}$$	Turbulent, entrance region, $10 < L/D < 400$, $0.7 \leq Pr \leq 16,700$
(10)	Chilton–Colburn analogy: $$j_H = St\,Pr^{2/3} = Nu\,Re^{-1}Pr^{-1/3} = \frac{f}{8}$$	Rough tubes, turbulent flow, St is based on the fluid bulk temperature while Pr and f are based on the film temperature
(11)	Skupinski equation for liquid metals: $$Nu_D = 4.82 + 0.0185\left(Re_D Pr\right)^{0.827}$$ $$= 4.82 + 0.0185 Pe_D^{0.827}$$	Liquid metals, $0.003 < Pr < 0.05$, turbulent, fully developed, constant Q_S, $3.6\times10^3 < Re_D < 9.05\times10^5$, $10^2 < Pe_D$ (Peclet number based on diameter) $< 10^4$
(12)	Seban and Shimazaki equation for liquid metals: $$Nu_D = 5.0 + 0.025\left(Re_D Pr\right)^{0.8}$$ $$= 5.0 + 0.025 Pe_D^{0.8}$$	Liquid metals, turbulent, fully developed, constant T_S, $Pe_D > 100$, properties at the average bulk temperature

Notes: Properties in Equations 2, 3, 7, 8, 11, and 12 are based on the mean fluid temperature T_m; properties in 1, 5, and 6 are based on the average film temperature, $T_f \equiv (T_S - T_m)/2$; properties in Equation 4 are based on the average of the mean temperature of fluids entering and leaving $\bar{T} = (T_{m,i} + T_{m,o})/2$. $Re_D = D_h V/\nu$; D_h = hydraulic diameter $\equiv 4A/P_w$; $V \equiv \dot{m}/\rho A$. UHF = uniform heat flow; UWT = uniform wall temperature.

provides a measure of the temperature change within a solid relative to the temperature change between the surface of the solid and the fluid.

Consider now the heat transfer process described in Figure 3.2. Under steady-state conditions, one may express the heat transfer rate as

$$Q = \frac{kA}{L}\left(T - T_S\right) = hA\left(T_S - T_M\right) \tag{3.20}$$

The above equation may be rearranged to give

$$\frac{T - T_S}{T_S - T_M} = \frac{hA}{kA/L} = \frac{L/kA}{1/hA} = \frac{R_{cond}}{R_{conv}} = \text{Bi} = \frac{hL}{k} \tag{3.21}$$

Figure 3.2 is useful when examining the range of $1.0 << \text{Bi} <<1.0$. Equation 3.21 indicates that for $\text{Bi} << 1.0$ (or $h << k$), one may assume that the temperature across the solid is approximately constant. When $\text{Bi} >> 1.0$ ($h>>k$), the temperature across the solid varies greatly. Finally, when $\text{Bi} = 1.0$, the resistances of the fluid and the solid are approximately equal.

The Buckingham π (Pi) Theorem: Forced Convection. To scale-up (or scale-down) a process, it is necessary to establish geometric and dynamic similarities between the model and the prototype. These two similarities are discussed below.

1. Geometric similarity implies using the same geometry of equipment. A circular pipe prototype should be modeled by a tube in the model. Geometric similarity establishes the scale of the model/prototype design. A 1/10th scale model means that the characteristic dimension of the model is 1/10th that of the prototype.
2. Dynamic similarity implies that certain important dimensionless numbers must be the same in the model and the prototype. For a flowing fluid, which is being heated in a tube of an exchanger, it has been shown that the friction factor, f, is a function of the dimensionless Reynolds number (defined earlier). By selecting the operating conditions such that Re in the model equals the Re in the prototype, then the friction factor in the prototype will equal the friction factor in the model.

It should now be apparent that dimensionless numbers, as well as dimensionless groups, play an important role in the engineering analysis of heat transfer phenomena. The Buckingham π (pi) theorem (see also Chapter 1) may be used to determine the number of independent dimensionless groups required to obtain a relation describing a physical phenomenon. The theorem requires that equations relating variables must be dimensionally homogenous (consistent). The approach can be applied to the problem of correlating experimental heat transfer data for a fluid flowing across (or through) a heated tube. This is defined as a convective process that requires expressing a heat transfer coefficient, h, in terms of its dependent variables. The units of this coefficient are energy/(time · area · temperature). For this process, it is reasonable to expect that the physical quantities listed below are pertinent to the description of this system. These have been expressed in terms of four primary dimensions – length (L), mass (m), time (t), and temperature (T).

The aforementioned heat transfer coefficient is now assumed a function of the following variables:

$$h = f\left(D, k, \overline{v}, \rho, \mu, c_p\right) \tag{3.22}$$

Each variable is expressed below in terms of the four aforementioned primary dimensions.

Tube diameter	D	L
Thermal conductivity	k	$m \cdot L/t^2 \cdot T$
Fluid velocity	$\overline{v}$	L/t
Fluid density	ρ	m/L^3
Fluid viscosity	μ	$m/L \cdot T$
Heat capacity	c_p	$L^2/t^2 \cdot T$
Heat transfer coefficient	h	$m/t^3 \cdot T$

Note that energy can be expressed as $m^2 \cdot L^2/t^2$, heat capacity is represented with c_p, and velocity with $\overline{v}$.

The following procedure is employed for the Buckingham π theorem, as applied to Equation (3.22).

$$\pi = D^a k^b \overline{v}^c \rho^d \mu^e c_p^f h^g \tag{3.23}$$

Substituting primary units gives

$$\pi = (L)^a \left(\frac{m \cdot L}{t^2 \cdot T}\right)^b \left(\frac{L}{t}\right)^c \left(\frac{m}{L^3}\right)^d \left(\frac{m}{L \cdot T}\right)^e \left(\frac{L^2}{t^2 \cdot T}\right)^f \left(\frac{m}{t^3 \cdot T}\right)^g \tag{3.24}$$

For π to be dimensionless, the exponents of each primary dimension *must* separately add up to zero, i.e.,

$$m: \qquad b+d+e+g=0$$

$$L: \quad a+b+c-3d-e+2f=0$$

$$t: \quad -3b-c-e-2f-3g=0$$

$$T: \qquad -b-f-g=0$$

There are seven unknowns but only four equations at this point. Since h is the dependent variable, set $g=1$. A trial-and-error procedure, based to some extent on experience, is now applied. Set $c=d=0$. For this condition (there are now four equations and four unknowns), resulting in

$$a=1 \qquad\qquad b=-1 \qquad\qquad e=f=0$$

The first dimensionless group, π, is then

$$\pi_1 = \frac{hD}{k} = \text{Nusselt number} = \text{Nu} \tag{3.25}$$

For π_2, once again set $g=0$, and $a=1$, $f=0$. The solution is given by

$$b=0 \qquad\qquad c=d=1 \qquad\qquad e=-1$$

$$\pi_2 = \frac{D\bar{v}\rho}{\mu} = \text{Reynolds number} = \text{Re} \tag{3.26}$$

For π_3, set $e=1$, and $c=g=0$. The solution now becomes

$$\pi_3 = \frac{c_p\mu}{k} = \text{Prandtl number} = \text{Pr} \tag{3.27}$$

Thus, the function relationship for h can be written as

$$\text{Nu} = f(\text{Re},\text{Pr}) \tag{3.28}$$

The reader should also note that the Peclet number, Pe, is given by the product of Pr and Re, i.e.,

$$Pe = (Pr)(Re) = \frac{D\bar{v}\rho c_p}{k} \qquad (3.29)$$

It is preferable to use the combination of Reynolds and Prandtl numbers (as seen in Equation (3.29)) because, unlike the Peclet number, the Prandtl number involves only the physical properties of the fluid and not $\bar{v}$ or D.

If a different selection of variables were chosen, one could have found that

$$\Pi = \frac{h}{\rho\bar{v}c_p} = \text{Stanton number} = \text{St} \qquad (3.30)$$

And, in place of Equation (3.28), one would have

$$\text{St} = f(\text{Re}, \text{Pr}) \qquad (3.31)$$

The Nusselt and Stanton numbers are alternative dimensionless heat-transfer coefficients for forced convection. Note that

$$\text{St} = \frac{(\text{Nu})}{(\text{Re})(\text{Pr})} \qquad (3.32)$$

It is found experimentally that for the case of fully developed *turbulent flow* in a long tube, with $\text{Re} \gg 2100$, and for fluids whose Prandtl numbers are in the range from about 0.5 and upwards, heat-transfer rates are given within an accuracy of about ± 10 percent by the simple empirical expression

$$\text{Nu} = 0.023\,\text{Re}^{0.8}\text{Pr}^{0.4} \qquad (3.33)$$

The heat-transfer factor h in the Nusselt number should be viewed as a mean value with physical properties μ, c_p, and k usually evaluated at the mean bulk temperature of the fluid, i.e., at the arithmetic mean of inlet and outlet temperatures. Minor variations in Equation (3.33) have been suggested, however, which claim to give better accuracy and involve the evaluation of the physical properties of the fluid at some hypothetical 'mean film temperature.' However, Equation (3.33) finds surprisingly wide application and covers must ordinary gases and liquids encountered in chemical engineering practice. It can be applied to heat-transfer calculations with passages or channels whose cross-sections are other than circular, provided appropriate values are taken for the equivalent diameter. However, the accuracy of the calculation is generally less satisfactory in such cases.

Physical Significance of the Nusselt, Stanton, and Prandtl Numbers. It is important to appreciate the physical significance of these dimensionless ratios employed in heat-transfer calculations.

1. The *Nusselt number* is a dimensionless heat-transfer coefficient number which provides a measure of the ratio of the heat-transfer rate to the rate at which heat would be conducted within the fluid under a temperature gradient.
2. The *Stanton number* is an alternative heat-transfer coefficient number and provides a measure of the ratio of the heat-transfer coefficient to the flow of heat along the pipe per unit temperature rise due to the velocity and heat capacity of the fluid, i.e., to $\rho \bar{v} c_p$.
3. The *Prandtl number* is represented as $\mu c_p / k = v / \alpha$; it is therefore simply the ratio of kinematic viscosity, v, to thermal diffusivity, α.

In all problems of forced convection, provided high velocities of flow are not involved, and provided the physical properties of the fluid (density, viscosity, and thermal conductivity) are essentially constant, one can assume that heat-transfer rates expressed in nondimensional form as a Nusselt or Stanton number are functions only of the Reynolds number and the Prandtl number. The form of the function will naturally depend on the geometry of the system and, although in a few special cases of laminar flow it may be calculated, in most cases it must be determined by experiment. Appropriate working formulae for other geometrical arrangements, such as flow across single tubes and tube banks, are available in the literature [2, 4].

If one is dealing with problems of heat transfer in which there are large temperature differences, it may be necessary to take account of variations of the physical properties of the fluid with temperature. Details here are also available in the literature [3, 4].

Example 3.5 – Dimensionless Groups. Identify the following three dimensionless groups:

1. $h_f L / k_f$ (subscript f refers to fluid)
2. $h_f L / k_s$ (subscript s refers to solid surface)
3. (Reynolds number)(Prandtl number), i.e., (Re)(Pr)

SOLUTION: As noted above,

1. $h_f L / k_f =$ Nusselt number $= Nu$

2. $h_f L / k_s =$ Biot number $= Bi$

3. $(Re)(Pr) = \left(\dfrac{L\bar{v}\rho}{\mu}\right)\left(\dfrac{c_p \mu}{k}\right) = \dfrac{L\bar{v}\rho c_p}{k} =$ Peclet number $= Pe$

Example 3.6 – Pipe Flow. Air with a mass rate of 0.075 kg/s flows through a tube of diameter $D = 0.225\,$m. The air enters at $100\,°$C and, after a distance of $L = 5\,$m, cools to $70\,°$C. Determine the heat transfer coefficient of the air. The properties of air at $85\,°$C are approximately, $c_p = 1010$ J/kg $\cdot$ K, $k = 0.030$ W/m $\cdot$ K, $\mu = 208 \times 10^{-7}$ N $\cdot$ s/m^2, and Pr $= 0.71$.

SOLUTION: The Reynolds number is (see Chapter 1)

$$\text{Re} = \frac{4\dot{m}}{\pi D \mu} = \frac{(4)(0.075)}{(\pi)(0.225)(208 \times 10^{-7}\,\text{N})}$$

Upon substitution,

$$\text{Re} = 20{,}400$$

Apply either Equation (7) in Tables 3.5 and/or Equation (3.15), with $n = 0.3$ for heating,

$$\text{Nu} = \frac{hD}{k} = 0.023\,\text{Re}^{0.8}\text{Pr}^{0.3}$$

$$= 0.023(20{,}400)^{0.8}(0.71)^{0.3} = 58.0$$

Thus,

$$h = \left(\frac{k}{D}\right)\text{Nu}$$

$$= \left(\frac{0.03}{0.225}\right)(58.0) = 7.73\,\text{W} / \text{m}^2 \cdot \text{K}$$

Example 3.7 – Average Film Heat Coefficient. Calculate the average film heat coefficient (Btu / hr $\cdot$ ft^2 $\cdot$°F) on the water side of a single pass steam condenser. The tubes are 0.902 inch inside diameter, and the cooling water enters at $60\,°$F and leaves at $70\,°$F. Employ the Dittus-Boelter equation and assume the average water velocity is 7 ft/s. Pertinent physical properties of water at an average temperature of 65 °F are:

$$\rho = 62.3\,\text{lb/ft}^3 \qquad c_p = 1.0\,\text{Btu/lb} \cdot {}^\circ\text{F}$$

$$\mu = 2.51\,\text{lb/ft} \cdot \text{hr} \qquad k = 0.340\,\text{Btu/hr} \cdot \text{ft} \cdot {}^\circ\text{F}$$

Solution: For heating, Equation (3.15) applies:

$$\text{Nu} = 0.023\,\text{Re}^{0.8}\,\text{Pr}^{0.4}$$

or

$$h = \left(\frac{k}{D}\right) 0.023 \left(\frac{D\bar{v}\rho}{\mu}\right)^{0.8} \left(\frac{c_p\mu}{k}\right)^{0.4}$$

The terms $\text{Re}^{0.8}$ and $\text{Pr}^{0.4}$ are given by

$$\text{Re}^{0.8} = \left[\frac{(0.902/12)(7)(62.4)}{(2.51/3600)}\right]^{0.8} = (47{,}091)^{0.8} = 5475$$

$$\text{Pr}^{0.4} = \left[\frac{(1.0)(2.51)}{0.340}\right]^{0.4} = (7.38)^{0.4} = 2.224$$

Therefore,

$$h = \left(\frac{0.340}{0.0752}\right)(0.023)(5475)(2.224)$$

$$= 1266\,\text{Btu/hr} \cdot \text{ft}^2 \cdot {}^\circ\text{F}$$

Example 3.8 – Constant Flux Application. Air at 1 atm and 300 °C is cooled as it flows at a velocity of 5.0 m/s through a tube with a diameter of 2.54 cm. Calculate the heat transfer coefficient if a constant heat flux condition is maintained at the wall and the wall temperature is 20 °C above the temperature along the entire length of the tube.

Solution: The density is first calculated using the ideal gas law in order to obtain the Reynolds number. Apply an appropriate value of R [4] for air on a *mass* basis.

$$\rho = \frac{P}{RT} = \frac{1\left(1.0132\times10^{5}\right)}{\left(287\right)\left(573\right)} = 0.6161\,\text{kg/m}^{3}$$

The following data is obtained from the Appendix assuming air to have the properties of nitrogen:

$$Pr = 0.713 \qquad\qquad c_{p} = 1.041\,\text{kJ/kg}\cdot\text{K}$$

$$\mu = 1.784\times10^{-5}\,\text{kg/m}\cdot\text{s} \qquad k = 0.0262\,\text{W/m}\cdot\text{K}$$

Thus,

$$Re = \frac{D\bar{v}\rho}{\mu} = \frac{\left(0.0254\right)\left(5\right)\left(0.6161\right)}{1.784\times10^{-5}} = 4386$$

Since the flow is turbulent, Equation (7) in Table 3.5 and/or Equation (3.15) applies:

$$Nu = \frac{hD}{k} = 0.023Re^{0.8}Pr^{0.3} = 0.023(4386)^{0.8}(0.713)^{0.3} = 17.03$$

Thus,

$$h = \left(\frac{k}{D}\right)Nu = \left(\frac{0.0262}{0.0254}\right)(17.03) = 17.57\,\text{W/m}^{2}\cdot\text{K}$$

Example 3.9 – Heat Transfer Coefficient Calculation. Water flows with an average velocity of 0.355 m/s through a long copper tube (inside diameter = 2.2 cm) in a heat exchanger. The water is heated by steam condensing at 150 °C on the outside of the tube. Water enters at 15 °C and leaves at 60 °C. Determine the heat transfer coefficient, h, for the water. (Adapted from Griskey [6].)

SOLUTION: First evaluate the average bulk temperature of the water which is

$$\bar{T} = \left(15+60\right)/2 = 37.5°C$$

Evaluating water properties (from the Appendix) at this temperature yields

$$\rho = 993\,\text{kg/m}^{3} \qquad\qquad c_{p} = 4.17\times10^{3}\,\text{J/kg}\cdot\text{K}$$

$$\mu = 0.000683\,\text{kg/m}\cdot\text{s} \qquad k = 0.630\,\text{W/m}\cdot\text{K}$$

Thus,

$$Re = \frac{D\bar{v}\rho}{\mu} = \frac{(0.022)(0.355)(993)}{6.83\times10^{-4}} = 11{,}350$$

$$Pr = \frac{c_p\mu}{k} = \frac{(4170)(6.83\times10^{-4})}{0.630} = 4.53$$

The flow is therefore turbulent and, since the tube is a long one (i.e., no L/D effect), one may use the Sieder and Tate turbulent relation [Equation (8), Table 3.5] for internal flow:

$$\frac{hD}{k} = 0.027\,Re^{0.8}Pr^{0.33}\left(\frac{\mu}{\mu_w}\right)^{0.14}$$

All of the quantities in the above equation are known except for μ_w. To obtain this value, the average wall temperature of the fluid must be determined. This temperature is between the fluid's bulk average temperature of 37.5 °C and the outside wall temperature of 150 °C. Once again, use a value of 93.75 °C (the average of the two). At this temperature (see Table AT.3 and AT.5 in the Appendix)

$$\mu_w = 0.000306\,\text{kg/m}\cdot\text{s}$$

Then

$$h = \left(\frac{0.630}{0.022}\right)0.027(11{,}350)^{0.8}(4.53)^{0.33}\left(\frac{0.000683}{0.000306}\right)^{0.14}$$
$$= 2498.1\,\text{W/m}^2\cdot\text{K}$$

Example 3.10 – Electric Current Application. The surface temperature T_s of a circular conducting rod is maintained at 250 °C by the passage of an electric current. The rod diameter is 10 mm, the length is 2.5 m, the thermal conductivity is 60 W/m·K, the density is 7850 kg/m³, and the heat capacity is 434 J/kg·K. The rod is in a fluid at temperature $T_f = 25°C$, and the convection heat transfer coefficient is 140 W / m^2 · K. The thermal conductivity of the fluid is 0.6 W / m · K.

1. What is the thermal diffusivity of the bare rod?
2. What is the Nusselt number of the fluid in contact with the bare rod?
3. What is the Biot number of the bare rod?
4. Calculate the heat transferred from the rod to the fluid.

SOLUTION:

1. The thermal diffusivity, α, of the bare rod is

$$\alpha = k / (\rho c_p) = 60 / [(7850)(434)] = 1.76 \times 10^{-5} \, \text{m}^2 / \text{s}$$

2. The Nusselt number of the fluid is

$$\text{Nu} = hD / k_f = (140)(0.01) / 0.6 = 2.33$$

3. The Biot number of the bare rod is

$$\text{Bi} = hD / k_s = (140)(0.01) / 60 = 0.0233$$

4. Finally, calculate Q for the bare rod:

$$\begin{aligned} Q_{\text{bare}} &= h(\pi DL)(T_s - T_f) \\ &= (140)(\pi)(0.01)(2.5)(250 - 25) \\ &= 2474 \, \text{W} \end{aligned}$$

3.4 Convection Heat Transfer: Microscopic Approach

The microscopic (transport) equations employed to describe heat transfer in a flowing fluid are presented in Table 3.6 [7, 8]. The equations describe the temperature profile in a moving fluid. Note that this table is essentially an extension of the microscopic material presented in the previous chapter on conduction. The illustrative examples that follow, drawn from the work of Theodore [8], involve the application of this table.

Table 3.6 Energy-transfer equations for incompressible fluids.

Rectangular coordinates:

$$\frac{\partial T}{\partial t}+\bar{v}_x\frac{\partial T}{\partial x}+\bar{v}_y\frac{\partial T}{\partial y}+\bar{v}_z\frac{\partial T}{\partial z}=\alpha\left[\frac{\partial^2 T}{\partial x^2}+\frac{\partial^2 T}{\partial y^2}+\frac{\partial^2 T}{\partial z^2}\right]+\frac{A}{\rho c_p} \tag{1}$$

Cylindrical coordinates:

$$\frac{\partial T}{\partial t}+\bar{v}_r\frac{\partial T}{\partial r}+\frac{\bar{v}_\phi}{r}\frac{\partial T}{\partial \phi}+\bar{v}_z\frac{\partial T}{\partial z}=\alpha\left[\frac{1}{r}\frac{\partial}{\partial r}\left(r\frac{\partial T}{\partial r}\right)+\frac{1}{r^2}\frac{\partial^2 T}{\partial \phi^2}+\frac{\partial^2 T}{\partial z^2}\right]+\frac{A}{\rho c_p} \tag{2}$$

Spherical coordinates:

$$\frac{\partial T}{\partial t}+\bar{v}_r\frac{\partial T}{\partial r}+\frac{\bar{v}_\theta}{r}\frac{\partial T}{\partial \theta}$$

$$+\frac{\bar{v}_\phi}{r\sin\theta}\frac{\partial T}{\partial \phi}$$

$$=\alpha\left[\frac{1}{r^2}\frac{\partial}{\partial r}\left(r^2\frac{\partial T}{\partial r}\right)+\frac{1}{r^2\sin\theta}\frac{\partial}{\partial \theta}\left(\sin\theta\frac{\partial T}{\partial \theta}\right)\right.$$

$$\left.+\frac{1}{r^2\sin^2\theta}\frac{\partial^2 T}{\partial \phi^2}+\frac{\partial^2 T}{\partial \phi^2}\right]+\frac{A}{\rho c_p} \tag{3}$$

Example 3.11 – Obtaining Describing Equations. An incompressible fluid enters an insulated tubular (cylindrical) heat exchanger at temperature T_0. A chemical reaction occurring in the tube causes a rate of energy per unit volume to be liberated. This rate is proportional to the temperature and is given by

$$(S)(T)$$

where S is a constant. Obtain the steady-state equations describing the temperature in the tube if the flow is laminar. Neglect axial diffusion.

SOLUTION: This problem is solved in cylindrical coordinates. The pertinent equation is a rearrangement of Equation (2) in Table 3.6. Based on the problem statement, the temperature is a function of *both z* and *r*, and the only velocity component is $\bar{v}_z$:

$$\bar{v}_z\frac{\partial T}{\partial z}=\alpha\left[\frac{1}{r}\frac{\partial}{\partial r}\left(r\frac{\partial T}{\partial r}\right)+\frac{\partial^2 T}{\partial z^2}\right]+\frac{A}{\rho c_p} \tag{1}$$

If one neglects axial diffusion

$$\frac{\partial^2 T}{\partial z^2}=0$$

The source term A is given by

$$A = (S)(T)$$

The resulting equation is

$$\overline{v}_z \frac{\partial T}{\partial z} = \alpha \left[\frac{1}{r} \frac{\partial}{\partial r} \left(r \frac{\partial T}{\partial r} \right) \right] + \left(\frac{S}{\rho c_P} \right) T$$

If the flow is laminar [2–5],

$$\overline{v}_z = \overline{v}_{max} \left[1 - \left(\frac{r}{a} \right)^2 \right]$$

where a = radius of cylinder and $\overline{v}_{max}$ = maximum velocity, located at $r = 0$. Equation (1) now takes the form

$$\overline{v}_{max} \left[1 - \left(\frac{r}{a} \right)^2 \right] \frac{\partial T}{\partial z} = \alpha \left[\frac{1}{r} \frac{\partial}{\partial r} \left(r \frac{\partial T}{\partial r} \right) \right] + \left(\frac{S}{\rho c_P} \right) T$$

Example 3.12 – Describing Equations: Plug Flow. Refer to the previous example. Obtain the describing equation if the flow is plug. Also obtain the temperature profile in the tube.

SOLUTION: If the flow is plug, $\overline{v}_z$ is constant [2] across the area of the tube. Based on physical grounds, the radial-diffusion term is zero and T is solely a function of z. Equation (1) in the previous example then becomes

$$\overline{v}_z \frac{dT}{dz} = \left(\frac{S}{\rho c_P} \right) T$$

The above equation is rewritten as

$$\frac{dT}{T} = \left(\frac{S}{\rho c_P \overline{v}_z} \right) dz$$

Integrating the above equation gives

$$\ln T = \left(\frac{S}{\rho c_P \overline{v}_z} \right) z + B$$

The BC is

$$T = T_0 \text{ at } z = 0$$

so that

$$B = \ln T_0$$

Therefore

$$\ln\left(\frac{T}{T_0}\right) = \left(\frac{S}{\rho c_p \bar{v}_z}\right) z$$

or

$$T = T_0 \exp\left(-\frac{S}{c_p \bar{v}_z} z\right)$$

3.5   Free Convection Principles and Applications

Convective effects, previously described as *forced convection*, are due to the bulk motion of the fluid. The bulk motion is caused by external forces, such as that provided by pumps, fans, compressors, etc., and is essentially independent of "thermal" effects.

Free convection is another effect that occasionally develops and was briefly discussed in the previous chapter. This effect is almost always attributed to buoyant forces that arise due to density differences within a system. It is treated analytically as another external force term in the momentum equation. The momentum (velocity) and energy (temperature) effects are therefore interdependent; consequently, both equations must be solved simultaneously in analytical analyses. This treatment is beyond the scope of this text but is available in the literature [9, 10].

Consider a heated body in an unbounded medium. In natural convection, the velocity is zero at the wall/surface of the heated body – commonly known as the no-slip boundary condition. The velocity increases rapidly in the thin boundary layer adjacent to the heated body. In reality, both natural convection and forced convection effects occur simultaneously so that one may be required to determine which is predominant. Both may therefore be required to be included in some analyses, even though one is often tempted to attach less significance to free convection effects. However, this temptation should be resisted since free convection occasionally plays the more important role in the design and/or performance of some heated system [9, 10].

As noted above, free convection fluid motion arises due to buoyant forces. Buoyancy arises due to the combined presence of a fluid density gradient and a body force that is proportional to density. The body force is usually gravity. Density gradients arise due to the presence of a temperature gradient. Furthermore, the density of gases and liquids depends on temperature, generally decreasing with increasing temperature, i.e., $(\partial \rho / \partial T)_p < 0$.

There are both industrial and environmental applications. Free convection influences industrial heat transfer from and within pipes. It is also important in transferring heat from heaters or radiators to ambient air and in removing heat from the coil of a refrigeration unit to the surrounding air. It is also relevant in the environmental sciences and engineering where it gives rise to both atmospheric and oceanic motion, a topic treated in the last section of this chapter.

In addition to this introduction, this section addresses the following three topics:

1. Key dimensionless numbers
2. Describing equations
3. Environmental applications

Key Dimensionless Numbers. If a solid surface temperature, T_s, is in contact with a gas or liquid at temperature T_∞, the fluid moves solely as a result of density variations in natural convection. It is the fluid motion that causes the so-called natural convection. The nature of the buoyant force is characterized by the coefficient of volumetric expansion, β. For an ideal gas, β is given by $(1.0 / T)$ where T is the *absolute* temperature; β is an important term in natural convection theory and applications.

Semi-theoretical equations for natural convection use the following key dimensionless numbers, some of which have been discussed earlier [3, 4].

$$\text{Gr} = \text{Grashof number} = \frac{L^3 g \beta \Delta T}{\nu^2} = \frac{L^3 \rho^3 g \beta \Delta T}{\mu^2} \qquad (3.34)$$

$$\text{Nu} = \text{Nusselt number} = \frac{hL}{k} \qquad (3.35)$$

$$\text{Ra} = \text{Rayleigh number} = (\text{Gr})(\text{Pr}) = \frac{L^3 g \beta \Delta T}{\nu \alpha} = \frac{L^3 g \beta \Delta T \rho^2 c_p}{\mu k} \qquad (3.36)$$

$$\text{Pr} = \text{Prandtl number} = \frac{\nu}{\alpha} = \frac{c_p \mu}{k} \qquad (3.37)$$

where,

v = kinematic viscosity

μ = absolute viscosity

α = thermal diffusivity $= k / \rho c_p$

ρ = fluid density

c_p = fluid heat capacity

k = thermal conductivity

L = characteristic length of a system

ΔT = temperature difference between the surface and the fluid $= |T_S - T_\infty|$

The above Rayleigh number is used to classify natural convection as either laminar or turbulent:

$$Ra < 10^9, \text{ laminar free convection;}$$

$$Ra > 10^9, \text{ turbulent free convection} \tag{3.38}$$

In the previous chapter on forced convection, the effects of natural convection were neglected, a valid assumption in many applications characterized by moderate-to-high-velocity fluids. Free convection may be significant with low-velocity fluids. A measure of the influence of each convection effect is provided by the ratio

$$\frac{Gr}{Re^2} = \frac{\text{bouyancy force}}{\text{inertia force}} = \frac{\rho g \beta L (\Delta T)}{\bar{v}^2}; \; \bar{v} = \text{fluid velocity} \tag{3.39}$$

The above dimensionless number is represented by LT by one of the authors [4], so that

$$\frac{Gr}{Re^2} = LT \tag{3.40}$$

For $LT > 1.0$, free convection is important. The regimes of these convection effects are:

Free convection predominates,

$$\text{i.e., } LT \gg 1.0 \text{ or } Gr \gg Re^2 \tag{3.41}$$

Forced convection predominates,

$$\text{i.e., } LT \ll 1.0 \text{ or } Gr \ll Re^2 \tag{3.42}$$

Both effects contribute; mixed free and forced convection,

$$\text{i.e., } LT \approx 1.0 \text{ or } Gr \approx Re^2 \tag{3.43}$$

Combining these three convection regimes with the two-flow regime – laminar and turbulent – produces six subregimes of potential interest.

Describing Equations [3, 4]. As stated earlier, free convection is an important consideration in the calculation of some heat transfer rates from pipes, transmission lines, electronic devices, and electric baseboards. The average Nusselt number, $\overline{Nu}$ and the Rayleigh number can be related through the following semi-theoretical correlation:

$$\overline{Nu} = \overline{h}L / k = CRa^m \tag{3.44}$$

with all the fluid properties evaluated at the film temperature, T_f,

$$T_f = (T_s + T_\infty) / 2 \tag{3.45}$$

Generally, but with some exceptions,

$$m = \frac{1}{4} \text{ for laminar free convection}$$

$$m = \frac{1}{3} \text{ for turbulent free convection} \tag{3.46}$$

As one might expect, the above characteristic length, L, depends on the geometry. For a vertical plate, L is the plate height and for a horizontal plate, the plate length. For a horizontal cylinder, L is the diameter, D, and for a horizontal disk, L is given by: $L = 0.9D$. The constants C and m to be used in Equation (3.44) are listed in Table 3.7 for several geometries and a wide range of Rayleigh numbers.

Another correlation that can be used to calculate the heat transfer coefficient for natural convection from spheres is Churchill's equation,

$$\overline{Nu} = 2 + \frac{0.589\,Ra^{0.25}}{\left[1 + \left(\dfrac{0.469}{Pr}\right)^{9/16}\right]^{4/9}} \tag{3.47}$$

Table 3.7 Coefficients for equation (3.44).

Geometry	$(\text{Gr})(\text{Pr}) = \text{Ra}$	C	m
Vertical planes and cylinders	10^4-10^9	0.59	0.25
	10^9-10^{13}	0.10	0.3333
Horizontal cylinders	$0-10^{-5}$	0.4	0
	$10^{-5}-10^4$	0.85	0.188
	10^4-10^9	0.53	0.25
	10^9-10^{12}	0.13	0.3333
Spheres	$0-10^{12}$	0.60	0.25
Upper surface of horizontal heated plates; plate is hotter than:			
surroundings $(T_S > T_\infty)$	$2 \times 10^4 - 8 \times 10^6$	0.54	0.25
lower surface of horizontal cooled plates $(T_S < T_\infty)$	$8 \times 10^6 - 10^{11}$	0.15	0.3333
Lower surface of horizontal heated plates:			
$(T_S > T_\infty)$ or upper surface of horizontal cooled plates $(T_S < T_\infty)$	10^5-10^{11}	0.58	0.2

The Rayleigh number, Ra, and the Nusselt number, Nu, in Equation (3.47) are based on the diameter of the sphere. Churchill's equation is valid for Pr ≥ 0.7 and Ra $\leq 10^{11}$.

There are also *simplified* correlations for natural (or free) convection in air at 1 atm. The correlations are dimensional and are based on the following SI units: h = heat transfer coefficient, W/m$^2 \cdot$K; $\Delta T = T_S - T_\infty$, °C; T_S = surface temperature, °C; T_∞ = surrounding temperature, °C; L = vertical or horizontal dimension, m, and D = diameter. These correlations are presented in Table 3.8.

The average heat transfer coefficient can be calculated once the Nusselt number has been determined. Rearranging Equation (3.44) gives

$$\overline{h} = \overline{\text{Nu}}k / L \tag{3.48}$$

The heat transfer rate, Q, is then given by the standard heat transfer equation:

$$Q = \overline{h}A\left(T_S - T_\infty\right) \tag{3.49}$$

Table 3.8 Free convection equation in air.

Geometry	$10^4 < (Gr)(Pr) = Ra < 10^9$	$(Gr)(Pr) = Ra > 10^9$
	Laminar	Turbulent
Vertical planes and cylinders	$h = 1.42\left(\dfrac{\Delta T}{L}\right)^{1/4}$	$h = 0.95(\Delta T)^{1/3}$
Horizontal cylinders	$h = 1.32\left(\dfrac{\Delta T}{D}\right)^{1/4}$	$h = 1.24(\Delta T)^{1/3}$
Upper surface of horizontal heated plates; plate is hotter than surroundings $(T_S > T_\infty)$ or lower surface of horizontal cooled plates $(T_S < T_\infty)$	$h = 1.32\left(\dfrac{\Delta T}{L}\right)^{1/4}$	$h = 1.43\left(\dfrac{\Delta T}{L}\right)^{1/3}$
Lower surface of horizontal heated plates $(T_S > T_\infty)$ or upper surface of horizontal cooled plates $(T_S < T_\infty)$	$h = 0.61\left(\dfrac{\Delta T}{L^2}\right)^{1/5}$	

If the air is at a pressure other than 1 atm, the following correction may be applied to the reference value at 1 atm:

$$h = h_{\text{ref}}\left(P \text{ in atmospheres}\right)^n \qquad (3.50)$$

where

$$n = \frac{1}{2} \text{ for laminar cases } (Ra < 10^9)$$

$$n = \frac{2}{3} \text{ for turbulent cases } (Ra > 10^9) \qquad (3.51)$$

The Buckingham π (Pi) Theorem and Free Convection. As noted earlier, in heat transfer by free convection, the motion of the fluid is caused by density changes. One may apply the Buckingham π theorem in a manner similar to that for forced convection.

Let the bulk fluid temperature be T_o and the corresponding density ρ_o. The buoyancy force per unit volume for an element of fluid at temperature T and density ρ will then be $(\rho_o - \rho)g$, i.e.,

$$\text{buoyancy force per unit mass} = \frac{(\rho_o - \rho)g}{\rho} \qquad (3.52)$$

If β is the coefficient of thermal expansion, referred to the bulk temperature of the fluid,

$$\frac{1}{\rho} = \frac{1}{\rho_o}(1 + \beta\Delta T) \qquad (3.53)$$

or

$$\rho_o = \rho(1 + \beta\Delta T) \qquad (3.54)$$

Note that the buoyancy force per unit mass is given by $\beta g(T - T_o)$ or $\beta g \Delta T$. For an ideal gas (neglecting pressure variation),

$$\frac{\rho_o}{\rho} = \frac{P_o T}{P T_o} \approx \frac{T}{T_o} \qquad (3.55)$$

Thus,

$$\text{buoyant force} = \frac{(T - T_o)g}{T_o} = \frac{\Delta T g}{T_o} \qquad (3.56)$$

Now assume that the heat transfer coefficient for the case of heat transfer by *free convection from a vertical plate* will depend on the following variables including the height of the plate L:

$$h = f\left(\Delta T, \beta g, \rho, k, c_p, \mu, L\right) \qquad (3.57)$$

The connection between eight quantities reduces to a relationship between three dimensionless ratios:

$$\Pi_1 = \frac{hL}{k} = \text{Nu} = \text{Nusselt number} \qquad (3.58)$$

$$\Pi_2 = \frac{\beta g \Delta T \rho^2 L^3}{\mu^2} = \text{Gr} = \text{Grashof number} \qquad (3.59)$$

$$\Pi_3 = \frac{c_p \mu}{k} = \text{Pr} = \text{Prandtl number} \qquad (3.60)$$

Therefore, for free convection from a vertical plate, one may write,

$$Nu = f(Gr, Pr) \tag{3.61}$$

A theoretical solution, assuming laminar motion in the free convection currents rising from the plate, gives

$$Nu = 0.52\,Gr^{1/4}Pr^{1/4} \tag{3.62}$$

It is found experimentally that, for laminar motion with the product $(Gr)(Pr)$ in the range from 10^4 to 10^8, the average heat-transfer coefficient is given by

$$Nu = 0.56\,Gr^{1/4}Pr^{1/4} \tag{3.63}$$

The motion is generally turbulent, for values of the product $(Gr)(Pr)$ greater than 10^9, and it is found that the Nusselt number is then proportional to $[(Gr)(Pr)]^{1/3}$. The following working formulae may then be applied:

$$\text{For gases: } Nu = 0.12\,Gr^{1/3}Pr^{1/3} \tag{3.64}$$

$$\text{For liquids: } Nu = 0.17\,Gr^{1/3}Pr^{1/3} \tag{3.65}$$

The product of Grashof number and Prandtl number which appears in all these expressions is occasionally referred to as the *Rayleigh number*, Ra.

For heat transfer by free convection from *horizontal cylinders,* the diameter is used in defining the Nusselt and Grashof numbers, and for values of the product $(Gr)(Pr) < 10^8$ the following expression is then employed:

$$Nu = 0.47\,Gr^{1/4}Pr^{1/4} \tag{3.66}$$

Another dimensionless number that appears in combined natural and forced convection systems as well as just natural convection systems is the Graetz number. There are several defining equations for this number:

$$Gz = (Re)(Pr)\left(\frac{D}{L}\right); \ L = \text{characteristic length} \tag{3.67}$$

and

$$Gz = \frac{4}{\pi}\frac{\dot{m}c_p}{kL}; \ \dot{m} = \text{mass flow rate} \tag{3.68}$$

However, it should also be noted that some define the Graetz number by

$$Gz = \frac{\dot{m}c_p}{kL} \tag{3.69}$$

There are several dimensionless numbers encountered in heat transfer applications. These dimensionless numbers, as well as others, will appear with regular consistency in both this and the next two Parts of this book. Some of the more important and most commonly used numbers are provided in alphabetical order in Table 1.1, with some of the terms to be defined later. With reference to Table 1.1, also note once again that

$$Nu = (St)(Re)(Pr) \tag{3.70}$$

Example 3.13 – Neglecting Free Convective Effects. The Grashof and Reynolds numbers for a system involved in a heat transfer process are approximately 100 and 50, respectively. Can free convection effects be neglected.

SOLUTION: Employ Equation (3.42).

$$LT = \frac{Gr}{Re^2}$$

Substituting

$$LT = \frac{100}{50^2} = 0.04$$

Since $LT \ll 1.0$, free convection effects can be neglected.

Example 3.14 – Determining the Grashof and Rayleigh numbers. The heat flux rate incident on a vertical plate at $110\,°C$ is 800 W/m². The plate is 2 m wide and 3.5 m high and is well insulated on the backside. The ambient air temperature is $30\,°C$. All the incident radiation (800 W/m²) on the plate is absorbed and dissipated by free convection to the ambient air at $30\,°C$. Determine the Grashof and Rayleigh numbers.

SOLUTION: Obtain v, k, and Pr from the Appendix for air at the film temperature, T_f.

Thus,

$$T_f = \left(T_S + T_\infty\right)/2$$
$$= \left(110 + 30\right)/2$$
$$= 70\,°C = 343\,K$$
$$v = 2.0 \times 10^{-5}\,m^2\,/\,s$$
$$k = 0.029\,W\,/\,m \cdot K$$
$$Pr = 0.7$$

Calculate the coefficient of expansion β.

$$\beta = \frac{1}{343} = 0.0029\,K^{-1}$$

Calculate the Grashof and Rayleigh numbers employing Equation (3.34–3.37) and (3.35), respectively.

$$Gr = \frac{L^3 g \beta \Delta T}{v^2}$$
$$= \frac{\left(9.807\right)\left(0.0029\right)\left(80\right)\left(3.5\right)^3}{\left(2.0 \times 10^{-5}\right)^2}$$
$$Gr = 2.44 \times 10^{11}$$
$$T_f = 70\,°C = 343K$$
$$Ra = \left(Gr\right)\left(Pr\right)$$
$$= \left(2.44 \times 10^{11}\right)\left(0.7\right)$$
$$= 1.71 \times 10^{11}$$

Since $Ra > 10^9$, the convection flow category is turbulent.

Example 3.15 – Determining Average Heat Transfer Coefficient. Refer to Example 3.14. Determine the average heat transfer coefficient.

SOLUTION: Calculate the average Nusselt number. Employ Equation (3.44). From Table 3.7 use $C = 0.1$ and $m = \dfrac{1}{3}$:

$$\overline{Nu} = c\,Ra^m$$

Substituting,

$$\overline{Nu} = (0.1)(1.71 \times 10^{11})^{1/3}$$
$$= 555.0$$

Calculate the average heat transfer coefficient employing Equation (3.48).

$$\overline{h} = \frac{\overline{Nu}k}{L}$$
$$= \frac{(555.0)(0.029)}{3.5}$$
$$= 4.6 \, W/m^2 \cdot K$$

Example 3.16 – Free Convection Coefficient for a Flat Plate. Calculate the free convection heat transfer coefficient for a plate 6 ft high and 8 ft wide at 120 °F that is exposed to nitrogen at 60 °F.

SOLUTION: The mean film temperature is

$$T_f = (120 + 60)/2 = 90°F = 550°R$$

From the Appendix,

$$\rho = 0.0713 \, lb/ft^3 \qquad \nu = 16.82 \times 10^{-5} \, ft^2/s$$
$$Pr = 0.713 \qquad k = 0.01514 \, Btu/hr \cdot ft \cdot °F$$

In addition,

$$\beta = 1/T = 1/550 = 1.818 \times 10^{-3} °R^{-1}$$

Employ Equation (3.34).

$$Gr = \frac{L^3 g \beta (T_s - T_\infty)}{\nu^2}$$

Substituting,

$$Gr = \frac{(32.2 \, ft/s^2)(1.818 \times 10^{-3} °R^{-1})[(120 - 60)°R](6 \, ft)^3}{(16.82 \times 10^{-5})^2 \, ft^4/s^2} = 2.682 \times 10^{10}$$

In addition, from Equation (3.36),

$$Ra = (Gr)(Pr) = (2.682 \times 10^{10})(0.713) = 1.912 \times 10^{10}$$

The flow is therefore turbulent (see Table 3.8). Equation (3.44) applies, with appropriate constants from Table 3.7, to give

$$\frac{\overline{h}L}{k} = 0.10\,\mathrm{Ra}^{1/3}$$

$$\overline{h} = \left(\frac{k}{L}\right)0.10\,\mathrm{Ra}^{1/3}$$

Substitution yields:

$$\overline{h} = \left(\frac{0.01514\,\mathrm{Btu/hr\cdot ft\cdot°F}}{6\,\mathrm{ft}}\right)(0.10)\left(1.912\times10^{10}\right)^{1/3}$$

$$\overline{h} = 0.675\,\mathrm{Btu/hr}$$

Example 3.17 – Flat Plate Heat Loss. Calculate the heat loss in the previous example.

SOLUTION: Apply Equation (3.49).

$$
\begin{aligned}
Q &= \overline{h}A\left(T_s - T_\infty\right) \\
&= \left(0.675\,\mathrm{Btu/hr\cdot ft^2\cdot°F}\right)\left[\left(6\times8\right)\mathrm{ft^2}\right]\left[\left(120-60\right)°F\right] \\
&= 1944\,\mathrm{Btu/hr}
\end{aligned}
$$

Example 3.18 – Light Bulb Heat Transfer. Calculate the heat transfer from a 100-W light bulb at 113°C to 31°C ambient air. Approximate the bulb as a 60 mm diameter sphere.

SOLUTION: For this example,

$$T_f = \left(T_s + T_\infty\right)/2 = 72°C$$

From the Appendix,

$$\nu = \left(22.38\times10^{-5}\,\mathrm{ft^2/s}\right)\left(\frac{0.0929\,\mathrm{m^2/s}}{\mathrm{ft^2/s}}\right) = 2.079\times10^{-5}\,\mathrm{m^2/s}$$

$$k = 0.01735\,\mathrm{Btu/hr\cdot ft\cdot°F}\left(\frac{1.729\,\mathrm{W/m\cdot K}}{\mathrm{Btu/hr\cdot ft\cdot°F}}\right) = 0.0300\,\mathrm{W/m\cdot K}$$

$$\mathrm{Pr} = 0.70$$

$$\beta = \frac{1}{T} = \frac{1}{345} = 2.899\times10^{-3}\,\mathrm{K^{-1}}$$

Employ the characteristic length as the diameter of the sphere, D, in Equation (3.34).

$$
\begin{aligned}
\mathrm{Gr} &= \frac{g\beta(T_s - T_\infty)D^3}{\nu^2} \\
&= \frac{(9.80\,\mathrm{m/s^2})(2.899\times 10^{-3}\,\mathrm{K^{-1}})[(113-31)\mathrm{K}](0.060\,\mathrm{m})^3}{(2.079\times 10^{-5}\,\mathrm{m^2/s})^2} \\
&= 1.16\times 10^6
\end{aligned}
$$

Apply Equation (3.44) with constants drawn from Table 3.7, i.e.,

$$
\begin{aligned}
\frac{\overline{h}D}{k} &= 0.60\,\mathrm{Ra}^{1/4} \\
\overline{h} &= \left(\frac{k}{D}\right)0.60\,\mathrm{Ra}^{1/4} \\
&= \left(\frac{0.0300\,\mathrm{W/m\cdot K}}{0.060\,\mathrm{m}}\right)(0.60)\left[(1.16\times 10^6)(0.7)\right]^{1/4} \\
&= 9.01\,\mathrm{W/m^2\cdot K}
\end{aligned}
$$

Example 3.19 – Light Bulb Free Convection Heat Loss. Refer to Illustrative Example 3.18. Calculate the heat transfer lost by free convection from the light bulb.

SOLUTION: Once again, apply Equation (3.49).

$$
\begin{aligned}
Q &= \overline{h}A(T_S - T_\infty) \\
&= (9.01\,\mathrm{W/m^2\cdot K})4\pi(0.060\,\mathrm{m})^2(82\,\mathrm{K}) \\
&= 33.4\,\mathrm{W}
\end{aligned}
$$

3.6   Environmental Applications

Two applications that involve the environment comprise the concluding section of this chapter. The first is concerned with lapse rates and the other with plume rise.

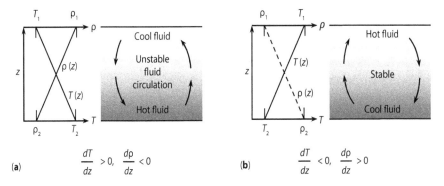

Figure 3.5 A fluid contained between two horizontal plates at different temperatures (a) Unstable temperature gradient (b) Stable temperature gradient.

The concept behind the so-called *lapse rate* that has found application in environmental science and engineering can be best demonstrated with the following example. Consider the situation where a fluid is enclosed by two horizontal plates of different temperature $(T_1 \neq T_2)$, see Figure 3.5. In case (*a*), the temperature of the lower plate exceeds that of the upper plate $(T_2 > T_1)$ and the density decreases in the (downward) direction of the gravitational force. If the temperature difference exceeds a particular value, conditions are termed *unstable* and buoyancy forces become important. In Figure 3.5 (a), the gravitational force of the cooler and denser fluid near the top plate exceeds that acting on the lighter hot fluid near the bottom plate and the circulation pattern as shown on the right-hand side of Figure 3.5 (a) will exist. The cooler heavier fluid will descend, being warmed in the process, while the lighter hot fluid will rise, cooling as it moves. However, for case (*b*) where $T_1 > T_2$, the density no longer decreases in the direction of the gravitational force. Conditions are now reversed and defined as *stable* since there is no bulk fluid motion. In case (*a*), heat transfer occurs from the bottom to the top surface by free convection; for case (*b*), any heat transfer (from top to bottom) occurs by conduction. These two conditions are similar to that experienced in the environment, particularly with regard to the atmosphere. The extension of the above development is now applied to the atmosphere but could also be applied to oceanographic systems [2].

Apart from the mechanical interference of the steady flow of air caused by buildings and other obstacles, the most important factor that determines the degree of turbulence, and hence how fast diffusion in the lower air occurs, is the variation of temperature with height above the ground (i.e., the aforementioned "lapse rate"). Air is a good insulator. Therefore, heat transfer in the atmosphere is caused by radiative heating or by mixing due to turbulence. If there is no mixing, an air parcel rises adiabatically (no heat transfer) in the atmosphere.

The Earth's atmosphere is normally treated as a perfect gas mixture. If a moving air parcel is chosen as a control volume, it will contain a fixed number of molecules. The volume of such an air parcel must be inversely proportional to the density. The ideal gas law is expressed as:

$$PV \propto RT$$

or

$$P \propto \rho RT \tag{3.71}$$

The air pressure at a fixed point is caused by the weight of the air above the point. The pressure is highest at the Earth's surface and decreases with altitude. This is a hydrostatic pressure distribution with the change in pressure proportional to the change in height [5].

$$dP = -\rho g\, dz \tag{3.72}$$

Since air is compressible, the density is also a decreasing function of height. An air parcel must have the same pressure as the surrounding air and so, as it rises, its pressure decreases. As the pressure drops, the parcel must expand adiabatically. The work done in the adiabatic expansion $(P\,dV)$ comes from the thermal energy of the air parcel [10]. As the parcel expands, the internal energy then decreases and the temperature decreases.

If a parcel of air is treated as a perfect gas rising in a hydrostatic pressure distribution, the rate of cooling by the adiabatic expansion can be calculated. The rate of expansion with altitude is fixed by the vertical pressure variation described in Equation (3.72). Near the Earth's surface, a rising air parcel's temperature normally decreases by 0.98 °C with every 100-m increase in altitude.

The vertical temperature gradient in the atmosphere (the amount the temperature changes with altitude, dT/dz) is defined as the *lapse rate*. The dry *adiabatic lapse rate* (DALR) is the temperature change for a rising parcel of dry air. The dry adiabatic lapse rate is approximately $-1°C/100$ m or $dT/dz = -10^{-2}°C/m$ or $-5.4°F/1000$ ft. Strongly stable lapse rates are commonly referred to as *inversions*; $dT/dz > 0$. The strong stability inhibits mixing. Normally, these conditions of strong stability only extend for several hundred meters vertically. The vertical extent of the inversion is referred to as the *inversion height*. Thus, a positive rate is particularly important in air pollution episodes because it limits vertical motion (i.e., the inversion traps the pollutants between the ground and the inversion layer).

Ground-level inversions inhibit the mixing of pollutants emitted from automobiles, smoke stacks, etc. This increases the ground-level concentrations of pollutants. At night the ground reradiates the solar energy that it received during the day. On a clear night with low wind speeds, the air near the ground is cooled and forms a ground-level inversion. By morning, the inversion depth may be 200–300 m with a 5–10 °C temperature from bottom to top. Clouds cut down the amount of heat radiated by the ground because they reflect the radiation back to the ground. Higher wind speeds tend to cause more mixing and spread the cooling effect over a larger vertical segment of the atmosphere, thus decreasing the change in lapse rate during the night [10].

Since temperature inversions arise because of solar radiation, the effects of nocturnal radiation often results in the formation of frost. When the Sun is down, some thermal radiation is still received by the Earth's surface from space, but the amount is small. Consequently, there is a net loss of radiation energy from the ground at night. If the air is relatively still, the surface temperature may drop below 32 °F. Thus, frost can form, even though the air temperature is above freezing. This frost is often avoided with the presence of a slight breeze or by cloud cover.

On to plume rise. Smoke from a stack will usually rise above the top of the stack for a certain distance (see Figure 3.6). The distance that the plume

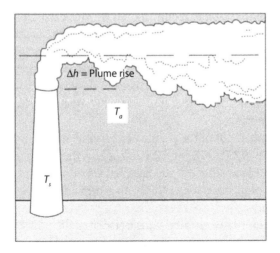

Figure 3.6 Plume rise.

rises above the stack is called the *plume rise*. It is actually calculated as the distance to the imaginary centerline of the plume rather than to the upper or lower edge of the plume. Plume rise, normally denoted Δh, depends on the stack's physical characteristics and on the effluent's (stack gas) characteristics. For example, the effluent characteristics of stack gas temperature T_s in relation to the surrounding air temperature T_a is more important than the stack characteristic of height. The difference in temperature between the stack gas and ambient air determines plume density, and this density affects plume rise. Therefore, smoke from a short stack could climb just as high as smoke from a taller stack.

Stack characteristics are used to determine momentum, and effluent characteristics are used to determine buoyancy. The *momentum* of the effluent is initially provided by the stack. It is determined by the speed of the effluent as it exits the stack. As momentum carries the effluent out of the stack, atmospheric conditions begin to affect the plume.

The condition of the atmosphere, including the winds and temperature profile along the path of the plume, will primarily determine the plume's rise. As the plume rises from the stack, the wind speed across the stack top begins to tilt (or bend) the plume. Wind speed usually increases with distance above the Earth's surface. As the plume continues upward, stronger winds tilt the plume even farther. This process continues until the plume may appear to be horizontal to the ground. The point where the plume appears level may be a considerable distance downwind from the stack.

Plume buoyancy is a function of temperature. When the effluent's temperature, T_s, is warmer than the atmosphere's temperature, T_a, the plume will be less dense than the surrounding air. In this case, the density difference between the plume and air will cause the plume to rise. The greater the temperature difference, ΔT, the more buoyant the plume. As long as the temperature of the pollutant remains warmer than the atmosphere, the plume will continue to rise. The distance downwind where the pollutant cools to atmospheric temperature may also be quite displaced from its original release point.

Buoyancy is taken out of the plume by the same mechanism that tilts the plume over–the wind. The faster the wind speed, the faster this mixing with outside air takes place. This mixing is called *entrainment*. Strong wind will "rob" the plume of its buoyancy rapidly and, on windy days, the plume will not climb significantly above the stack.

Many individuals have studied plume rise over the years. The most popular plume rise formulas in use are those of Briggs [11]. Most plume rise equations are used on plumes with temperatures greater than the ambient air temperature. Plume rise formulas determine the imaginary *centerline* of the plume; the centerline is located where the *greatest concentration* of pollutant occurs at a given downward distance. Finally, plume rise is a linear measurement, usually expressed in feet or meters.

Briggs [11] used the following equations to calculate the plume rise:

$$\Delta h = 1.6F^{1/3}u^{-1}x^{2/3}; \ x < x_f \qquad (3.73)$$

$$= 1.6F^{1/3}u^{-1}x_f^{2/3}; \ \text{if } x \geq x_f \qquad (3.74)$$

$$x^* = 14F^{5/8}; \ \text{when } F < 55\text{m}^4/\text{s}^3 \qquad (3.75)$$

$$= 34F^{5/8}; \ \text{when } F \geq 55\text{m}^4/\text{s}^3 \qquad (3.76)$$

$$x_f = 3.5x^* \qquad (3.77)$$

where,

Δh = plume rise, m
F = buoyancy flux, m^4/s^3 = $(3.7 \times 10^{-5})Q_H$
u = wind speed m/s
x^* = downward distance, m
x_f = distance of transmission from the first stage of rise to the second stage of rise, m
Q_H = heat emission rate, kcal/s

If the term Q_H is not available, the term F may be estimated by

$$F = q\left(\frac{g}{\pi}\right)\left(\frac{T_S - T}{T_S}\right) \qquad (3.78)$$

where

g = gravity term, 9.8 m^2/s
q = stack gas volumetric flow rate, m^3/s (actual conditions)
T_S, T = stack gas and ambient air temperature, K, respectively

Example 3.20 [6, 7] – Plume Rise Application. If a waste source emits a gas with a buoyancy flux of 50 m^4/s^3, and the wind speed averages 4 m/s, find the plume rise, under unstable atmospheric conditions, at a distance of 750 m downward from a stack that is 50 m high. Although several plume rise equations are available, use the equation proposed by Briggs.

SOLUTION: Calculate x_f to determine which plume equation applies.

$$x^* = 14F^{5/8}, \quad \text{since } F \text{ is less than } 55\text{m}^4/\text{s}^3$$
$$= 14(50)^{5/8}$$
$$= 161.43\,\text{m}$$

$$x_f = 3.5x^*$$
$$= 3.5(161.43)$$
$$= 565.0\,\text{m}$$

The plume rise is therefore

$$\Delta h = 1.6F^{1/3}u^{-1}x_f^{2/3}, \quad \text{since } x \geq x_f$$
$$= 1.6(50)^{1/3}(4)^{-1}(565.0)^{2/3}$$
$$= 101\,\text{m}$$

Example 3.21 – Plume Rise Equations. Briefly discuss other plume rise equations.

SOLUTION: Many other plume rise equations may be found in the literature [14]. The Enviromental Protection Agency (EPA) is mandated to use Briggs's equations to calculate plume rise. In past years, industry has often chosen to use the Holland or Davidson-Bryant equation. The Holland equation is [12–14]

$$\Delta h = D_s\left(\frac{\bar{v}_s}{u}\right)\left[1.5 + \left(2.68 \times 10^{-3}\right)P\left(\frac{\Delta T/T_s}{d_s}\right)\right] \qquad (3.79)$$

where

$\quad D_s$ = inside stack diameter, m
$\quad \bar{v}_s$ = stack exit velocity, m/s
$\quad u$ = wind speed, m/s
$\quad P$ = atmospheric pressure, mbar
$\quad T_s, T$ = stack gas and ambient air temperature, K, respectively
$\quad \Delta T = T_s - T$
$\quad \Delta h$ = plume rise, m

The Davidson-Bryant equation is [13]

$$\Delta h = D_s\left(\frac{v_s}{u}\right)^{1.4}\left[1.0 + \frac{T_s + T}{T_s}\right] \qquad (3.80)$$

The reader should also note that the "plume rise" may be negative in some instances due to surrounding structures, topography, and so on.

References

1. L. Theodore, personal notes, East Williston, NY, 1992.
2. D. Kern, *Process Heat Transfer*, McGraw-Hill, New York City, NY, 1950.
3. I. Farag and J. Reynolds, *Heat Transfer*, A Theodore Tutorial, Theodore Tutorials, East Williston, NY, originally published by USEPA/APTI, RTP, NC, 1996.
4. L. Theodore (adapted form), *Heat Transfer Applications for the Practicing Engineers*, John Wiley & Sons, Hoboken, NJ, 2011.
5. P. Abulencia and L. Theodore, *Fluid Flow for the Practicing Chemical Engineer*, John Wiley & Sons, Hoboken, NJ, 2009.
6. R. Griskey, *Transport Phenomena and Unit Operations*, John Wiley & Sons, Hoboken, NJ, 2002.
7. R. Bird, W. Stewart, and E. Lightfoot, *Transport Phenomena*, 2nd edition, John Wiley & Sons, Hoboken, NJ, 2002.
8. L. Theodore, *Transport Phenomena for Engineers*, Theodore Tutorials, East Williston, NY, originally published by International Textbook Co., Scranton, PA, 1971.
9. L. Theodore: personal notes, East Williston, NY, 1980.
10. L. Theodore and A.J. Buonicore, *Air Polluting Control Equipment: Gases*, CRC Press, Boca Raton, FI, 1976.
11. G. Briggs, *Plume Rise*, AEC Critical Review Series, USAEC, Division of Technical Information, date and location unknown.
12. D. Turner, *Workbook of Atmospheric Estimates*, Publ. No. AP-626, USEPA, RTP, NC, 1970.
13. W. Davidson, *Trans. Conf. Ind. Wastes*, 14th Annual Meet Ind. Hyg. Foundation, pg 38, location unknown, 1949.
14. L. Theodore, *Air Pollution Control Equipment Calculations*, John Wiley & Sons, Hoboken, NJ, 2008.

4

Radiation

Introduction

In addition to conduction and convection, heat can be transmitted by radiation. Conduction and convection both require the presence of molecules to "carry" or pass along the energy. Unlike conduction or convection, radiation does not require the presence of any medium between the heat source and the heat sink since the thermal energy travels as electromagnetic waves. This radiant energy (thermal radiation) phenomena is emitted by every body having a temperature greater than absolute zero. Quantities of radiation emitted by a body are a function of both temperature and surface conditions, details of which will be presented later in this chapter. Applications of thermal radiation include industrial heating, drying, energy conversion, solar radiation, and combustion [1, 2].

The amount of thermal radiation emitted is not always significant. Its importance in a heat transfer process depends on the quantity of heat being transferred simultaneously by the other aforementioned mechanisms. The reader should note that the thermal radiation of systems operating at or below room temperature is almost always negligible. In contrast, thermal radiation tends to be the principal mechanism for heat transfer for systems operating in excess of 1200 °F. When systems operate between room temperature and 1200 °F, the amount of heat transfer contributed by radiation depends on such variables as the convection film coefficient and the nature of the radiating surface [1, 2].

In the heat transfer mechanisms of conduction and convection discussed in Chapters 2 and 3, respectively, the movement of energy in the form of heat takes place through a material medium—a fluid in the case of convection. Since a transfer medium is not required for this third mechanism, the energy is carried by the aforementioned electromagnetic radiation. Thus, a piece of steel plate heated in a furnace until it is glowing red and then placed several inches away from a cold piece of steel plate will cause the temperature of the cold steel to rise, even if the process takes place in an evacuated container.

Several additional examples of heat transfer mechanisms by radiation have appeared in the literature. Badger and Banchero [3] provided the following explanation of radiation: "If radiation is passing through empty space, it is not transformed to heat or any other form of energy and it is not diverted from its path. If, however, matter appears in its path it is only the absorbed energy that appears as heat, and this transformation is quantitative. For example, fused quartz transmits practically all the radiation which strikes it; a black surface will absorb most of the radiation received by it (as one can experience on a sunny day while wearing a black shirt) and will transform such absorbed energy quantitatively into heat." Bennett and Meyers [4] provide an additional example involving the operation of a steam "radiator." The relationship between the energy transmitted, reflected, and absorbed is discussed in a later section.

Characteristic wavelengths of radiation are provided in Table 4.1. Note that light received from the Sun passes through the Earth's atmosphere which absorbs some of the energy and thus affects the quality of visible light as it is received. The units of wavelength may be expressd in

Table 4.1 Characteristic wavelengths [3].

Type of radiation	$\lambda \times 10^8$
Gamma rays	0.01–0.15 cm
X-rays	0.06–1000 cm
Ultraviolet	100–35,000 cm
Visible	3500–7800 cm
Infrared	7800–4,000,000 cm
Radio	0.01–0.15 cm

meters (m), centimeters (cm), micrometers (μm), or Angstroms (1.0 Å = 10 μm), with the centimeter being the usual unit of choice. The speed of electromagnetic radiation is approximately 3×10^8 m/s in a vacuum. This velocity (c) is given by the product of the wavelength (λ) and the frequency (v) of the radiation, i.e.,

$$c = \lambda v; \text{ consistent units} \tag{4.1}$$

As noted earlier, the energy emitted from a "hot" surface is in the form of electromagnetic waves, λ. One of the types of electromagnetic waves is thermal radiation. Thermal radiation is defined as electromagnetic waves falling within the following range:

$$0.1 \text{ μm} < \lambda < 100 \text{ μm}; \quad 1.0 \text{μm} = 10^{-6} \text{ m} = 10^{-4} \text{ cm}$$

However, most of this energy is in the interval from 0.1 to 10 μm. The visible range of thermal radiation lies within the narrow range of 0.4 μm (violet) $< \lambda < 0.8$ μm (red).

Example 4.1 – Heat Transfer Processes. Identify the pertinent heat transfer processes for the following three systems.

1. A heat exchanger made of metal tubing with the fluid inside the tube hotter than the outside fluid
2. Air flowing across a heated radiator
3. Asphalt pavement on a breezy summer's day

SOLUTION.

1. In scenario 1, metallic solids transfer heat by conduction. The metal tube wall conducts heat from the hot fluid, through the metal wall, to the cold fluid.
2. In scenario 2, heat is transferred by natural convection. If the air currents are caused by an external force, the heat transfer from the radiator surface to the air is by forced convection.
3. Pathways for energy to and from the asphalt paving for scenario 3 include incident solar radiation (a large portion of which is absorbed by the asphalt). Asphalt emits heat by radiation to the surroundings and also loses heat by convection to the air.

The remainder of the chapter consists of six additional sections:

4.1   The Origin of Radiant Energy
4.2   The Distribution of Radiant Energy
4.3   Radiant Exchange Principles
4.4   Kirchoff's Law
4.5   Emissivity Factors and Energy Interchange
4.6   View Factors

4.1   The Origin of Radiant Energy

Radiant energy is believed to originate within the molecules of the radiating body, the atoms of such molecules vibrating in a simple harmonic motion as linear oscillators. The emission of radiant energy is believed to represent a decrease in the amplitudes of the vibrations within the molecules, while an absorption of energy represents an increase. In its essence, quantum theory postulates that for every frequency of radiation there is a small minimum pulsation of energy which may be emitted. This is the *quantum*, and a smaller quantity cannot be emitted although many such quanta may be emitted. The total radiation of energy of a given frequency emitted by a body is an integral number of quanta and thus, the total energy may be different. Planck showed that the energy associated with a quantum is proportional to the frequency of the vibration and inversely proportional to the wavelength. Thus, radiant energy of a given frequency may be pictured as consisting of successive pulses of radiant energy, each pulse of which has the value of the quantum for a given frequency.

The picture of the atom proposed by Bohr is helpful in providing a clearer understanding of one possible origin of radiant energy. Electrons are presumed to travel about the nucleus of an atom in elliptical orbits at varying distances from the nucleus. The orbital electrons possess definite energies comprising their kinetic and potential energies by virtue of their rotation about the nucleus. The potential energy is the energy required to remove an electron from its orbit to an infinite distance from the nucleus. A given electron in an orbit at a given distance from the nucleus will have a certain energy. Should a disturbance occur such as the collision of the atom with another atom or electron, the given electron may be displaced from its orbit and may (1) return to its original orbit, (2) pass to another orbit whose electrons possess a different energy, or (3) entirely leave the system influenced by the nucleus. If the transition is from an orbit of high energy to one of lower energy, the readjustment is affected by the radiation of the excess energy.

Another origin of radiant energy may be attributed to the changes in the energies of atoms and molecules themselves without reference to their individual electrons. If two or more nuclei of the molecule are vibrating with respect to each other, a change in the amplitude or amplitudes of vibration will cause a change in energy content. As noted, a decrease in amplitude is the result of an emission of radiant energy, while an increase is the result of the absorption of radiant energy. The energy of a molecule may be changed by an alteration of its kinetic energy by translation or rotation, and this will likewise result in the emission of radiant energy. A decrease in particle velocity corresponds to the emission of radiant energy, while an increase corresponds to the absorption of radiant energy.

Since temperature is a measure of the average kinetic energy of molecules, the higher the temperature the higher the average kinetic energy both of translation and of vibration. It can therefore be expected that the higher the temperature the greater the quantity of radiant energy emitted from a substance. Since the molecular movement ceases completely at the absolute zero of temperature, it may be concluded that all substances are above absolute zero.

For radiant energy to be emitted from the interior of a solid, it has to do so without being dissipated by producing other energy changes within its molecules. There is little probability that radiant energy generated in the interior of a solid will reach its surface without encountering other molecules, and therefore all radiant energy emitted from the surfaces of solid bodies is generated by energy-level changes in molecules near and on their surfaces. The quantity of radiant energy emitted by a solid body is consequently a function of the surface of the body, and conversely, radiation incident on a solid body is absorbed at the surface. The probability that internally generated radiant energy will reach the surface is far greater for hot radiating gases than for solids, and the radiant energy emitted by a gas is a function of the gas volume rather than the surface of the gas shape. The situation with liquids is intermediate between gases and solids, and radiation may originate somewhat below the surface, depending on the nature of the liquid.

4.2   The Distribution of Radiant Energy

A body at a given temperature will emit radiation over a range of wavelengths, and not a single wavelength. This is attributed to the existence of an infinite variety of linear oscillators. The energy emitted at each wavelength can be determined through the use of a dispersing prism and thermopiles. Such measurements on a given body will produce curves as shown in Figure 4.1 for each given temperature. The curves are plots of the intensities of the radiant energy, I, Btu/hr·ft²·μm against the wavelength, λ, (in μm) as determined at numerous wavelengths and connecting , i.e., data is taken for a constant temperature while varying wavelength and the data points are "connected" to form each curve. Each curve possesses a wavelength at which the amount of spectral energy given off is a maximum. This maximum intensity of radiation is obviously less for the same body at a lower temperature. The maximum intensity falls between 0.75 and 400 microns, indicating that red heat is a far better source of energy than white heat. Were it not for this fact, the near-white incandescent lamp would require more energy for illumination and give off uncomfortable quantities of heat.

It is necessary to differentiate between two kinds of properties when dealing with the properties of radiation: monochromatic and total. A monochromatic property, such as the maximum value of I in Figure 4.1 refers to a single wavelength. Atmospheric data indicates that the maximum intensity, I,

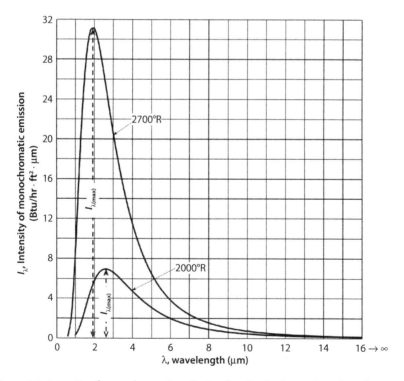

Figure 4.1 Intensity of monochromatic radiation for a hot body as a Function of temperature.

of the Sun is experienced around 0.25 μm wavelength. This accounts for the predominance of blue in the visible spectrum and the high ultraviolet content of the Sun's rays. A total property indicates that it is the algebraic sum of the monochromatic values of the property. Monochromatic radiation literally means, "one color" or one wavelength, but experimentally it actually refers to a group or band of wavelengths since wavelengths cannot be resolved individually. Monochromatic values are not important to the direct solution of engineering problems but are necessary for the derivation of basic radiation relationships.

Since the $I - \lambda$ curve for a single temperature depicts the amount of energy emitted at a given wavelength, the sum of all the energy radiated by the body at all its wavelengths is simply the area under a plot of I vs. λ. This quantity of radiant energy (of all wavelengths) emitted by a body per unit area and time is defined as the *total emissive power E* (Btu/hr·ft²). Given the intensity of the radiation at any wavelength, I, one may therefore calculate the total emissive power, E, from

$$E = \int_0^\infty I d\lambda \tag{4.2}$$

Maxwell Planck was the first to fit the I vs. λ relationship to equation form, as

$$I_\lambda = \frac{C_1 \lambda^{-5}}{e^{(C_2/\lambda T)} - 1} \tag{4.3}$$

where

I_λ = intensity of emission at $\lambda \left(\text{Btu} / \text{hr} \cdot \text{ft}^2 \cdot \mu\text{m} \right)$

λ = wavelength (μm)

C_1 = 1.16×10^5 (dimensionless)

C_2 = = 25,740 (dimensionless)

T = temperature of the body (°R)

It was later shown that the product of the wavelength of the maximum value of the intensity of emission and the absolute temperature is a constant. This is referred to as Wien's displacement law, where

$$\lambda T = 2884 \, \mu\text{m} \cdot °\text{R} \approx 5200 \, \mu\text{m} \cdot \text{K} \tag{4.4}$$

The reader can derive Equation (4.4) from (4.3) as follows. Since $dI / d\lambda$ at the maximum value of the intensity is zero,

$$\frac{dI_\lambda}{d\lambda} = \frac{d\left[\dfrac{C_1 \lambda^{-5}}{e^{(C_2/\lambda T)} - 1}\right]}{d\lambda} = 0 \qquad (4.5)$$

differentiation of Equation (4.5) leads to

$$\left(-5C_1\lambda^{-6}\right)\left(e^{(C_2/\lambda T)} - 1\right) + C_1\lambda^{-5}\left(e^{(C_2/\lambda T)}\right) / \left(e^{(C_2/\lambda T)} - 1\right) = 0$$

which can be reduced to

$$\left(-5C_1\lambda^{-6}\right)\left(e^{(C_2/\lambda T)} - 1\right) + C_1\lambda^{-5}\left(e^{(C_2/\lambda T)}\right)\left(\frac{C_2}{\lambda^2 T}\right) = 0$$

This ultimately simplifies to

$$\left(-5 + \frac{C_2}{\lambda T}\right)e^{(C_2/\lambda T)} + 5 = 0 \qquad (4.6)$$

with

$$C_2 = 25{,}740$$

The reader is left with the exercise of showing that the first term in Equation (4.6) equals −5 when $\lambda = 2884$.

Example 4.2 – The Temperature of the Sun. Estimate the Sun's temperature. Employ Equation (4.4).

SOLUTION: Assuming a wavelength of 0.25 μm, substitute into Equation (4.4):

$$\lambda T = 2884\,\mu m\,°R, \cdot \lambda = 0.25\,\mu m$$

Therefore,

$$T = 2884\,\mu m \cdot °R / 0.25\,\mu m$$

$$T = 11{,}500°R \approx 11{,}000°F$$

Example 4.3 – Emissive Power Calculation. The intensity of radiation as a function of wavelength (λ = μm) is specified as

$$I = 40e^{-\lambda^2}; \quad \text{Btu} / \text{hr} \cdot \text{ft}^2 \cdot \mu m$$

Calculate the total emissive power.

SOLUTION: Apply Equation (4.2):

$$E = \int_0^\infty I \, d\lambda \tag{4.2}$$

Substitute for I,

$$E = \int_0^\infty 40e^{-\lambda^2} \, d\lambda = 40 \int_0^\infty e^{-\lambda^2} \, d\lambda$$

The above integral is calculated as follows [4]. Set

$$\lambda^2 = x; \quad \lambda = x^{1/2}$$

so that

$$2\lambda d\lambda = dx$$

$$d\lambda = \frac{dx}{2\lambda}$$

$$= \frac{1}{2} x^{-1/2} dx$$

Insertion into the above integral gives

$$E = \frac{40}{2} \int_0^\infty e^{-x} x^{-1/2} dx$$

This integral is the gamma function [2] of ½, i.e., $\Gamma(\frac{1}{2})$. Since [2]

$$\Gamma\left(\frac{1}{2}\right) = \pi^{1/2} = \sqrt{\pi}$$

$$E = 20\sqrt{\pi}$$

$$= 35.5 \frac{\text{Btu}}{\text{hr} \cdot \text{ft}^2}$$

4.3   Radiant Exchange Principles

The conservation law of energy indicates that any radiant energy incident on a body will partially absorb, reflect, or transmit stored energy. An energy balance around a receiving body on which the total incident energy is assumed to be unity gives

$$\alpha + \rho + \tau = 1 \qquad (4.7)$$

where the absorptivity α is the fraction absorbed (and no longer the thermal diffusivity employed in the previous chapters), the reflectivity ρ is the fraction reflected, and the transmissivity τ is the fraction transmitted [1, 4]. It should be noted that the majority of engineering applications involve opaque substances having transmissivities approaching zero (i.e., $\tau = 0$). This topic receives additional treatment below in a later paragraph.

When an ordinary body emits radiation to another body, it will have some of the emitted energy returned to itself by reflection. Equation (4.7) assumes that none of the emitted energy is returned; this is equivalent to assuming that bodies having zero transmissivity also have zero reflectivity. This introduces the concept of a perfect "black body" for which $\alpha = 1$.

Not all substances radiate energy at the same rate at a given temperature. The theoretical substance to which most radiation discussions refer to is called a "black body." This is defined as a body that radiates the maximum possible amount of energy at a given temperature. Much of the development to follow is based on this concept.

To summarize, when radiation strikes the surface of a semi-transparent material such as a glass plate or a layer of water, three types of interactive effects occur. Some of the incident radiation is reflected off the surface, some of it is absorbed within the material, and the remainder is transmitted through the material. Examining the three fates of the incident radiation, one can see that α and ρ depend on inherent properties of the material; it is for this reason that they are referred to as *surface* properties. The transmissivity, τ, on the other hand, depends on the amount of the material in question; it is therefore referred to as a *volumetric* property.

It is appropriate to examine a few common surfaces. An opaque surface, the most commonly encountered surface type, has $\tau \approx 0$. Because of this, Equation (4.7) becomes:

$$\alpha + \rho = 1 \qquad (4.8)$$

or

$$\rho = 1 - \alpha \qquad (4.9)$$

Equation (4.9) may also be applied to gases; this may be counterintuitive since most gases are invisible. With respect to the reflectivity term, ρ, surfaces may have either *specular* reflection, in which the angle of incidence of the radiation is equal to the angle of reflection, or *diffuse* reflection, in which their reflected radiation scatters in all directions. In addition, a *gray* surface is one for which the absorptivity is the same as the emissivity, ε, at the temperature of the radiation source, For this case,

$$\alpha = \varepsilon \qquad (4.10)$$

$$\rho = 1 - \varepsilon \qquad (4.11)$$

Reflectivity and transmissivity are characteristics experienced in the everyday world. Polished metallic surfaces have high reflectivities and granular surfaces have low reflectivities. Reflection from a surface depends greatly on the characteristics of the surface. If a surface is very smooth, the angles of incidence and reflection are essentially the same. However, most surfaces encountered in engineering practice are sufficiently rough so that some reflection occurs in all directions. Finally, one may state that a system in thermal equilibrium has its absorptivity equal to the emissivity.

4.4   Kirchoff's Law

Consider a body of given size and shape placed within a hollow sphere of constant temperature and assume that the air has been evacuated. After *thermal equilibrium* has been reached, the temperature of the body and that of the enclosure (the sphere) will be the same, inferring that the body is absorbing and radiating heat at equal rates [1, 4]. Let the total intensity of radiation falling on the body be I (now with units of Btu/hr·ft^2), the fraction absorbed a_1, and the total emissive power E_1 (Btu/hr·ft^2). Kirchoff noted that the energy emitted by a body of surface A_1 at thermal equilibrium is equal to that received, so that:

$$E_1 A_1 = I \alpha_1 A_1 \qquad (4.12)$$

or simply,

$$E_1 = I \alpha_1 \qquad (4.13)$$

If the body is replaced by another of identical shape, then:

$$E_2 = I \alpha_2 \qquad (4.14)$$

If a third body that is a black body is introduced, then:

$$E_b = I \tag{4.15}$$

Since the absorptivity, a, of a black body is 1.0, one may write:

$$I = \frac{E_1}{\alpha_1} = \frac{E_2}{\alpha_2} = E_b \tag{4.16}$$

Thus, at thermal equilibrium, the ratio of the total emissive power to the absorptivity for all bodies is the same. This is referred to as Kirchoff's law. Since $a = \varepsilon$, the above equation may also be written

$$\frac{E_1}{E_b} = \alpha_1 = \varepsilon_1 \tag{4.16}$$

and

$$\frac{E_2}{E_b} = \alpha_2 = \varepsilon_2 \tag{4.17}$$

where the ratio of the actual emissive power to the black-body emissive power is defined as the aforementioned emissivity, ε. Selected values of ε for various bodies and surfaces are given in Table 4.2. Additional values are available in the literature [5–7].

Table 4.2 Total emissivity of various sources.

Surface	T, °F	Emissivity, ε
Alumina, effect of mean grain size (μm)		
10		0.18 – 0.30
50		0.28 – 0.39
100		0.40 – 0.50
Aluminum		
Anodized		0.76
Highly polished plate, 98.3% pure	400 – 1070	0.039 – 0.057
Commercial sheet	212	0.09
Heavily oxidized	299 – 940	0.20 – 0.31
Al-surfaced roofing	100	0.216
Asbestos		
Board	74	0.96

(*Continued*)

Table 4.2 Cont.

Surface	T, °F	Emissivity, ε
Brass		
Highly polished		
73.2% Cu, 26.7% Zn	476 – 674	0.028 – 0.031
62.4% Cu, 36.8% Zn, 0.4% Pb, 0.3% Al	494 – 710	0.033 – 0.037
82.9% Cu, 17.0% Zn	530	0.030
Hard-rolled, polished, but with the direction of polishing visible	70	0.038
Dull plate	120 – 660	0.22
Brick		
Red, rough, but no gross irregularities	70	0.96
Fireclay	1832	0.75
Carbon		
T-Carbon 0.9% ash, started with an emissivity of 0.72 at 260°F but on heating changed to specified values	260 – 1160	0.79 – 0.81
Filament	1900 – 2560	0.526
Rough plate	212 – 608	0.77
Lampblack, rough deposit	212 – 932	0.78 – 0.84
Chromium		
Polished	100 – 2000	0.08 – 0.38
Concrete tiles	1832	0.63
Copper		
Polished	242	0.028
Plate, heated a long time, and covered with thick oxide layer	77	0.78
Enamel		
White fused, on iron	66	0.90
Glass		
Smooth	72	0.94
Pyrex, lead and soda	500 – 1000	0.85 – 0.95
Gold		
Pure, highly polished	440 – 1160	0.018 – 0.035

(Continued)

Table 4.2 Cont.

Surface	T, °F	Emissivity, ε
Iron and steel (not including stainless)		
Steel, polished	212	0.066
Iron, polished	800 – 1800	0.14 – 0.38
Cast iron, newly turned	72	0.44
Turned and heated	1620 – 1810	0.60 – 0.70
Milled steel	450 – 1950	0.20 – 0.32
Lead		
Unoxidized, 99.96% pure	260 – 440	0.057 – 0.075
Gray oxidized	75	0.28
Oxidized at 300°F	390	0.63
Magnesium		
Magnesium oxide	530 – 1520	0.055 – 0.20
Molybdenum		
Filament	1340 – 4700	0.096 – 0.202
Massive, polished	212	0.071
Monel metal		
Oxidized at 1110°F	390 – 1110	0.41 – 0.46
Nickel		
Polished	212	0.072
Nickel oxide	1200 – 2290	0.59 – 0.86
Nickel Alloys		
Copper-nickel, polished	212	0.059
Nichrome wire, bright	120 – 1830	0.65 – 0.79
Nichrome wire, oxidized	120 – 930	0.95 – 0.98
Oxidized surfaces		
Iron plate, pickled, then rusted red	68	0.61
Iron, dark gray surface	212	0.31
Rough ingot iron	1700 – 2040	0.87 – 0.95
Sheet steel with strong, rough, oxide layer	75	0.80
Platinum		
Polished plate, pure	440 – 1160	0.054 – 0.104
Rubber		
Hard, glossy plate	74	0.94

(Continued)

Table 4.2 Cont.

Surface	T, °F	Emissivity, ε
Silver		
Polished, pure	440 – 1160	0.020 – 0.032
Polished	100 – 700	0.022 – 0.031
Stainless steels		
Polished	212	0.074
Type 301	450 – 1725	0.54 – 0.63
Tin		
Bright tinned iron	76	0.043 – 0.064
Tungsten		
Filament	6000	0.39
Zinc		
Galvanized, fairly bright	82	0.23
Water	32 – 212	0.95 – 0.963

If a *black* body radiates energy, the total radiation may be determined from the aforementioned Planck's law:

$$I_\lambda = \frac{C_1 \lambda^{-5}}{e^{(C_2/\lambda T)} - 1} \qquad (4.3)$$

Integration over the entire spectrum at a particular temperature yields:

$$F_b = \int_0^\infty \frac{C_1 \lambda^{-5}}{e^{(C_2/\lambda T)} - 1} d\lambda \qquad (4.18)$$

The evaluation of the integral can be shown to be:

$$E_b = \left(0.173 \times 10^{-8}\right) T^4 = \sigma T^4; \, T[=]°R \qquad (4.19)$$

Thus, the total radiation from a perfect black body is proportional to the fourth power of the absolute temperature of the body. This is also referred to as the Stefan-Boltzmann law. The constant 0.173×10^{-8} Btu/hr·ft^2·°R [2] is known as the Stefan-Boltzmann constant, usually designated by σ. Its counterpart in SI units is 5.669×10^{-8} W/m^2·K^4. However, note that this equation was derived for a *black* body.

If the body is *non-black*, the emissivity is given by

$$E = E_b \varepsilon \tag{4.16}$$

Substituting Equation (4.16) into Equation (4.19) gives

$$E = \varepsilon \sigma T^4 \tag{4.20}$$

or, since $E = Q / A$,

$$\frac{Q}{A} = \varepsilon \sigma T^4 \tag{4.21}$$

Thus, when the law is applied to a real surface with emissivity, ε, the total emissive power of a real body is given by Equation (4.21). As noted, typical values for the emissivity are provided in Table 4.2.

Now consider the energy transferred *between* two *black* bodies. Assume the energy transferred from the hotter body and the colder body is E_H and E_C, respectively. All of the energy that each body receives is absorbed since they are black bodies. Then, the net exchange between the two bodies maintained at two constant temperatures, T_H and T_C, is therefore

$$\frac{Q}{A} = E_H - E_C = \sigma\left(T_H^4 - T_C^4\right) \tag{4.22}$$

$$\frac{Q}{A} = 0.173\left[\left(\frac{T_H}{100}\right)^4 - \left(\frac{T_C}{100}\right)^4\right] \tag{4.23}$$

Example 4.4 – Radiation Between Two Black Bodies/Large Planes. Two large walls are required to be maintained at constant temperatures of 800 °F and 1200 °F. Assuming the walls are black bodies, how much heat must be removed from the colder wall to maintain a steady-state, constant temperature?

SOLUTION: For this application,

$$T_1 = 1200\ °F = 1660\ °R$$

$$T_2 = 800\ °F = 1260\ °R$$

Apply Equation (4.23) and substitute:

$$\frac{Q}{A} = 0.173\left[\left(\frac{1660}{100}\right)^4 - \left(\frac{1260}{100}\right)^4\right]$$

$$= 0.173 \, [75,900 - 25,200]$$

$$= 8770 \text{ Btu/hr} \cdot \text{ft}^2$$

Example 4.5 – Increases in Radiative Heat Transfer. Estimate the increase in heat transferred by radiation of a black body at 1500 °F relative to one at 1000 °F.

SOLUTION: Convert to *absolute* temperatures:

$$T_1 = 1500 \, °F = 1960 \, °R$$

$$T_2 = 1000 \, °F = 1460 \, °R$$

The percent increase in the rate of heat transfer from a black body at $T_1 = 1000 \, °F$ to $T_2 = 1500 \, °F$ is calculated as follows:

$$\% \text{ Increase} = \frac{Q_2 - Q_1}{Q_1} \times 100$$

$$= \frac{\left(\varepsilon \sigma T_2^4 / A\right) - \left(\varepsilon \sigma T_1^4 / A\right)}{\varepsilon \sigma T_1^4 / A} \times 100$$

$$= \frac{T_2^4 - T_1^4}{T_1^4} \times 100$$

$$= \frac{\left[1960°R\right]^4 - \left[1460°R\right]^4}{\left[1460°R\right]^4} \times 100$$

$$\approx 225\%$$

This represents a 225% increase in heat transfer.

4.5   Emissivity Factors and Energy Interchange

If the two bodies referred to the previous section are not black bodies, and instead (each) have an emissivity ε, then the net interchange of radiant energy is given by

$$\frac{Q}{A} = \varepsilon \sigma \left(T_H^4 - T_C^4\right) \tag{4.24}$$

This equation can be "verified" as follows [4]. If the two bodies discussed above are not black bodies and have *different* emissivities, the net exchange of energy will be different. Some of the energy emitted from the first body will be absorbed and the remainder radiated back to the other source. For two parallel bodies (planes) of infinite size, the radiation of each body can be accounted for. If the energy emitted from the first body is E_H with emissivity ε_H, the second body will absorb $E_H \varepsilon_C$ and reflect $1 - \varepsilon_C$ of it, i.e., $E_H(1 - \varepsilon_C)$. The first body will then receive $E_H(1 - \varepsilon_C)\varepsilon_H$ and again radiate to the cold body, but in the amount $E_H(1 - \varepsilon_C)(1 - \varepsilon_H)$. The exchanges for the two bodies are therefore:

Hot body
Radiated: E_H
Reflected back: $E_H(1 - \varepsilon_C)$
Radiated: $E_H(1 - \varepsilon_C)(1 - \varepsilon_H)$
Reflected back: $E_H(1 - \varepsilon_C)(1 - \varepsilon_H)(1 - \varepsilon_C)$
etc.

Cold body
Radiated: E_C
Reflected back: $E_C(1 - \varepsilon_H)$
Radiated: $E_C(1 - \varepsilon_H)(1 - \varepsilon_C)$
Reflected back: $E_C(1 - \varepsilon_H)(1 - \varepsilon_C)(1 - \varepsilon_H)$
etc.

For a non-black body, ε is not unity and must be included in Equation (4.19), i.e.,

$$E_b = 0.173\left(\frac{T}{100}\right)^4 \tag{4.25}$$

When Equation (4.25) is applied to the above infinite series analysis, one can show that Equation (4.26) results [5]:

$$E = \frac{Q}{A} = \frac{\sigma}{\left[\left(\dfrac{1}{E_H}\right) + \left(\dfrac{1}{E_C}\right) - 1\right]}\left(T_H^4 - T_C^4\right) \tag{4.26}$$

The radiation between a sphere and an enclosed sphere of radii R_H and R_C, respectively, may be treated in a manner similar to that provided above. The radiation emitted initially by the inner sphere is $E_H A_H$, all of which fall on A_C. Of this total, however, $(1 - \varepsilon_C)E_H A_H$ is reflected back to the hot body. If this analysis is similarly extended as before, the radiant exchange will again be represented by an infinite series whose solution may be shown to give [5]

$$E = \frac{Q}{A} = \frac{\sigma_H \left(T_H^4 - T_C^4 \right)}{\left[\frac{1}{\varepsilon_H} + \left(\frac{A_H}{A_C} \right) \left(\frac{1}{\varepsilon_C} - 1 \right) \right]}$$

$$= \frac{\sigma_H \left(T_H^4 - T_C^4 \right)}{\left[\frac{1}{\varepsilon_H} + \left(\frac{R_H}{R_C} \right)^2 \left(\frac{1}{\varepsilon_C} - 1 \right) \right]} \tag{4.27}$$

A similar relation applies for infinitely long concentric cylinders except that A_H/A_C is replaced R_H/R_C, not R_H^2/R_C^2 [5].

In general, an emissivity correction factor, F_ε, is introduced to account or the exchange of energy between different surfaces of different emissivities. The describing equation takes the form

$$E_H = \frac{Q}{A_H} = F_\varepsilon \sigma \left(T_H^4 - T_C^4 \right) \tag{4.28}$$

Values of F_ε for the interchange between surfaces are provided in Table 4.3 for the three cases already considered, i.e., (a), (b), (c), plus three additional cases (d), (e), (f) [5].

Table 4.3 Values of F_ε.

Condition	F_ε
a. Surface A_H is small compared with the totally enclosing surface A_C	ε_H
b. Surface A_C and A_H of infinite parallel planes or surface A_H of a completely enclosed body that is small compared with A_H	$\dfrac{1}{\left(\dfrac{1}{\varepsilon_H} + \dfrac{1}{\varepsilon_C} \right) - 1}$
c. Concentric spheres or infinite concentric cylinders with surfaces A_H and A_C	$\dfrac{1}{\dfrac{1}{\varepsilon_H} + \left(\dfrac{A_H}{A_C} \right) \left(\dfrac{1}{\varepsilon_C} - 1 \right)}$
d. Surfaces A_H and A_C of parallel disks, squares, 2:1 rectangles, long rectangles (see Figures 4.3 and 4.5 later)	$\varepsilon_H \varepsilon_C$
e. Surfaces A_H and A_C of perpendicular rectangles having a common side (see Figure 4.4 later)	$\varepsilon_H \varepsilon_C$
f. Surfaces A_H and parallel rectangular surface A_C with one corner of rectangle above A_H (see Figure 4.3 later)	$\varepsilon_H \varepsilon_C$

Finally, when a heat source is small compared to its enclosure, it is customary to assume that some of the heat radiated from the source is reflected back to it. Such is often the case on the loss of heat from a pipe to surrounding air. For these applications, it is convenient to represent the net radiation heat transfer in the same form employed for convection, i.e.,

$$Q = h_r A(T_H - T_C) \qquad (4.29)$$

where h_r is the *effective* radiation heat transfer coefficient. When $T_H - T_C$ is less than 120 °C (120 K or 216 °R), one may calculate the radiation heat transfer coefficient using

$$h_r = 4\varepsilon\sigma T_{avg}^3 \qquad (4.30)$$

where $T_{avg} = (T_H + T_C)/2$ [5].

SOLUTION: Apply Equation (4.26) and substitute:

$$\frac{Q}{A} = \frac{0.173}{\left[\left(\dfrac{1}{0.5}\right) + \left(\dfrac{1}{0.75}\right) - 1\right]}\left[(16.6)^4 - (12.6)^4\right]$$

$$= 3760 \text{ Btu/hr} \cdot \text{ft}^2$$

$$\% \text{ difference} = \left(\frac{8770 - 3760}{8770}\right)100$$

$$= 57.1\%$$

Note once again that the above solution is similar to Example 4.5 and applies to black bodies.

Example 4.6 – Radiation Between Two Large Planes with Different Emissivities. If the two bodies from Example 4.5 have emissivities of 0.5 and 0.75, respectively, what is the net energy exchange (per unit area)? Assume that the temperatures remain constant at 1660 °R and 1260 °R, and the two bodies are of infinite size. Also compare the energy exchange for the two examples.

SOLUTION: Apply Equation (4.26) and substitute:

$$\frac{Q}{A} = \frac{0.173}{\left[\left(\dfrac{1}{0.5}\right) + \left(\dfrac{1}{0.75}\right) - 1\right]}\left[(16.6)^4 - (12.6)^4\right]$$

$$= 3760 \text{ Btu / hr} \cdot \text{ft}^2$$

$$\% \text{difference} = \left(\frac{8770 - 3760}{8770}\right)100$$

$$= 57.1\% \text{ (decrease)}$$

Note once again that the above solution is similar to Example 4.5 and applies to black bodies.

Example 4.7 – Pipe Radiation. Calculate the radiation from a 2-inch IPS cast iron pipe (assumed polished) carrying steam at 300°F and passing through the center of a 1 ft × 1 ft galvanized zinc duct at 75°F and whose outside is insulated.

SOLUTION: Base the calculation on 1 ft of pipe/duct. For a 2-inch pipe A_H is 0.622 ft² of external surface per foot of pipe (see Appendix). The emissivity of oxidized steel from Table 4.2 is $\varepsilon_H = 0.44$. The surface of the duct is $A_C = 4(1)(1) = 4.0 \text{ ft}^2$, and for galvanized zinc, $\varepsilon_C = 0.23$. Assume condition (c) in Table 4.3 applies with the physical representation of the duct replaced by a cylinder of the same area. Therefore,

$$F_\varepsilon = \cfrac{1}{\left[\cfrac{1}{\varepsilon_H} + \left(\cfrac{A_H}{A_C}\right)\left(\cfrac{1}{\varepsilon_C} - 1\right)\right]} = \cfrac{1}{\left[\cfrac{1}{0.44} + \left(\cfrac{0.622}{4.0}\right)\left(\cfrac{1}{0.23} - 1\right)\right]} = 0.358$$

Apply Equation (4.28):

$$Q_H = Q = F_\varepsilon A \sigma \left(T_H^4 - T_C^4\right) \tag{4.28}$$

Substituting,

$$Q = (0.358)(0.622)(0.173 \times 10^{-8})(760^4 - 535^4)$$
$$= 97.85 \text{ Btu / hr} \cdot \text{ft}^2$$

Example 4.8 – Heat Loss Due to Radiation. The outside temperature of a 10 ft² hot insulated pipe is 140°F and the surrounding atmosphere is 60°F. The heat loss by free convection and radiation is 13,020 Btu/hr·ft²·°F. How much of the heat loss is due to radiation? Assume the pipe emissivity is approximately 0.9.

SOLUTION: For this example,

$$T_H = 140 + 460 = 600°R$$
$$T_C = 60 + 460 = 520°R$$

Apply Equation (4.21) and substitute:

$$Q = Q_{rad} = (0.9)(10)(0.173)\left[\left(\frac{600}{100}\right)^4 - \left(\frac{520}{100}\right)^4\right] = 880 \text{ Btu / hr}$$

Thus, radiation accounts for about 880 / 13,000 or 6.7% of the heat loss.

Example 4.9 – Radiation Heat Transfer Coefficient. With reference to Example 4.8, calculate the radiation heat transfer coefficient, h_r.

SOLUTION: Apply Equation (4.29):

$$h_r = \frac{Q}{A(T_H - T_C)} \qquad (4.29)$$

Substituting,

$$h_r = \frac{880}{10(600 - 520)} = 1.10 \text{ Btu / hr} \cdot \text{ft}^2 \cdot {}^\circ\text{F}$$

Example 4.10 – Insulated Pipe Application. The outside temperature of a lagged pipe carrying steam at 300 °F is 125 °F and the surrounding atmosphere is at 70 °F. The heat loss by free convection and radiation is estimated to be 100 Btu/hr·ft, and the combined coefficient of heat transfer is 2.10 Btu/hr·ft² · °F. How much of the heat loss is due to radiation, and what is the equivalent coefficient of heat transfer for the radiation alone?

SOLUTION:

$$\text{Area / lin. ft} = (\pi)\left(\frac{3.375}{12}\right)(1) = 0.88 \text{ ft}^2 / \text{ft}$$

In addition,

$$T_H = 125 + 460 = 585°\text{R}$$
$$T_C = 70 + 460 = 530°\text{R}$$

From Table 4.2, the emissivity is approximately 0.90. Apply Equation (4.24). Substituting,

$$Q = (0.9)(0.88)(0.173)\left[\left(\frac{585}{100}\right)^4 - \left(\frac{530}{100}\right)^4\right] = 52.5 \text{ Btu / hr}$$

Apply Equation (4.29).

$$h_r = \frac{Q}{A(T_H - T_C)} = \frac{52.5}{0.88(125 - 70)} = 1.08 \text{ Btu / hr} \cdot \text{ft}^2 \cdot {}^\circ\text{F}$$

Example 4.11 – Light Bulb Surface. The filament of a light bulb is at a temperature of 900 °C and emits 5 W of heat toward the glass bulb. The interior of a light bulb can be considered a vacuum and the temperature of the glass bulb is 150 °C. Ignore heat transfer to the room and assuming the emissivity of the filament is 1.0, calculate the surface of the filament in cm².

SOLUTION: Write the equation for radiation heat transfer. See Equation (4.24).

$$Q = A\varepsilon\sigma\left(T_H^4 - T_C^4\right)$$

Solve for the unknown surface, A, and substitute.

$$A = \frac{Q}{\varepsilon\sigma\left(T_H^4 - T_C^4\right)}$$

Substitution:

$$A = \frac{5.0}{(1)\left(5.669\times10^{-8}\right)\left[(273+900)^4 - (273+150)^4\right]}$$

$$A = 4.74\times10^{-5}\,m^2$$

$$A = 0.47\,cm^2$$

Example 4.12 – Uninsulated Steam Pipe. A system consists of an uninsulated steam pipe made of anodized aluminum with a diameter D = 0.06 m and a length L = 100 m. The surface temperature is T_1 = 127 °C and the surface emissivity of anodized aluminum is ε = 0.76. The pipe is in a large room with a wall temperature T_2 = 20 °C. The air in the room is at a temperature T_3 = 22 °C. The pipe convective heat transfer coefficient is h = 15 W/m²·K.

Estimate the emissive power, the total heat transfer by convection and radiation, the radiation heat transfer coefficient and the percent radiation relative to the total heat transfer. Assume steady-state operation, constant properties, and a room surface area much larger than the pipe surface area.

SOLUTION: Calculate the emissive energy of the pipe surface assuming it is a black body:

$$T_1 = 127°C = 400K$$

$$E_b = \sigma T_1^4 = \left(5.669\times10^{-8}\right)(400)^4 = 1451\ W/m^2$$

Employ Equation (4.17), and calculate the emissive power from the surface of the pipe.

$$E = \alpha E_b = (0.76)(1451) = 1103\ W/m^2$$

The total heat transfer, Q, from the pipe to the air and walls is

$$Q = Q_c + Q_r$$

With reference to the convection equation,

$$Q_c = hA(T_1 - T_3)$$

calculate the surface area of the pipe

$$A = \pi(0.06)(100) = 18.85 \text{ m}^2$$

The convective heat transfer to the air is therefore

$$Q_c = (15)(18.85)(127 - 22) = 29,700 \text{ W} = 29.7 \text{ kW}$$

The total radiative heat transfer rate is

$$Q_r = \varepsilon\sigma A(T_1^4 - T_2^4)$$
$$= (0.76)(5.669 \times 10^{-8})(18.85)(400^4 - 293^4)$$
$$= 14,800 \text{ W} = 14.8 \text{ kW}$$

Since $(T_1 - T_2) = 107°C$, which is less than $120°C$, it is valid to use the approximation formula in Equation (4.30).

$$h_r = 4\varepsilon\sigma T_{avg}^3 \qquad (4.30)$$

In order to calculate the radiation heat transfer coefficient, first calculate T_{avg}.

$$T_{avg} = (T_H + T_C)/2$$

Substituting,

$$T_{avg} = \left[(127 + 273) + (20 + 273)\right]/2 = 346.5 \text{ K}$$

Substituting into Equation (4.30), and solving for h_r

$$h_r = 4\varepsilon\sigma T_{avg}^3 = 4(0.76)(5.669 \times 10^{-8})(346.5)^3 = 7.2 \text{ W} / \text{m}^2 \cdot \text{K}$$

Since the convection heat transfer coefficient is $15.0 \text{ W/m}^2 \cdot \text{K}$ and the radiation heat transfer coefficient is $7.2 \text{ W/m}^2 \cdot \text{K}$,

$$\% \text{ heat transfer by radiation} = \frac{7.2}{15.0 + 7.2} \times 100 = 32.4\%$$

Therefore, radiation accounts for approximately one-third of the total heat transfer.

4.6 View Factors

As indicated earlier, the amount of heat transfer between two surfaces depends on geometry *and* orientation of the two surfaces. Again, it is assumed that the intervening medium is non-participating. The previous analyses were concerned with sources that were situated so that every point on one surface could be "connected" with every surface on the second... in effect possessing a perfect view. This is very rarely the case in real-world engineering applications, particularly in the design of boilers and furnaces. Here, the receiving surface, such as a bank of tubes, is cylindrical and may partially obscure some of the surfaces from "viewing" the source. These systems are naturally difficult to evaluate. The simplest cases are addressed below; however, many practical applications must resort to the use of empirical methods.

To introduce the subject of view factors, the reader should note that the flow of radiant heat is analogous to the flow of light (i.e., one may follow the path of radiant heat as one may follow the path of light). If any object is placed between a hot and cold body, light from the hot body would cast a shadow on the cold body and prevent it from receiving all the light leaving the hot body.

As noted above, the computation of actual problems involving "viewing" difficulty is beyond the scope of this book. One simple problem may

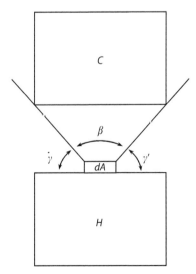

Figure 4.2 View factor illustration.

illustrate the nature of these complications. Contemplate two black bodies, with surfaces consisting of parallel planes of finite size and separated by a finite distance (see Figure 4.2). Then, a small differential unit of surface dA can "see" the colder body only through solid angel β, and any radiation emitted by it through solid angles γ and γ' will fall elsewhere. To evaluate the heat lost by radiation from body H, one must integrate the loss from element dA over the whole surface of A on body H.

The previously mentioned problem is sufficiently complicated. However, it can become more complicated. If the colder body is not a black body, it will then reflect some of the energy imparted upon it. Some of the reflected energy will return to the hot body. Since the hot body is black, it will absorb the reflected energy, tending to raise its temperature. One can envision even more complicated scenarios.

In order to include the effect of "viewing," Equation (4.28) is expanded to

$$\frac{Q}{A} = F_v F_\varepsilon \sigma \left(T_H^4 - T_C^4 \right) \tag{4.31}$$

with the inclusion of a view factor, F_v. View factors for the six cases considered in Table 4.3 are presented in Table 4.4 in conjunction with

Table 4.4 Values of F_v.

Condition	F_v
a. Surface A_H small compared with the totally enclosing surface A_C	1.0
b. Surface A_H and A_C of infinite parallel planes or surface A_H of a completely enclosed body is small compared to A_H	1.0
c. Concentric spheres or infinite concentric cylinders with surfaces A_H and A_C	1.0
d. Surfaces A_H and A_C of parallel disks, squares, 2:1 rectangles, long rectangles	Figure 4.3, 4.5
e. Surfaces A_H and A_C of perpendicular rectangles having a common side	Figure 4.4
f. Coaxial parallel disks	Figure 4.5

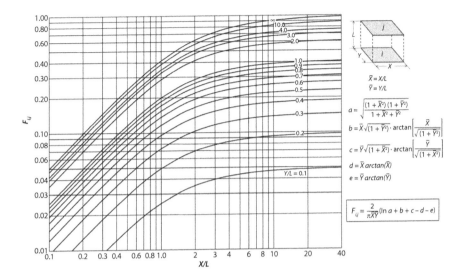

Figure 4.3 View factors; parallel aligned rectangles [7].

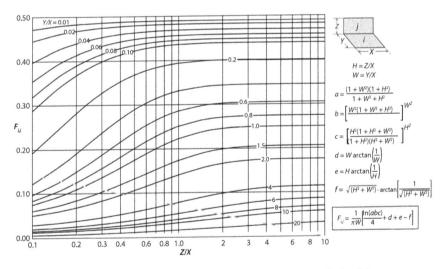

Figure 4.4 View factors; perpendicular rectangles with a connected edge [7].

Figures 4.3 – 4.5. View factors for parallel disks are provided in Figure 4.5. Note that $F_{i,j}$ represents the fraction of energy leaving surface i that strikes surface j. The graphical results for these figures are presented in equation form on the bottom right side of the figure. Other information and applications involving view factors are available in the literature [1, 6, 7].

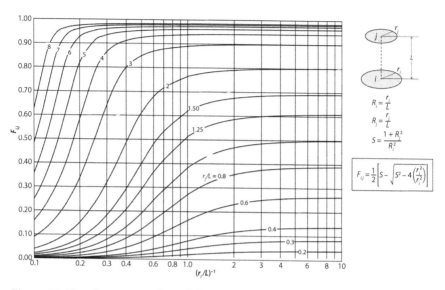

Figure 4.5 View factors; coaxial/parallel disks [7].

Example 4.13 – Cast Iron Pipe Heat Transfer Rate. Refer to Example 4.7. Calculate the heat transfer rate if $F_v = 1.0$.

SOLUTION: Apply Equation (4.31).

$$Q = F_v F_\varepsilon \sigma A \left(T_H^4 - T_C^4 \right)$$

Since $F_v = 1.0$, the solution remains unchanged.

$$Q = 97.85 \text{ Btu / hr}$$

Example 4.14 – Radiative Heat Transfer Between Two Plates. Two parallel rectangular *black* plates 0.5 m by 2.0 m are spaced 1.0 m apart. One plate is maintained at $1000\,°C$ and the other at $2000\,°C$. What is the net radiant heat exchange between the two plates?

SOLUTION: Figure 4.3 is to be employed. For the application, $L = 1.0$, $X = 0.5$, and $Y = 2.0$. Therefore,

$$\frac{Y}{L} = \frac{2.0}{1.0} = 2.0$$

and

$$\frac{X}{L} = \frac{0.5}{1.0} = 0.5$$

From Figure 4.3,

$$F_v \cong 0.18$$

The net heat transfer rate is calculated employing Equation (4.31), noting that $F_\varepsilon = 1.0$ and $\sigma = 5.669 \times 10^{-8}$ (when employing temperatures in K).

$$
\begin{aligned}
Q &= F_v F_\varepsilon \sigma A \left(T_H^4 - T_C^4 \right) \\
&= (0.18)(1.0)\left(5.669 \times 10^{-8}\right)(0.5)(2.0)\left(2273^4 - 1273^4\right) \\
&= 245{,}000 \text{ kW} = 8.37 \times 10^8 \text{ Btu / hr}
\end{aligned}
$$

References

1. I. Farag and J. Reynolds, *Heat Transfer, A Theodore Tutorial*, Theodore Tutorials, East Williston, NY, originally published by USEPA/APTI, RTP, NC, 1996.
2. L. Theodore, *Heat Transfer Applications for the Practicing Engineer*, John Wiley & Sons, Hoboken, NJ, 2013.
3. W. Badger and J. Banchero, *Introduction to Chemical Engineering*, McGraw-Hill, New York City, NY, 1955.
4. C. Bennett and J. Meyer, *Momentum, Heat and Mass Transfer*, McGraw-Hill, New York City, NY, 1962.
5. D. Kern, *Process Heat Transfer*, McGraw-Hill, New York City, NY, 1950.
6. J. Holman, *Heat Transfer*, McGraw-Hill, New York City, NY, 1981.
7. F. Incropera and D. Dewitt, *Fundamentals of Heat Transfer* (adapted form), John Wiley and Sons, Hoboken, NJ, 2002.

Part II
HEAT EXCHANGERS

5

The Heat Transfer Equation

Introduction

A modification of Fourier's general equation for heat transfer from Chapter 3 is often referred to simply as *the heat transfer equation*. Given that $Q = \Delta T / \Sigma R$ and $1 / \Sigma R = UA$, then

$$Q = UA\Delta T \qquad (5.1)$$

The variable U is defined as the *overall heat transfer coefficient*.

This chapter examines two main parameters of the heat transfer equation $Q = UA\Delta T$. Specifically, it includes an analysis of:

1. ΔT, the temperature difference driving force
2. U, the overall heat transfer coefficient

But first, an introduction to heat exchangers is warranted before discussing the describing equation that is employed in the heat exchanger predictive and design equations. *Heat exchangers* are defined as equipment that effect the transfer of thermal energy in the form of heat from one fluid to another. The simplest exchangers involve the direct mixing of hot and cold fluids. Most industrial exchangers are those in which the fluids are separated by a wall. The latter type, referred by some as a *recuperator*, can range from a simple plane wall between two flowing fluids to more complex configurations involving multiple passes, fins, or baffles. Conductive and convective heat transfer principles are required to describe and design these units; radiation effects are generally neglected [1, 2].

Heat exchangers for the chemical, petroleum, petrochemical, paper, power, etc., industries encompass a wide variety of designs that are available for many manufacturers. Equipment design practice first requires the selection of safe operable equipment. The selection and design process must also seek a cost-effective balance between initial (capital) installation costs, operating costs, maintenance costs, and energy considerations that include energy conservation concerns.

The proper application of heat exchange principles can significantly minimize both the initial cost of a plant and the daily operating and/or utility costs. Each heat exchange application may be accomplished by the use of many types of heat exchange equipment. To perform these applications, their design and materials of construction must be suitable for the desired operating conditions; the selection of materials of construction is primarily influenced by the operating temperature and the corrosive nature of the fluids being handled [1, 2].

The presentation in this chapter keys not only on the pertinent energy relationships that specify the heat transferred and the various heat exchanger equipment (and their classification) but also on the log mean temperature difference driving force, ΔT_{LM}, and the overall heat transfer coefficient, U. The development of ΔT_{LM} and U ultimately leads to the aforementioned classic heat transfer equation employed for heat exchangers. Topics addressed in this chapter include:

5.1 Heat Exchanger Equipment Classification
5.2 Energy Relationships
5.3 The Log Mean Temperature Difference Driving Force (LMTD)
5.4 The Overall Heat Transfer Coefficient (U)
5.5 The Heat Transfer Equation

5.1 Heat Exchanger Equipment Classification

There is a near infinite variety of heat exchange equipment [1]. These can vary from a simple electric heater in the home to a giant boiler in a utility power plant. A limited number of heat transfer devices likely to be encountered by the practicing engineer have been selected for description in this Part. Most of those units transfer heat from one fluid to another fluid, with the heat passing through a solid interface such as a tube wall that separates the two (or more) fluids. The size, shape, and material(s) employed to separate the two fluids is of course important. Another problem in a heat exchanger is the method of confining one or both of the two fluids involved in the heat transfer process.

As noted above, heat exchangers are devices used to transfer heat from a hot fluid to a cold fluid. They are classified by their functions, as shown in the Table 5.1. In a general sense, heat exchangers are classified into three broad types:

1. Recuperators or through-the-wall non-storing exchangers (e.g., double-pipe heat exchangers and shell-and-tube heat exchangers to be discussed in the next two chapters)
2. Direct contact non-storing exchangers
3. Regenerators, accumulators, or heat storage exchangers

Through-the-wall non-storing exchangers can be further classified into:

1. Double pipe heat exchangers
2. Shell-and-tube heat exchangers
3. Cross-flow exchangers

Exchangers (1) and (2) will be discussed with more detail in this portion of the text (Part II) in Chapters 6 and 7, respectively.

Flow considerations, or the directions to which the fluid is flowing with respect to the other fluid's direction of flow, are often included in the classification of a heat exchanger. This will be discussed in more detail in Section 5.4.

5.2   Energy Relationships

The flow of heat from a hot fluid to a cooler fluid through a solid wall is a situation often encountered in engineering equipment; examples of such equipment are the aforementioned heat exchangers, condensers, evaporators, boilers, and economizers [1, 2]. The heat absorbed by the cool fluid or given by the hot fluid may be sensible heat, causing a temperature change in the fluid, or it may be latent heat, causing a phase change such as vaporization

Table 5.1 Heat exchanger equipment.*

Equipment	Function
Chiller	Cools a fluid to a temperature below that obtainable if only water were used as a coolant. It often uses a refrigerant such as ammonia or freon.
Condenser	Condenses a vapor or mixture of vapors, either alone or in the presence of a non-condensable gas.
Cooler	Cools liquids or gases by means of water.
Exchanger	Performs a double function: (1) heats a cold fluid and (2) cools a hot fluid. Little or none of the transferred heat is normally lost.
Final condenser	Condenses the vapors to a final storage temperature of approximately 100 °F. It uses water cooling, which means the transferred heat is often lost in the process.
Forced-circulation reboiler	A pump is used to force liquid through the reboiler (see reboiler below).
Heater	Imparts sensible heat to a liquid or a gas by means of condensing steam or some other hot fluid (e.g., Dowtherm).
Partial condenser	Condenses vapors at a temperature high enough to provide a temperature difference sufficient to preheat a cold stream of process fluid. This saves heat and eliminates the need for providing a separate preheater.
Reboiler	Connected to the bottom of a fractionating tower, it provides the reboil heat necessary for distillation. The heating medium may be either steam (usually) or a hot process fluid.
Steam generator	Generates steam for use elsewhere in the plant by using available high-level heat, e.g., from tar or a heavy oil.
Superheater	Heats a vapor above its saturation temperature.
Thermosiphon Reboiler	Natural circulation of the boiling medium is obtained by maintaining sufficient liquid heat to provide for circulation (see reboiler).
Vaporizer	A heater which vaporizes all or part of the liquid.
Waste-heat boiler	Produces steam; similar to a steam generator, except that the heating medium is a hot gas or liquid produced in a chemical reaction.
Heat pipe	It utilizes both boiling and condensation processes in its operation.

*Most of the above equipment are discussed in later chapters.

or condensation. In a typical waste heat boiler, for example, the hot flue gas gives up heat to water through thin metal tubes walls separating the two fluids. As the flue gas loses heat, its temperature drops. As the water gains heat, its temperature quickly reaches the boiling point where it continues to absorb heat with no further temperature rise as it changes into steam. The rate of heat transfer between the two streams, assuming no heat loss due to the surroundings, may be calculated by the enthalpy change of either fluid.

$$Q = \dot{m}_h \left(h_{h1} - h_{h2} \right) = \dot{m}_c \left(h_{c1} - h_{c2} \right) \tag{5.2}$$

where Q is the rate of heat flow (Btu/hr), $\dot{m}_h$ is the mass flow rate of hot fluid (lb/hr), $\dot{m}_c$ is the mass flow rate of cold fluid (lb/hr), $h_{h,1}$ is the enthalpy of entering hot fluid (Btu/lb), $h_{h,2}$ is the enthalpy of exiting hot fluid (Btu/lb), $h_{c,1}$ is the enthalpy of entering cold fluid (Btu/lb), and $h_{c,2}$ is the enthalpy of exiting cold fluid (Btu/lb).

Equation (5.2) is applicable to the heat exchange between two fluids whether a phase change is involved or not. In the above waste heat boiler example, the enthalpy change of the flue gas is calculated from its sensible temperature change.

$$Q = \dot{m}_h \left(h_{h,1} - h_{h,2} \right) = \dot{m}_c c_{p,h} \left(T_{h,1} - T_{h,2} \right) \tag{5.3}$$

where $c_{p,h}$ is the heat capacity of the hot fluid (Btu/lb·°F), $T_{h,1}$ is the temperature of the entering hot fluid (°F), and $T_{h,2}$ is the temperature of the exiting hot fluid (°F). The enthalpy change of the water, on the other hand, involves a small amount of sensible heat to bring the water to its boiling point plus a considerable amount (usually) of latent heat to vaporize the water. Assuming all of the water is vaporized and no superheating of the steam occurs, the enthalpy change is

$$Q = \dot{m}_c \left(h_{c,2} - h_{c,1} \right) = \dot{m}_c c_{p,c} \left(t_{c,2} - t_{c,1} \right) + \dot{m}_c \Delta h_{vap} \tag{5.4}$$

where $c_{p,c}$ is the heat capacity of the cold fluid (Btu/lb·°F), $t_{c,1}$ is the temperature of the entering cold fluid (°F), $t_{c,2}$ is the temperature of the exiting cold fluid (°F), and Δh_{vap} is the heat vaporization of the cold fluid (Btu/lb).

Example 5.1 – Sensible Enthalpy Change. If 20,000 scfm (32 °F, 1 atm) of an air stream is heated from 200 °F to 2000 °F, calculate the heat transfer rate required to bring about this change in temperature. Use the following enthalpy and average heat capacity data:

$$H_{200°F} = 1{,}170 \, \text{Btu/lbmol}$$

$$H_{2000°F} = 14,970\,\text{Btu/lbmol}$$

$$\overline{C}_{p,\text{avg}} = 7.53\,\text{Btu/lbmol} \cdot °F \text{ (over the 200–2000 °F range)}$$

SOLUTION: Convert the flow rate of air to a molar flow rate, $\dot{n}$:

$$\dot{n} = \frac{\text{scfm}}{359} = \frac{20,000}{359} = 55.7\,\text{lbmol/min}$$

Note that at standard conditions of 32 °F and 1 atm, 1.0 lbmol of an ideal gas occupies 359 ft³ or 22.4 L (see also Chapter 1). Calculate the heat transfer rate, Q, using enthalpy data:

$$Q = \dot{n}\Delta H = 55.7(14,970 - 1,170) = 7.69 \times 10^5\,\text{Btu / min}$$

Also, calculate the heat transfer rate using the average heat capacity data:

$$Q = \dot{n}\overline{C}_{p,\text{avg}}\Delta T = 55.7(7.53)(2000 - 200) = 7.55 \times 10^5\,\text{Btu / min}$$

As one would expect, both approaches provide near identical results.

5.3   Log Mean Temperature Difference (LMTD) Driving Force

A temperature difference between two materials is the driving force by which heat is transferred from a source to a receiver. In contrast, when the temperature of two materials is the same, the temperature driving force is nonexistent because $\Delta T = 0$; as a result, $Q = 0$ in Equation (5.1).

For the sake of understanding ΔT, first examine heat flow in a composite rod instead of a typical heat exchange equipment–refer to Figure 5.1. When one views heat flow (conduction) through a composite pipe wall, the heat flow in the radial direction is constant throughout the entire distance from the center line ($r = 0$) of the pipe to the exterior ($r = r_3$) of the pipe wall for steady state conditions. With respect to a unit length of "solid" pipe in Figure 5.1, the area for heat transfer through the pipe wall increases from r_1 to r_3.

Note that the sample composite rod in Figure 5.1 is solid (with no fluid flowing through it) and not to be confused with a hollow pipe. Heat transfer situations in hollow pipes with fluids running through the inner pipe cavity will be discussed in the next section.

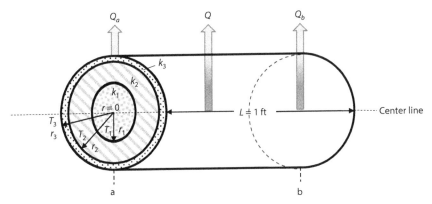

r_1 = radius of solid inner layer T_1 = temperature of solid inner layer k_1 = thermal of solid inner layer

r_2 = radius of solid middle layer T_2 = temperature of solid middle layer k_2 = thermal of solid middle layer

r_3 = radius of solid outer layer T_3 = temperature of solid outer layer k_3 = thermal of solid outer layer

Figure 5.1 Solid composite rod.

When comparing the heat transfer, Q, through the composite pipe wall at points a and b, the heat transfer can be described by one of the following equations:

$$Q_a = \frac{T_1 - T_3}{\Sigma R} \text{ at point } a, \quad Q_b = \frac{T_1 - T_3}{\Sigma R} \text{ at point } b \qquad (5.5)$$

In this scenario, Q, the heat transferred at each terminal end of the pipe at points a and b, is the same. More to the point, the ΔT at point a and point b, as well as the temperature difference at all points in between points a and b along the length of the exchanger (at any given radius, r) is the same, i.e., $\Delta T_a = \Delta T_b = T_1 - T_3$.

Flow Directions. Unlike the pipe with a composite wall in Figure 5.1, fluid flow through a hollow pipe (or double pipe) changes the way in which ΔT is calculated. The cool stream heats up, and the hot stream cools down as they move through the pipe. In other words, the temperature of each fluid changes from one end to the other. The variations of fluid temperature within the heat exchanger depend on whether the flow is cocurrent or countercurrent.

Plots of temperature vs. pipe length for a system of *two concentric pipes* (pipe inside a pipe) are shown in Figure 5.2. When both fluids enter at point (a) and flow in the same direction towards point (b), the flow regime is known as *cocurrent, concurrent,* or *parallel.* When the fluids enter at the opposite ends of the concentric pipes, the flow regime is known as *counter-current* or *counterflow*. From this point forward, the terminology *cocurrent* and *countercurrent* will be used.

The flow area between the inner pipe and the outer pipe is called the *annulus*. The cross-sectional diagram given in Figure 5.2(c) outlines the terminology commonly used to describe concentric pipes. Thus, the temperature of each fluid, whether located in the inner pipe or annulus, changes as it flows through the pipe from (a) to (b). In this case, the entrance to the concentric pipes is defined as position *a*, while the fluid exits at position *b*. The temperature difference, or temperature driving force, ΔT, between the two fluids can be represented as the vertical distance between the two curves at any point, as shown in Figures 5.2(a) and 5.2(b).

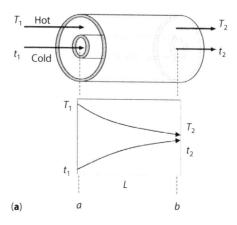

(a)

Figure 5.2(a) Cocurrent flow. Fluids flowing in the same direction is described as *cocurrent, concurrent,* or *parallel*.

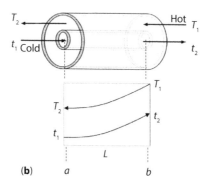

(b)

Figure 5.2(b) Countercurrent flow. Fluids flowing in the opposite direction is described as either *countercurrent* or *counterflow*.

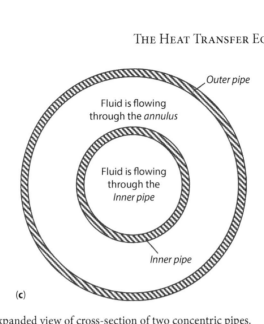

Figure 5.2(c) Expanded view of cross-section of two concentric pipes.

It is not ordinarily possible in industrial equipment to measure the average pipe-wall temperature or the temperature difference driving force across the entire length of the heat exchanger. Only the inlet and outlet temperatures of the hot and cold fluids are known or can be measured, and these are referred to as the *process temperatures*. For the remainder of this book the subscript "1" will usually denote the inlet and the subscript "2" the outlet. The definitions below are employed throughout the book:

t_1 = temperature of fluid entering the inside tube
t_2 = temperature of fluid exiting the inside tube
T_1 = temperature of fluid entering the annulus (space between the two tubes)
T_2 = temperature of fluid exiting the annulus

In a cocurrent heat exchanger, both hot and cold fluids enter on the same side and flow through the exchanger in the same direction. The *temperature approach* is defined as the temperature difference driving force at the heat exchanger entrance, ΔT_1 or $(T_1 - t_1)$. This driving force drops as the streams approach the exit of the exchanger. At the exit, the temperature difference driving force is ΔT_2 or $(T_2 - t_2)$. The heat exchanger is more effective at the entrance than at the exit because of the greater driving force as shown by the temperature differential represented by the dashed lines in Figure 5.3. The dashed lines grow smaller in length towards the exit of the heat exchanger.

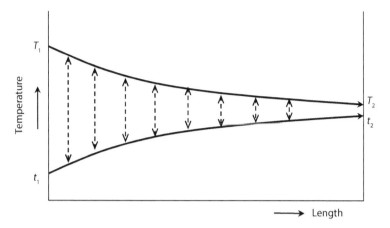

Figure 5.3 Cocurrent flow temperature profile from Figure 5.2(a).

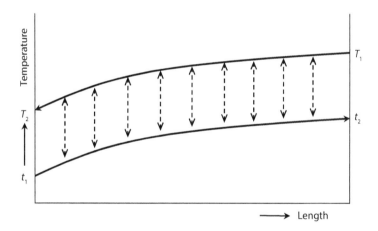

Figure 5.4 Countercurrent flow temperature profile from Figure 5.2(b).

In a countercurrent flow exchanger, the two fluids flow in opposite directions. The temperature approach at the tube entrance end, ΔT_1 or $\left(T_1 - t_2\right)$, and at the annulus entrance end, ΔT_2 or $\left(T_2 - t_1\right)$ are usually roughly the same. The thermal driving force is normally relatively constant over the length of the exchanger, as shown by the temperature differential represented by the dashed lines in Figure 5.4. The lengths of the dashed lines remain relatively constant throughout the heat exchanger.

Recall that for a composite rod from Figure 5.1, $\Delta T_a = \Delta T_b$. However, when there is flow of fluid, it is clear from Figure 5.2(a) and 5.2(b) that

$\Delta T_a \neq \Delta T_b$. The question becomes, "Which ΔT should one use for $Q = UA\Delta T$? The one at the beginning? The one at the end? The average of the two?"

One solution is to take the average (or mean) of ΔT_a and ΔT_b. The most commonly used methods are the arithmetic mean, logarithmic mean, and geometric mean. Example 5.2 briefly demonstrates the difference in the way each is calculated.

It should be noted that the temperature profiles of the two fluids, whether in cocurrent or countercurrent flow arrangements, are curved – not straight lines. For the purposes of heat transfer calculations, it will be shown that the exponential profile of the logarithmic mean most accurately represents ΔT over the entire length of a heat exchanger since ΔT varies along the flow path. It is referred to as the *Log Mean Temperature Difference Driving Force* (LMTD), indicated by ΔT_{LM}, and calculated as follows:

$$\Delta T_{LM} = \frac{\Delta T_a - \Delta T_b}{\ln\left(\dfrac{\Delta T_a}{\Delta T_b}\right)} \tag{5.6}$$

As noted earlier, ΔT_a is the temperature difference between the two fluids at the entrance, and ΔT_b is the temperature difference at the exit. The formula for ΔT_{LM} is independent of flow configuration, i.e., cocurrent or countercurrent.

Example 5.2 – Comparison of Arithmetic Mean, Logarithmic Mean, and Geometric Mean Temperature Difference Driving Force. A hot fluid enters the annulus of a concentric pipe at 147 °F and exits at 102 °F. A cold fluid enters the inner pipe at 68 °F and exits at 91 °F. The two fluids flow *cocurrently*. Compare the mean ΔT from the entrance to exit, from points a to b, respectively, using the formulae for arithmetic mean, logarithmic mean, and geometric mean.

SOLUTION: The temperature profile for the system is provided in Figure 5.5.

Shortcut Notations. Before delving further into the chapter, the reader should note that, whenever possible, lowercase t and uppercase T will be employed to represent the cooler fluid temperature and hotter fluid temperature, respectively; in addition, "lowercase c / uppercase C," "lowercase $\dot{m}$ / uppercase $\dot{M}$," and "lowercase h / uppercase H" will be employed to represent the cold and hot fluid, respectively. Although this is contrary to

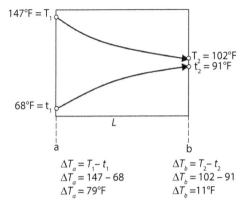

$$\Delta T_a = T_1 - t_1 \qquad\qquad \Delta T_b = T_2 - t_2$$
$$\Delta T_a = 147 - 68 \qquad\qquad \Delta T_b = 102 - 91$$
$$\Delta T_a = 79°F \qquad\qquad \Delta T_b = 11°F$$

Pertinent calculations for the three ΔT's follows.

Arithmetic mean	Logarithmic mean	Geometric mean
$\Delta T_{AM} = \dfrac{\Delta T_a - \Delta T_b}{2}$	$\Delta T_{LM} = \dfrac{\Delta T_a - \Delta T_b}{\ln\left(\dfrac{\Delta T_a}{\Delta T_b}\right)}$	$\Delta T_{GM} = \sqrt{\Delta T_a \cdot \Delta T_b}$
$\Delta T_{AM} = \dfrac{79 - 11}{2}$	$\Delta T_{LM} = \dfrac{79 - 11}{\ln\left(\dfrac{79}{11}\right)}$	$\Delta T_{GM} = \sqrt{79 \cdot 11}$
$\Delta T_{AM} = 45°F$	$\Delta T_{LM} = 34.5°F$	$\Delta T_{GM} = 29.5°F$

Figure 5.5 Temperature difference driving force comparison; for Example 5.2.

the notation defined in Chapter 1, where capital-case notations referred to molar basis while lower-case notations referred to mass basis, the authors decided that this notation distinguishing cold and hot fluid is more convenient and "less cluttered," mainly because it involves less subscripts in writing the notation. This shortcut notation will primarily be used throughout the book.

Derivation of ΔT_{LM}. For the derivation of the temperature difference between the two fluids of Figure 5.2(b) in countercurrent flow, the following assumptions must be made:

1. The overall coefficient of heat transfer U is constant over the entire length of path.
2. The mass flow rate, $\dot{m}$ is constant, obeying the steady-state requirement.
3. The heat capacity is constant over the entire length of path.
4. There are no partial phase changes in the system, i.e., vaporization or condensation. The derivation is applicable for

sensible-heat changes and when vaporization or condensation is isothermal over the entire length of path.
5. Heat losses to the surroundings are negligible.

With respect to countercurrent flow in Figure 5.2(b), ΔT_a is the driving
force at the entrance to the heat exchanger, $(T_1 - t_1)$ and ΔT_b is the temperature driving force at the exit, $(T_2 - t_2)$. In a perfectly insulated steady-state
system, the heat lost by the hot fluid is theoretically equal to the amount
of energy, or heat, gained by the cold fluid. The heat lost or gained by an
individual fluid is described as the sensible heat – defined as a change in
temperature, with no phase change. The sensible heat, also known as the
duty of the heat exchanger, is calculated as follows:

$$Q = \dot{M}C(\Delta T) = \dot{m}c(\Delta t) \tag{5.7}$$

where,

1. $\dot{M}, \dot{m}$ are the mass flow rates of the hot and cold fluids,
   respectively [lb/hr or kg/hr]
2. C, c, are the heat capacities of the hot and cold fluids, respectively [Btu/lb·°F or kJ/kg·K]

These are assumed to be constant for the purposes of this derivation. Under
the above conditions, the heat lost by the hot fluid is equal to the heat
gained by the cold fluid. This is also equal to the heat transferred between
the two fluids. Mathematically, this is written as,

$$\dot{M}C(\Delta T) = \dot{m}c(\Delta t) = UA\Delta T_x \tag{5.8}$$

where ΔT_x is the temperature difference between the hot and cold fluid
along the heat exchanger, i.e., $a \leq x \leq b$,

$$\Delta T_x = T - t \tag{5.9}$$

Equation (5.9) in differential form is written as follows:

$$d\Delta T_x = dT - dt \tag{5.10}$$

The term $d\Delta T_x$ denotes the differential change in temperature between the
hot and cold fluid at any point, x, along the length of the heat exchanger.

From a differential perspective,

HOT FLUID	COLD FLUID

$(a)^*$ $(UdA)\Delta T_x = -(\dot{M}C)dT$ (b) $(UdA)\Delta T_x = -(\dot{m}c)dt$ (5.11)

(a) $dT = -\dfrac{(UdA)\Delta T_x}{\dot{M}C}$ (b) $dt = -\dfrac{(UdA)\Delta T_x}{\dot{m}c}$ (5.12)

*(−) the negative sign denotes loss of heat

If Equations [5.11(a)] and [5.11(b)] are substituted into Equation (5.10) (and simplifying),

$$d\Delta T_x = -\frac{(UdA)\Delta T_x}{\dot{M}C} - \frac{(UdA)\Delta T_x}{\dot{m}c} \tag{5.13}$$

$$d\Delta T_x = -(UdA)\Delta T_x\left[\frac{1}{\dot{M}C} - \frac{1}{\dot{m}c}\right] \tag{5.14}$$

$$\frac{d\Delta T_x}{\Delta T_x} = -(UdA)\left[\frac{1}{\dot{M}C} - \frac{1}{\dot{m}c}\right] \tag{5.15}$$

However, the value ΔT_x changes at different points along the pipe. For a fixed radius, $d\Delta T_x$ changes with dL. Since $dA = \pi dL$, $d\Delta T_x$ is a function of πdL, i.e., $f(\pi dL)$. Therefore, $d\Delta T_x$ is a function of dL, i.e., $d\Delta T_x = f(dL)$. Also note that

$$A = a''x \tag{5.16}$$

$$dA = a''dx \tag{5.17}$$

where a'' is the cross-sectional area of the pipe per unit length [ft^2/lin ft or m^2/lin m].

Integrating both sides of Equation (5.15), using *boundary conditions* (BC),

$$BC1: T = \Delta T_a \text{ at } x = 0;$$

$$BC2: T = \Delta T_b \text{ at } x = L,$$

leads to

$$\int_{\Delta T_a}^{\Delta T_b} \frac{d\Delta T_x}{\Delta T_x} = -Ua''\left[\frac{1}{\dot{M}C} + \frac{1}{\dot{m}c}\right]\int_0^L dx \qquad (5.18)$$

$$\ln\Delta T_b - \ln\Delta T_a = -Ua''\left[\frac{1}{\dot{M}C} + \frac{1}{\dot{m}c}\right]L \qquad (5.19)$$

$$\ln\frac{\Delta T_b}{\Delta T_a} = -Ua''\left[\frac{1}{\dot{M}C} + \frac{1}{\dot{m}c}\right]L \qquad (5.20)$$

$$A = a''L \qquad (5.21)$$

Equation (5.20 becomes: $\ln\dfrac{\Delta T_a}{\Delta T_b} = UA\left[\dfrac{1}{\dot{M}C} + \dfrac{1}{\dot{m}c}\right]$ \qquad (5.22)

Terms $\dfrac{1}{\dot{M}C}$ and $\dfrac{1}{\dot{m}c}$ for the hot and cold streams, respectively, may be solved for in the following manner:

$$Q = \dot{M}C(T_1 - T_2) = \dot{m}c(t_2 - t_1) \qquad (5.23)$$

$$\frac{1}{\dot{M}C} = \frac{(T_1 - T_2)}{Q} \qquad (5.24)$$

$$\frac{1}{\dot{m}c} = \frac{(t_2 - t_1)}{Q} \qquad (5.25)$$

Substituting Equations (5.24) and (5.25) into Equation (5.22) and simplifying:

$$\ln\frac{\Delta T_a}{\Delta T_b} = \frac{UA}{Q}\left[(T_1 - T_2) + (t_2 - t_1)\right] \qquad (5.26)$$

$$\ln\frac{\Delta T_a}{\Delta T_b} = \frac{UA}{Q}\left[(T_1 - t_1) - (T_2 - t_2)\right] \qquad (5.27)$$

$$\ln\frac{\Delta T_a}{\Delta T_b} = \frac{UA}{Q}\left[\Delta T_a - \Delta T_b\right] \qquad (5.28)$$

Rearranging Equation (5.28) and solving for Q provides an expression for ΔT_{LM}:

$$Q = UA \left[\frac{\Delta T_a - \Delta T_b}{\ln\left(\dfrac{\Delta T_a}{\Delta T_b}\right)} \right]$$

where,

$$\Delta T_{LM} = \frac{\Delta T_a - \Delta T_b}{\ln\left(\dfrac{\Delta T_a}{\Delta T_b}\right)} \tag{5.29}$$

Alternative Derivation of ΔT_{LM}. The concept of a log-mean temperature can also be developed in the following derivation [1, 2]. Consider an absolute temperature profile $T(x)$ that is continuous and where T_1 is the absolute temperature at some point x_1 with T_2 at x_2. By definition, the mean value of the reciprocal temperatures $(1/T)$ between x_1 and x_2 is given by

$$\left(\frac{\overline{1}}{T}\right) = \frac{\int_{T_1}^{T_2} \dfrac{1}{T} dT}{\int_{T_1}^{T_2} dT} = \left(\frac{1}{T_2 - T_1}\right) \int_{T_1}^{T_2} \frac{1}{T} dT = \frac{\ln(T_2 / T_1)}{T_2 - T_1} \tag{5.30}$$

If the average of the reciprocal is a reasonable approximation to the reciprocal of the average, i.e.,

$$\left(\frac{\overline{1}}{T}\right) \approx \frac{1}{\overline{\overline{T}}} \tag{5.31}$$

then one can combine the above two equations to give

$$\frac{1}{\overline{T}} = \frac{\ln(T_2 / T_1)}{T_2 - T_1}$$

or,

$$\overline{T} = \frac{T_2 - T_1}{\ln(T_2 / T_1)} \tag{5.32}$$

The right-hand side of Equation (5.32) is *defined* to be the log-mean temperature between T_1 and T_2, i.e.,

$$T_{LM} = \frac{T_2 - T_1}{\ln(T_2 / T_1)} \tag{5.33}$$

Note once again that absolute temperatures must be employed.

An overall energy balance is first applied to a fluid in a conduit (e.g., a pipe with inlet and exit temperatures t_1 and t_2, respectively), and heated by a source at temperature T_S (see also Figure 5.6):

$$Q = \dot{m}c_p(t_2 - t_1) = \dot{m}c_p\left[(T_S - t_1) - (T_S - t_2)\right] = \dot{m}c_p(\Delta T_1 - \Delta T_2) \tag{5.34}$$

where ΔT_1, or $(T_S - t_1)$, is the temperature difference driving force (also termed the *approach*) at the fluid entrance, and ΔT_2, or $(T_S - t_2)$, is the temperature driving force, or approach, at the fluid exit. The differential energy balance is

$$dQ = \dot{m}c_p dt = \dot{m}c_p d(\Delta T) \tag{5.35}$$

One may also apply an energy balance to a differential fluid element of cross-sectional area, A, and a thickness of dx. Set $x = 0$ at the pipe entrance and $x = L$ at the pipe exit. The pipe diameter is D, so

$$A = \frac{\pi}{4}D^2 \tag{5.36}$$

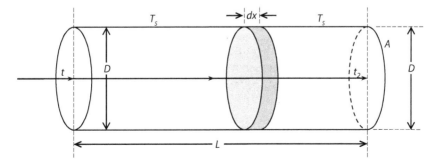

Figure 5.6 Energy balance of a differential element of a pipe.

The outer surface differential area of the element is $dA = \pi D(dx)$. One may also apply the energy balance across the wall of the unit

$$dQ = h(T_S - t)(\pi D)(dx) = \dot{m}c_p dt \qquad (5.37)$$

where h is the heat transfer coefficient of the fluid defined earlier in Chapter 3. Separation of variables in Equation (5.37) yields

$$\frac{dt}{T_S - t} = \frac{h}{\dot{m}c_p}(\pi D)(dx) \qquad (5.38)$$

Integrating from $x = 0$ where $t = t_1$, to $x = L$, where $t = t_2$, yields

$$\int_{t_1}^{t_2} \frac{dt}{T_S - t} = \frac{h}{\dot{m}c_p}(\pi D)\int_0^L dx \qquad (5.39)$$

$$\ln\left(\frac{T_S - t_1}{T_S - t_2}\right) = \frac{h}{\dot{m}c_p}(\pi D)L \qquad (5.40)$$

Letting $\Delta T_1 = T_S - t_1$ and $\Delta T_2 = T_S - t_2$ in Equation (5.40) leads to

$$\ln\left(\frac{\Delta T_1}{\Delta T_2}\right) = \frac{h}{\dot{m}c_p}(\pi D)L \qquad (5.41)$$

Thus,

$$\dot{m}c_p = \frac{h(\pi D)L}{\ln(\Delta T_1 / \Delta T_2)} \qquad (5.42)$$

Combining Equations (5.34) and (5.42) gives

$$Q = h(\pi D)L\left[\frac{\Delta T_1 - \Delta T_2}{\ln(\Delta T_1 / \Delta T_2)}\right] = hA\Delta T_{LM} \qquad (5.43)$$

where ΔT_{LM} is the aforementioned log mean temperature difference driving force, or log mean temperature approach, and is defined as

$$\Delta T_{LM} = \frac{\Delta T_1 - \Delta T_2}{\ln\left(\Delta T_1 / \Delta T_2\right)} \tag{5.44}$$

For the special case of $\Delta T_1 = \Delta T_2$,

$$\Delta T_{LM} = \Delta T_1 = \Delta T_2 \tag{5.45}$$

Thus, the arithmetic and logarithm means approach equality for this condition.

Difference Between ΔT_{LM} for Cocurrent and Countercurrent Flow.
Although two fluids may transfer heat in a concentric pipe heat exchanger in either countercurrent flow or cocurrent flow, *the relative direction* of the two fluids influences the value of the temperature difference. This point cannot be overemphasized: the flow pattern of two fluids has a direct impact on its unique temperature difference.

In general, the *average* temperature driving force, ΔT_{LM}, between points *a* and *b* is *greater* for countercurrent flow than for cocurrent flow. While cocurrent flow starts with a large ΔT at *a*, the driving force continually decreases down the length of the exchanger and is at a minimum at point *b*. In contrast, the ΔT for countercurrent flow remains reasonably large over the entire length of the exchanger. As a result, when given the same two fluids, a countercurrent arrangement will usually give a larger ΔT_{LM} and therefore, a greater heat transfer rate, Q, compared to cocurrent flow for the same heat exchanger.

With the exception of when one fluid is isothermal (such as condensing steam), the examples that follow demonstrate that there is a distinct thermal *disadvantage* to the use of cocurrent flow.

Example 5.3 – Calculation of ΔT_{LM}. A hot fluid enters a concentric-pipe apparatus at a temperature of 300 °F and is to be cooled to 200 °F by a cold fluid entering at 100 °F and heated to 150 °F. Should they be directed in cocurrent or countercurrent flow?

SOLUTION:

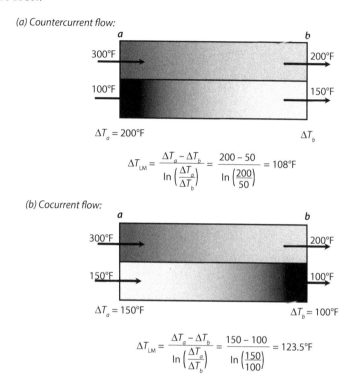

(a) Countercurrent flow:

$\Delta T_a = 200°F$ ΔT_b

$$\Delta T_{LM} = \frac{\Delta T_a - \Delta T_b}{\ln\left(\frac{\Delta T_a}{\Delta T_b}\right)} = \frac{200 - 50}{\ln\left(\frac{200}{50}\right)} = 108°F$$

(b) Cocurrent flow:

$\Delta T_a = 150°F$ $\Delta T_b = 100°F$

$$\Delta T_{LM} = \frac{\Delta T_a - \Delta T_b}{\ln\left(\frac{\Delta T_a}{\Delta T_b}\right)} = \frac{150 - 100}{\ln\left(\frac{150}{100}\right)} = 123.5°F$$

The ΔT_{LM} for the same two fluids is higher for countercurrent flow than for cocurrent flow. Therefore, employ countercurrent flow.

Example 5.4 – Calculation of ΔT_{LM} with Equal Outlet Temperatures.
A hot fluid enters a concentric-pipe apparatus at 300 °F and is to be cooled
to 200 °F by a cold fluid entering at 150 °F and heated to 200 °F.
SOLUTION:

(a) Countercurrent flow:

$\Delta T_a = 100°F$ $\Delta T_b = 50°F$

$$\Delta T_{LM} = \frac{\Delta T_a - \Delta T_b}{\ln\left(\frac{\Delta T_a}{\Delta T_b}\right)} = \frac{100 - 50}{\ln\left(\frac{100}{50}\right)} = 72.1°F$$

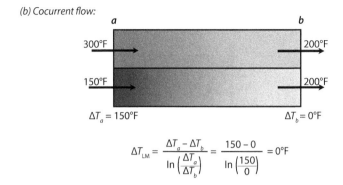

(b) Cocurrent flow:

$$\Delta T_{LM} = \frac{\Delta T_a - \Delta T_b}{\ln\left(\frac{\Delta T_a}{\Delta T_b}\right)} = \frac{150 - 0}{\ln\left(\frac{150}{0}\right)} = 0°F$$

In cocurrent flow, the lowest temperature theoretically attainable by the hot fluid is that of the outlet temperature of the cold fluid, t_2. If this temperature were attained, ΔT_{LM} would be zero. Since Q and U are finite in the heat transfer equation, $Q = UA\Delta T_{LM}$, the heat-transfer surface area, A, would have to be infinite. In other words, in order to have the two streams approach the same temperature, the heat exchanger would have to be infinitely long. This is not feasible.

The inability of the hot fluid in cocurrent flow to fall below the outlet temperature of the cold fluid has a marked effect upon the ability of cocurrent flow apparatus to *recover* heat. Suppose it is desired to recover as much heat as possible from the hot fluid in Example 5.3 by using the same quantities of hot and cold fluid as before but by assuming that more heat-transfer surface is available. In a countercurrent flow apparatus, it is possible to have the hot fluid outlet, T_2, fall to within perhaps 5 or 10 °F of the cold-fluid inlet, t_1, say 110 °F. In a cocurrent flow apparatus, the heat transfer would be restricted by the cold fluid outlet temperature rather than the cold fluid inlet and the difference would be the loss in recoverable heat. In other words, the hot fluid would not get as cold and the cold fluid would not get as hot when using cocurrent flow compared to countercurrent flow. Cocurrent flow is used for cold viscous fluids, however, since the arrangement occasionally enables a higher value of U to be obtained.

Consider the next example where ΔT (hot stream) does not have the greater temperature difference.

Example 5.5 – Calculation of ΔT_{LM} when Δt (cold stream) > ΔT (hot stream). (Note: Δt does not have to be equal to ΔT. Only Q for the hot fluid must be equal to q for the cold fluid, or rather, $\dot{M}C\Delta T = \dot{m}c\Delta t$. A hot fluid is cooled from 300 to 200 °F in countercurrent flow while a cold fluid is heated from 100 to 275 °F. Assuming the system is perfectly insulated, calculate ΔT_{LM}.

SOLUTION:

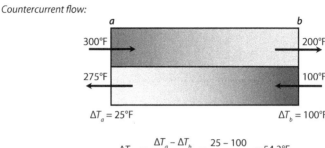

Countercurrent flow:

$$\Delta T_{LM} = \frac{\Delta T_a - \Delta T_b}{\ln\left(\frac{\Delta T_a}{\Delta T_b}\right)} = \frac{25 - 100}{\ln\left(\frac{25}{100}\right)} = 54.3°F$$

Example 5.6 – Calculation of ΔT_{LM} with One Isothermal Fluid. A flowing cold fluid is heated from 100 to 275 °F by steam at 300 °F. Calculate ΔT_{LM} for countercurrent and cocurrent flow conditions.

SOLUTION:

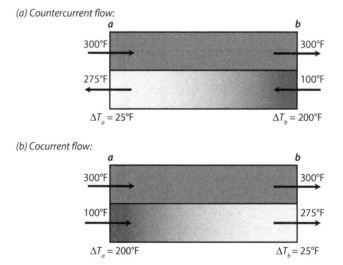

The results are identical and independent of whether flow is cocurrent or countercurrent.

Note: The color gradient for the red-shaded hot fluid does not change for this exchanger. This situation arises because steam is transferring heat from its latent heat that is released during condensation at a constant temperature. Sensible heat changes occur whenever a fluid experiences a change in temperature in the same phase as with the cold blue-shaded fluid in the drawing.

Hereafter, unless specified, all temperature arrangements will be assumed in countercurrent flow. Many industrial heat exchangers are actually a compromise between cocurrent and countercurrent flow and receive additional study in later chapters.

Example 5.7 – Rate of Heat Transfer. A heavy hydrocarbon oil with heat capacity, $c_p = 0.55\,\text{Btu/lb} \cdot °\text{F}$, is being cooled from $T_1 = 210°\text{F}$ to $T_2 = 170°\text{F}$. The oil flows inside a tube at a rate of 8000 lb/hr and the tube surface temperature is maintained at $60°\text{F}$. Calculate the heat transfer rate and the LMTD in $°\text{F}$.

SOLUTION: Calculate the heat transferred from the heavy oil:

$$Q = \dot{m}c_p \Delta T = (8000)(0.55)(170 - 210) = -176{,}000\,\text{Btu/hr}$$

The negative sign is a reminder that heat is lost by the oil. The thermal driving forces, or the temperature approaches, at the pipe entrance and exit are

$$\Delta T_1 = T_1 - t = 210 - 60 = 150°\text{F}$$

$$\Delta T_2 = T_2 - t = 170 - 60 = 110°\text{F}$$

The LMTD is therefore:

$$\Delta T_{LM} = \frac{150 - 110}{\ln(150/110)} = 129°\text{F}$$

Note that the use of ΔT_{LM} is valid for heating or cooling.

5.4   The Overall Heat Transfer Coefficient (U)

In order to design heat transfer equipment and calculate the required energy, one must know more than just the heat transfer rate calculated by the enthalpy (energy) balances described previously. The rate at which heat can travel from the hot fluid T_H, through the tube walls, into the cold fluid at t_C, must also be considered in the calculation of certain design variables (e.g., the contact area). The slower this rate is, for given hot and cold fluid rates, the more contact area is required. The rate of heat transfer through a unit of contact area was referred to earlier as the *heat flux* and, at any point along the area or the tube length, is given by

$$\frac{dQ}{dA} = U\left(T_H - t_C\right) \tag{5.46}$$

where dQ/dA is the local heat flux (Btu/hr·ft²) and U is defined as the local *overall heat transfer coefficient* (Btu/hr·ft²·°F), a term that provides a measure (inversely) of the resistance to heat transfer [1, 2].

The use of the above overall heat transfer coefficient (U) is a simple, yet powerful concept. In most applications, it combines both conduction and convection effects, although heat transfer by radiation can also be included. In actual practice, it is not uncommon for vendors to provide a numerical value for U. For example, a typical value for U for estimating heat losses from an incinerator is approximately 0.1 Btu/hr·ft²·°F. Methods for calculating the overall heat transfer coefficient are presented later in this section. The following assumptions apply in the development to follow.

1. The heat transfer coefficient is constant over the entire length of path.
2. The mass fluid flow is constant, obeying the steady-state requirement.
3. The heat capacity is constant over the entire length of path.
4. There are no partial phase changes in the system, i.e., vaporization or condensation. The derivation is applicable for sensible-heat changes or when vaporization or condensation is isothermal over the whole length of the path.
5. Heat losses are negligible.

With reference to Equation (5.46), the temperatures T_H and t_C are actually local average values. As described in the previous section, when a fluid is being heated or cooled, the temperature will vary throughout the cross-section of the stream. If the fluid is being heated, its temperature will be highest at the tube wall and will decrease with increasing distance from the tube wall. The average temperature across the stream cross-section is denoted as t_C, i.e., the temperature that would be achieved if the fluid at this cross-section was suddenly mixed to a uniform temperature. If the fluid is being cooled, on the other hand, its temperature will be lowest at the tube wall and will increase with increasing distance from the wall.

In order to apply Equation (5.46) to an entire heat exchanger, the equation must be integrated. This cannot be accomplished unless the geometry of the exchanger is first defined. For simplicity, one of the simplest geometries will be assumed here—the double pipe heat exchanger to be discussed in Chapter 6, Part II. As described earlier, this device consists of two parallel concentric pipes. The outer surface of the outer pipe is well insulated so that no heat exchange with the surroundings may be assumed. One of the fluids flows through the center pipe and the other flows through the annular channel (known as the annulus) between the pipes. The fluid flows may be either cocurrent where the two fluids flow in the same directions, or countercurrent where the flows are in the opposite directions; however, and as noted earlier, the countercurrent arrangement is more efficient and is more commonly used. For this heat exchanger, integration of Equation (5.46) along the exchanger area or length, and applying several simplifying assumptions noted earlier, yields:

$$Q = \frac{T_H - T_C}{\dfrac{1}{h_i A} + \dfrac{\Delta x}{kA} + \dfrac{1}{h_o A}} \tag{5.47}$$

This is defined as the aforementioned *heat transfer equation*.

The above equation was previously applied to heat transfer across a plane wall in Chapter 2, and it was shown that:

$$Q = UA\Delta T_{LM} \tag{5.48}$$

where the h's represent the individual heat transfer coefficients (see also Chapter 2). Since $Q = UA\Delta T$,

$$U = \frac{1}{\dfrac{1}{h_i} + \dfrac{\Delta x}{k} + \dfrac{1}{h_o}} \tag{5.49}$$

and since A is constant,

$$\frac{Q}{\Delta T} = UA = \frac{1}{\dfrac{1}{h_i A} + \dfrac{\Delta x}{kA} + \dfrac{1}{h_o A}} \tag{5.50}$$

For a tubular unit, $Q = UA\Delta T$ still applies, but the surface area available for heat transfer, A, is no longer constant for a cylinder, i.e., the inside surface area of the inner pipe is not equal to the outside surface area of the inner pipe because the radius increases going from inside to outside.

Equation (5.50) is rewritten for this geometry as (see also Chapter 3 for additional details).

$$\frac{Q}{\Delta T} = \frac{1}{\dfrac{1}{h_i A_i} + \dfrac{\Delta D}{2k A_{LM}} + \dfrac{1}{h_o A_o}}; \quad \Delta D = D_o - D_i = 2\left(r_o - r_i\right) \quad (5.51)$$

where the subscript i and o refer to the inside and outside of the tube, respectively. In addition,

$$A_{LM} = \frac{A_o - A_i}{\ln\left(A_o / A_i\right)} = \frac{\pi L \Delta D}{\ln\left(D_o / D_i\right)} \quad (5.52)$$

and L is the tube length (ft). Thus, Equation (5.51) may also be written as:

$$\frac{Q}{\Delta T} = UA = \frac{1}{\dfrac{1}{h_i A_i} + \dfrac{\ln\left(D_o / D_i\right)}{2\pi L k} + \dfrac{1}{h_o A_o}} \quad (5.53)$$

Equation (5.53) can also be written as

$$\frac{Q}{\Delta T} = UA = \frac{1}{R_i + R_w + R_o} = \frac{1}{R} \quad (5.54)$$

The term U in Equation (5.53) may be based on the inner area (A_i) or the outer area (A_o) so that

$$\frac{Q}{\Delta T} = U_i A_i = U_o A_o = \frac{1}{\dfrac{1}{h_i A_i} + \dfrac{\ln\left(D_o / D_i\right)}{2\pi L k} + \dfrac{1}{h_o A_o}} \quad (5.55)$$

Dividing by either A_i or A_o yields an expression for U_i and U_o:

$$U_i = \frac{1}{\dfrac{1}{h_i} + \dfrac{A_i \ln\left(D_o / D_i\right)}{2\pi Z k} + \dfrac{A_i}{h_o A_o}} \quad (5.56)$$

$$U_o = \frac{1}{\dfrac{A_o}{h_i A_i} + \dfrac{A_o \ln\left(D_o / D_i\right)}{2\pi Z k} + \dfrac{1}{h_o}} = \frac{1}{\dfrac{D_o}{h_i D_i} + \dfrac{D_o \ln\left(D_o / D_i\right)}{2k} + \dfrac{1}{h_o}} \quad (5.57)$$

Numerical values of U can range from as low as 2 Btu/hr·ft²·°F (10 W/m²·K) for gas-to-gas heat exchangers to as high as 250 Btu/hr·ft²·°F for water-to-water units. Additional values are provided in the literature [1–3]. Kern [4] provides an expanded list of values of U in Table AT.11 in the Appendix. Some typical values of the resistance, R, are provided below in Table 5.2.

Relationship between U and R. In order to understand the heat exchanged between two streams in concentric pipes, it is necessary to understand the resistances encountered between the two streams. These resistances include the pipe fluid-film resistance, the pipe wall resistance, and the annulus fluid-film resistance. The overall heat transfer coefficient, U, summarizes the resistances in a system and is defined as:

$$U_i A_i = U_o A_o = \frac{1}{\Sigma R} \tag{5.58}$$

ΣR is the total resistance in the system. U_i and U_o are the overall heat transfer coefficients calculated based on the inside area of the inner pipe and outside area of the inner pipe, respectively. In the absence of any assumptions, it is important to note:

$$U_i A_i = U_o A_o = UA, \underline{but}\ U_i \neq U_o$$

Example 5.8 – Overall Heat Transfer Coefficient for a Plane Surface. A rectangular plane window glass panel is mounted on a house. The glass is 0.125 thick and has a surface area of 1.0 m²; its thermal conductivity, k_2, is 1.4 W/m·K. The inside house temperature, T_1, and outside air temperature, t_4, are 25 °C and –14 °C, respectively. The heat transfer coefficient inside the room, h_1, is 11.0 W/m²·K and the heat transfer coefficient from the window to the surrounding cold air, h_3, is 9.0 W/m²·K. Calculate the overall heat transfer coefficient in W/m²·K.

Table 5.2 Typical R values.

Fluids	Usual Range for R hr·ft²·°F/Btu	Suggested Values for R hr·ft²·°F/Btu
River water	0.002 – 0.003	0.0025
Boiler feed water (treated)	0.0005 – 0.001	0.00075
Fuel oils	0.004 – 0.010	0.007
Light oils and other organic liquids	0.001 – 0.004	0.0025

SOLUTION: Calculate the internal convection resistance, R_1:

$$R_1 = \frac{1}{h_1 A} = \frac{1}{(11)(1)} = 0.0909 \, \text{K/W}$$

Calculate the conduction resistance through the glass panel:

$$L_2 = \Delta x = 0.125 \, \text{in.} \left(\frac{0.0254 \, \text{m}}{\text{in.}} \right) = 0.00318 \, \text{m}$$

$$R_2 = \frac{\Delta x}{k_2 A} = \frac{0.00318}{(1.4)(1)} = 0.00227 \, \text{K/W}$$

Also calculate the outside convection resistance:

$$R_3 = \frac{1}{h_3 A} = \frac{1}{(9)(1)} = 0.111 \, \text{K/W}$$

The total thermal resistance is therefore,

$$\Sigma R = R_1 + R_2 + R_3 = 0.0909 + 0.00227 + 0.111 = 0.204 \, \text{K/W}$$

The overall heat transfer coefficient may now be calculated employing Equation (5.58):

$$U = \frac{1}{\Sigma R \cdot A} \tag{5.58}$$

Substituting,

$$U = \frac{1}{(0.204)(1)} = 4.9 \, \text{W/m}^2 \cdot \text{K}$$

Example 5.9 – Copper Plate Heat Exchanger. A heat exchanger wall consists of a copper plate 0.049 inch thick. If the two surface film coefficients are 208 and 10.8 Btu/hr·ft²·°F, respectively, calculate the overall heat transfer coefficient in Btu/hr·ft²·°F.

SOLUTION: Since information on the internal and external area is not given, neglect the effect of area. Equation (5.49) may be applied:

$$U = \frac{1}{\dfrac{1}{h_1} + \dfrac{\Delta x}{k} + \dfrac{1}{h_2}}$$

Substituting,

$$U = \cfrac{1}{\cfrac{1}{208} + \cfrac{(0.049/12)}{220} + \cfrac{1}{10.8}}$$

$$= 10.26\,\text{Btu/hr} \cdot \text{ft}^2 \cdot {}^\circ\text{F}$$

Expressing the Overall Heat Transfer Coefficient, U, in terms of the Resistance, R. For two fluids flowing through a concentric pipe, assume that the outer pipe wall is perfectly insulated (see Figure 5.7).
(i) Based on *Inside Area of the Inner Pipe*:

$$\frac{1}{U_i A_i} = \frac{T_i - T_o}{Q} = \Sigma R \tag{5.59}$$

(ii) Based on *Outside Area of the Inner Pipe*:

$$\frac{1}{U_o A_o} = \frac{T_i - T_o}{Q} = \Sigma R \tag{5.60}$$

From first principles, the heat transfer and the resistances to heat transfer through a concentric pipe are computed in the following manner:

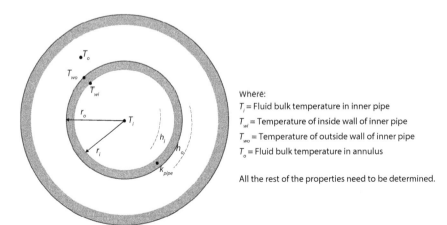

Where:
T_i = Fluid bulk temperature in inner pipe
T_{wi} = Temperature of inside wall of inner pipe
T_{wo} = Temperature of outside wall of inner pipe
T_o = Fluid bulk temperature in annulus

All the rest of the properties need to be determined.

Figure 5.7 Diagrammatic representation of a concentric pipe and process fluid properties used to calculate overall heat transfer coefficients.

Inside Film:

$$Q_i = h_i A_i \left(T_i - T_{wi} \right) \tag{5.61}$$

$$Q_i = h_i \left(2\pi r_i L \right) \left(T_i - T_{wi} \right) \tag{5.62}$$

and,

$$R_i = \frac{1}{h_i A_i} = \frac{1}{h_i \left(2\pi r_i L \right)} \tag{3.8}$$

Pipe Wall:

$$Q_i = \frac{k_{pipe} A_{LM} \left(T_{wi} - T_{wo} \right)}{\Delta x_w} \tag{5.63}$$

where,

$$R_w = \frac{\Delta x_w}{k_{pipe} A_{LM}} \tag{2.4}$$

and,

$$\Delta x_w = r_o - r_i \tag{5.64}$$

and,

$$A_{LM} = \frac{A_o - A_i}{\ln\left(\dfrac{A_o}{A_i}\right)} = 2\pi \left[\frac{r_o - r_i}{\ln\left(\dfrac{r_o}{r_i}\right)} \right] L \tag{5.65}$$

Outside Film:

$$Q_o = h_o A_o \left(T_{wo} - T_o \right) \tag{5.66}$$

$$Q_o = h_o \left(2\pi r_o L \right) \left(T_{wo} - T_o \right) \tag{5.67}$$

and

$$R_o = \frac{1}{h_o A_o} = \frac{1}{h_o \left(2\pi r_o L \right)} \tag{3.8}$$

When the resistances are substituted into the expression for *overall heat transfer coefficient*,

$$\frac{1}{U_o A_o} = \Sigma R = R_i + R_w + R_o \tag{5.68}$$

and

$$\frac{1}{U_o A_o} = \frac{1}{h_i A_i} + \frac{\Delta x_w}{k A_{LM}} + \frac{1}{h_o A_o} \tag{5.69}$$

The above expression is the most general expression used to calculate the overall heat transfer coefficient, U, when there is heat transfer between two fluids in a concentric pipe. It is common practice to choose the outside area of the inner pipe, A_o, as the basis for the calculation of U.

There are many situations where assumptions can be made that will simplify Equation (5.69). Two of these situations are presented below:

Assumption 1: Metal Inner Pipe

    (i) Due to the high thermal conductivity, k, the resistance of the pipe wall is relatively small compared to the resistance of the two fluids.

    (ii) R_w is considered negligible and can be neglected.

Equation (5.69) is then reduced to:

$$\frac{1}{U_o A_o} = \frac{1}{h_i A_i} + \frac{1}{h_o A_o} \tag{5.70}$$

or,

$$\frac{1}{U_o} = \frac{1}{h_i \left(A_i / A_o \right)} + \frac{1}{h_o} \tag{5.71}$$

Assumption 2: Very Thin Inner Pipe Wall

    (i) This implies that $r_i \approx r_o$; $\Delta x_w = r_o - r_i \approx$ very small ; $R_w \approx$ very small compared to R_i and R_o and is considered negligible.

    (ii) Assume $A_i \approx A_o \approx A$

Equation (5.71) then becomes:

$$\frac{A}{U_o A} = \frac{A}{h_i A} + \frac{A}{h_o A}$$

(5.72)

or

$$\frac{1}{U} = \frac{1}{h_i} + \frac{1}{h_o}$$

(5.73)

Example 5.10 – Overall Heat Transfer Coefficient based on Inner Area.
Steam at 247 °F is flowing through a pipe exposed to air but covered with
1.5 in. thick insulation. The following data is provided:

1. Pipe diameter, inside = 0.825 in.
2. Pipe diameter, outside = 1.05 in.
3. Surrounding air temperature = 60 °F
4. Thermal conductivity, pipe = 26 Btu/hr·ft². °F
5. Thermal conductivity, insulation = 0.037 Btu/hr·ft². °F
6. Steam film coefficient = 800 Btu/hr·ft². °F
7. Air film coefficient = 2.5 Btu/hr·ft². °F

Calculate the overall heat transfer coefficient based on the inside area
of the pipe.

SOLUTION: Obviously, cylindrical coordinates are to be employed here.
Assume a basis 1.0ft of pipe length. The inside (i) and outside (o) areas of
the pipe (P) plus the insulation (I) areas are

$$A_i = 0.2157 \, \text{ft}^2$$

$$A_o = 0.2750 \, \text{ft}^2$$

$$A_I = 1.060 \, \text{ft}^2$$

The log mean area for the steel pipe (P) and insulation (I) are, therefore,

$$A_{P,LM} = \frac{A_o - A_i}{\ln(A_o / A_i)} = \frac{0.2750 - 0.2157}{\ln(0.2750 / 0.2157)} = 0.245 \, \text{ft}^2$$

$$A_{I,LM} = \frac{A_I - A_o}{\ln(A_I / A_o)} = \frac{1.060 - 0.2750}{\ln(1.060 / 0.2750)} = 0.582 \, \text{ft}^2$$

A slightly modified form of Equation (5.51) and (5.56) is employed to calculate the overall heat coefficient based on the inside area:

$$U_i = \frac{1}{A_i \Sigma R} = \frac{1}{A_i \left(R_i + R_P + R_o + R_I \right)}$$

$$= \frac{1}{A_i \left(\dfrac{1}{h_i A_i} + \dfrac{r_o - r_i}{k_P A_{P,LM}} + \dfrac{1}{h_o A_I} + \dfrac{r_1 - r_o}{k_I A_{I,LM}} \right)} \tag{5.74}$$

$$= \cfrac{1}{0.2157\left(\cfrac{1}{(800)(0.2157)} + \cfrac{(0.525 - 0.412)}{(26)(0.245)} + \cfrac{1}{(2.5)(1.060)} + \cfrac{(2.025 - 0.525)/12}{(0.037)(0.582)} \right)}$$

$$= \frac{1}{0.2157 \left(0.005795 + 0.001478 + 0.37736 + 5.80477 \right)}$$

$$= 0.749 \, \text{Btu/hr} \cdot \text{ft}^2 \cdot {}^\circ\text{F}$$

Note that the resistance from the insulation is much higher than the resistances of other materials, as one can see in the denominator of the resistance values: 0.005795, 0.001475, and 0.37736 $\lll$ 5.80477.

Example 5.11 – Rate of Heat Transfer. Calculate the rate of heat transfer in the previous example.

SOLUTION: The above result can be extended to calculate the heat transfer rate.

$$A_i = \pi D_i L$$

$$A_i = \pi \left(0.825/12 \right)\left(1 \right) = 0.2157 \, \text{ft}^2$$

$$Q = U_i A_i \left(T_s - T_a \right)$$

$$Q = \left(0.7492 \right)\left(0.2157 \right)\left(247 - 60 \right) = 30.2 \, \text{Btu/hr}$$

Controlling Film Coefficient and Controlling Resistance [5]. Consider two fluids in a concentric pipe system. When the resistance of the pipe metal, R_w, is small by comparison with the other resistances, and it usually is, it may be neglected. Consider the layout provided in Figure 5.8.

Total resistance can be calculated in one of two ways for the given pipe wall shown in Figure 5.8. In one method, one can calculate the individual resistances and simply sum them up:

$$R_i = 1/10 = 0.1$$
$$R_o = 1/1000 = 0.001$$

Therefore,

$$R = R_i + R_o$$
$$= 0.1 + 0.001 = 0.101 \, \mathrm{hr \cdot ft^2 \cdot {}^\circ F/Btu}$$

In another method, one can utilize Equation (5.71).

$$h_i \left(A_i / A \right) \approx 10$$

$$h_o = 1000$$

Therefore,

$$R = \frac{1}{U_o} = \frac{1}{h_i \left(A_i / A \right)} + \frac{1}{h_o} \tag{5.71}$$

$$= \frac{1}{10} + \frac{1}{1000} = 0.101 \, \mathrm{hr \cdot ft^2 \cdot {}^\circ F/Btu}$$

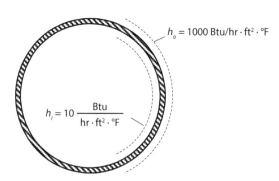

Figure 5.8 Cross-section of a pipe.

Ultimately, both methods should give the same total resistance. Now consider if one of the convective heat transfer coefficients, i.e., h_i or h_o, drastically changed in its value. How would this change affect the total resistance?

If h_o suddenly becomes 500 instead of 1000 Btu/hr·ft·°F, a variation of 50 percent, the resulting R would change very little. Using the same calculation procedure just demonstrated, R increases from 0.101 to 0.102 hr·ft²·°F/Btu. h_o is comparatively much larger than h_i, which also means that the resistance R_o is comparatively much smaller than R_i. In other words, what limits the rate of heat transfer through the pipe wall is the material with the greatest resistance. That is why changing h_i had very little impact on the overall resistance; it did not affect the component of the system that offered the largest resistance, which in this case was R_i.

On the other hand, if h_i was changed by 50 percent, to a new value of 5 Btu/hr·ft·°F, the resulting R would change significantly. R changes from 0.101 to 0.201, a nearly 100 percent increase in overall resistance. Since R_i is the component of the system that offers the largest bulk of resistance, R_i determines or *controls* the nature of the overall resistance. Therefore, the resistance of the system can be approximated as

$$R = \frac{1}{h_{smaller} A} \qquad (5.74)$$

where the other components of R are considered negligible. As h gets smaller↓, R gets larger↑.

The smaller of the two coefficients is defined as the *controlling film coefficient* and the large resistance it creates is called the *controlling resistance*. This resistance tends to dominate the magnitude of the overall heat transfer coefficient, as shown in the following example.

Example 5.12 – Identifying the Controlling Resistance. Hot water flows through a 3-inch Schedule-40 steel pipe. The convective heat transfer coefficients for the hot water and air outside the pipe are 1000 and 0.2 Btu/hr·ft²·°F, respectively. The thermal conductivity for the steel pipe is 26 Btu/hr·ft·°F. Determine which resistance, if any, is the controlling resistance.

SOLUTION: Refer to Figure 5.9. As noted above, in many heat transfer situations involving resistances in series, one resistance will be larger than all others, and will therefore control the rate of heat transfer.

$$R_i = \frac{1}{h_i A_i} = \frac{1}{(1000)(0.803)} = 0.00124\,\text{ft}^2 \cdot °\text{F} \cdot \text{hr/Btu}$$

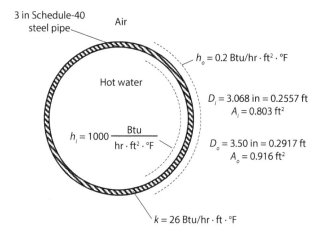

Figure 5.9 Pipe resistances; Example 5.12.

$$R_w = \frac{\Delta x_w}{kA_{LM}}$$

$$\Delta x_w = r_o - r_i = \frac{D_o - D_i}{2} = \frac{0.2917 - 0.2557}{2} = 0.018\,\text{ft}$$

$$A_{LM} = \frac{A_o - A_i}{\ln\left(\dfrac{A_o}{A_i}\right)} = \frac{0.916 - 0.803}{\ln\left(\dfrac{0.916}{0.803}\right)} = 0.8583\,\text{ft}^2$$

Therefore,

$$R_w = \frac{\Delta x_w}{kA_{LM}} = \frac{0.018}{(26)(0.8583)} = 0.0008\,\text{ft}^2\cdot{}^\circ\text{F}\cdot\text{hr/Btu}$$

and,

$$R_o = \frac{1}{h_o A_o} = \frac{1}{(0.2)(0.916)} = 5.4585\,\text{ft}^2\cdot{}^\circ\text{F}\cdot\text{hr/Btu}$$

therefore,

$$R = R_i + R_w + R_o = 5.4605\,\text{ft}^2\cdot{}^\circ\text{F}\cdot\text{hr/Btu}$$

It is clear from the above example that the magnitude of the total resistance is dominated by the resistance of the air. R_o is considered the controlling resistance in this system and the others are considered negligible.

Example 5.13 – Steel Pipe Resistance and Overall Heat Transfer Coefficient. A coolant flows through a steel pipe (inside diameter of 0.05 m, outside diameter of 0.06 m) at a velocity of 0.25 m/sec. What is the overall heat transfer coefficient for the system based on the pipe's outside surface area if the inside and outside coefficients are 2000 W/m²·K and 8.25 W/m²·K, respectively. The "resistance" term for the steel is specified as

$$R = 1.33 \times 10^{-4} \, \text{m}^2 \cdot \text{K/W}$$

SOLUTION: For this example, Equation (5.57) applies. Therefore,

$$U_o = \cfrac{1}{\cfrac{D_o}{h_i D_i} + \cfrac{D_o \ln(D_o / D_i)}{2k} + \cfrac{1}{h_o}}$$

After substitution,

$$U_o = \cfrac{1}{\cfrac{0.06}{(2000)(0.05)} + 1.33 \times 10^{-4} + \cfrac{1}{8.25}}$$

$$U_o = \frac{1}{0.0006 + 1.33 \times 10^{-4} + 0.121} = 8.21 \, \text{W/m}^2 \cdot \text{K}$$

Example 5.14 – Identifying the Controlling Resistance. With reference to the previous example, determine the controlling resistance.

SOLUTION: Obviously, the controlling resistance is located outside (external to) the pipe ($1/h_o = 0.121$).

Varying Overall Heat Transfer Coefficients. The calculation of ΔT_{LM} is valid for heating or cooling, i.e., whether ΔT_1 and ΔT_2 are both positive or both negative. It is not possible that ΔT_1 and ΔT_2 have opposite signs since it would constitute a violation of the Second Law of Thermodynamics. In addition, the LMTD should not be used when the overall heat transfer coefficient, U, *changes appreciably* through the unit. When U varies lineally with the temperature difference over the entire heating surface, the rate of heat transfer may be estimated from the following derived equation [6]:

$$Q = A \left[\frac{U_2 \Delta T_1 - U_1 \Delta T_2}{\ln\left(\dfrac{U_2 \Delta T_1}{U_1 \Delta T_2}\right)} \right] \tag{5.75}$$

where U_1 is the overall heat transfer coefficient at the fluid entrance side and U_2 is the overall heat transfer coefficient at the fluid exit side. Equation (5.75) requires the use of a log mean value of the *product* of U and ΔT. Note that the U at one end of the exchanger is multiplied by the temperature difference at the other end. This equation was derived by Colburn [6] assuming that:

1. the variation of U was linearly dependent on temperature
2. constant mean flow
3. constant heat capacity
4. no partial phase changes

Interestingly, if Equation (5.75) is rewritten as

$$Q = \bar{U}A\Delta T_{LM} \qquad (5.76)$$

the term $\bar{U}$ may be viewed as an average (or mean) overall heat transfer coefficient. The entire surface area A can then be regarded as transferring heat at ΔT_{LM} rather than at some average value.

Example 5.15 – Approach Temperature Calculation with Varying U.
A brine solution at 10 °F in a food processing plant is heated by flowing through a heated pipe. The pipe surface is maintained at 80 °F. The pipe surface area for heat transfer is 2.5 ft². The brine solution (with a density of 62.4 lb/ft² and a heat capacity of 0.99 Btu/lb· °F) flows at a rate of 20 lb/min. Assume the overall heat transfer coefficient varies in a linear fashion with temperature, with values of 150 Btu/hr·ft²· °F at the brine solution entrance (where the brine temperature is 10 °F) and 140 Btu/hr·ft²· °F at the brine solution exit. Determine the temperature approach at the brine inlet side, the exit temperature of the brine solution, and the rate of heat transfer, Q.

SOLUTION: Set T as the surface temperature and t_1 and t_2 as the brine inlet and outlet temperatures, respectively. Calculate the temperature approach at the pipe entrance:

$$\Delta T_1 = T - t_1$$

$$\Delta T_1 = 80 - 10 = 70\,°F$$

$$\Delta T_1 = 70/1.8 = 38.9\,°C$$

Note that ΔT_2, which is equal to $(T - t_2)$, cannot be calculated since t_2 is not known. Apply an energy balance to the brine solution across the full length of the pipe.

$$Q = \dot{m}c_p\left(t_2 - t_1\right)$$
$$= \dot{m}c_p\left(\Delta T_1 - \Delta T_2\right)$$
$$= \left(1200\right)\left(0.99\right)\left(70 - \Delta T_2\right)$$
$$= 1188\left(70 - \Delta T_2\right)$$

The corresponding equation for the LMTD is

$$\Delta T_{LM} = \frac{\left(70 - \Delta T_2\right)}{\ln\left(70/\Delta T_2\right)}$$

Write the equation for the heat transfer rate. Note that U varies lineally with ΔT. See Equation (5.75):

$$Q = A\left[\frac{U_2\Delta T_1 - U_1\Delta T_2}{\ln\left(\dfrac{U_2\Delta T_1}{U_1\Delta T_2}\right)}\right] \tag{5.75}$$

Substitution yields,

$$Q = 2.5\left[\frac{\left(140\right)\left(70\right) - \left(150\right)\Delta T_2}{\ln\left[\dfrac{\left(140\right)\left(70\right)}{\left(150\right)\Delta T_2}\right]}\right]$$

Combining the previous equations and eliminating Q:

$$1188\left(70 - \Delta T_2\right) = 2.5\left[\frac{\left(140\right)\left(70\right) - \left(150\right)\Delta T_2}{\ln\left[\dfrac{\left(140\right)\left(70\right)}{\left(150\right)\Delta T_2}\right]}\right]$$

This equation is non-linear with one unknown (ΔT_2). This equation may be solved by trial-and-error. Note that $0 \le \Delta T_2 \le 70°F$. Solution gives $\Delta T_2 = 51.6°F = 28.7°C$.

Calculate the discharge temperature of the brine solution:

$$\Delta T_2 = T - t_2$$

$$t_2 = T - \Delta T_2 = 80 - 51.6 = 28.5°F$$

$$t_2 = \frac{28.4 - 32}{1.8} = -2°C$$

Finally, calculate the heat transfer rate, Q. Using the earlier equations for Q,

$$Q = 1188(70 - 51.6) = 21{,}860\,\text{Btu/hr}$$

$$Q = 21{,}860/3.412 = 6{,}407\,\text{W}$$

Determination of Unknown Exit Temperatures in Countercurrent Flow. Very often, a countercurrent flow heat exchanger is available which has a given length, L, and therefore a fixed surface area, A.

Two process streams are available with inlet temperatures (T_1 and t_1), flow rates ($\dot{M}$ and $\dot{m}$), and heat capacities (C and c). A common question asked of the newly hired engineer: "What outlet temperatures will be attained by the exchanger?" This problem requires an estimate of U which can be obtained in the Appendix.

Since,

$$\dot{m}c(t_2 - t_1) = \dot{M}C(T_1 - T_2) \tag{5.77}$$

The ratio of the temperature ranges can be established as a unique ratio, R, where

$$R = \frac{\dot{m}c}{\dot{M}C} = \frac{T_1 - T_2}{t_2 - t_1} \tag{5.78}$$

Substituting $UA\Delta T_{LM} = \dot{M}C(T_1 - T_2)$ and rewriting Equation (5.77),

$$\dot{m}c(t_2 - t_1) = UA\frac{(T_1 - t_2) - (T_2 - t_1)}{\ln\left(\dfrac{T_1 - t_2}{T_2 - t_1}\right)} \tag{5.79}$$

Rearranging Equation (5.79) leads to:

$$\ln\left(\frac{T_1-t_2}{T_2-t_1}\right) = \frac{UA}{\dot{m}c}\left[\frac{(T_1-t_2)-(T_2-t_1)}{(t_2-t_1)}\right] \tag{5.80}$$

$$\ln\left(\frac{T_1-t_2}{T_2-t_1}\right) = \frac{UA}{\dot{m}c}\left[\frac{(T_1-T_2)}{(t_2-t_1)}-\frac{(t_2-t_1)}{(t_2-t_1)}\right] \tag{5.81}$$

$$\ln\left(\frac{T_1-t_2}{T_2-t_1}\right) = \frac{UA}{\dot{m}c}\left[\frac{(T_1-T_2)}{(t_2-t_1)}-1\right] \tag{5.82}$$

Substitute the ratio, R, into Equation (5.82),

$$\ln\left(\frac{T_1-t_2}{T_2-t_1}\right) = \frac{UA}{\dot{m}c}\left[R-1\right] \tag{5.83}$$

The above equation may also be written as

$$\frac{T_1-t_2}{T_2-t_1} = e^{\omega}, \quad \omega = \frac{UA}{\dot{m}c}\left[R-1\right] \tag{5.84}$$

Solving for the exit temperature, T_2,

$$T_2 = t_1 + \frac{T_1-t_2}{e^{\omega}} \tag{5.85}$$

One may now develop an expression for t_2 from R in order to obtain a relationship for T_2 in terms of T_1 and t_1 from Equation (5.78).

$$t_2 = t_1 + \frac{T_1-T_2}{R} \tag{5.86}$$

Substituting t_2 into T_2, and solving for T_2 ultimately yields for a countercurrent flow exchanger

$$T_2 = \frac{(1-R)T_1 + \left(1-e^{\frac{UA}{\dot{m}c}\left[R-1\right]}\right)Rt_1}{1-e^{\frac{UA}{\dot{m}c}\left[R-1\right]}R} \tag{5.87}$$

For cocurrent flow, it can be shown that

$$
T_2 = \frac{\left(R + e^{\frac{UA}{\dot{m}c}[R+1]}\right)T_1 + \left(e^{\frac{UA}{\dot{m}c}[R+1]} - 1\right)Rt_1}{(R+1)e^{\frac{UA}{\dot{m}c}[R+1]}}
\tag{5.88}
$$

The final exit temperature, t_2, may be obtained by applying the heat balance:

$$
\dot{m}c(t_2 - t_1) = \dot{M}C(T_1 - T_2)
$$

Fouling Factors. When a double pipe heat exchanger has been in service for some time, dirt and scale builds up on the inside and outside of the inner pipe, adding two more resistances than were originally included in the calculation of the heat transfer coefficient, U. This accumulation may consist of rust, boiler scale, or silt, etc. These additional resistances reduce the original value of U and the results are as follows:

1. The required amount of heat is no longer transferred by the original surface A.
2. The outlet temperatures of the hot and cold streams, T_2 and t_2, do not reach the desired temperatures
3. h_i and h_o remain substantially constant over the length of the exchanger.

To overcome this, it is customary in designing equipment to anticipate the deposition of dirt and scale by incorporating a resistance, R_d, called the *dirt, scale,* or *fouling factor.*

Let R_{di} represent the dirt factor for the inner pipe fluid at its inside diameter, and R_{do} the dirt factor for the annulus fluid at the outside diameter of the inner pipe as seen in Figure 5.10. These may be considered very thin for dirt but may be appreciably thick for scale, which has a higher thermal conductivity than dirt. The resistance from dirt is calculated from the fouling coefficients as follows:

$$
R_d = R_{di} + R_{do}
\tag{5.89}
$$

The value of U obtained using Equation (5.73) from $1/h_i$ and $1/h_o$ may only be considered the *clean overall coefficient*, designated by U_C to show that dirt has not been taken into account. The coefficient which includes

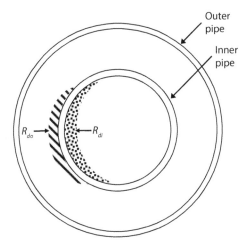

Figure 5.10 Location of fouling factors.

the dirt resistance is called the *design* or *dirty overall coefficient* U_D. The relationship between the two overall coefficients, U_C and U_D is

$$\frac{1}{U_D} = \frac{1}{U_C} + R_d \tag{5.90}$$

The value of the area, A, corresponding to U_D rather than U_C provides the basis on which equipment is ultimately designed and built. (Note that this is only true for $A_i \approx A_o$.)

The variable h_f is a heat transfer coefficient for fouling – similar to the convective film heat transfer coefficients, h_i and h_o with units **Btu/hr·ft²·°F**.

Table 5.3 Common fouling coefficients, h_f.

Fluid	h_f, Btu/(hr·ft²·°F)
Alcohol Vapors	0.0005
Boiler Feed Water	0.001
Fuel Oil	0.005
Industrial Air	0.002
Quench Oil	0.004
Refrigerating Liquids	0.001
Seawater	0.0008
Steam, Non Oil-Bearing	0.0006

Common fouling coefficients, h_f, are shown in Table 5.3. For example, suppose that for a double pipe exchanger, h_i and h_o have been computed to be 300 and 100 respectively, with units of Btu/(hr·ft²·°F). A typical question that the practicing engineer needs to answer is: "For what overall coefficient should the surface be calculated so that the equipment need be cleaned only once a year?" Assume a very thin, metal pipe in the analysis, so that

$$\frac{1}{U_C} = \frac{1}{h_i} + \frac{1}{h_o} = \frac{1}{300} + \frac{1}{100} = 0.0033 + 0.01 = 0.0133$$

Thus,

$$U_C = \frac{1}{0.0133} = 75.2\,\text{Btu/hr} \cdot \text{ft}^2 \cdot {}^\circ\text{F}$$

From experience it has been found that a thermal dirt resistance of $R_{di} = 0.001\,\text{ft}^2 \cdot {}^\circ\text{F} \cdot \text{hr/Btu}$ will deposit inside the inner pipe and $R_{do} = 0.0015\,\text{ft}^2 \cdot {}^\circ\text{F} \cdot \text{hr/Btu}$ will deposit on the outside of the inner pipe annually. The *allowable* dirt factor, R_d *(allowable)* is computed as follows,

$$R_d = R_{di} + R_{do} \tag{5.91}$$

U_D, the *design* or *dirty overall heat transfer coefficient* is then calculated. Employ Equation (5.91).

$$R_d = R_{di} + R_{do} = 0.0025$$

U_D may now be calculated by employing Equation (5.90):

$$\frac{1}{U_D} = \frac{1}{U_C} + R_d \tag{5.90}$$

substitution:

$$\frac{1}{U_D} = \frac{1}{75.2} + 0.0025 = 0.0158\,\text{ft}^2 \cdot {}^\circ\text{F} \cdot \text{hr/Btu}$$

$$U_D = \frac{1}{0.0158} = 63.3\,\text{Btu/hr} \cdot \text{ft}^2 \cdot {}^\circ\text{F}$$

Note that U_D is nearly 20% less than U_C for this application.

The heat transfer equation for a designed surface area, A, on which dirt will be deposited now becomes

$$A = \frac{Q}{U_D \Delta T_{LM}} \tag{5.92}$$

If ΔT is calculated from *observed* temperatures instead of desired process temperatures, then Equation (5.93) may be used to determine U_D, and R_d *(deposited)* from Equation (5.94) for a given fouling period. Rearranging Equation (5.92):

$$U_D = \frac{Q}{A \Delta T_{LM}} \tag{5.93}$$

If fouling is to be included,

$$\frac{1}{U_D} - \frac{1}{U_C} = R_d \left(\text{deposited}\right) \tag{5.94}$$

Because U_C is constant, as U_D decreases $\downarrow$ with fouling, $1/U_D$ increases$\uparrow$, and R_d*(deposited)* increases$\uparrow$. Therefore, when R_d*(deposited)* $> R_d$*(allowable)* — as after a period of service — the exchanger no longer delivers a quantity of heat equal to the process requirements and must be cleaned – whether it is scheduled or not.

Numerical values of dirt or fouling factors for a variety of process services are provided in Appendix Table AT.12. The tabulated fouling factors are intended to protect the exchanger from delivering less than the required process heat load for a period of about a year to a year and a half. Actually, the purpose of the tabulated fouling factors should be considered from another point of view. In designing a process plant containing many heat exchangers but without alternate or spare pieces of heat-transfer equipment, the process must be discontinued and the equipment cleaned as soon as the first exchanger becomes fouled. It is impractical to shut down every time one exchanger or another is fouled, and by using the tabulated fouling factors, it can be arranged so that all the exchangers in the process become dirty at the same time regardless of service. At that time all can be dismantled and cleaned during a single shutdown. The tabulated values may differ from those encountered by experience in particular services. If too frequent cleaning is necessary, a greater value of R_d should be kept in mind for future design.

It is to be expected that heat-transfer equipment will transfer more than the process requirements when newly placed in service and that it will deteriorate through operation as a result of dirt until it just meets the process requirements. The calculation of the temperatures delivered initially by a clean exchanger whose surface has been designed for U_D, but which is operating without dirt is not difficult. Referring to Equation (5.47), use U_C for U and the actual surface area of the exchanger, A (which is based on U_D). This calculation is also useful in checking whether or not a clean exchanger will be able to deliver the process heat requirements when it becomes dirty.

Example 5.16 – Neglecting Wall Resistance. Using the following information provided for a heat exchanger, calculate the heat transfer rate for the exchanger if the wall resistance can be neglected.

$$h_w(\text{water}) = 200 \text{ Btu/hr·ft}^2\cdot{}^\circ\text{F}$$

$$h_o(\text{oil}) = 50 \text{ Btu/hr·ft}^2\cdot{}^\circ\text{F}$$

$$h_f(\text{fouling}) = 1000 \text{ Btu/hr·ft}^2\cdot{}^\circ\text{F}$$

$$\Delta T_{LM} = 90\,^\circ\text{F}$$

$$A = 15.0\,\text{ft}^2$$

SOLUTION: Apply a modified form of Equation (5.71).

$$\frac{1}{U} = \frac{1}{h_w} + \frac{1}{h_o} + \frac{1}{h_f}$$

Substituting,

$$\frac{1}{U} = \frac{1}{200} + \frac{1}{50} + \frac{1}{1000}$$
$$= 0.005 + 0.02 + 0.001$$
$$= 0.026$$

Therefore,

$$U = \frac{1.0}{0.026} = 38.46\,\text{Btu/hr}\cdot\text{ft}^2\cdot{}^\circ\text{F}$$

The heat transfer rate is

$$Q = UA\Delta T_{LM}$$
$$= (38.46)(15.0)(90)$$
$$= 51,920\,\text{Btu/hr}$$

Using Caloric Temperature to Account for Varying Overall Heat Transfer Coefficients. Of the four assumptions used in the derivation of Equation (5.6) for ΔT_{LM} – the one which is subject to the *largest deviation* – is that of a constant overall heat-transfer coefficient U. The film coefficients were computed for the properties of the fluid at the arithmetic mean temperatures between inlet and outlet. This assumption is not always accurate – *specifically for very viscous fluids*.

In fluid-fluid heat exchangers where the viscosity is a strong function of temperature, the hot fluid possesses a viscosity on entering which becomes greater as the fluid cools. However, the cold countercurrent flow fluid enters with a viscosity which decreases as it is heated. Therefore, the values of h_o and $h_i \left(A_i / A \right)$ vary over the length of the pipe to produce a larger U at the hot terminal than at the cold terminal. Under actual conditions, the variation of U may be even greater than the variation of h_i alone since the outside film coefficient h_o will vary at the same time and in the same direction as h_i.

The variation of U can be taken into account by numerical integration of dQ, the heat transferred over incremental lengths of the pipe $a''dL = dA$, and using the average values of U from point to point in the differential equation $dQ = U_{avg} dA \Delta T$. The summation from point to point then approximates $Q = UA\Delta T$. This is a time-consuming method, and the increase in the accuracy of the result normally does not warrant the effort.

Colburn [6], as described earlier by Kern [4], has undertaken an alternate solution to heat transfer between two highly viscous fluids. First, a Caloric Temperature, tc, of each fluid is calculated using Figure 17 in the appendix of Kern's First Edition. The properties of each fluid are then found at the caloric temperature – as opposed to the arithmetic mean of inlet and exit. These corrected fluid properties are then used to calculate h_i and h_o, which are then used to calculate a new U – defined as U_x. Colburn chose to obtain a single, overall coefficient, U_x, at which it can be assumed that the entire surface area, A, is transferring heat at the LMTD.

In summary, whenever there is heat transfer between highly viscous fluids, there is a sizeable difference between $U_{entrance}$ and U_{exit}. ΔT_{LM}, is not *the true temperature difference* for countercurrent flow. ΔT_{LM} according to Kern/Coburn [4], may be retained, if a suitable value of U is calculated based on caloric temperature.

It should be noted that this effect can be safely neglected in most real-world applications. While caloric temperature is not used in the example problems sited in this text, there are a variety of examples that used caloric temperature correction in Chapters 5 and 6 of Kern's original text.

5.5   The Heat Transfer Equation

The chapter concludes with a short section entitled, *The Heat Transfer Equation*, which has been expressed in the following manner:

$$Q = UA\Delta T; \quad \Delta T = \Delta T_{LM} \tag{5.1}$$

This equation will find extensive use in the remainder of the text, particularly in the remaining chapters in this section.

The reader will recall that for a plane wall:

$$UA = \frac{1}{\dfrac{1}{h_i A} + \dfrac{\Delta x}{kA} + \dfrac{1}{h_o A}} \tag{5.96}$$

Because the area of a plane wall remains constant, this equation may be rewritten as,

$$U = \frac{1}{R_i + R_w + R_o} = \frac{1}{\dfrac{1}{h_i} + \dfrac{\Delta x}{k} + \dfrac{1}{h_o}} \tag{5.95}$$

Similarly, *UA* for a pipe wall is expressed in the following manner:

$$UA = \frac{1}{\dfrac{1}{h_i A_i} + \dfrac{\Delta x}{kA_{LM}} + \dfrac{1}{h_o A_o}} \tag{5.97}$$

where:

$$A_{LM} = \frac{A_o - A_i}{\ln(A_o / A_i)} = \frac{2\pi L \Delta r}{\ln(r_o / r_i)} \tag{5.98}$$

Therefore,

$$UA = \frac{1}{\dfrac{1}{h_i A_i} + \dfrac{\ln(r_o / r_i)}{2\pi kL} + \dfrac{1}{h_o A_o}} \tag{5.99}$$

The overall heat transfer coefficient, *U*, may be calculated based on the inside or outside area of pipe. For example, $U_i A_i$:

$$U_i A_i = \cfrac{1}{\cfrac{1}{h_i A_i} + \cfrac{\ln(r_o / r_i)}{2\pi kL} + \cfrac{1}{h_o A_o}} \tag{5.100}$$

$$U_i = \cfrac{1}{\cfrac{1}{h_i} + \cfrac{A_i \ln(r_o / r_i)}{2\pi kL} + \cfrac{1}{h_o}\left(\cfrac{A_i}{A_o}\right)} \tag{5.101}$$

$$U = \cfrac{1}{\cfrac{1}{h_i} + \cfrac{\Delta r}{k} + \cfrac{1}{h_o}} \tag{5.102}$$

A similar expression can be written for U_o based on A_o, the outside area of the inner pipe. *However*, it is critically important that the reader be aware of the following differences.

For a *plane wall*, A_i is always equal to A_o. For a cylindrical pipe wall (or a square duct), A_i is *NOT* equal to A_o. Therefore, the following exists:

This chapter would be incomplete without a brief discussion of design considerations as they relate to the heat transfer equation. There are two design categories of interest to the practicing engineer: process design and equipment design. In the heat exchanger industry, process "design" can involve the calculation of the required heat transfer area for a particular application, adopting an existing exchanger to operate under a new set of conditions, designing a new exchanger to solve an existing problem, or predicting the performance of an "existing" unit with known existing process conditions. Each involve calculations associated with the aforementioned heat transfer equation $Q = UA\Delta T_{LM}$. Equipment design, as its

Table 5.4 Notable Differences between U_i and U_o for a Plane Wall compared to a Cylindrical Pipe Wall.

	Plane wall	Thick pipe wall	Thin pipe wall
Relationship:	$U_i A_i = U_o A_o$	$U_i A_i = U_o A_o$	$U_i A_i = U_o A_o$
Assumption:	$A_i = A_o$	$A_i \neq A_o$	$A_i \approx A_o$
Result:	$U_i = U_o = U$	$U_i \neq U_o$	$U_i \approx U_o = U$

name implies, is concerned with mechanical details of the equipment itself, e.g., the size (diameter and length) and number of tubes, the materials of construction, the number of passes (see Chapter 7), the nuts and bolts employed, etc. However, equipment design is only occasionally addressed in this *process* heat transfer book.

In addition, a factor of safety, similar to that used in structural design, may be applied to the U value in heat exchanger design to account for uncertainties. If a heat exchanger is selected as an off-the-shelf item, the heat exchanger supplier may provide the necessary U value so that a suitable exchanger can be selected for a specific application.

It should be noted that current process heat exchanger design practices can be categorized as state-of-the-art and pure empiricism. Past experience with similar applications is commonly used as the sole basis for the design procedure. The vendor (seller) maintains proprietary files on past heat exchangers, and these files are periodically revised and expanded as new exchangers are evaluated. During the design of a new unit, the files are consulted for similar applications and old designs are heavily relied on; thus, a new design is rarely required. However, industry in general, and the heat exchanger industry in particular, have developed fairly well-defined procedures for the design, construction, and operation of heat exchangers. These techniques are routinely used by today's practicing engineers. The same procedures have also been used in the design of other equipment.

Note that the bulk of the material to follow in this section has been drawn from one of the author's personal notes [7], from Shen *et al.* [8], and Theodore and Dupont [9]. Finally, the reader should note that no attempt is made to provide extensive coverage of this topic; only general procedures and concepts are presented and discussed.

Regarding new designs, one is occasionally required in more specialized applications, such as fundamental research projects with the aerospace and electronics industries. Whether the heat exchanger is selected as an off-the-shelf item or designed especially for a new application, the five conceptual steps to be initially considered for heat exchangers are:

1. Identification of the parameters that must be specified.
2. Application of the fundamentals underlying theoretical equations or concepts.
3. Enumeration, explanation, and application of simplifying assumptions.
4. Possible use of correction factors for non-ideal behavior.
5. Identification of other factors that must be considered for adequate equipment specification.

This should be followed by an examination of the 8 key factors listed below.

1. Heat-Exchanger Type
2. Heat Transfer Rate
3. Flow Direction
4. Physical Size and Weight
5. Materials of Construction
6. Maintenance
7. Capital Cost
8. Operating Cost

The relative importance of each of the above need also be included in the selection process for the exchanger.

Although all engineers approach heat exchanger design problems somewhat differently, Theodore [7] expanded the above and detailed the six major steps that are generally required. These six steps are discussed below and may also be applied to the design of most (other) equipment.

1. The first step is to conceptualize and define the heat exchange process. Some of the answers to a host of questions pertaining to the process operation is often known from past experience.
2. After a heat exchanger application has been defined, a method of solution must be sought. Although a method is seldom obvious, a good starting point is the preparation of a process flowchart for the exchanger. This effort can produce valuable results. It is an efficient way to become familiar with process and information that is initially lacking.
3. The third step involves the sometimes numerous calculations needed to arrive at specifications of operating conditions, equipment geometry, size, materials of construction, controls, instrumentation, monitors, safety equipment (automatic feed cutoff), etc. As part of this step, capital costs must be established. Cost-estimating precision is dependent on the desired accuracy of the estimate.
4. An overall economic analysis (including operating costs) must also be performed in order to determine if the application is "feasible." To answer this, labor, equipment, and other processing costs are estimated to provide an accurate economic forecast for the proposed exchanger.

5. In a case where alternate heat exchanger possibilities exist, economics and engineering optimization are necessary. Since this is occasionally the case, optimization calculations are usually applied several times during the process design calculations.

6. The final step of this design scheme is the compilation of a report. If applicable, the report should utilize the latest computer-generated layouts and graphics.

These six activities are prominent steps in traditional project analysis. Today, safety and regulatory (if applicable) concerns have also been integrated into the approach [9]. These two aspects are briefly discussed.

The safe operation of equipment requires that some of the operational parameters be constrained within specific bounds. Each system has parameters that must remain within the appropriate bounds to assure that the system is stable. All system should have safety equipment to prevent the system from being operated at a condition outside of the safe limits. Insurance companies such as Industrial Risk Insurers (IRI) and Factory Mutual (FM) and national groups such as the National Fire Protection Agency (NFPA) have recommended specific training requirements; the most important is to assume that all personal are properly trained on operational limitations of the heat exchanger.

Consideration should be given to conducting a hazard and operability (HAZOP) study (see Chapter 12, Part III) to examine and identify any possible safety issues [9]. It is sometimes necessary to apply safety factors to calculations and/or design; attempting to justify these as a responsible engineer is often a difficult task.

Any environmental regulation requires that each operational limit be monitored to ensure that the system is not operated when the parameters have been exceeded, i.e, when excursion arises. Variation in the steady-state conditions should be included. As part of any environmental regulatory review, the flow rate, composition, system size, and other physical and chemical characteristics of all discharged streams must be reviewed to assess any potential problems. In addition, concerns regarding pollution prevention and sustainability should be addressed where applicable [10].

Unlike many of the problems encountered and solved by the practicing engineer, there is absolutely no correct solution to a heat exchanger design. However, there is usually a *better* solution. Many alternative solutions [11] when properly implemented will function satisfactorily, but one alternative will usually prove to be more economical [12], more efficient and/or more attractive than the others [12].

It is common practice for the heat exchanger engineer to predict the performance and use computer software to design heat exchangers. Because of the uncertainties and variabilities discussed above, the engineer should not select an exchanger based solely on a lowest first – cost basis.

On to the specifics associated with what one would define as engineering calculations. As noted above, the calculation procedure for a heat exchanger is naturally based on the experimental data available. The six key process variables are generally:

$$\text{Hot side (h): } \dot{M}, T_1, T_2$$

$$\text{Cold side (c): } \dot{m}, t_1, t_2$$

Other variables usually assumed to be known include the physical properties of both the hot (h) and cold (c) fluids, plus the pressure drop and the fouling heat transfer coefficient (dirt factor) on both the hot (h) and cold (c) side. Kern's design methodology [4] allows one to assume that all six variables above are known/specified. His design methodology calculations are provided in several illustrative examples for double pipe, shell-and-tube, and finned units in Chapters 6, 7, and 8, respectively, to follow. These six variables permit the heat exchanger engineer to check on the consistency of the data via an energy balance on both the cold and hot side.

The reader should definitely note that the design procedure prepared by Kern — which in many respects has withstand the test of time—may not apply in many real-world applications. On examining the process information required for the calculation proposed by Kern, one comes to realize that many of the process conditions required are unknown. In fact, other than T_1 and $\dot{M}$ plus the average values of the mass properties if a hot fluid is to be cooled (or t_1 and $\dot{m}$ if a cold fluid is to be heated), the remaining process conditions may not be known. When this be the case, these other conditions must somehow be estimated or obtained by some other means.

The Kern methodology changes if one or more of the process variables is *not* known. His procedure can be employed if the above 6 conditions are known since the sixth can be obtained directly from the energy balance equation. Two unknowns could give rise to a trial-and-error procedure where other constraints would enter into the calculation; three unknowns could lead to double trial-and-error calculations. Other considerations can include capital costs, operating costs, environmental issues, the recovery of "quality" energy (see Chapter 10, Part III), etc. It should also be noted that for many industrial applications, all but t_2 is known, which would permit employing Kern's methodology. As noted by Theodore [7], the problem is then one of selecting an appropriate heat exchanger type and determining the size, i.e., the heat transfer area A, required to achieve the desired outlet temperature.

Consider an application for which $\dot{M}$, $\dot{m}$, T_1, and t_1 are known, and the objective is to specify a heat exchanger that will provide a desired value of t_2. The corresponding values of Q and T_2 may be computed from the energy balance equations, and the value of ΔT_{LM} may be found from its definition. It is then a simple matter to determine the required value of A from the heat transfer equation $Q = UA\Delta T_{LM}$. Alternatively, the heat exchanger type and size A may be known and the objective is to determine the heat transfer rate Q and the fluid outlet temperatures, T_2 and t_2, but this calculation would require iteration.

Finally, Theodore [7] has condensed some of the above to include the following consideration. As described earlier, there are two heat transfer calculations encountered in practice.

1. Predicting the performance of an existing exchanger.
2. Designing a new exchanger; this calculation can on occasion also include those involving the optimum design based on minimum cost (usually), minimum weight (rarely), and minimum volume (rarely).

Calculation (2) is usually one order of magnitude more difficult than (1) when the physical unit (the exchanger) is clearly specified in what function it needs to fulfill, e.g., duty. But, respective of whether the calculation is to be concerned with (1) or (2), there are two specific steps in the analysis that would usually follow. One is concerned with basic calculations and the other delves into physical design aspects. Details on both are as follows.

Basic calculations:

1. Application of the conservation law of energy and determining the heat load (or rate of heat transfer)
2. Calculating the inlet and/or outlet exchanger temperatures
3. Calculating the log mean temperature difference driving force
4. Calculating the individual heat transfer coefficients and/or the overall heat transfer coefficient
5. Applying the heat transfer equation to calculate the exchanger area

Physical design aspects:

1. Type of exchanger
2. Flow directions of the fluids
3. Tube size
4. Internal physical arrangement of tubes
5. Baffle arrangement (if applicable)
6. Specifying vents
7. Specifying drains
8. Eliminate any potential health and physical hazards

Once the above have been determined, the process engineer must then include all pertinent economic, energy, and environmental considerations.

The last several paragraphs introduced the reader to Kern's design methodology. All of the above design aspects and calculation demonstrations will be revisited in the Kern's Design Methodology sections in Chapters 6, 7, and 8.

The Role of Computers in Heat Transfer Calculation. It is no secret that some process engineers and/or vendors have a computer program available that will optimize the calculations associated with the selection of a heat exchanger from a cost perspective. Heat exchanger selection usually refers to the decision related to choosing an existing, or off-the-shelf, exchanger for a specific application. As one might suppose, there are thousands of these units of a variety of types that are available for most industrial applications. However, in some cases, the aforementioned exchangers will not satisfy the needs of a process, and so an exchanger must be custom-designed. A heat exchanger may also have to be designed (or manufactured) if the units are not available. The above term, *design*, usually refers to the creation of a new heat exchanger for an application in which an existing unit is either unsuitable or not available.

Computers play an important role with heat exchangers, particularly in the design of unique and/or specialty units. The reader should also note that it may be difficult to justify detailed trial-and-error and/or numerical solutions for many applications to obtain the optimum design since the overall heat-transfer coefficient, U, is a critical parameter. Because the conduction component of U is often negligible compared to the convection component(s), the heat-transfer coefficients for the fluids significantly affect heat transfer. As pointed out earlier, heat-transfer coefficients for both external and internal forced flow can be calculated for applications that correspond to laboratory conditions under which experimental correlations were developed. Due to heat-transfer surface complexities and other geometrical considerations in heat exchangers, however, actual flow characteristics and heat transfer mechanisms may not closely match those on which the correlations are based. Hence, heat-transfer coefficients employed in practice should be regarded as approximations and a factor of safety, similar to that used in many engineering calculations [12], may be applied to the U value in heat-exchanger design to account for these uncertainties. To compliment this topic, a short writeup on computer-aided design of heat exchangers is provided at the conclusion of Chapter 7 — Shell-and-Tube Heat Exchangers. The decision to place this material in Chapter 7 was influenced by the fact that most computer-designed heat exchangers find primary application with tube-and-shell exchangers.

References

1. I. Farag and J. Reynolds, *Heat Transfer*, A Theodore Tutorial, Theodore Tutorials, East Williston, NY, originally published by USEPA/APTI, RTP, NC, 1996.
2. L. Theodore, *Heat Transfer Applications for the Practicing Engineer*, John Wiley & Sons, Hoboken, NJ, 2011.
3. D. Green and R. Perry (editors), *Perry's Chemical Engineers Handbook*, 8th edition, McGraw-Hill, New York City, NY, 2008.
4. D. Kern, *Process Heat Transfer*, McGraw-Hill, New York City, NY, 1950.
5. A.M. Flynn, personal notes, Manhattan College, Bronx, NY, 1989.
6. A.P. Colburn, Ind. Eng. Chem., 35, 873–877, New York City, NY, 1933.
7. L. Theodore, personal notes, East Williston, NY, 1967.
8. T. Shen, Y. Choi and L. Theodore, *Hazardous Waste Incineration Manual*. USEPA/APTI, Research Triangle Park, NC, 1985.
9. L. Theodore and R. Dupont, *Environmental Health and Hazard Risk Assessment: Principles and Calculations*, CRC Press/Taylor & Francis Group, Boca Raton, FL, 2012.
10. R. Dupont, K. Ganesan and L. Theodore, *Pollution Prevention: Sustainability, Industrial Ecology, and Green Engineering*, 2nd edition, CRC Press/Taylor & Francis Group, Boca Raton, FL, 2017.
11. P. Abulencia and L. Theodore, *Open-ended Problems: A Future Chemical Engineering Approach*, Scrivener-Wiley Publishing, Salem, MA, 2015.
12. L. Theodore, *Chemical Engineering: The Essential Reference*, McGraw-Hill, New York City, NY, 2014.
13. L. Theodore and K. Behan, *Introduction to Optimization for Environmental and Chemical Engineers*, CRC Press/Taylor & Francis Group, Boca Raton, FL, 2018.

6

Double Pipe Heat Exchangers

Introduction

Of the various types of heat exchangers that are employed in industry, perhaps the two most important are the *double pipe* and the *shell and tube*. Despite the fact that shell and tube heat exchangers (see next chapter) generally provide greater surface area for heat transfer with a more compact design, greater ease of cleaning, and less probability of leakage, the double pipe heat exchanger still finds use in practice [1–3].

As will be discussed in the next section, the double pipe unit consists of two concentric pipes, two connecting tees (*T's*), a return head, a return feed, and packing glands that support the inner pipe within the outer pipe. Each of two fluids–hot and cold–flow either through the inside of the inner pipe or through the annulus formed between the outside of the inner pipe and the inside of the outer pipe. Generally, it is more economical (from a heat efficiency perspective) for the hot fluid to flow through the inner pipe and the cold fluid through the annulus, thereby reducing heat losses to the surroundings. In order to ensure sufficient contacting times, pipes longer than approximately 20 ft are extended by connecting them to *return bends*. The length of pipe is generally kept to a maximum of 20 ft because the weight of

217

the piping may cause the pipe(s) to sag. Sagging may allow the inner pipe to touch the outer pipe, distorting the annulus flow region and disturbing proper operation. When two pipes are connected in a "U" configuration by a return bend, the bend is referred to as a *hairpin*. In some instances, several hairpins may be connected in series. Additional equipment details follow in the next section.

Double pipe heat exchangers have been used in the chemical process industry for over 90 years. The first patent on this unit appeared in 1923. The unique (at that time) design provided "the fluid to be heated or cooled and flow longitudinally and transversely around a tube containing the cooling or heating liquid," etc. Interestingly, and perhaps unfortunately, the original design has not changed significantly since that time [1–3].

Although this unit is not extensively employed in industry (the heat transfer area is small relative to other heat exchangers), it serves as an excellent starting point from an academic and/or training perspective in the treatment of all the various heat exchangers reviewed in this Part.

Topics covered in this chapter include:

6.1   Equipment Description and Details
6.2   Key Describing Equations
6.3   Calculation of Exit Temperatures
6.4   Pressure Drop in Pipes and Annuli
6.5   Open-Ended Problems
6.6   Kern's Design Methodology

The reader should note that since each chapter in Part II has been written in a stand-alone fashion, there is overlapping material, not only between the chapters, but also with Part I.

6.1   Equipment Description and Details

As discussed in the Introduction to this chapter, the simplest kind of heat exchanger, which confines both the hot and cold fluids, consists of two concentric tubes, or pipes. When conditions are such that only a few tubes per pass are required, the simplest construction is the double pipe heat interchanger shown in Figure 6.1. This consists of special fittings that are attached to standard iron (typically) pipe so that one fluid (usually liquid) flows through the *inside* pipe and the second fluid (also usually a liquid) flows through the *annular* space between the two pipes. Such a heat interchanger will usually consist of a number of passes which are almost

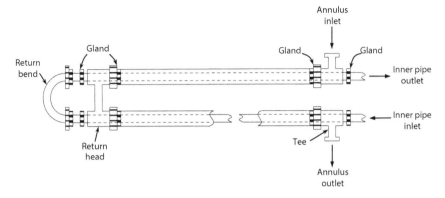

Figure 6.1 Double pipe exchanger.

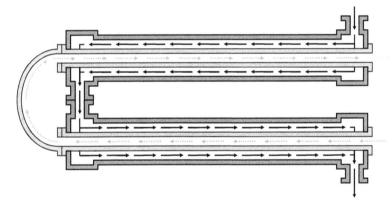

Figure 6.2 Fluid flow inside a double pipe exchanger.

invariably arranged in a vertical stack; when arranged in this manner it is referred to as a hairpin unit. If more than one pipe per pass is required, the proper number of such stacks is connected in parallel. Although they do not provide a large surface area for heat transfer, double pipe heat exchangers are at times used in industrial settings.

Figure 6.2 illustrates a more detailed perspective on the nature of flow inside a typical double pipe exchanger. Solid black arrows represent fluid flow in the annulus while dotted gray arrows represent fluid flow in the inner pipe.

A schematic of this unit and flow classification(s) is provided in Figure 6.3. Note that shades of red represent the hot fluid while shades of blue represent cold fluid. As heat transfer takes place, the hot fluid loses heat becoming less "red," while the cold fluid gains heat becoming less "blue."

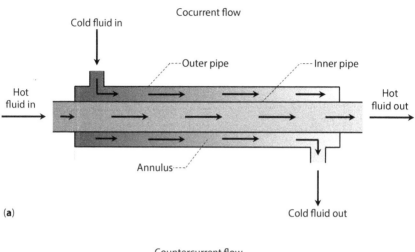

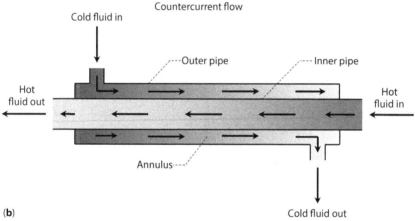

Figure 6.3 Double pipe schematic.

Recommended standard fittings for double pipe interchangers are provided in the literature [1, 3]. Additional details were provided in Part I, Chapter 1. Other types of tubes or pipes available include:

1. Plain tubes
2. Duplex tubes
3. Finned tubes

Plain tubes are used when the corrosion potential on either the tube or the shell side (see next chapter) are approximately the same. Duplex tubes have walls of two different metals and are used when it is impossible to find a single metal that can adequately handle corrosion on both sides of the unit. Finned tubes (to be discussed in Chapter 8) are usually employed for heat transfer to gases where film coefficients are very low. Tube materials include: carbon steel, carbon alloy steels, stainless steels, brass and alloys, cupro-nickel, nickel, monel, glass, reinforced fiberglass plastic (RFP), etc.

The double pipe exchanger is extremely useful because it can be assembled in any pipe-fitting shop from standard parts and provides an inexpensive heat-transfer surface. The standard sizes of tees and return heads are given in Table 6.1. Note that Iron Pipe Size (IPS) refers to a pipe sizing system traditionally used in the U.S. that references pipes according to its outside diameter in inches. In many cases, schedule numbers and Birmingham Wire Gauge (BWG) numbers are mentioned with the pipe size to indicate pipe wall thickness (refer to Table AT.7 and AT.8 in the Appendix).

This class of exchanger is usually assembled in 12-, 15-, or 20-ft effective lengths, where the effective length is defined as the distance in *each* leg over which heat transfer occurs and excludes inner pipe protruding beyond the exchanger section. In other words, a *20 ft hairpin is 40 ft long*. As noted earlier, hairpins should not be designed for pipes in excess of 20 ft in effective length. Whenever the linear pipe distance of each hairpin exceeds 40 ft, the inner pipe tends to sag and touch the outer pipe in the middle of the leg, thereby causing a poor flow distribution in the annulus. The principal disadvantage to the use of double pipe exchangers lies in the small amount of heat-transfer surface contained with a single hairpin. In an industrial process, a very large number of hairpins are required. These require considerable space, and each double pipe exchanger introduces no fewer than 14 points at which leakage might occur. The time and expense required, such as for dismantling and periodically cleaning, are excessive compared with other types of equipment. However, the double pipe exchanger is of greatest use where the total required heat-transfer surface is small, e.g., 100 to 200 ft² or less [2].

An important physical description of this exchanger is the representation of the so-called diameter of the annular region of the unit. When a fluid flows in a conduit having other than a circular cross section, such

Table 6.1 Double pipe exchanger fittings.

Outer pipe, IPS	Inner pipe, IPS
2	1 ¼
2 ½	1 ¼
3	2
4	3

as an annulus, it is convenient to express heat-transfer coefficients and friction factors by the same types of equations and curves used for pipes and tubes. To permit this type of representation for annulus heat transfer heat transfer, it has been found advantageous to describe this space by a characteristic length that is represented by a diameter term. As noted in Chapter 1, Part I, calculations for flow in the annular region referred to above require the use of the aforementioned characteristic or *equivalent diameter*. By definition, this diameter, D_e, is given by 4 times the area available for flow divided by the "wetted" perimeter. Note that there are two wetted perimeters in an annular area. This is given by,

$$D_e = 4\left(\frac{\pi}{4}\right)\left[\frac{D_{o,i}^2 - D_{i,o}^2}{\pi\left(D_{o,i} - D_{i,o}\right)}\right] \tag{6.1}$$

where $D_{o,i}$ is the inside diameter of the outer pipe and $D_{i,o}$ is the outside diameter of the inner pipe. This equation reduces to,

$$D_e = D_{o,i} - D_{i,o} \tag{6.2}$$

which is four times the hydraulic radius, r_H, i.e.,

$$D_e = 4r_H \tag{6.3}$$

with,

$$r_H = \frac{1}{4}\left(D_{o,i} - D_{i,o}\right) \tag{6.4}$$

The flow in a double pipe heat exchanger may be countercurrent or parallel (co-current). In countercurrent flow, the fluid in the pipe flows in a direction opposite to the fluid in the annulus. In parallel flow, the two fluids flow in the same direction. The variations of fluid temperature within the heat exchanger depend on whether the flow is parallel or countercurrent - as depicted in Figure 6.4.

The definitions below are employed in the development to follow (see also Figure 6.4):

T_1 = temperature of the hot fluid entering the inside pipe/tube

T_2 = temperature of the hot fluid exiting the inside pipe/tube

t_1 = temperature of the cold fluid entering the annulus

t_2 = temperature of the cold fluid exiting the annulus

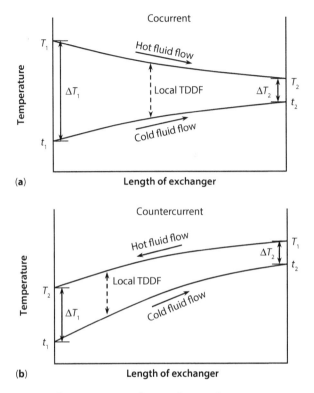

Figure 6.4 Cocurrent and countercurrent flow in a heat exchanger.

The difference between the temperature of the tube side fluid and that of the annulus side is the *temperature difference driving force* (TDDF), ΔT.

As described in earlier chapters, in a cocurrent flow heat exchanger, both hot and cold fluids enter on the same side and flow through the exchanger in the same direction. The *temperature approach* is defined as the temperature difference driving force at the heat exchanger entrance, ΔT_1 or $(T_1 - t_1)$. This driving force drops as the streams approach the exit of the exchanger. At the exit, the temperature difference driving force is ΔT_2 or $(T_2 - t_2)$. Thus, the heat exchanger is more effective at the entrance than the exit. In a countercurrent flow exchanger, the two fluids exchange heat while flowing in opposite directions. The temperature approach at the tube entrance end ΔT_1 or $(T_1 - t_2)$ and at the annular entrance end, ΔT_2 or $(T_2 - t_1)$ is usually roughly the same. In addition, the thermal driving force is normally relatively constant over the length of the exchanger. The temperature profile for both parallel and countercurrent systems are presented in Figure 6.4.

Summarizing, the double pipe heat exchanger employed in practice consists of two pipes: an inner and outer pipe. Hot fluid normally flows in the inner pipe and the cold fluid flows in the annulus between the outer diameter of the inner pipe and inner diameter of the outer pipe. Heat transfer occurs from the inner pipe to the outer pipe as the fluids flow through the piping system. By recording both inner and outer fluid temperatures at various points along the length of the exchanger, it is possible to calculate heat exchanger duties and heat transfer coefficients (to be discussed shortly) [1, 3].

Kern [2] has also provided design calculation details for series-parallel arrangements for double pipe heat exchangers. The difference of this class of exchangers and standard exchangers in series is demonstrated on Figures 6.5 and 6.6. However, series-parallel arranged exchangers are rarely employed in practice.

Example 6.1 – Fluid Flow Location. Explain why the cold fluid flows in the annular region in nearly all double pipe heat exchanger applications.

SOLUTION: As discussed earlier, heat losses to the environment are lower (except in cryogenic applications) with the hot fluid flowing in the inside pipe. Any losses are thus transferred to the cold fluid and not the surroundings.

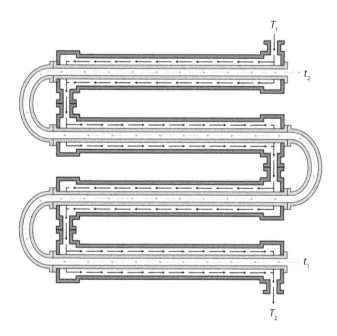

Figure 6.5 Double pipe exchangers in series.

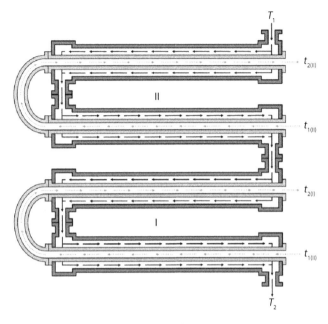

Figure 6.6 Double pipe exchangers in series-parallel arrangement.

6.2   Key Describing Equations

When it is necessary to heat or cool a process fluid, the heat transfer equation is employed to find a suitable heat exchanger, using one or more rearrangements of the following steps:

(i) Solve for unknown variables in the heat balance.

$$\Rightarrow \qquad Q = \dot{m}c\Delta t = \dot{M}C\Delta T$$

(ii) Calculate ΔT_{LM} from known and calculated temperatures.

$$\Rightarrow \qquad \Delta T_{LM} = \frac{\Delta T_a - \Delta T_b}{\ln\left(\Delta T_a / \Delta T_b\right)}$$

(iii) Calculate the film coefficients from fluid properties and U_C (as U_o or U_i).

$$\Rightarrow \qquad \frac{1}{U_o A_o} = \frac{1}{h_i A_i} + \frac{\Delta x_w}{k A_{LM}} + \frac{1}{h_o A_o}$$

(iv) When applicable, neglect pipe-wall resistance; assume $A_i \approx A_o$.

$$\Rightarrow \qquad \frac{1}{U_C} = R_i + R_o = \frac{1}{h_i} + \frac{1}{h_o}$$

(v) Assume an allowable dirt factor in order to calculate U_D.

$$\Rightarrow \qquad \frac{1}{U_D} = \frac{1}{U_C} + R_d$$

(vi) Use the heat transfer equation to
solve for unknown variables such $\Rightarrow$ $Q = U_D A \Delta T_{LM}$
as heat transfer surface area, A.

From the student's and practicing engineer's perspectives, there are two calculations associated with heat exchangers.

1. Design of a new heat exchanger to heat or cool process fluids.
2. Use of an existing heat exchanger to heat or cool process fluids. Determining the suitability of an existing heat exchanger for given process conditions is known as *rating an exchanger*. An example that analyzes the performance of an existing exchanger is provided in Chapter 7.

Details on the above follow.

Film Coefficient Calculations. Individual coefficients, h_i and h_o, can be calculated by using empirical equations provided in Chapter 3, Part I, some of which are repeated below. Except for the viscosity term at the wall temperature, all of the physical properties in the equations that follow are evaluated at bulk temperatures. For the hot stream in the inner tube, the bulk temperature is $T_{H,bulk} = (T_1 + T_2)/2$. For the cold stream in the annulus, the bulk temperature is $t_{C,bulk} = (t_1 + t_2)/2$. Employing viscosity values at the wrong temperature can occasionally lead to substantial errors; however, the density and thermal conductivity of liquids do not vary significantly with temperature [1, 3].

The Reynolds number for both the cold and hot process streams must be calculated in order to determine whether the flow rate for each stream is in the laminar, turbulent, or transition region. Flow regimes for various Reynolds numbers appear in Table 6.2 [3].

In all of the equations that follow, calculations for the annulus require that the aforementioned equivalent or hydraulic diameter, $D_e = (D_{o,i} - D_{i,o})$, replace the tubular diameter, D. Thus, the Reynolds numbers, Re_i and Re_o, are defined as follows:

Table 6.2 Reynolds number values vs. type of flow.

Reynolds number, Re	Flow region
$Re < 2100$	Laminar
$2100 < Re < 10,000$	Transitional
$Re > 10,000$	Turbulent

Inner pipe:
$$\text{Re}_i = \frac{4\dot{m}}{\pi D_{i,i} \mu_i} \tag{6.5}$$

Annulus between pipes:
$$\text{Re}_o = \frac{4\dot{m}_o}{\pi \left(D_{o,i} - D_{i,o}\right)\mu_o} = \frac{D_e \dot{m}_o}{\mu_o S} \tag{6.6}$$

where,

$D_{i,i}$ = inside diameter of inner pipe, ft

$D_{i,o}$ = outside diameter of inner pipe, ft

$D_{o,i}$ = inside diameter of outer pipe, ft

D_e = equivalent diameter, ft

S = cross-sectional annular area, ft^2

μ = viscosity of hot or cold fluid at bulk temperature, lb/ft · hr

Similarly, the Nusselt numbers, Nu_i and Nu_o, are defined by the following equations:

Inner pipe:
$$\text{Nu}_i = \frac{h_i D_{i,i}}{k_i} \tag{6.7}$$

Annulus between pipes:
$$\text{Nu}_o = \frac{h_o \left(D_{o,i} - D_{i,o}\right)}{k_o} \tag{6.8}$$

where,

k = thermal conductivity at the bulk temperature of the hot or cold fluid, Btu/(hr·ft·°F)

Nu = Nusselt number, inside or outside

For *laminar flow*, the Nusselt number equals 4.36 for uniform surface heat flux $\left(Q/A\right)$ or 3.66 for constant surface temperature [1, 3]; this value should only be used for Graetz numbers, Gz, less than 10. For laminar flow with Graetz numbers from 10 to 1000, the following equation applies [1, 3].

$$\text{Nu} = 2.0 \text{Gz}^{1/3} \left(\frac{\mu}{\mu_{wall}}\right); \quad \text{Gz} = \frac{\dot{m}c_p}{kL} \tag{6.9}$$

where,

μ_{wall} = viscosity at wall temperature, lb/ft · hr
L = total length of tubular exchanger, ft
Gz = Graetz number

For *turbulent flow* (Re > 10,000), the Nusselt number may be calculated from the Dittus-Boelter equation if $0.7 \leq$ Pr ≤ 160 or the Sieder -Tate equation if $0.7 \leq$ Pr $\leq 16{,}700$ where Pr is the Prandtl number. Both equations are valid for L/D greater than 10 [1, 3].

Dittus-Boelter equation:

$$Nu = 0.023 \, Re^{0.8} Pr^{n}$$

$$St = 0.023 \, Re^{0.8} Pr^{-2/3}$$

$$Pr = \frac{c_p \mu}{k} \tag{6.10}$$

$$n = 0.4 \text{ for heating or } 0.3 \text{ for cooling}$$

Sieder-Tate equation:

$$Nu = 0.023 \, Re^{0.8} Pr^{1/3} \left(\frac{\mu}{\mu_{wall}} \right)^{0.14} \tag{6.11}$$

Note that Equation (6.10) is often written in terms of the Stanton number, St (occasionally referred to as the modified Nusselt number), where $St = Nu/(Re \cdot Pr) = h / \rho \overline{v} c_p$.

The Dittus-Boelter equation should only be used for small to moderate temperature differences. The Sieder-Tate equation applies for larger temperature differences [1, 3]. Errors as large as 25% are associated with both equations [4]. Other empirical equations with more complicated formulas and less error are available in the literature [1–6].

It should also be noted that there are numerous other correlations for calculating film heat transfer correlations. The reader is referred to Perry's [5] for some of these equations. Both Equation (6.10) or (6.11) requires a trail-and-error solution between the hot and cold streams, as each equation includes a viscosity term that is evaluated at the wall temperature, T_{wall}. An educated guess can be made for the wall temperature, or it can be estimated to be average of the four known temperatures:

$$\Delta T_i = \frac{t_{C,i} + t_{C,o} + T_{H,i} + T_{H,o}}{4} = \frac{t_1 + t_2 + T_1 + T_2}{4} \tag{6.12}$$

After the individual heat transfer coefficients are calculated, a new wall temperature can be calculated with the following equation [1, 3].

$$\Delta T_i = \frac{1/h_i}{(1/h_i)+(D_{i,i}/D_{i,o})h_o}\left(T_{H,bulk}-t_{C,bulk}\right) \tag{6.13}$$

with,

$$T_{wall} = T_{H,bulk} - \Delta T_i \tag{6.14}$$

The entire calculation is repeated several times until T_{wall} converges. As previously stated, all of the terms in the operation above may be evaluated at the bulk temperature of the hot and cold streams, except for μ_{wall} which is estimated at the wall temperature. Note that this correction factor is included to account for the distortion to the velocity profile that arises because of the viscosity variation with temperature.

Equation (6.15) below is based on experimental data, and was obtained for heating several oils in a pipe:

$$\frac{h_i D}{k} = 0.0115\left(\frac{DG}{\mu}\right)^{0.90}\left(\frac{c_p \mu}{k}\right)^{1/2} \tag{6.15}$$

Sieder and Tate made a later correlation by both heating and cooling a number of fluids, principally petroleum fractions in horizontal and vertical tubes, and arrived at Equation (6.16) for streamline flow when $DG/\mu < 2100$ [2].

$$\begin{aligned}
\frac{h_i D}{k} &= 1.86\left[\left(\frac{DG}{\mu}\right)\left(\frac{c\mu}{k}\right)\left(\frac{D}{L}\right)\right]^{1/3}\left(\frac{\mu}{\mu_w}\right)^{0.14} \\
&= 1.86\left[\frac{4}{\pi}\left(\frac{\dot{m}c_p}{kL}\right)\right]^{1/3}\left(\frac{\mu}{\mu_w}\right)^{0.14}
\end{aligned} \tag{6.16}$$

Here, L is the total length of the heat-transfer path before mixing occurs. Equation (6.16) gave maximum mean deviations of approximately ±12 percent from $Re = 100$ to $Re = 2100$, except for water. Beyond the transition range, the data may be extended to turbulent flow, resulting in Equation (6.17).

$$\frac{h_i D}{k} = 0.027 \left(\frac{DG}{\mu} \right)^{0.8} \left(\frac{c_p \mu}{k} \right)^{1/3} \left(\frac{\mu}{\mu_w} \right)^{0.14} \qquad (6.17)$$

Equation (6.15) gave maximum mean deviations of approximately ±15 and ±10 percent for the Reynolds numbers above 10,000. While Equation (6.16) and (6.17) were obtained for tubes, they have been used indiscriminately for pipes. Pipes are rougher than tubes and produce more turbulence at equal Reynolds numbers. Coefficients calculated from tube-data correlations are actually lower and more accurate than corresponding calculations based on pipe data. Equations (6.16) and (6.17) are applicable for organic liquids, aqueous solutions, and gases [2]. They are not consistent for water. Additional data for water will be provided later.

Equivalent Diameter and Film Coefficients for Fluids in Annuli. When a fluid flows in a conduit having other than a circular cross section, such as an annulus, it is convenient to express heat-transfer coefficients and friction factors by the same types of equations and curves used for flow in pipes and tubes. To permit this type of representation for annulus heat transfer it has been found advantageous to employ the aforementioned *equivalent diameter*, D_e. (See also the Introduction to this chapter.)

The equivalent diameter is four times the hydraulic radius, and the hydraulic radius is, in turn, the radius of a pipe equivalent to the annulus cross section. The hydraulic radius is defined as the ratio of the flow area to the wetted perimeter. For a fluid flowing in an annulus as shown in Figure 6.7, the flow area is $(\pi/4)(D_2^2 - D_1^2)$. It should be noted that in the development to follow the wetted perimeters for heat transfer and pressure drops are *different* [2].

For heat transfer, the wetted perimeter is the outer circumference of the inner pipe with diameter D_1. Therefore, for *heat transfer* in annuli,

$$D_e = 4r_h = \frac{4 \times flow\, area}{wetted\, perimeter}$$

$$= \frac{4\pi \left(D_2^2 - D_1^2 \right)}{4\pi D_1} = \frac{D_2^2 - D_1^2}{D_1} \qquad (6.18)$$

However, in pressure-drop calculations (which will be discussed more in a later section of this chapter) the friction not only results from the resistance

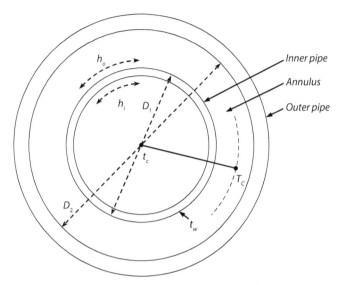

Figure 6.7 Annulus diameters and location of heat transfer coefficients.

of the outer pipe but is also affected by the outer surface of the inner pipe. Therefore, the total wetted perimeter is $4\pi\left(D_2 + D_1\right)$. As shown earlier, for the *pressure drop* in annuli,

$$D'_e = 4r_h = \frac{4 \times flow\ area}{frictional\ wetted\ perimeter}$$

$$= \frac{4\pi\left(D_2^2 - D_1^2\right)}{4\pi\left(D_2 + D_1\right)} = D_2 - D_1 \qquad (6.19)$$

The above development leads to the anomalous result that the Reynolds numbers for the same flow conditions, $\dot{m}$, G, and μ, are *different* for heat transfer and pressure drop since D_e might result in $Re > 2100$ while D'_e might result in $Re < 2100$. Actually, both Reynolds numbers should be considered only approximations, since the sharp distinction between streamline and turbulent flow at a Reynolds number in the range of 2100 is not completely valid in annuli.

Once the equivalent diameter has been obtained, the outside or annulus coefficient, h_o, is found in the same manner as h_i described earlier.

Area Requirements. In designing a double pipe heat exchanger, mass balances, heat balances, and the applicable heat transfer equation(s) are used. The steady-state heat balance equation is,

$$
\begin{aligned}
Q = \Delta \dot{h}_H &= \Delta \dot{h}_C \\
&= \Delta \dot{m}_h c_{p,h} \left(T_1 - T_2 \right) \\
&= \Delta \dot{m}_c c_{p,c} \left(t_2 - t_1 \right)
\end{aligned}
\tag{6.20}
$$

This equation assumes steady state, no heat loss, no viscous dissipation, and no heat generation [1]. The rate equation used to design an exchanger is the heat transfer equation, which includes the previously developed log mean temperature difference (ΔT_{LM} or LMTD) and overall heat transfer coefficient U,

$$
Q = U A \Delta T_{LM}
\tag{6.21}
$$

where Q is the heat load, U is the overall heat transfer coefficient, A is the heat transfer area, and ΔT_{LM} is the log mean (or global) temperature difference driving force. If the temperature difference driving forces are ΔT_1 and ΔT_2 at the entrance and exit of the heat exchanger, respectively, then (see also Chapter 5, Part I),

$$
\Delta T_{LM} = \frac{\Delta T_1 - \Delta T_2}{\ln \left(\Delta T_1 / \Delta T_2 \right)}
\tag{6.22}
$$

Note: if $\Delta T_1 = \Delta T_2$, then $\Delta T_{LM} = \Delta T_1 = \Delta T_2$.

The overall heat transfer coefficient U is usually based on the *inside* area of the tube; the A term in Equation (6.21) should then be based on the inside surface area. Typical values of U for new, clean exchangers are available in the literature and are often assigned the symbol, U_{clean}.

If the heat duties are known (or have been calculated), values for the overall heat transfer coefficient, U, can be calculated as follows:

$$
U_o = \frac{Q_C}{A_o \Delta T_{LM}}
\tag{6.23}
$$

$$
U_i = \frac{Q_H}{A_i \Delta T_{LM}}
\tag{6.24}
$$

where,

A = surface area for heat transfer (outside or inside), ft^2

Q = average heat duty (cold or hot), Btu/hr

U_o = overall heat transfer coefficient based on the outside area of the inner pipe, Btu/ft^2·hr·°F

U_i = overall heat transfer coefficient based on the inside area of the inner pipe, Btu/ft^2·hr·°F

In addition, one may write,

$$\frac{1}{UA} = \frac{1}{U_i A_i} = \frac{1}{U_o A_o} \qquad (6.25)$$

As described in Chapter 5 the overall heat transfer coefficient, U, for flow in a tube is related to the individual coefficients by the following equation:

$$\frac{1}{UA} = \frac{1}{h_i A_i} + \frac{R_{d,i}}{A_i} + \frac{\ln\left(D_{i,i} / D_{o,i}\right)}{2\pi kL} + \frac{R_{d,o}}{A_o} + \frac{1}{h_o A_o} \qquad (6.26)$$

where,

h_i = inside heat transfer coefficient, Btu/ft^2 · hr · °F

h_o = outside heat transfer coefficient, Btu/ft^2 · hr · °F

$R_{d,i}$ = fouling factor based on inner pipe surface

$R_{d,o}$ = fouling factor based on outer pipe surface

$D_{i,i} / D_{o,i}$ = (inside diameter/outside diameter) of inner pipe, ft

k = pipe thermal conductivity, Btu/ft · hr · °F

L = tube length, ft

If the resistance of the metal wall is small compared to the other resistances, and the fouling resistances are negligible, the above equation may be rewritten as:

$$\frac{1}{U} = \frac{A}{h_i A_i} + \frac{A}{h_o A_o} \qquad (6.27)$$

If the outside area of the inner pipe is used, A_o, then the relationship is simplified as:

$$\frac{1}{U_o} = \frac{D_o}{h_i D_i} + \frac{1}{h_o}$$

When the inside area, A_i, is used, then the relationship is simplified as:

$$\frac{1}{U_i} = \frac{1}{h_i} + \frac{D_i}{h_o D_o} \qquad (6.28)$$

Furthermore, if $D_0 \cong D_i$, the diameter ratio in Equation (6.28) may be assumed equal to unity, so that,

$$\frac{1}{U} = \frac{1}{h_i} + \frac{1}{h_o} \qquad (6.29)$$

or,

$$U = \frac{h_i h_o}{h_i + h_o} \qquad (6.30)$$

The locations of the coefficients and temperatures are shown in Figure 6.7. When U has been obtained from values of h_i and h_o, and Q and ΔT_{LM} are calculated from the process conditions, the surface A required for the process can be computed. Kern refers to the calculation of A as *design* [2].

Example 6.2 – Area Calculation. The following data is provided for a double pipe exchanger,

$$Q = 12{,}000 \text{ Btu / hr}$$

$$U = 48.0 \text{ Btu / hr} \cdot \text{ft}^2 \cdot {}^\circ\text{F}$$

$$\Delta T_{LM} = 50{}^\circ\text{F}$$

Calculate the area of the exchanger.

SOLUTION: Calculate the area of the exchanger by applying the heat transfer equation.

$$Q = UA\Delta T_{LM} \qquad (6.21)$$

Substitute the data provided and solve for A.

$$A = \frac{Q}{U \Delta T_{LM}}$$
$$= \frac{12,000}{(48.0)(50)}$$
$$= 5.0 \text{ ft}^2$$

Example 6.3 – Outlet Temperature Calculation. Calculate the outlet cold water temperature, $t_{C,o}$, flowing at a rate of 14.6 lb/min in a double pipe heat exchanger given the following data. Assume co-current operation.

$$t_1 = t_{C,i} = 63°F$$
$$T_1 = T_{H,i} = 164°F$$
$$T_2 = T_{H,o} = 99°F$$
$$Q = 56,760 \text{ Btu} / \text{hr}$$
$$U = 35.4 \text{ Btu} / (\text{hr} \cdot \text{ft}^2 \cdot °F)$$
$$A = 32.1 \text{ ft}^2$$

SOLUTION: Calculate ΔT_{LM} by applying Equation (6.21).

$$Q = UA\Delta T_{LM} \qquad (6.21)$$
$$\Delta T_{LM} = \frac{Q}{UA}$$
$$\Delta T_{LM} = \frac{56,760}{(32.1)(35.4)}$$
$$\Delta T_{LM} = 50.0°F$$

Apply Equation (6.22) and substitute for ΔT_1 and ΔT_2.

$$\Delta T_{LM} = \frac{\Delta T_2 - \Delta T_1}{\ln(\Delta T_2 / \Delta T_1)}$$
$$\Delta T_{LM} = \frac{(T_1 - t_1) - (T_2 - t_2)}{\ln[(T_2 - t_2)/(T_2 - t_2)]}$$

$$50 = \frac{(164-63)-(99-t_{C,o})}{\ln\left[(164-63)/(99-t_{C,o})\right]}$$

$$50 = \frac{101-(99-t_{C,o})}{\ln\left[101/(99-t_{C,o})\right]}$$

Solve for $t_{C,o}$ through a trial-and-error process or by using Microsoft Excel software's data solver (goal-seek).

$$t_{C,o} = t_2 = 79°F$$

Example 6.4 – Data Consistency. As Shakespeare once said: "Something is rotten in the state of…" Comment on whether the information provided in the previous example is "consistent."

SOLUTION: In addition to satisfying the heat transfer rate equation, the information provided must also satisfy the energy transfer (or conservation) equation. For this example,

$$Q = \dot{m}_C c_{p,C}\left(t_{C,o} - t_{C,i}\right) = \dot{m}c\left(t_2 - t_1\right)$$
$$= (14.6)(1.0)(79-63)$$
$$= 233.6 \text{ Btu / min}$$
$$= 14{,}016 \text{ Btu / hr}$$

This result does not agree with the Q provided in the problem statement. Shakespeare is right, something is indeed rotten. Also, as noted in Chapter 5, the heat capacity term c_p will more often be replaced by c throughout the rest of the book, mainly for the sake of keeping the notation simple. Lowercase c pertains to the cold fluid while capital-case C pertains to the hot fluid.

Example 6.5 – Heavy Hydrocarbon Oil and Area Calculation. A heavy hydrocarbon with a heat capacity, $C = 0.55$ Btu / lb·°F , is being cooled in a double pipe heat exchanger from $T_1 = 210°F$ to $T_2 = 170°F$. The oil flows inside a tube at a rate of 8000 lb/hr and the tube surface temperature is maintained at $60°F$. The overall heat transfer coefficient $U = 63$ Btu / $\left(\text{hr} \cdot \text{ft}^2 \cdot °F\right)$. Calculate the required heat transfer area, A, in ft².

SOLUTION: First calculate the LMTD.

$$\Delta T_1 = 210 - 60 = 150°F$$
$$\Delta T_2 = 170 - 60 = 110°F$$

Apply Equation (6.22).

$$\Delta T_{LM} = \frac{\Delta T_1 - \Delta T_2}{\ln(\Delta T_1 / \Delta T_2)} = \frac{150 - 110}{\ln(150/110)} = 129°F$$

Calculate the heat transferred or the duty. Apply Equation (6.20).

$$Q = \dot{M}C\Delta T = (8000)(0.55)(210 - 170) = 176{,}000 \text{ Btu / hr}$$

Write the describing equation for the area and substitute.

$$Q = UA\Delta T_{LM}$$

$$A = \frac{Q}{U\Delta T_{LM}}$$

$$= \frac{176{,}000}{(63)(129)} = 21.65 \text{ ft}^2$$

Example 6.6 – Cocurrent and Countercurrent Flow Calculation. Two variations of the design of a heat exchanger have been proposed. The unit is to cool a hot water stream from 140 °F to 110 °F through indirect contact with a cold water stream that is being heated from 60 °F to 90 °F. The water flow rate is 100 lb/min and the overall heat transfer coefficient may be assumed equal to 750 Btu/(hr · ft² · °F). Calculate the area requirements for the following two exchanger conditions:

1. Double pipe cocurrent flow
2. Double pipe countercurrent flow

Solution:

1. For co-current flow,

$$\Delta T_{LM} = \frac{\Delta T_1 - \Delta T_2}{\ln(\Delta T_1 / \Delta T_2)}$$

$$= \frac{(140 - 60) - (110 - 90)}{\ln[(140 - 60)/(110 - 90)]} = 43.3°F$$

The rate of heat transfer is,

$$Q = \dot{m}c\Delta t = (100)(1)(90-60)$$
$$= 3000 \text{ Btu / min}$$
$$= 180,000 \text{ Btu / hr}$$

The area for parallel (or cocurrent) flow is therefore,

$$A_{cocurrent} = \frac{Q}{U\Delta T_{LM}} = \frac{180,000}{(750)(43.3)} = 5.55 \text{ ft}^2$$

2. For countercurrent flow,

$$\Delta T_{LM} = 50°F \text{ (a constant*)}$$

and,

$$A_{countercurrent} = \frac{Q}{U\Delta T_{LM}} = \frac{180,000}{(750)(50)} = 4.80 \text{ ft}^2$$

*Note that whenever the temperature driving force on both the inlet and out-let is the same, ΔT_{LM} will return an error because the denominator will be 0. In this case, one should recall the reason why the log mean average was used in the first place, i.e., the ΔT at one end of the pipe varied from the ΔT at the other end of the pipe. When ΔT is constant from one end of the pipe to the other, there is no reason to use ΔT_{LM}, and a simple ΔT will suffice.

Example 6.7 – Flow Direction and Area Comparison. Compare and comment on the area requirements for the two flow conditions in the previous example.

SOLUTION: As expected, the countercurrent exchanger yields the smaller (more compact) design due to the higher driving force.

Example 6.8 – Trombone Heat Exchanger [7]. The double pipe heat exchanger shown in Figure 6.8 is used for heating 2500 lb/hr benzene from 60 °F to 120 °F. Hot water at 200 °F from a boiler is available for heating purposes in amounts up to 4000 lb/hr. Schedule-40 brass pipe and an integral number of 15-foot long sections are to be employed in the design of the exchanger.

The design is to be based on achieving a Reynolds number of 13,000 (turbulent flow) in both the inner pipe and the annular region. As an

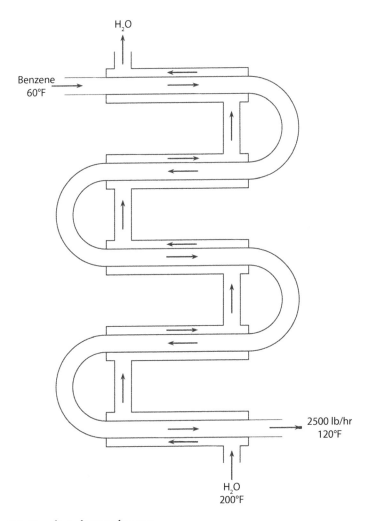

Figure 6.8 Trombone heat exchanger.

additional constraint, due to pressure drop considerations, the *width* of the annular region must be at least equal to one fourth the outside diameter of the inner pipe.

Note that one can obtain Q by calculating the duty of the benzene stream, and a heat balance will yield the exit temperature of the water, T_2. However, the area or the overall heat transfer coefficient is not known. Obtaining U becomes more complicated since information on flow rates/velocities, or the pipe diameter is not given.

Data is provided below for this problem, but fluid properties can be obtained from the literature or estimated from Tables in the Appendix.

Benzene at 90 °F:

$$\mu = 0.55 \text{ cP} = 3.70 \times 10^{-4} \text{ lb / ft} \cdot \text{s}$$
$$SG = 0.879 \text{ at } 68.8°F$$
$$\rho = 0.879(62.4) = 54.8 \text{lb / ft}^3$$
$$c = 0.415 \text{ Btu / lb} \cdot °F$$
$$k = 0.092 \text{ Btu /} \left(\text{hr} \cdot \text{ft}^2 \cdot °F \right) \text{ at } 86°F$$

Water at 200 °F:

$$\mu = 0.305 \text{cP} = 2.05 \times 10^{-4} \text{ lb / ft} \cdot \text{s}$$
$$\rho = 60.13 \text{lb / ft}^3$$
$$C = 1.0 \text{Btu / lb} \cdot °F$$
$$k = 0.392 \text{Btu /} \left(\text{hr} \cdot \text{ft}^2 \cdot °F \right)$$

SOLUTION: For flow inside tubes (pipes),

$$Re = \frac{4\dot{m}}{\pi D_{i,i} \mu} \tag{6.5}$$

Substituting,

$$D_{i,i} = \frac{4\dot{m}}{Re\, \pi\mu} = \frac{4(2500 / 3600)}{(13,000)(\pi)(3.7 \times 10^{-4})} = 0.184 \text{ft} = 2.20 \text{in}$$

The closest standard pipe with $D_{i,i} = 2.20$ in is the 2 in schedule-40 pipe (see also AT.7 in the Appendix). Thus,

$$D_{i,i} = 2.067 \text{in}$$

This results in,

$$Re = 13,874$$

For 2-inch schedule-40 pipe,

$$A_i = 0.541 \, \text{ft}^2 / \text{ft}$$

$$A_o = 0.622 \, \text{ft}^2 / \text{ft}$$

$$D_i = 2.067 \, \text{in}$$

$$D_o = 2.375 \, \text{in}$$

$$S_i = 0.0233 \, \text{ft}^2 \, (\text{inside flow area})$$

$$\Delta x_w = 0.154 \, \text{in} = 0.0128 \, \text{ft}$$

For flow in the annular region, the equivalent form of Equation (6.6) is,

$$\text{Re} = \frac{D_e \bar{v}_p \rho}{\mu} = \frac{D_e \dot{M}}{\mu S}$$

with,

$$D_e = 4r_H = \frac{4S}{L_p} = \frac{4S}{\pi \left(D_{o,i} + D_{i,o} \right)}$$

Therefore,

$$\text{Re} = \frac{4\dot{M}}{\pi \left(D_{o,i} + D_{i,o} \right) \mu}$$

However, note the constraint:

$$\frac{1}{2} \left(D_{o,i} + D_{i,o} \right) \geq \frac{1}{4} D_{o,i}$$

or,

$$D_{i,o} \geq \frac{3}{2} D_{o,i} = 3.562 \, \text{in}$$

The smallest pipe, which satisfies this constraint, is 4 in schedule 40 pipe $\left(D_{i,o} = 4.026 \, \text{in} \right)$. The outer pipe is therefore (see Appendix once again) 4 in schedule-40 pipe.

For Re = 13,000, the water flow is:

$$\dot{M} = \frac{\text{Re} \, \pi \mu \left(D_{o,i} + D_{i,o} \right)}{4}$$

$$= \frac{(13,000)(\pi)(2.05\times10^{-4})\left[(2.375/12)+(4.206/12)\right]}{4}$$

$$= 4019\,\text{lb}/\text{hr}$$

This is close enough to 4000 lb/hr.
The outlet water temperature is obtained from an energy balance:

$$Q = \dot{m}c\Delta t = \dot{M}C\Delta T$$
$$= (2500)(0.415)(60)$$
$$= 62,250\,\text{Btu}/\text{hr}$$

From the water energy balance, with $C = 1.0$,

$$\Delta T = \Delta T_H = 62,250/4000 = 15.6°\text{F}$$
$$T_2 = T_{H,\text{out}} = 200 - 15.6 = 184.4°\text{F}$$
$$\bar{T} = T_{H,\text{avg}} = (200 + 184.4)/2 = 192.2°\text{F}$$

Properties of water at 192 °F are estimated from the data at 200 °F. At 200 °F,

$$\mu = 0.322\,\text{cP} = 2.16\times10^{-4}\,\text{lb}/\text{ft}\cdot\text{s}$$
$$\rho = 60.13\text{lb}/\text{ft}^3$$
$$C = 1.0\,\text{Btu}/\text{lb}\cdot°\text{F}$$
$$k = 0.390\,\text{Btu}/\left(\text{hr}\cdot\text{ft}^2\cdot°\text{F}\right)$$

The Dittus-Boelter equation is employed to determine the heat transfer coefficients.

$$St = 0.023\,Re^{-0.2}\,Pr^{-0.667}$$

For the inside pipe (benzene):

$$Re = 13,000$$

$$Pr = \frac{c\mu}{k} = \frac{(0.415)(3.70\times10^{-4})}{0.092/3600} = 6.01$$

$$St = (0.023)(13,000)^{-0.2}(6.01)^{-0.667} = 0.00105$$

$$h_i = \frac{St\,c\dot{m}}{s} = \frac{(0.00105)(0.415)(2500)}{0.0233} = 46.7\,\text{Btu}/\left(\text{hr}\cdot\text{ft}^2\cdot{}^\circ\text{F}\right)$$

For the annular region (water):

$$Re = \frac{4\dot{M}}{\pi\mu\left(D_{o,i}+D_{i,o}\right)} = \frac{4(4000/3600)}{\pi(2.16\times10^{-4})\left[(2.375/12)+(4.206/12)\right]}$$

$$= 12,279$$

$$Pr = \frac{C\mu}{k} = \frac{(1.0)(2.16\times10^{-4})}{0.390/3600} = 1.99$$

$$St = (0.023)(12,279)^{-0.2}(1.99)^{-0.667} = 0.00221$$

$$h_i = \frac{St\,C\dot{M}}{S_{annulus}} = \frac{(0.00221)(1.0)(4000)}{0.0576} = 153\,\text{Btu}/\left(\text{hr}\cdot\text{ft}^2\cdot{}^\circ\text{F}\right)$$

Neglect fouling and employ a modified form of Equation (6.26).

$$\frac{1}{U_o} = \frac{D_o}{D_i h_i} + \frac{\Delta x_w D_o}{k_w D_{LM}} + \frac{1}{h_o}$$

For the pipe,

$$D_{LM} = \frac{2.375-2.067}{\ln(2.375/2.067)} = 2.217\,\text{in}$$

Substituting,

$$\frac{1}{U_o} = \frac{2.375}{(2.067)(46.7)} + \frac{(0.0128)(2.375)}{(26)(2.217)} + \frac{1}{153}$$

$$= 0.0246 + 0.0005274 + 0.00654$$

$$= 0.0316\,\text{hr}\cdot\text{ft}^2\cdot{}^\circ\text{F}/\text{Btu}$$

Therefore,

$$U_o = 31.6\,\text{Btu}/\left(\text{hr}\cdot\text{ft}^2\cdot{}^\circ\text{F}\right)$$

Finally, to obtain the length use,

$$Q = U_o A_o \Delta T_{LM} \tag{6.21}$$

$$\Delta T_{LM} = \frac{124.4 - 80}{\ln(124.4 / 80)} = 100.6°F$$

The heat exchanger equation is again employed to calculate the length:

$$A_o = 0.622L$$

$$62,250 = (31.6)(0.622 \text{ L})(100.6)$$

$$L = 31.5 \text{ ft}$$

Example 6.9 – Trombone Sections. With reference to the previous example, how many sections should be required in the design?

SOLUTION: Two sections would do the job. However, three sections might be recommended; the unit would then be somewhat over-designed.

6.3 Calculations of Exit Temperatures

The equations describing exit temperatures for parallel flow heat exchangers follows [7]. These equations are based on three assumptions:

1. Streams do not experience a phase change from inlet and exit.
2. Heat capacities of both streams are constant.
3. A single average U applies for the entire exchanger.

For parallel (p) flow, one notes (with capital and lowercase letters representing the hot and cold fluid, respectively, as per Famularo's [7] notation),

$$\Delta T_1 = T_1 - t_1$$
$$\Delta T_2 = T_2 - t_2 \tag{6.31}$$

In addition,

$$Q_p = \dot{M}C(T_1 - T_2) = \dot{m}c(t_2 - t_1) \tag{6.32}$$

or,

$$\frac{T_1 - T_2}{B} = \frac{t_2 - t_1}{b}$$

with,

$$B = \frac{1}{\dot{M}C}$$

$$b = \frac{1}{\dot{m}c}$$

Combining the above equations with the heat exchanger equation $\left(Q = UA\Delta T_{LM}\right)$ leads to,

$$\ln\left(\frac{\Delta T_2}{\Delta T_1}\right) = -\left(B + b\right)UA \tag{6.33}$$

Combining Equations (6.32) and (6.33) gives,

$$T_2 = \frac{\left[\left(b/B\right) + e^{-UA(B+b)}\right]T_1 + \left[1 - e^{-UA(B+b)}\right]t_1}{1 + \left(b/B\right)} \tag{6.34}$$

The cold fluid outlet temperature may be calculated by replacing T_2, T_1, and t_1 in Equation (6.35) by t_2, t_1, and T_1, respectively [7].

For countercurrent (c) flow, one notes,

$$\Delta T_1 = T_1 - t_2$$
$$\Delta T_2 = T_2 - t_1 \tag{6.35}$$

Once again,

$$Q_c = \dot{M}C\left(T_1 - T_2\right) = \dot{m}c\left(t_2 - t_1\right) \tag{6.36}$$

or,

$$\frac{T_1 - T_2}{B} = \frac{t_2 - t_1}{b}$$

Combining the above equations with the heat transfer equations leads to,

$$\ln\left(\frac{\Delta T_2}{\Delta T_1}\right) = -(B-b)UA \tag{6.37}$$

Combining Equation (6.34) and (6.36) gives,

$$T_2 = \frac{\left[(b/B)-1\right]T_1 + \left[(B/b)e^{UA(B-b)} - (B/b)\right]t_1}{(B/b)e^{UA(B-b)} - 1} \tag{6.38}$$

The cold fluid outlet temperature may be calculated by replacing T_2, T_1, and t_1 in Equation (6.38) by t_2, t_1, and T_1, respectively [7].

Another application involves a simplified case in which the fluid temperature on one side of the dividing wall (tube/pipe) is constant. If one can assume that,

$$\dot{M}C \gg \dot{m}c$$

or,

$$\frac{1}{\dot{M}C} \ll \frac{1}{\dot{m}c} \tag{6.39}$$

and $T_2 = T_1 = T$ and $B \ll b$, the co-current case reduces to,

$$\ln\left(\frac{T-t_2}{T-t_1}\right) = -bUA \tag{6.40}$$

In addition, the countercurrent case reduces to,

$$\ln\left(\frac{T-t_1}{T-t_2}\right) = bUA \tag{6.41}$$

If the cold stream is flowing inside the inner pipe, then $UA = U_i A_i$, so that,

$$t_2 = T - (T-t_1)e^{-U_i A_i / \dot{m}c} \tag{6.42}$$

This equation applies to both the co-current and countercurrent case. There are two other situations in which the fluid temperature is constant and the above equation applies:

1. Condensation of a vapor at T_s without sub-cooling. Here, set $T = T_s$ in Equation (6.42).
2. Outside flow at T_∞ normal to a tube. Here, set $T = T_\infty$ in Equation (6.42).

Example 6.10 – Exit Temperature Calculation - Countercurrent Flow. Calculate T_2 and t_2 (in consistent units) for the following countercurrent flow system.

$$\dot{M}C = 2000 \, \text{Btu} \, / \, \text{hr} \, \cdot \, °\text{F}$$

$$\dot{m}c = 1000 \, \text{Btu} \, / \, \text{hr} \, \cdot \, °\text{F}$$

$$U = 2000 \, \text{Btu} \, / \, \text{hr} \, \cdot \text{ft}^2 \cdot \, °\text{F}$$

$$A = 10 \, \text{ft}^2$$

$$T_1 = 300 \, °\text{F}$$

$$t_1 = 60 \, °\text{F}$$

Solution: Employ Equation (6.38).

$$B = 0.001, \quad b = 0.0005, \quad \text{and} \quad B/b = 2.0$$

In addition,

$$UA(B-b) = 2000(0.001 - 0.0005) = 1.0$$

so that,

$$e^{-UA(B-b)} = e^{1.0} = 2.7183$$

Substituting into Equation (6.38) gives,

$$T_2 = \frac{(2-1)(300) + 2(2.7183 - 1)(60)}{2(2.7183) - 1}$$

$$= \frac{506.2}{4.4366} = 114.1°\text{F} \approx 114°\text{F}$$

Employ a revised form of Equation (6.38) or an overall energy balance to generate t_2,

$$t_2 = t_1 + \frac{T_1 - T_2}{(B/b)} = 60 + \frac{300 - 114.1}{2}$$

$$= 152.95°F \approx 153°F$$

Example 6.11 – Exit Temperature Calculation - Cocurrent Flow. The following information is provided for a cocurrent flow double pipe heat exchanger. The pipe consists of 200 ft of 2-inch schedule 40 pipe with $k = 25\,\text{Btu}/(\text{hr}\cdot\text{ft}\cdot°F)$. The hot and cold film coefficients are 1200 and 1175 $\text{Btu}/(\text{hr}\cdot\text{ft}^2\cdot°F)$, respectively. Calculate T_2 and t_2 given the following data.

$$\dot{m}c = 22,300\,\text{Btu}/\text{hr}\cdot°F$$

$$T_1 = 300°F$$

$$t_1 = 60°F$$

Solution: First calculate the overall inside heat transfer coefficient. For the pipe (see also Table AT.7 in the Appendix),

$$D_i = 2.067\,\text{in}$$

$$D_o = 2.375\,\text{in}$$

$$\Delta x = 0.154\,\text{in}$$

$$A_i' = 0.541\,\text{ft}^2 / \text{ft}$$

$$k = 25\,\text{Btu}/(\text{hr}\cdot\text{ft}\cdot°F)$$

Neglect fouling and apply a modified form of Equation (6.26).

$$\frac{1}{U_i} = \frac{1}{h_i} + \frac{\Delta x}{k} + \frac{1}{h_o(D_o/D_i)}$$

$$U_i = \frac{1}{\dfrac{1}{h_i} + \dfrac{\Delta x}{k} + \dfrac{1}{h_o(D_o/D_i)}}$$

$$U_i = \frac{1}{\dfrac{1}{1200} + \dfrac{(0.154)(1\,\text{ft}/12\,\text{in.})}{(25)} + \dfrac{1}{(1175)(2.375/2.067)}}$$

$$= 482\,\text{Btu}/(\text{hr}\cdot\text{ft}^2\cdot°F)$$

In addition,

$$A_i = A_i' L$$
$$= (0.541)(200)$$
$$= 108.2\,\text{ft}^2$$

Thus, there are two key equations–one based on an energy balance and one based on rate considerations. These two equations are:

$$Q_H = Q = \dot{M}C(300 - T_2)$$
$$= (30,000)(300 - T_2)$$
$$Q_C = 22,300(t_2 - 60)$$

with,

$$Q_C = Q_H$$

and,

$$Q = U_i A_i \Delta T_{\text{LM}}$$
$$= (482)(108.2)\left\{ \frac{(300 - 60) - (T_2 - t_2)}{\ln\left[(300 - 60)/(T_2 - t_2)\right]} \right\}$$

There are also the two unknowns: T_2 and t_2. These can be solved by any suitable numerical method. By trail-and-error, one obtains,

$$T_2 = 200°F$$
$$t_2 = 195°F$$

with,

$$Q = 3.0 \times 10^6\,\text{Btu / hr}$$

Example 6.12 - Applying Famularo's Procedure. Solve the previous example employing the above procedure provided by Famularo [7]. Also calculate the discharge temperature, the LMTD, and Q for the *countercurrent* case.

SOLUTION: Equation (6.34) applies for cocurrent case,

$$T_2 = \frac{\left[(b/B) + e^{-UA(B+b)}\right]T_1 + \left[1 - e^{-UA(B+b)}\right]t_1}{1 + (b/B)} \qquad (6.34)$$

Using the expressions for "B" and "b" developed from Equation (6.32): $B = 3.33 \times 10^{-5}$, $b = 4.48 \times 10^{-5}$, and $A = 108.2$ and $U = 482$, one obtains:

$$T_2 = 199.4°F \approx 200°F$$

Apply an energy balance to both fluids.

$$(30{,}000)(300 - T_2) = (22{,}300)(t_2 - 60)$$

For $T_2 = 200°F$, one obtains:

$$t_2 = 195°F$$

The agreement between both methods is excellent. Note that the temperature t_2 may also be calculated directly from Equation (6.34) with the temperature(s) appropriately reversed. Substitute into Equation (6.38) to obtain T_2 for the countercurrent case:

$$T_2 = \frac{\left[(b/B) - 1\right]T_1 + \left[(B/b)e^{UA(B-b)} - (B/b)\right]t_1}{(B/b)e^{UA(B-b)} - 1} \tag{6.38}$$

Once again, for $B = 3.33 \times 10^{-5}$, $b = 4.48 \times 10^{-5}$, $A = 108.2$ and $U = 482$, one obtains,

$$T_2 = 164.20°F \approx 164°F$$

Substitute into Equation (6.38) with the temperature reversed to generate t_2:

$$t_2 = \frac{(b/B)\left[e^{UA(B-b)} - 1\right]T_1 + \left[(B/b) - 1\right]t_1}{(B/b)e^{UA(B-b)} - 1}$$

Substituting gives,

$$t_2 = 242.85°F \approx 243°F$$

For LMTD (see Equation 6.22),

$$\Delta T_{LM} = \frac{(T_2 - t_1) - (T_1 - t_2)}{\ln\left(\dfrac{T_2 - t_1}{T_1 - t_2}\right)}$$

Substituting,

$$T_2 - t_1 = 164 - 60 = 104$$
$$T_1 - t_2 = 300 - 243 = 57$$
$$\Delta T_{LM} = \frac{104 - 57}{\ln(104/57)} = 78.2°F$$

Apply the heat transfer equation to calculate Q.

$$Q = UA\Delta T_{LM}$$
$$= \left(482\,\text{Btu}/\text{hr}\cdot\text{ft}^2\cdot°F\right)\left(108.2\,\text{ft}^2\right)\left(78.2°F\right)$$
$$= 4{,}078{,}000\,\text{Btu}/\text{hr}$$

This result may be checked by employing the heat conservation equation.

$$Q = \dot{M}C\left(T_1 - T_2\right)$$
$$= \left(30{,}000\,\text{Btu}/\text{hr}\cdot°F\right)\left(300°F - 164.20°F\right)$$
$$= 4{,}074{,}000\,\text{Btu}/\text{hr}$$

The two results are once again in reasonable agreement.

6.4   Pressure Drops in Pipes and Annuli

The *pressure drop* allowance in an exchanger is the static fluid pressure which may be expended to drive the fluid through the exchanger. The pump selected for the circulation of a process fluid is one which develops sufficient head at the desired capacity to overcome the frictional losses caused by connecting piping, fittings, control regulators, and the pressure drop in the exchanger itself. To this head must be added the static pressure at the end of the line such as the elevation or pressure of the final receiving vessel [6]. Once a definite pressure drop allowance has been designated for an exchanger as a part of a pumping circuit, it should always be utilized as completely as possible in the exchanger since it will otherwise be blown off or expanded through a pressure reducer.

The pressure drop in pipes can be computed from the Fanning Equation [Equation (1.54)] using an appropriate value of f from Table 3.5, depending upon the type of flow. For the pressure drop in fluids flowing in annuli, replace D_e in the Reynolds number by D_e' to obtain f. The Fanning equation may then be modified to give [6],

$$\Delta P_{\text{Fanning}} = \Delta F = \frac{4 f L \bar{v}^2}{2 g D_e'} = \frac{4 f G^2 L}{2 g \rho^2 D_e'} \tag{6.43}$$

The reader should note that in the calculation demonstrations that will be shown throughout the remainder of this book, pressure drop values obtained directly from the Fanning equation [Equation (6.43)] will use a notation of ΔF, which typically has a unit of psf. ΔF can then be converted to ΔP which typically has a unit of psi. In other words, ΔF and ΔP are both pressure drops, but with different units.

Furthermore, for turbulent flow, the pressure drop and mass velocity can have a proportional relationship to each other in the following manner:

$$\Delta P \propto G^2 \tag{6.44}$$

However, Equation (6.44) shows an approximate (not fully accurate) relationship between ΔP and G since f varies somewhat with DG / μ for turbulent flow. Regardless of the possible variations, it is apparent that the best use of available pressure is to increase the mass velocity, which also increases h_i and reduces the size and cost of the apparatus. Standard allowable pressure drops are as follows [2]:

1. 5 to 10 psi is a customary allowable pressure drop for an exchanger or battery of exchangers fulfilling a single process service (except where the flow is by gravity).
2. 10 psi is fairly standard for each pumped system.
3. For gravity flow, the allowable pressure drop is determined by the elevation of the storage vessel above the final outlet, z_h, in feet of fluid. The feet of fluid may be converted to pounds per square inch by multiplying z_h by 144ρ, where ρ is the density of the fluid.

When several pipe exchangers are connected in series, annulus to annulus and pipe to pipe as in Figure 6.9, the length in Equation (6.43) is the total for the entire path.

The pressure drop computed by Equation (6.43) does not include the pressure drop encountered when the fluid enters or leaves exchangers. For the inner pipes of double pipe exchangers connected in series, the entrance loss is usually negligible, but for annuli it may be significant. The allowance of

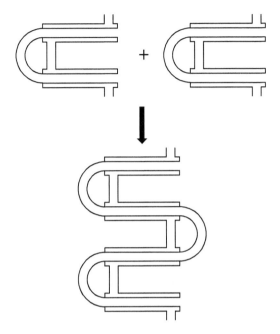

Figure 6.9 Connecting exchangers in series.

pressure drop of one velocity head, $\bar{v}^2 / 2g'$, per hairpin will ordinarily suffice - where $g' = 32.2$ ft/sec^{-2} [2].

Consider the following example. Water flows in an annulus with a mass velocity of 720,000 lb/hr·ft^2. Since $\rho = 62.5$ lb/ft^3 (approximately),

$$\bar{v} = \frac{G}{3600\rho} = \frac{720,000}{(3600)(62.4)} = 3.2\,\text{fps}$$

The pressure-drop per hairpin would then be,

$$\frac{\bar{v}^2}{2g'} = \frac{(3.2)^2}{(2)(32.2)} = 0.159\,\text{ft}\,H_2O \text{ or } 0.07\,\text{psi}$$

Unless the velocity is well above 3 fps, the entrance and exit losses may be neglected.

The reader should note that Equation (6.43) — Fanning's Equation employing Fanning's friction factor — is recommended by Kern to calculate the pressure drop. However, his approach to calculating pressure drop for shell and tube exchangers (see Chapter 7) uses a modified form of Darcy's friction factor. This is noted for the reader in the next chapter to help reduce the confusion created by this change.

6.5   Open-Ended Problems

This section introduces the concept of open-ended problems. The authors decided to include open-ended material since they believe chemical engineers in the future will have to be innovative and creative in order to succeed in their careers. One approach to developing the chemical engineer's ability to solve unique problems is by employing open-ended problems. Although the term "open-ended problem" has come to mean different things to different people, it describes an approach to the solution of a problem and/or situation where there is usually not a unique solution. Two of the authors of this text have applied this approach by including numerous open-ended problems in several of their courses. Although the literature is inundated with texts emphasizing theory and theoretical derivations, the goal of this section is to present the subject of open-ended problems from a pragmatic point-of-view in order to better prepare the reader for the future when dealing with unique heat exchanger applications. This section also includes a review of Wilson's method [9]. Six examples in this section complement the presentation.

Example 6.13 – Troubleshooting with a Heat Exchanger: Discharge Temperature of Cold Stream. It was recently determined that a double pipe heat exchanger at a facility was no longer operating within design specifications. The cold stream (on the annular side, surprisingly) was only being heated to 125 °F, not 135 °F as required for the process. A 180 °F hot fluid was providing the heat and was *cocurrently* travelling through the inner tube. Justify what steps one can take as the Plant Manager to increase the discharge temperature of the cold fluid.

SOLUTION: Since the flow is cocurrent, the first recommendation is to attempt to convert the flow pattern to countercurrent. This will increase ΔT_{LM}, as noted earlier in Chapter 5.

If a flow direction change is not possible, or does not provide the required increase in temperature, the flow rate of the heating medium should be increased. The increased flowrate will increase both U *and* ΔT_{LM}, resulting in Q increasing. Although any increase in the discharge temperature of the cold fluid could decrease ΔT_{LM} it would automatically increase Q via the energy equation. Thus, any increase in the hot fluid flowrate will produce a higher Q.

Example 6.14 – Troubleshooting with a Heat Exchanger: Discharge Temperature of Hot Stream. Corenza Partners designed a countercurrent double pipe heat exchanger to operate with a maximum discharge temperature of a hot stream to be cooled to 90 °F. Once the unit was purchased, installed, tested, and running, the exchanger operated with a discharge temperature of 105 °F. Rather than purchase a new exchanger, what operation options are available to bring the unit into compliance with the specified design temperature?

SOLUTION: Depending on the type of heat exchanger, the approaches to bring the heat exchanger back to design specification are different. However, most of them include the following:

1. Cleaning of the heat exchanger tubes/pipes
2. Installation or removal of insulation
3. Correction of any leakage problems
4. Upgrading or replacement of gaskets
5. Increasing the mass flow rate of the cooling fluid (see also Example 6.13)
6. Change the state of the cooling medium (e.g., have colder water pass through the heat exchanger).

Example 6.15 – Analysis of a Double-Pipe Heat Exchanger. The following is an open-ended heat transfer final exam question prepared by one of the authors [8].

A double-pipe heat exchanger is to be designed to condense a specified quantity of saturated steam at T_c. This T_c in the annular region of the exchanger sets the value of Q. U was estimated from a correlation in the literature, and a 20 °F approach temperature $(T_c - t_2)$ for the coolant outlet was specified. The required exchanger area, A, was then calculated. The appropriate exchanger was purchased, installed, tested, and put into operation. However, it soon became apparent that the exchanger was *not* condensing the required quantity of steam.

You have been asked, as part of your final exam, to provide a *qualitative* technical analysis of the describing equation(s) that will produce recommendation(s) to alleviate this deficiency and satisfy the original design specification.

SOLUTION: The key equation is,

$$Q = UA\Delta T_{\text{LM}} \tag{6.24}$$

The objective is to somehow increase Q, given that A — the area available for heat transfer — is a constant. Options available include:

1. Increasing U
2. Increasing ΔT_{LM}
3. Increasing both U and ΔT_{LM}
4. Increasing either U or ΔT_{LM} more than a corresponding decrease in ΔT_{LM} or U, respectively.

Whichever option is to be selected, it is assumed that any change in the pressure drop associated with the option will not affect either the operation or performance of the exchanger

Example 6.16 – Quantitative Discussion of Heat Exchanger Characteristics. Refer to the operation/performance aspect of the heat exchanger in the previous example. Paul Farber, an authority in the heat exchanger field, has suggested that doubling the coolant flow rate will increase Q and perhaps satisfy the original design criteria. Comment on Paul's recommendation.

SOLUTION: There are three equations that arise in the development that follows (employing earlier notation).

1. The continuity or conservation law for mass equation:

$$\dot{m} = \rho A_c \bar{v}; \quad A_c = \text{area available for flow} \tag{6.45}$$

2. The conservation law for energy:

$$Q = \dot{m}c_p \left(t_2 - t_1 \right) = \dot{m}c_p \Delta t \tag{6.46}$$

3. The heat transfer equation:

$$Q = UA\Delta T_{\text{LM}} \tag{6.47}$$

where,

$$\Delta T_{\text{LM}} = \frac{t_2 - t_1}{\ln\left(\dfrac{T_c - t_1}{T_c - t_2} \right)}; \quad T_c = \text{condensing temperature}$$

The term U in Equation (6.47) compounds this problem/analysis since there are three terms that affect the numerical value of U — the condensing medium, the wall, and the coolant fluid. In addition, the term U can be constrained by two values:

1. The resistance is present only in coolant fluid (i.e., $h_o \rightarrow \infty$ and $k \rightarrow \infty$). For this condition, the Dittus-Boelter equation is employed to estimate h_i:

$$U = h_i = a \cdot \bar{v}^{0.8} = f(\bar{v})$$

(6.48)

This will be referred to as Condition (1) in the developments that follow in this and the next chapter.

2. The resistance of the coolant fluid is negligible so that $h_i \rightarrow \infty$ and

$$U = h_i \neq f(\bar{v})$$

(6.49)

In effect U is no longer a function of the coolant's velocity (i.e., it does *not* increase as $\bar{v}$ increases); it is constant. This will be referred to as Condition (2) in the developments that follow.

Combining Equations (6.45) thru (6.48) above for Condition (1) results in the following equation:

$$Q = \rho \bar{v} A_c c_p \left(t_2 - t_1\right) = a\left(\bar{v}^{0.8}\right) A \left\{ \frac{t_2 - t_1}{\ln\left[\left(T_c - t_1\right)/\left(T_c - t_2\right)\right]} \right\}$$

(6.50)

Equation (6.50) may be rearranged to

$$\ln\left(\frac{T_c - t_1}{T_c - t_2}\right) = \left(\frac{a\left(\bar{v}^{-0.2}\right)}{\rho c_p}\right)\left(\frac{A}{A_c}\right) = \frac{B}{\bar{v}^{0.2}}$$

(6.51)

where,

$$B = \frac{a}{\rho c_p}\left(\frac{A}{A_c}\right)$$

(6.52)

Equation (6.51) may also be rearranged to yield,

$$\frac{T_c - t_1}{T_c - t_2} = e^{B/\bar{v}^{0.2}} \tag{6.53}$$

The above analysis may now be applied to this example. For this case, as $\bar{v}$ increases, Equation 6.53 predicts that the resulting cold fluid outlet temperature t_2, will decrease below its initial value. Although Δt has decreased, $\dot{m}$ has doubled — producing an increase in Q.

With reference to the problem question, first note that since the coolant film coefficient is proportional to the (velocity)$^{0.8}$ or $\bar{v}^{0.8}$, U would increase. In addition, the increased flow rate would produce a lower cold fluid outlet temperature resulting in a larger ΔT_{LM}. Now, examine the heat transfer equation.

$$Q = UA\Delta T_{LM}$$

Since A is constant, and both U and ΔT_{LM} have increased, Q will correspondingly increase.

Farber [10] computerized the above equation for a specified set of conditions and the results are presented in Figure 6.10. One notes that doubling the flow has produced only a small decrease in Δt, not nearly compensating for the

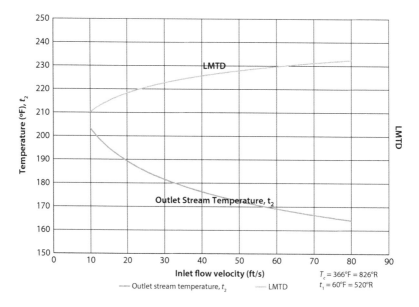

Figure 6.10 Outlet Temperature and LMTD for Example 6.16.

doubling (increasing) of the mass flow rate (see also Equation (6.46)). The same conclusion is drawn from Equation (6.47) where U and ΔT_{LM} have increased.

The same conclusion can also be drawn for Condition (2). Here, the RHS (right-hand side) of Equation (6.49) is given by a constant since $\bar{v}$ is constant. Once again, doubling the velocity produces a decrease in t_2 but an increase in ΔT_{LM}. Thus, Q also increases for Condition (2) since this decrease in t_2 (or t) results in a greater ΔT_{LM}.

To help the reader visualize ΔT_{LM} of Condition (2), one could represent ΔT_{LM} via the area between two curves, similar to the temperature profile graphs shown earlier in this chapter. In Figure 6.11, the shaded region represents the extra (or additional) ΔT_{LM} gained from the increased flow rate of the coolant.

Wilson's Method [1, 3]. There is a procedure for evaluating the *outside* film coefficient for a double pipe unit. Wilson's method [9] is a graphical technique for evaluating this coefficient. The inside coefficient is a function

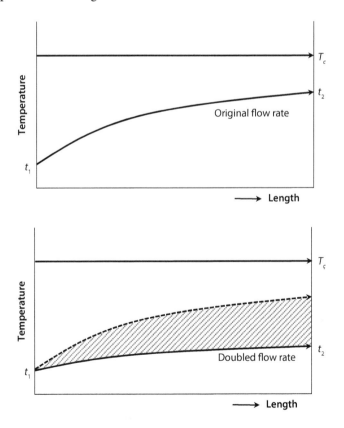

Figure 6.11 Comparison of ΔT_{LM}; Example 6.16.

of the Reynolds and Prandtl numbers via the Dittus-Boelter equation presented in Equation (6.10), i.e.,

$$h_i = f\left(\mathrm{Re}^{0.8}\mathrm{Pr}^{0.3}\right) \tag{6.54}$$

A series of experiments can be carried out on a double pipe exchanger where all conditions are held relatively constant except for the velocity (v) of the cooling (in this case) inner stream. Therefore, for the proposed experiment:

$$h_i = f\left(\mathrm{Re}^{0.8}\right) = f\left(\bar{v}^{0.8}\right) = f\left(\dot{m}\right); \ \ \mathrm{Re} = \frac{D\bar{v}\rho}{\mu} \tag{6.55}$$

or, in equation form,

$$h_i = a\bar{v}^{0.8} \tag{6.56}$$

where a is a constant. Equation (6.56) can be substituted into the overall coefficient equation,

$$\frac{1}{U_o A_o} = R_o + R_w + R_i$$

so that,

$$\frac{1}{U_o A_o} = \frac{1}{h_o A_o} + \frac{\Delta x_w}{k A_{LM}} + \frac{1}{a\bar{v}^{0.8} A_i} \tag{6.57}$$

Data can be generated at varying velocities. By plotting $1/U_o A_o$ versus $1/\bar{v}^{0.8}$, a straight line should be obtained since the first two terms on the right hand side of Equation (6.57) are constants. The intercept of this line corresponds to an infinite velocity and an inside resistance of zero. Thus, the above equation may be rewritten as,

$$\left(\frac{1}{U_o A_o}\right)_{\mathrm{intercept}} = \frac{1}{h_o A_o} + \frac{\Delta x_w}{k A_{LM}} \tag{6.58}$$

The second term on the right-hand side of Equation (6.58) is known and/or can be calculated, and h_o can then be evaluated from the intercept (details on this calculational scheme are provided in the illustrative

Table 6.3 Wilson Method Information: Example 6.17.

Run	$1/U_oA_o$, °F·hr/Btu	T, °F average temperature	$1/\bar{v}$, hr/ft
1	1.2176×10^{-3}	124.5	2.29×10^{-4}
2	0.9454×10^{-3}	125	1.65×10^{-4}
3	0.9366×10^{-3}	129	1.29×10^{-4}
4	0.8618×10^{-3}	121.5	1.117×10^{-4}
5	0.7966×10^{-3}	122	0.962×10^{-4}

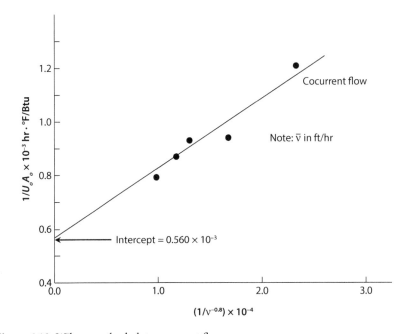

Figure 6.12 Wilson method plot: cocurrent flow.

example to follow). Fouling coefficients, h_f, can be estimated by the Wilson method if the outside film coefficient, h_o can be predicted or is negligible. Note that the fouling resistance is normally included in the intercept value.

Example 6.17 - Application of Wilson's Method. In 1975, Arthur Andrews [11]–a chemical engineering senior at Manhattan College–conducted a double pipe heat exchanger experiment in the Unit Operations Laboratory. Some of the *cocurrent* flow data and calculations submitted are present in Table 6.3. Employing Andrew's data and Wilson's method, estimate h_o for the exchanger fluid. For this lab unit, $A_o = 1.85 \text{ ft}^2$.

SOLUTION: A "best" straight line representation plot of $1/U_o A_o$ versus $1/\overline{v}^{0.8}$ is provided in Figure 6.12. The intercept $(1/h_o A_o)$ is approximately 0.560×10^{-3}. This value may now be used to estimate h_o:

$$h_o = \frac{1}{(0.560 \times 10^{-3})(1.85)} = 965 \text{ Btu} / (\text{hr} \cdot \text{ft}^2 \cdot {}^\circ\text{F})$$

Example 6.18 – Neglecting the Wall Resistance. With reference to the Wilson method, discuss why the resistance of the wall is often neglected.

SOLUTION: As noted earlier, the resistance of the wall is usually small (due to the high conductivity of the metal) in comparison to the resistances of the two flowing fluids.

6.6   Kern's Design Methodology

On to the specifics associated with what one would define as process heat exchanger engineering calculations [2]. As noted above, the calculational procedures for a heat exchanger is dependent on the information/data available. Kern's design methodology is also provided in several illustrative examples for shell-and-tubes, and finned units in Chapters 7 and 8, respectively. This section, however, is concerned with Kern's design methodology for double pipe heat exchangers.

All the equations developed previously can be combined to outline the design of a double pipe exchanger. Kern's calculation consists simply of

Table 6.4 Flow Areas and Equivalent Diameters in Double Pipe Exchangers.

Exchanger, IPS	Flow area, in²		Annulus, in	
	Annulus	Pipe	D_e	D_e'
$2 \times 1\,\frac{1}{4}$	1.19	1.50	0.915	0.40
$2\,\frac{1}{2} \times 1\,\frac{1}{4}$	2.63	1.50	2.02	0.81
3×2	2.93	3.35	1.57	0.69
4×3	3.14	7.38	1.14	0.53

computing h_o and h_{io} to obtain U_C. Allowing a reasonable fouling resistance, a value of U_D is calculated from which the surface can be found with the use of the Heat Transfer Equation $Q = U_D A \Delta T$.

Usually, the first problem is to determine which fluid should be placed in the annulus and which in the inner pipe. This is simplified by establishing the relative sizes of the flow areas for both streams. For equal allowable pressure drops on both the hot and cold streams, the decision rests in the arrangement producing the most nearly equal mass velocities and pressure drops. For the standard arrangements, the flow areas are given in Table 6.4.

It should also be noted that double pipe heat exchangers come in standard sizes so the practicing engineer may have to modify his/her calculation to accommodate the exchangers that are commercially available. However, the procedure used to design double pipe heat exchangers vary with the type of problem. Some prefer to apply a method referred to as *rating an exchanger*. Here, the practicing engineer assumes the existence of an exchanger and performs calculations to determine if the exchanger can handle the process requirements under existing or near-existing conditions. If not, a different exchanger is selected, and the previous calculations are repeated until a suitable design is developed. Repeated trials, i.e., a trial-and-error calculation procedure, is employed.

Kern's Original Design Methodology. In the outline below, hot- and cold-fluid temperatures are once again represented by upper and lower-case letters, respectively. All fluid properties are indicated by lower case letters to eliminate the requirement for new nomenclature. The diameter of the pipes must be given or assumed. With references to Kern's design methodology from the first edition, the following calculation procedure involves an application where a hot fluid is to be cooled by contact with a cooler fluid [2].

Process conditions (variables) that needs to be given, obtained, or calculated:

Hot fluid: $T_1, T_2, \dot{M}, C, S, \rho_H, \mu_H, k_H, \Delta P, R_{d,o}$ or $R_{d,i}$

Cold fluid: $t_1, t_2, \dot{m}, c, s, \rho_C, \mu_C, k_C, \Delta P, R_{d,o}$ or $R_{d,i}$

It should be noted that in the example problems throughout most of the remainder of the book, worked-out calculations and solutions will not use h, c, i, o, p, or a subscripts (hot, cold, inside, outside, pipe, and annulus, respectively) for μ, k, or ρ since the calculation of each of the process variables is such that the reader should be easily able to identify what stream each variable pertains to. Most process variables are calculated together in one portion of the system at a time, so that it should not be overly chaotic for the reader to follow the calculational procedure.

Process conditions that are usually specified or estimated:

   Hot fluid: ΔP_{max}, $R_{d,o}$ or $R_{d,i}$ (from past design and/or experience)

   Cold fluid: ΔP_{max}, $R_{d,o}$ or $R_{d,i}$ (from past design and/or experience)

A convenient order of calculation provided by Kern [2] follows:

(1)   From $T_1, T_2, t_1,$ *and* t_2 check the heat balance, Q, using C at T_{avg} and c at t_{avg}

$$T_{avg} = \frac{T_1 + T_2}{2} \quad and \quad t_{avg} = \frac{t_1 + t_2}{2}$$

$$Q = \dot{M}C(T_2 - T_1) = \dot{m}c(t_2 - t_1)$$

Radiation losses from the exchanger are usually insignificant compared with the heat load transferred in the exchanger.

(2)   LMTD, assuming countercurrent flow (as stated in Chapter 5, countercurrent flow arrangements tend to produce a larger temperature driving force, ΔT_{LM}).

(3)   Viscosity considerations: If the fluid is nonviscous and Newtonian, then ϕ can be assumed to equal 1, where $\phi = (\mu / \mu_w)^{0.14}$. However, if the fluid is viscous like that of petroleum or heavy hydrocarbons, the viscosity may behave differently depending on where the fluid is because it is often sensitive to temperature, i.e., the viscosity of the fluid at the inner pipe wall can differ from that of the bulk flow area of the fluid.

Inner pipe (p):

(4)   Flow area, $a_p = \dfrac{D^2 \pi}{4}$;   ft^2

(5)   Mass velocity, $G_p = \dfrac{\dot{m}_p}{a_p}$; $lb/(hr \cdot ft^2)$

(6) Obtain μ at T_{avg} or t_{avg} depending upon which flows through the inner pipe.

$$\mu[=]\text{lb}/(\text{ft}\cdot\text{hr})$$

(7) From D ft, G_p lb/(hr · ft²), and μ lb/(ft · hr), obtain the Reynolds number, Re_p.

$$Re_p = \frac{DG_p}{\mu}$$

(8) From c_p Btu /(lb·°F), μ lb/(ft·hr), and k Btu /(hr·ft²·°F/ft), all obtained at T_{avg} or t_{avg}, compute h_i using the appropriate equation based on the range of Re_p.

For Laminar Flow: $\dfrac{h_i D}{k} = 1.86\left[\left(\dfrac{DG_p}{\mu}\right)\left(\dfrac{c_p \mu}{k}\right)\left(\dfrac{D}{L}\right)\right]^{\frac{1}{3}}\left(\dfrac{\mu}{\mu_w}\right)^{0.14}$

For Turbulent Flow: $\dfrac{h_i D}{k} = 0.027\left(\dfrac{DG_p}{\mu}\right)^{0.8}\left(\dfrac{c_p \mu}{k}\right)^{\frac{1}{3}}\left(\dfrac{\mu}{\mu_w}\right)^{0.14}$

Annulus (a):

(4') Flow area, $a_a = \dfrac{\pi}{4}\left(D_2^2 - D_1^2\right)[=]\text{ft}^2$

Equivalent diameter, $D_e = 4\times\dfrac{Flow\ Area}{Wetted\ Perimeter} = \dfrac{D_2^2 - D_1^2}{D_1}[=]\text{ft}$

(5') Mass velocity, $G_a = \dfrac{\dot{m}_a}{a_a}[=]\text{lb}/\left(\text{hr}\cdot\text{ft}^2\right)$

(6') Obtain μ of the annulus's fluid at T_{avg} or t_{avg} depending upon which flows through the inner pipe.

$$\mu[=]\text{lb}/(\text{ft}\cdot\text{hr})$$

(7') From D_e ft, G_a lb/(hr · ft²), and μ lb/(ft · hr), obtain the Reynolds number, Re_a.

$$Re_a = \frac{D_e G_a}{\mu}$$

(8′) From $c_p \text{Btu}/(\text{lb}\cdot{}^\circ F), \mu \text{ lb}/(\text{ft}\cdot\text{hr})$, and $k \text{ Btu}/(\text{hr}\cdot\text{ft}^2\cdot{}^\circ F/\text{ft})$, all obtained at T_{avg} or t_{avg}, compute h_o using the appropriate equation based on the range of Re_p.

$$\text{For Laminar Flow: } \frac{h_o D_e}{k} = 1.86\left[\left(\frac{D_e G_a}{\mu}\right)\left(\frac{c_p \mu}{k}\right)\left(\frac{D_e}{L}\right)\right]^{1/3}\left(\frac{\mu}{\mu_w}\right)^{0.14}$$

$$\text{For Turbulent Flow: } \frac{h_o D_e}{k} = 0.027\left(\frac{D_e G_a}{\mu}\right)^{0.8}\left(\frac{c_p \mu}{k}\right)^{1/3}\left(\frac{\mu}{\mu_w}\right)^{0.14}$$

Overall coefficients:

(9)   Compute U_C,
$$\frac{1}{U_o A_o} = \frac{1}{h_i A_i} + \frac{\Delta x_w}{k A_{LM}} + \frac{1}{h_o A_o}$$

$$= \frac{1}{h_i A_i} + \frac{1}{h_o A_o}, \text{ neglecting pipe wall resistance}$$

$$= \frac{h_i h_o}{h_i + h_o}, \text{ assuming } A_i \approx A_o$$

(10) Compute U_D,
$$\frac{1}{U_D} = \frac{1}{U_C} + R_d$$

(11) Compute A from $Q = U_D A \Delta T_{LM}$ which may be translated into length. This length can be thought of as the "design" of the exchanger since it would help one determine how much material pipe of a certain size that would be needed to make an exchanger that will meet the desired process requirement. Although optional, an additional task in this step is to recalculate the dirt factor. If the length should not correspond to an integral number of hairpins, a change in dirt factor will result. The recalculated dirt factor should equal or exceed the required dirt factor by using the next larger integral number of hairpins.

Calculation of ΔP [12]. This requires a knowledge of the total length of the path satisfying the heat-transfer requirements. It also serves as a check to confirm whether the proposed heat exchanger design can be implemented. For example, if the calculated pressure drop, ΔP, is too high, the rate of heat transfer may be too low for it to operate as specified. This situation arises because the fluid would require more energy for it to be flowing through the system. If the pump forcing the fluid through the system does not provide sufficient flow rate, this may result in a low h, low U, and therefore a reduced rate of heat transfer, Q.

ΔP for Inner pipe (p):

(1)   Compute Re_p, $Re_p = \dfrac{DG}{\mu}$, identify type of flow

(2)   Compute f and ΔF_p, $f = \dfrac{16}{(DG/\mu)}$,   for laminar flow

$$f = 0.0035 + \dfrac{0.264}{\left(DG/\mu^{0.42}\right)},\quad \text{for turbulent flow}$$

$$\Delta F_p = \dfrac{4 f G^2 L}{2 g \rho^2 D}[=]\text{ft}$$

(3)   Compute ΔP_p,* $\Delta P_p = \Delta F_p \left(\dfrac{\rho}{144}\right)[=]\text{psi}$

*Note that 144 converts units from ft² to in²

ΔP for Annulus (a):

(1′)   *Compute D_e' and Re_a', $D_e' = \dfrac{4\pi\left(D_2^2 - D_1^2\right)}{4\pi\left(D_2 + D_1\right)} = D_2 - D_1$

$$Re_a' = \dfrac{D_e' G_a}{\mu},\ \text{identify type of flow}$$

*Note that D_e' for pressure drop differs from D_e used for the heat transfer rate calculation. The Reynolds number must be calculated based on D_e'.

(2′)   Compute f, ΔF_a, and ΔF_i, $f = \dfrac{16}{(D_e' G_a/\mu)}$,   for laminar flow

$$f = 0.0035 + \dfrac{0.264}{\left(D_e' G_a/\mu^{0.42}\right)},$$
$$\text{for turbulent flow}$$

$$\Delta F_a = \frac{4fG_a^2 L}{2g\rho^2 D'_e}[=]\,ft$$

$$\Delta F_i = \frac{\bar{v}^2}{2g'}[=]\,ft/hairpin$$

(3') Compute ΔP_a, $\Delta P_a = \dfrac{(\Delta F_a + \Delta F_i)\rho}{144}[=]\,psi$

There is an advantage if both fluids are computed side by side, as per Kern's recommendation, and the use of the outline in this manner will be demonstrated in Example 6.19.

Example 6.19 – Double Pipe Benzene-Toluene Exchanger [2]. It is desired to heat 9820 lb/hr of cold benzene from 80 to 120 °F using hot toluene which is cooled from 160 to 100 °F. Assume countercurrent flow. The specific gravities at 68 °F are 0.88 and 0.87, respectively. The other fluid properties can be found either in the Appendix of this book or in *Perry's Chemical Engineers' Handbook* [5]. A fouling factor of 0.001 is provided for each stream, and the allowable pressure drop on each stream is 10.0 psi. A number of 20-ft hairpins of 2- by 1¼in IPS pipe are available. How many hairpins are required?

SOLUTION:
(1)   Heat Balance:

$$Q = \dot{M}C\Delta T = \dot{m}c\Delta t$$

Cold Benzene,

$$t_{avg} = \frac{80+120}{2} = 100°F$$
$$c = 0.425\,Btu/(lb \cdot °F)\ \text{at}\,100°F$$
$$Q = (9820)(0.425)(120-80) = 167,000\,Btu/hr$$

Hot Toluene,

$$T_{avg} = \frac{160+100}{2} = 130°F$$
$$C = 0.44\ Btu/(lb \cdot °F)\ \text{at}\,130°F$$
$$\dot{M} = \frac{167,000}{(0.44)(160-100)} = 6330\ lb/hr$$

(2)   LMTD, ΔT_{LM} :

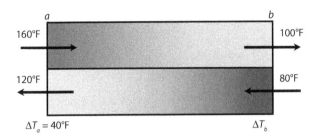

$$\mathrm{LMTD} = \frac{\Delta T_b - \Delta T_a}{\ln\left(\Delta T_b / \Delta T_a\right)} = \frac{20}{\ln\left(20 / 40\right)} = 28.8°\mathrm{F}$$

Proceed now to the inner pipe. A check of Table 6.4 indicates that the flow area of the inner pipe is greater than that of the annulus. Place the larger stream (benzene) in the inner pipe.

Hot Fluid (Toluene) – *Annulus (a)*	*Cold Fluid (Benzene) – Inner* *pipe (p)*
(4′) Flow Area, a_a	(4) Flow Area, a_p

$$D_2 = 2.067 \text{ in}\left(\frac{1\,\mathrm{ft}}{12\,\mathrm{in}}\right)$$
$$= 0.1725 \text{ ft}$$

$$D = 1.38 \text{ in}\left(\frac{1\,\mathrm{ft}}{12\,\mathrm{in}}\right)$$
$$= 0.115 \text{ ft}$$

$$D_1 = 1.66 \text{ in}\left(\frac{1\,\mathrm{ft}}{12\,\mathrm{in}}\right)$$
$$= 0.138 \text{ ft}$$

$$a_a = \frac{\pi}{4}\left(D_2^2 - D_1^2\right)$$
$$= \frac{\pi}{4}\left[\left(0.1725\right)^2 - \left(0.138\right)^2\right]$$
$$= 0.00826 \,\mathrm{ft}^2$$

$$a_p = \frac{D^2\pi}{4}$$
$$= \frac{\left(0.115\right)^2\pi}{4}$$
$$= 0.0104 \,\mathrm{ft}^2$$

Equivalent Diameter, D_e

$$D_e = \frac{D_2^2 - D_1^2}{D_1}$$

$$= \frac{(0.1725)^2 - (0.138)^2}{0.138}$$

$$= 0.0762 \, \text{ft}$$

(5') Mass Velocity, G_a

$$G_a = \frac{\dot{M}}{a_a} = \frac{6330}{0.00826}$$

$$= 767,000 \text{lb} / (\text{hr} \cdot \text{ft}^2)$$

(5) Mass Velocity, G_p

$$G_p = \frac{\dot{m}}{a_p} = \frac{9820}{0.0104}$$

$$= 943,000 \text{lb} / (\text{hr} \cdot \text{ft}^2)$$

(6') At 130 °F, $\mu = 0.41$ cP

$$\mu = 0.41 \text{cP} \left(\frac{2.42 \text{lb} / (\text{ft} \cdot \text{hr})}{1 \text{cP}} \right)$$

$$= 0.99 \text{lb} / (\text{ft} \cdot \text{hr})$$

(6) At 100 °F, $\mu = 0.50$ cP

$$\mu = 0.50 \text{ cP} \left(\frac{2.42 \text{lb} / (\text{ft} \cdot \text{hr})}{1 \text{ cP}} \right)$$

$$= 1.21 \text{ lb} / (\text{ft} \cdot \text{hr})$$

(7') Reynolds Number, Re_a

$$\text{Re}_a = \frac{D_e G_a}{\mu}$$

$$= \frac{(0.0762)(767,000)}{0.99}$$

$$= 59,000$$

(5) Reynolds Number, Re_p

$$\text{Re}_p = \frac{D G_p}{\mu}$$

$$= \frac{(0.115)(943,000)}{1.21}$$

$$= 89,500$$

(8') Solve for h_o

$$k = 0.085 \, \text{Btu} / \text{hr} \cdot \text{ft}^2 \cdot °\text{F/ft}$$

$$\mu_w = 0.51 \, \text{cP at } 130°\text{F}$$

$$= 0.51 \text{ cP} \left(\frac{2.42 \text{lb} / (\text{ft} \cdot \text{hr})}{1 \text{ cP}} \right)$$

$$= 1.23 \, \text{lb} / (\text{ft} \cdot \text{hr})$$

$$C = 0.44 \, \text{Btu} / \text{lb} \cdot °\text{F at } 130°\text{F}$$

$$\frac{h_o D_e}{k} = 0.027 \left(\frac{D_e G_a}{\mu} \right)^{0.8} \left(\frac{C\mu}{k} \right)^{1/3} \left(\frac{\mu}{\mu_w} \right)^{0.14}$$

(8) Solve for h_i

$$k = 0.091 \, \text{Btu} / \text{hr} \cdot \text{ft}^2 \cdot °\text{F/ft}$$

$$\mu_w = 0.68 \, \text{cP at } 100°\text{F}$$

$$= 0.68 \text{ cP} \left(\frac{2.42 \text{lb} / (\text{ft} \cdot \text{hr})}{1 \text{ cP}} \right)$$

$$= 1.65 \, \text{lb} / (\text{ft} \cdot \text{hr})$$

$$= 0.425 \, \text{Btu} / \text{lb} \cdot °\text{F at } 100°\text{F}$$

$$\frac{h_i D}{k} = 0.027 \left(\frac{D G_p}{\mu} \right)^{0.8} \left(\frac{c\mu}{k} \right)^{1/3} \left(\frac{\mu}{\mu} \right)^{0.14}$$

(8') $\quad h_o = 0.027 \left(\dfrac{D_e G_a}{\mu} \right)^{0.8} \left(\dfrac{C\mu}{k} \right)^{\frac{1}{3}} \left(\dfrac{\mu}{\mu_w} \right)^{0.14} \left(\dfrac{k}{D_e} \right)$

$\qquad = 0.027 \left(\dfrac{G_a}{\mu} \right)^{0.8} \left(C\mu k^2 \right)^{\frac{1}{3}} \left(\dfrac{\mu}{\mu_w} \right)^{0.14} \left(\dfrac{1}{D_e} \right)^{0.2}$

$\qquad = 0.027 \left(\dfrac{767,000}{0.99} \right)^{0.8} \left((0.44)(0.99)(0.085)^2 \right)^{\frac{1}{3}} \left(\dfrac{0.99}{1.23} \right)^{0.14} \left(\dfrac{1}{0.0762} \right)^{0.2}$

$\qquad \approx 330 \text{ Btu / hr} \cdot \text{ft}^2 \cdot {}^\circ\text{F}$

(8) $\quad h_i = 0.027 \left(\dfrac{D G_p}{\mu} \right)^{0.8} \left(\dfrac{c\mu}{k} \right)^{\frac{1}{3}} \left(\dfrac{\mu}{\mu_w} \right)^{0.14} \left(\dfrac{k}{D} \right)$

$\qquad = 0.027 \left(\dfrac{G_p}{\mu} \right)^{0.8} \left(c\mu k^2 \right)^{\frac{1}{3}} \left(\dfrac{\mu}{\mu_w} \right)^{0.14} \left(\dfrac{1}{D} \right)^{0.2}$

$\qquad = 0.027 \left(\dfrac{943,000}{1.21} \right)^{0.8} \left((0.425)(0.99)(0.091)^2 \right)^{\frac{1}{3}} \left(\dfrac{1.21}{1.65} \right)^{0.14} \left(\dfrac{1}{0.115} \right)^{0.2}$

$\qquad \approx 312 \text{ Btu / hr} \cdot \text{ft}^2 \cdot {}^\circ\text{F}$

(9) Clean overall coefficient, U_C, neglecting pipe wall resistance and assuming $A_i \approx A_o$:

$$U_C = \frac{h_i h_o}{h_i + h_o} = \frac{(330)(312)}{330 + 312} \approx 160 \text{ Btu / hr} \cdot \text{ft}^2 \cdot {}^\circ\text{F}$$

(10) Design overall coefficient, U_D:

$$R_d = R_{d,i} + R_{d,o} \qquad\qquad \frac{1}{U_D} = \frac{1}{U_C} + R_d$$
$$= 0.001 + 0.001 \qquad\qquad \frac{1}{U_D} = \frac{1}{160} + 0.002$$
$$- 0.002 \qquad\qquad\qquad U_D \approx 121 \text{ Btu / hr} \cdot \text{ft}^2 \cdot {}^\circ\text{F}$$

Summary

Annulus/Toluene	h	Inner Pipe/Benzene
330	Btu / hr $\cdot$ ft^2 $\cdot$ $^\circ$F	312
$U_C = 160$		
$U_D = 121$		

(11) Required surface, A:

$$Q = U_D A \Delta T_{LM} \quad \rightarrow \quad A = \frac{Q}{U_D \Delta T_{LM}}$$

$$A = \frac{167{,}000}{(121)(28.8)} \approx 47.9 \text{ ft}^2$$

From Table AT.8 from the Appendix, for 1¼in IPS standard pipe there are 0.435 ft² of external surface per foot length. Thus,

$$\text{Required length}, L = \frac{47.9}{0.435} \approx 110.1 \text{ lin. ft}$$

This may be fulfilled by connecting three 20-ft hairpins in series. Note that each 20-ft hairpin has 40 ft of heat exchanger length (each leg of the "U" shaped hairpin is 20 ft long).

(12) The surface supplied will actually be $(120)(0.435) = 52.2 \text{ ft}^2$. The dirt factor will accordingly be greater than required, and is calculated from the actual U_D:

$$U_D = \frac{167{,}000}{(52.2)(28.8)} \approx 111 \text{Btu} / \text{hr} \cdot \text{ft}^2 \cdot {}^\circ\text{F}$$

$$R_d = \frac{U_C - U_D}{U_C U_D} = \frac{160 - 111}{(160)(111)} \approx 0.0028 \text{ hr} \cdot \text{ft}^2 \cdot {}^\circ\text{F} / \text{Btu}$$

Pressure Drop

(1') D_e' for pressure drop differs from D_e for heat transfer.

$$D_e' = D_2 - D_1$$
$$= 0.1725 - 0.138$$
$$= 0.0345 \text{ ft}$$

$$Re_a' = \frac{D_e' G_a}{\mu}$$
$$= \frac{(0.0345)(767{,}000)}{0.99}$$
$$\approx 26{,}700$$

(1) $Re_p = 89{,}500$ from (7) above.

(2′) Compute ΔF_a and ΔF_i

$$f = 0.0035 + \frac{0.264}{Re_a^{'\,0.42}}$$

$$= 0.0035 + \frac{0.264}{(26,700)^{0.42}}$$

$$\approx 0.0072$$

$$S = 0.87$$

$$\rho = (62.5)(0.87) = 54.3\,\frac{\text{lb}}{\text{ft}^3}$$

$$\Delta F_a = \frac{4\,fG_a^2 L}{2g\rho^2 D_e'}$$

$$= \frac{4(0.0072)(767,000)^2(120)}{2(4.18\times10^8)(54.3)^2(0.0345)}$$

$$\approx 23.5\,\text{ft of toluene}$$

$$\bar{v} = \frac{G_a}{3600\rho} = \frac{767,000}{(3600)(54.3)}$$

$$\approx 3.92\,\text{fps}$$

$$\Delta F_i = 3\left(\frac{\bar{v}^2}{2g'}\right) = 3\left[\frac{(3.92)^2}{2(32.2)}\right]$$

$$\approx 0.7\,\text{ft of toluene}$$

(3′) Compute ΔP_a

$$\Delta P_a = \frac{(\Delta F_a + \Delta F_i)\rho}{144}$$

$$= \frac{(23.5+0.7)(54.3)}{144} \approx 9.2\,\text{psi}$$

$\Delta P_a < 10\,\text{psi as allowable } \Delta P$

---

(2) Compute ΔF_p

$$f = 0.0035 + \frac{0.264}{Re_p^{0.42}}$$

$$= 0.0035 + \frac{0.264}{(89,500)^{0.42}}$$

$$\approx 0.0061$$

$$s = 0.88$$

$$\rho = (62.5)(0.88) = 55.0\,\frac{\text{lb}}{\text{ft}^3}$$

$$\Delta F_p = \frac{4\,fG^2 L}{2g\rho^2 D}$$

$$= \frac{4(0.0061)(943,000)^2(120)}{2(4.18\times10^8)(55.0)^2(0.115)}$$

$$\approx 9.0\,\text{ft of benzene}$$

(3) Compute ΔP_p

$$\Delta P_p = \Delta F_p\left(\frac{\rho}{144}\right)$$

$$= \frac{(9.0)(55.0)}{144} \approx 3.4\,\text{psi}$$

$\Delta P_p < 10\,\text{psi as allowable } \Delta P$

It should be noted that if the pressure drops in Example 6.19 were greater than 10 psi, the possibility of reversing the location of the streams should always be examined *first* whenever the allowable pressure drop cannot be met. In the case of Example 6.19, a cocurrent flow arrangement would be investigated through another work of calculations if the pressure drop exceeded 10 psi under countercurrent flow arrangement. Unfortunately, reversing the location of the streams by placing the benzene in the annulus does not provide a solution in this case, since the flow of benzene is larger than that of the toluene.

It should also be noted that Kern [2] provides other illustrative examples concerned with the design of a double pipe oil/crude-oil heat exchangers. The properties of different oils can often be found in the literature according to the appropriate grades or degrees API gravity of oil. (API gravity indicates a measure of how light or heavy the given oil is compared to water.)

Heat Exchanger Calculation with a Viscosity Correction, ϕ. For heating or cooling fluids, the use of $(\mu / \mu_w)^{0.14} = 1.0$ assumes a negligible deviation of viscosity with temperature, i.e., the viscosity at the wall, μ_w, is the same as the viscosity of the bulk fluid – even though the temperature is not the same. When the pipe-wall temperature differs appreciably from the bulk temperature, the value of $\phi = (\mu / \mu_w)^{0.14}$ must be taken into account for very viscous fluids.

In cases where neither ϕ nor t_w is provided, one may need to utilize trial-and-error calculations and repeatedly refer to viscosity or heat transfer coefficient graphs/data table upon each trial (i.e., a new t_w is employed at each trial) to find the necessary correction. Referring to Figure 6.13 with the necessary variables, multiple expressions for Q can be made assuming a steady state heat transfer condition.

$$Q = \frac{\Delta t}{\sum R} = \frac{t_o - t_i}{R_o + R_w + R_i} = \frac{t_{w,i} - t_i}{R_w + R_i} = \frac{t_o - t_{w,o}}{R_o + R_w} \qquad (6.59)$$

If one can also assume that $t_{w,i} \approx t_{w,o} \approx t_w$ and a high value of k for pipe wall material (such as copper), then Equation (6.59) can be rearranged algebraically as follows:

$$\frac{R_o (t_o - t_i)}{R_o + R_i} = t_w - t_i$$

$$t_i + \frac{R_o}{R_o + R_i}(t_o - t_i) = t_w$$

$$t_i + \frac{h_o A_o}{\dfrac{1}{h_o A_o} + \dfrac{1}{h_i A_i}}(t_o - t_i) = t_w \qquad (6.60)$$

Assuming $A_i \approx A_o$ due to the small thickness of the pipe wall, Δx_w,

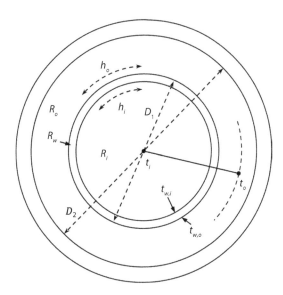

Figure 6.13 Location of variables for double pipe exchanger.

$$t_w = t_i + \frac{\dfrac{1}{h_o}}{\dfrac{1}{h_o} + \dfrac{1}{h_i}}\left(t_o - t_i\right) \qquad (6.61)$$

Therefore,

$$t_w = t_i + \left(\frac{h_i}{h_i + h_o}\right)\left(t_o - t_i\right) \qquad (6.62)$$

With the above assumptions and derived expressions, the following trial-and-error calculation procedure can be applied to determine an approximate temperature at the pipe wall as well as the viscosity correction. Note that if the fluid is nonviscous, this procedure would not have been necessary in the first place. It should only arise in situations where the viscosity of the fluid is very sensitive to temperature changes within the heat exchanger.

1. Initially, assume that $\left(\mu / \mu_w\right)^{0.14} = 1.0$ and obtain the values of h_i and h_o from the Sieder Tate Equations.

For laminar flow:

$$\frac{hD}{k} = 1.86\left[\left(\frac{DG}{\mu}\right)\left(\frac{c\mu}{k}\right)\left(\frac{D}{L}\right)\right]^{1/3}\left(\frac{\mu}{\mu_w}\right)^{0.14} \tag{6.16}$$

$$= 1.86\left[\left(\mathrm{Re}\right)\left(\mathrm{Pr}\right)\left(\frac{D}{L}\right)\right]^{1/3}\phi$$

For turbulent flow:

$$\frac{hD}{k} = 0.027\left(\frac{DG}{\mu}\right)^{0.8}\left(\frac{c\mu}{k}\right)^{1/3}\left(\frac{\mu}{\mu_w}\right)^{0.14}$$

$$= 0.027\left(\mathrm{Re}\right)^{0.8}\left(\mathrm{Pr}\right)^{1/3}\phi \tag{6.17}$$

2. Calculate t_w with h_i and h_o obtained from step 1 by employing Equation (6.59).
3. Use appropriate data table or graphs from available literature to find μ_w at t_w obtained from step 2 (if step 3 was never done before) or step 5.
4. Employing the Sieder Tate equations, calculate new values of h_i and h_o using μ_w obtained from step 3.
5. Calculate a new value for t_w with h_i and h_o obtained from step 4 by employing Equation (6.62).
6. Compare the old and new values of t_w. If $\left|t_{w,new} - t_{w,old}\right| < 0.1$, then an approximate t_w was determined, and thus, ϕ can be determined. If $\left|t_{w,new} - t_{w,old}\right| > 0.1$, repeat steps 3, 4, and 5 until the correction is better fine-tuned with each trial of calculation.

One of Kern's examples involving the viscosity correction factor is demonstrated in Example 6.20 to follow [2]. Although the trail-and-error calculation is not shown (at step 8), the reader is strongly encouraged to practice this trial-and-error calculation. *Perry's Chemical Engineers' Handbook* [5] is a good literature source to obtain viscosity values of many types of fluids at varying temperatures. For most given cases, 2 to 4 iterations should suffice in obtaining an acceptable viscosity correction for each fluid.

Example 6.20 – Double Pipe Lube Oil-Crude Oil Exchanger. 6,900 lb/hr of a 26°API lube oil must be cooled from 450 to 350 °F by an unspecified amount of 34°API mid-continent crude oil. The crude oil is to be heated from 300 to 390 °F. A fouling factor of 0.003 should be provided for each stream, and the allowable pressure drop on each stream will be 10 psi. A number of 20-ft hairpins of 3-by 2-in IPS pipes are available. How many must be used, and how shall they be arranged? For the lube oil, viscosities are 1.4 centipoises at 500 °F, 3.0 at 400 °F, and 7.7 at 300 °F. These viscosity variations are sufficient enough to introduce error if one assumes that $(\mu/\mu_w)^{0.14} = 1.0$.

SOLUTION:

(1) Heat Balance: $Q = \dot{M}C\Delta T = \dot{m}c\Delta t$

Hot Lube Oil, $T_{avg} = \dfrac{450 + 350}{2} = 400°F$

$C = 0.62\,\text{BTU}/(\text{lb}\cdot°\text{F})$ for $26°$API at $400\,°F$

$Q = (6900)(0.62)(450 - 350) = 427,000\,\text{Btu/hr}$

Cold Crude Oil, $t_{avg} = \dfrac{300 + 390}{2} = 345°F$

$c = 0.538\,\text{BTU}/(\text{lb}\cdot°\text{F})$ for $34°$API at $345°F$

$\dot{m} = \dfrac{Q}{c\Delta t} = \dfrac{427,000}{(0.538)(390 - 300)} = 8,800\,\text{lb}/\text{hr}$

(2) LMTD, ΔT_{LM}: (assume countercurrent flow)

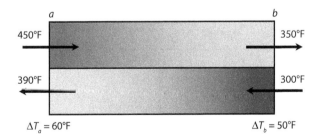

$$\text{LMTD} = \dfrac{\Delta T_b - \Delta T_a}{\ln(\Delta T_b / \Delta T_a)} = \dfrac{60 - 50}{\ln(60/50)} = 54.8°F$$

Proceed now to the inner pipe. A check of Table 6.4 indicates that the flow area of the inner pipe is greater than that of the annulus. Place the larger stream (crude oil) in the inner pipe.

Hot Fluid (Lube Oil) – Annulus (a)	Cold Fluid (Crude Oil) – Inner Pipe (p)
(4') Flow Area, a_a	(4) Flow Area, a_p

$D_2 = 3.068 \, \text{in} \left(\dfrac{1 \text{ft}}{12 \text{ in}} \right) = 0.256 \, \text{ft}$	$D = 2.067 \text{ in} \left(\dfrac{1 \text{ft}}{12 \text{ in}} \right) = 0.172 \text{ ft}$
$D_1 = 2.38 \, \text{in} \left(\dfrac{1 \text{ft}}{12 \text{ in}} \right) = 0.199 \, \text{ft}$	
$a_a = \dfrac{\pi}{4} \left(D_2^2 - D_1^2 \right)$	$a_p = \dfrac{D^2 \pi}{4}$
$\quad = \dfrac{\pi}{4} \left[(0.256)^2 - (0.199)^2 \right]$	$\quad = \dfrac{(0.172)^2 \pi}{4}$
$\quad = 0.0206 \, \text{ft}^2$	$\quad = 0.0233 \, \text{ft}^2$

Equivalent Diameter, D_e

$$D_e = \frac{D_2^2 - D_1^2}{D_1}$$

$$= \frac{(0.256)^2 - (0.199)^2}{0.199}$$

$$= 0.13 \, \text{ft}$$

(5') Mass Velocity, G_a	(5) Mass Velocity, G_p
$G_a = \dfrac{\dot{M}}{a_a} = \dfrac{6900}{0.0206}$	$G_p = \dfrac{\dot{m}}{a_p} = \dfrac{8800}{0.0233}$
$\quad = 335,000 \, \text{lb} / \left(\text{hr} \cdot \text{ft}^2 \right)$	$\quad = 378,000 \, \text{lb} / \left(\text{hr} \cdot \text{ft}^2 \right)$

(6') At 400 °F, $\mu = 3.0$ cP	(6) At 345 °F, $\mu = 0.81$ cP
$\mu = 3.0 \text{ cP} \left(\dfrac{2.42 \text{lb} / (\text{ft} \cdot \text{hr})}{1 \text{cP}} \right)$	$\mu = 0.81 \text{cP} \left(\dfrac{2.42 \text{lb} / (\text{ft} \cdot \text{hr})}{1 \text{cP}} \right)$
$\quad = 7.25 \text{lb} / (\text{ft} \cdot \text{hr})$	$\quad = 1.96 \text{lb} / (\text{ft} \cdot \text{hr})$

(7') Reynolds Number, Re_a	(7) Reynolds Number, Re_p
$\text{Re}_a = \dfrac{D_e G_a}{\mu}$	$\text{Re}_p = \dfrac{D G_p}{\mu}$
$\quad = \dfrac{(0.13)(335,000)}{7.25}$	$\quad = \dfrac{(0.172)(378,000)}{1.96}$
$\quad = 6,000$	$\quad = 33,200$

(8′) Solve for h_o (through trial and error)	(8) Solve for h_i (through trial and error)
$k = 0.067 \text{Btu} / \text{hr} \cdot \text{ft}^2 \cdot {}^\circ\text{F/ft}$	$k = 0.073 \text{ Btu} / \text{hr} \cdot \text{ft}^2 \cdot {}^\circ\text{F/ft}$
$\mu_w = 4.1 \text{cP at } t_w = 395{}^\circ\text{F}$	$\mu_w = 0.77 \text{cP at } t_w = 395{}^\circ\text{F}$
$= 4.1 \text{cP} \left(\dfrac{2.42 \text{lb} / (\text{ft} \cdot \text{hr})}{1 \text{cP}} \right)$	$= 0.77 \text{cP} \left(\dfrac{2.42 \text{lb} / (\text{ft} \cdot \text{hr})}{1 \text{cP}} \right)$
$= 16.0 \text{lb} / (\text{ft} \cdot \text{hr})$	$= 1.86 \text{lb} / (\text{ft} \cdot \text{hr})$
$C = 0.62 \text{ Btu} / \text{lb} \cdot {}^\circ\text{F at } T_{\text{avg}} = 400{}^\circ\text{F}$	$c = 0.538 \text{Btu} / \text{lb} \cdot {}^\circ\text{F at } t_{\text{avg}} = 345{}^\circ\text{F}$

(8′) $\quad h_o = 0.027 \left(\dfrac{D_e G_a}{\mu} \right)^{0.8} \left(\dfrac{C\mu}{k} \right)^{\frac{1}{3}} \left(\dfrac{\mu}{\mu_w} \right)^{0.14} \left(\dfrac{k}{D_e} \right)$

$\qquad = 0.027 \left(\dfrac{G_a}{\mu} \right)^{0.8} \left(C\mu k^2 \right)^{\frac{1}{3}} \left(\dfrac{\mu}{\mu_w} \right)^{0.14} \left(\dfrac{1}{D_e} \right)^{0.2}$

$\qquad = 0.027 \left(\dfrac{335{,}000}{7.25} \right)^{0.8} \left((0.62)(7.25)(0.067)^2 \right)^{\frac{1}{3}} \left(\dfrac{7.25}{16.0} \right)^{0.14} \left(\dfrac{1}{0.13} \right)^{0.2}$

$\qquad \approx 53.4 \text{ Btu} / \text{hr} \cdot \text{ft}^2 \cdot {}^\circ\text{F}$

(8) $\quad h_i = 0.027 \left(\dfrac{DG_p}{\mu} \right)^{0.8} \left(\dfrac{c\mu}{k} \right)^{\frac{1}{3}} \left(\dfrac{\mu}{\mu_w} \right)^{0.14} \left(\dfrac{k}{D} \right)$

$\qquad = 0.027 \left(\dfrac{G_p}{\mu} \right)^{0.8} \left(c\mu k^2 \right)^{\frac{1}{3}} \left(\dfrac{\mu}{\mu_w} \right)^{0.14} \left(\dfrac{1}{D} \right)^{0.2}$

$\qquad = 0.027 \left(\dfrac{378{,}000}{1.96} \right)^{0.8} \left((0.538)(1.96)(0.073)^2 \right)^{\frac{1}{3}} \left(\dfrac{1.96}{1.86} \right)^{0.14} \left(\dfrac{1}{0.172} \right)^{0.2}$

$\qquad \approx 116 \text{ Btu} / \text{hr} \cdot \text{ft}^2 \cdot {}^\circ\text{F}$

(9) Clean overall coefficient, U_C:

$$U_C = \frac{h_i h_o}{h_i + h_o} = \frac{(53.4)(116)}{53.4 + 116} \approx 36.6 \text{ Btu} / \text{hr} \cdot \text{ft}^2 \cdot {}^\circ\text{F}$$

(10) Design overall coefficient, U_D :

$$R_d = 0.003 + 0.003 = 0.006 \, \text{hr} \cdot \text{ft}^2 \cdot {}^\circ\text{F} / \text{Btu}$$

$$\frac{1}{U_D} = \frac{1}{U_C} + R_d$$

$$\frac{1}{U_D} = \frac{1}{36.6} + 0.006$$

$$U_D \approx 30.0 \, \text{Btu} / \text{hr} \cdot \text{ft}^2 \cdot {}^\circ\text{F}$$

Summary

Annulus/Lube Oil	h	Inner Pipe/Crude Oil
53.4	Btu / hr · ft² · °F	116
$U_C = 36.6$		
$U_D = 30.0$		

(11) Required surface, A:

$$Q = U_D A \Delta T_{LM} \;\; \rightarrow \;\; A = \frac{Q}{U_D \Delta T_{LM}}$$

$$A = \frac{427{,}000}{(30.0)(54.8)} \approx 260 \, \text{ft}^2$$

External surface/lin. ft, $a'' = 0.622 \, \text{ft} \leftarrow \left[\text{See Appendix AT.8} \right]$

$$\text{Required length}, L = \frac{260}{0.622} \approx 418 \, \text{lin. ft}$$

This may be satisfied by connecting eleven 20-ft hairpins at a total of 440 lin. ft of heat transfer area. The corrected U_D will be:

$$U_D = Q / A \Delta T_{LM} = 427{,}000 / \left(440 \times 0.622 \times 54.8 \right)$$
$$= 28.5 \, \text{Btu} / \text{hr} \cdot \text{ft}^2 \cdot {}^\circ\text{F}$$

The corrected dirt factor will be:

$$R_d = 1 / U_D - 1 / U_C = 1 / 28.5 - 1 / 36.6 = 0.0078 \, \text{hr} \cdot \text{ft}^2 \cdot {}^\circ\text{F} / \text{Btu}$$

Pressure Drop

(1') D'_e for pressure drop differs from D_e	(1) $Re_p = 89,500$ from (7) above.

(1') D'_e for pressure drop differs from D_e

$$D'_e = D_2 - D_1$$
$$= 0.256 - 0.199$$
$$= 0.057 \text{ ft}$$

$$Re'_a = \frac{D'_e G_a}{\mu}$$
$$= \frac{(0.057)(335,000)}{7.25}$$
$$\approx 2630$$

---

(1) $Re_p = 89,500$ from (7) above.

(2') Compute ΔF_a and ΔF_i.

$$f = 0.0035 + \frac{0.264}{Re'^{0.42}_a}$$
$$= 0.0035 + \frac{0.264}{(2630)^{0.42}}$$
$$\approx 0.0132$$
$$S = 0.775$$
$$\rho = (62.5)(0.775) = 48.4 \frac{\text{lb}}{\text{ft}^3}$$
$$\Delta F_a = \frac{4 f G_a^2 L}{2 g \rho^2 D'_e}$$
$$= \frac{4(0.0132)(335,000)^2 (440)}{2(4.18 \times 10^8)(48.4)^2 (0.057)}$$
$$\approx 23.4 \text{ ft}$$
$$\overline{v} = \frac{G_a}{3600\rho} = \frac{335,000}{(3600)(48.4)}$$
$$\approx 1.92 \text{ fps}$$
$$\Delta F_i = 6\left(\frac{\overline{v}^2}{2g}\right) = 6\left[\frac{(1.92)^2}{2(32.2)}\right]$$
$$\approx 0.172 \text{ ft}$$

---

(2) Compute ΔF_p.

$$f = 0.0035 + \frac{0.264}{Re^{0.42}_p}$$
$$= 0.0035 + \frac{0.264}{(33,200)^{0.42}}$$
$$\approx 0.00683$$
$$s = 0.76$$
$$\rho = (62.5)(0.76) = 47.5 \frac{\text{lb}}{\text{ft}^3}$$
$$\Delta F_p = \frac{4 f G_p^2 L}{2 g \rho^2 D}$$
$$= \frac{4(0.00683)(33,200)^2 (440)}{2(4.18 \times 10^8)(47.5)^2 (0.172)}$$
$$\approx 0.041 \text{ ft}$$

(3′) Compute ΔP_a .	(3) Compute ΔP_p .
$\Delta P_a = \dfrac{(\Delta F_a + \Delta F_i)\rho}{144}$ $= \dfrac{(23.4 + 0.172)(48.4)}{144}$ $\approx 7.9 \text{ psi}$ $\Delta P_a < 10 \text{ psi as allowable } \Delta P$	$\Delta P_p = \Delta F_p \left(\dfrac{\rho}{144} \right)$ $= \dfrac{(0.041)(47.5)}{144} \approx 0.014 \text{ psi}$ $\Delta P_p < 10 \text{ psi as allowable } \Delta P$

Theodore's Analysis. Theodore [13] has provided a revised version Kern's design methodology. Although Example 6.19 gave a detailed demonstration of Kern's design methodology, other approaches may be required and/ or employed. A common scenario of given process conditions for a heat exchanger is shown in the following presentation.

Given Process Conditions:

Hot fluid: $T_1, T_2, \dot{M}, C, S, \rho_H, \mu_H, k_H, \Delta P, R_{d,o}$ or $R_{d,i}$

Cold fluid: $t_1, t_2, \dot{m}, c, s, \rho_C, \mu_C, k_C, \Delta P, R_{d,v}$ or $R_{d,i}$, where $x =$ unknown

What remains to be calculated is,

1. t_2 and $\dot{m}$
2. Exchanger area requirement
3. Physical design

There are obviously an infinite number of solutions if some or all of the above specified terms are treated as variables. The simplest approach is to *initially* use a "guessed value" of the temperature driving force of 20 °F — assumed at the hot inlet so that $t_2 = T_1 - 20$ (assuming countercurrent flow) — and proceed as noted in the outline of Kern's methodology, i.e. calculate $\dot{m}$, A, and select a physical design. This would reduce the number of unknown process conditions from two to one. However, the physical design can include such variables as,

1. Pipe diameter
2. Number of tube turns (passes), which can increase ΔP
3. Material of construction/equipment/heat transfer surface

4. Tube layout
5. Baffles (if applicable)

Thus, there also are a near infinite number of physical designs but the design engineer will probably settle on one that has been extensively used in the past. In many cases in industry, the physical design can often be limited by what the supplying vendor has in stock and the design engineer will utilize those physical specifications and determine which items to purchase while considering the financial cost involved with the design. More details regarding tube layout and baffles will be presented later in Chapters 7 and 8. Also note that steady-state conditions are assumed to apply and constraints associated with quality energy considerations (see Part III, Chapter 11) are assumed *not* to exist; these generally do *not* usually arise in real-world applications.

If the approach temperature in the above example is not specified, a trial-and-error calculation arises since there are now two unknown process conditions, t_2 and $\dot{m}$. The reader should also note that if the number of process variables above is increased to three (rather than two), a double trial-and-error calculation can be required.

Finally, the original premise of this design was based on the cooling of a hot fluid. Any changes and/or modification to this premise will almost definitely require a different (but not significantly different) design calculation methodology.

Despite the forewarning in the previous paragraph, the readers should realize that heat exchanger designs are usually based on past experience so that many of the above concerns do not arise in a real-world design. The degree of simplification will naturally vary with both the application and the methodology employed by the heat exchanger engineer. For example, pressure drop concerns often do not arise, there are usually no phase changes of the fluids (except in condensers), and health and safety factors are not present.

Here is a simple summary of Kern's basic design methodology applied to the scenario above. The main difference from the general methodology is that in this proposed outline of calculation procedure, pressure drop is either specified or not a concern.

1. Set the approach temperature, i.e., t_2, and calculate the log mean temperature difference driving force (LMTD).
2. Apply a heat balance and calculate $\dot{m}$.

3. With the tube/pipe diameter set, calculate the mass velocity of both streams.
4. Calculate the film coefficients for both streams and evaluate the overall heat transfer coefficient *or* obtain a numerical value for the overall heat transfer coefficient from literature sources, experience, data from industrial sources, etc.
5. Apply the heat transfer equation and calculate the area of heat transfer of the exchanger.
6. Calculate the length of pipe and tubing from step 5.
7. Determine the number of passes from step 6, after specifying the length for each pass.

The above design procedure can be "reversed" for an existing exchanger when the area is specified. The above outline, for all intents and purposes, reflect the calculational procedures provided in Example 6.19. The pressure drop can also be treated as a variable – as opposed to a specified constraint.

In terms of optimum design, the design equation is often concerned with the trade-off associated with fluid velocity and pressure drop. In general, for a given rate of heat transfer, increased fluid velocities result in larger heat-transfer coefficients and, consequently, less-heat transfer area and lower exchanger cost. However, the increased fluid velocities cause an increase in pressure drop and greater pumping costs. Thus, the optimum economic design occurs at the conditions where the total cost (of both equipment and operating cost) is a minimum, and this occurs when the sum of the annual costs for the exchanger and its operation are minimized. By choosing various conditions, the process design engineer could ultimately arrive at a final design that would give the least total cost for fixed charges and operation [3].

6.7   Practice Problems from Kern's First Edition

1. What is the fouling factor when (a) $U_C = 30$ and $U_D = 20$, (b) $U_C = 60$ and $U_D = 50$, and (c) $U_C = 110$ and $U_D = 100$? Which do you consider reasonable to specify between to moderately clean streams?
2. A double pipe exchanger was oversized because no data were available on the rate at which dirt accumulated. The exchanger was originally designed to heat 13,000 lb/hr of 100 percent

acetic acid from 150 to 250 °F by cooling 19,000 lb/hr of butyl alcohol from 157 to 100 °F. A design coefficient $U_D = 85$ was employed, but during initial operation a hot-liquid outlet temperature of 117 °F was obtained. It rose during operation at the average rate of 3 °F per month. What dirt factor should have been specified for a 6-month cleaning cycle?

3. O-xylene coming from storage at 100 °F is to be heated to 150 °F by cooling 18,000 lb/hr of butyl alcohol from 170 to 140 °F. Available for the purpose are five 20-ft hairpin double pipe exchangers with annuli and pipes each connected in series. The exchangers are 3- by 2-in IPS. What is (a) the dirt factor, (b) pressure drops? (c) If the hot and cold streams in (a) are reversed with respect to the annulus and inner pipe, how does this justify or refute your initial decision where to place the hot stream?

4. 10,000 lb/hr of 57°API gasoline is cooled from 150 to 130 °F by heating 42° API kerosene from 70 to 100 °F. Pressure drops of 10 psi are allowable with a minimum dirt factor of 0.004. (a) How many 2½- by 1¼-in IPS hairpins 20 ft long are required? (b) How shall they be arranged? (c) What is the final fouling factor?

5. 12,000 lb/hr of 26°API lube oil (see Example 6.3 in text for viscosities) is to be cooled from 450 to 350 °F by heating 42° API kerosene from 325 to 375 °F. Pressure drops of 10 psi is permissible on both streams, and a with a minimum dirt factor of 0.004 should be provided. (a) How many 20-ft hairpins of 2½- by 1¼-in IPS are required? (b) How shall they be arranged? (c) What is the final dirt factor?

6. 7,000 lb/hr of aniline is to be heated from 100 to 150 °F by cooling 10,000 lb/hr of toluene with an initial temperature of 185 °F in 2- by 1-in IPS double pipe hairpin exchangers 15 ft long. Pressure drops of 10 psi are allowable, and a dirt factor of 0.005 is required. (b) How many hairpin sections are required? (c) How shall they be arranged? (c) What is the final dirt factor?

7. 24,000 lb/hr of 35°API distillate is cooled from 400 to 300 °F by 50,000 lb/hr of 34° API crude oil heated from an inlet temperature of 250 °F. Pressure drops of 10 psi are allowable, and a dirt factor of 0.006 is required. Using 20-ft hairpins of 4- by 3-in IPS (a) How many are required? (b) How shall they be arranged? (c) What is the final fouling factor?

8. 6330 lb/hr of toluene is cooled from 160 to 100 °F by heating amyl acetate from 90 to 100 °F using 15-ft hairpins. The exchangers are 2½- by 1¼-in IPS. Allowing 10 psi pressure drops and providing a minimum dirt factor of 0.004 (a) How many hairpins are required? (b) How shall they be arranged? (c) What is the final dirt factor?

9. 13,000 lb/hr of 26°API gas oil (see Example 6.3 in text for viscosities) is to be cooled from 450 to 350 °F by heating 57° API gasoline under pressure from 220 to 230 °F in as many 3- by 2-in IPS double pipe 20-ft hairpins as are required. Pressure drops of 10 psi are permitted along with a minimum dirt factor of 0.004 should be provided. (a) How many hairpins of are required? (b) How shall they be arranged? (c) What is the final dirt factor?

10. 100,000 lb/hr of nitrobenzene is to be cooled from 325 to 275 °F by benzene heated from 100 to 300 °F. Twenty-foot hairpins of 4- by 3-in IPS double pipe will be employed, and pressure drops of 10 psi are permissible. A minimum dirt factor of 0.004 is required. (a) How many hairpins of are required? (b) How shall they be arranged? (c) What is the final dirt factor?

References

1. I. Farag and J. Reynolds, *Heat Transfer*, a Theodore Tutorial, Theodore Tutorials, East Williston, NY, originally published by USEPA/APTI, RTP, NC,1996.

2. D. Kern, *Process Heat Transfer*, McGraw-Hill, New York City, New York, 1950.

3. L. Theodore, *Heat Transfer Applications for the Practicing Engineering*, John Wiley & Sons, Hoboken, NJ, 2011.

4. E. Incropera and D. De Witt, *Fundamentals of Heat and Mass Transfer*, 4th edition, Chapters 8, 11, John Wiley & Sons, Hoboken, NJ, 1996.

5. R. Perry and D. Green (editors), *Perry's Chemical Engineers' Handbook*, 8th edition, 11–67 to 11–137, McGraw-Hill, New York City, NY, 2008.

6. W. McCabe, J. Smith, and P. Harriott, *Unit Operations of Chemical Engineering*, 5th edition, Chapters 11–12. McGraw-Hill, New York City, NY, 1993.

7. J. Famularo: private communication to L. Theodore, Englewood Cliffs, NJ, 1980.

8. P. Farber, personal communication to L. Theodore, Chicago, IL, 2018.

9. E. Wilson, *Trans. ASME*, 34, 47, New York City, NY, 1915.

10. L. Theodore, personal notes, Manhattan College, Bronx, NY, 1963.
11. A. Andrews, report submitted to L. Theodore, Manhattan College, Bronx, NY, date unknown.
12. P. Abluencia and L. Theodore, *Fluid Flow for the Practicing Chemical Engineer*, John Wiley & Sons, Hoboken, NJ, 2009.
13. L. Theodore, personal notes, East Williston, NY, 2017.

7

Shell-and-Tube Heat Exchangers

Introduction

Shell-and-tube (also referred to as tube-and-bundle) heat exchangers provide a large heat transfer area both economically and practically. The tubes are placed in a bundle and the ends of the tubes are mounted in tube sheets. The tube bundle is enclosed in a cylindrical shell through which the second fluid flows. Most shell-and-tube exchangers used in practice are of welded construction. The shells are built as a piece of pipe with flanged ends and associated/necessary branch connections. The shells are made of seamless pipe up to 24 in. in diameter; they consist of bent and welded steel plates if above 24 in. Channel sections are usually of built-up construction, with welding-neck forged-steel flanges, rolled-steel barrels and welded-in pass partitions. Shell covers are either welded directly to the shell or are built-up constructions of flanged and dished heads plus welding-neck forged-steel flanges. The tube sheets are usually nonferrous castings in which the holes for inserting the tubes have been drilled and reamed before assembly.

Baffles can be employed to both control the flow of the fluid outside the tubes and provide turbulence [1–3].

There are vast industrial uses of shell-and-tube heat exchangers. These units are used to heat or cool process fluids, either through a single-phase heat exchanger or a two-phase heat exchanger. In single-phase exchangers, both the tube-side and shell-side fluids remain in the same phase that they enter. In two-phase exchangers (examples include condensers and boilers), the shell-side fluid is usually condensed to a liquid or heated to a gas, while the tube-side fluid usually remains in the same phase.

Generally, shell-and-tube exchangers are employed when double pipe exchangers do not provide sufficient area for heat transfer. Shell-and-tube exchangers usually require less materials of construction and are consequently more economical when compared to double pipe and/or multiple double pipe heat exchangers in parallel. Furthermore, the fulfillment of many industrial services require the use of a large number of double pipe hairpins. These consume considerable ground area and also entail a large number of points at which leakage may occur. Thus, when large heat-transfer surfaces are required, they can usually be best obtained by means of shell-and-tube exchangers [2].

Chapter topics include:

7.1   Equipment Description and Details
7.2   Key Describing Equations
7.3   Open-Ended Problems
7.4   Kern's Design Methodology
7.5   Other Design Procedures and Applications
7.6   Computer Aided Heat Exchanger Design

This is the longest chapter in the book, and it is so for good reason since shell-and-tube heat exchangers are the workhorse of the chemical/petrochemical industry.

7.1   Equipment Description and Details

There were reportedly more than 280 different types of shell-and-tube heat exchangers that had been defined earlier by the Tubular Exchanger Manufacturers Association (TEMA). The simplest shell-and-tube heat exchanger has a single pass through the shell and a single pass through the tubes. This is termed a 1–1 shell-and-tube heat exchanger. A schematic (line diagram) of 1–2 unit can be found in Figure 7.1. (It should be noted

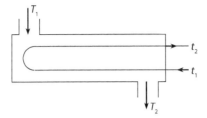

Figure 7.1 One shell pass, two tube passes schematic.

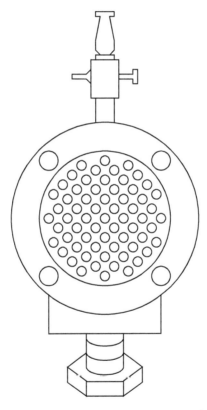

Figure 7.2 Side view of tube layout in the shell-and-tube heat exchanger.

that the bulk of the treatment in this chapter is concerned with 1–2 units since they are often the common choice in practice.) A side view of the tubes in a typical exchanger is shown in Figure 7.2. An exchanger with one pass on the shell-side and four tube passes is termed a 1–4 shell-and-tube heat exchanger. It is also possible to increase the number of passes on the shell-side by using dividers. A 2–8 shell-and-tube heat exchanger has two passes on the shell-side and 8 passes on the tube-side. Additional details are presented later in this section.

Fluids that flow through tubes at low velocity result in low heat transfer coefficients and low pressure drops. To increase the heat transfer rates, multi-pass operations may be used. Baffles are used to divert the fluid within the distribution header.

There is no debating that the shell-and-tube exchanger is the workhorse of the chemical process industries. As such, a significant amount of material on this unit is presented in this chapter, particularly in this equipment description section. There are ten specific equipment details that are reviewed. The ten subsection titles include:

1. The Tubular Element
2. Heat Exchanger Tubes
3. Tube Pitch
4. Shells
5. Stationary Tube-sheet Exchangers
6. Baffles
7. Fixed-tube-sheet Exchanger with Integral Channels
8. Fixed-tube-sheet 1–2 Exchangers
9. Removal-bundle Exchangers
10. Design Layout of Inner Tube Bundles

Several other topics are also briefly reviewed [2].

The Tubular Element. Fabricating shell-and-tube equipment involves expanding a tube into a tube sheet and forming a seal which does not leak under reasonable operating conditions. A simple and common example of an expanded tube is shown in Figure 7.3. A tube hole is drilled in a tube sheet with a slightly greater diameter than the outside diameter of the tube, and two or more groves are cut in the wall of the hole. The tube is placed inside the tube hole, and a tube roller is inserted into the end of the tube. The roller is a rotating mandril having a slight taper. It is capable of exceeding the elastic limit of the tube metal and transforming it into a semi-plastic condition so that it flows into the grooves and forms an extremely tight seal. Tube rolling is a skill, since a tube may be damaged by rolling it to paper thinness and leaving a seal with little structural strength.

In some industrial uses it is desirable to install tubes in a tube sheet so that they can be removed easily as shown in Figure 7.4. The tubes are actually packed in the tube sheet by means of ferrules using a soft metal packing ring.

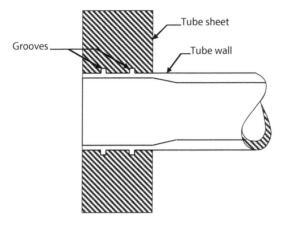

Figure 7.3 Tube roll.

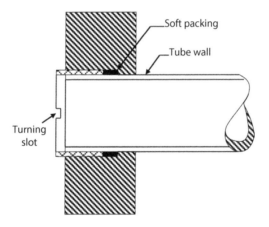

Figure 7.4 Ferrule.

Heat Exchanger Tubes. Heat-exchanger tubes are also referred to as condenser tubes and should not be confused with steel pipes or other types of pipes which are extruded to iron pipe sizes. The outside diameter of heat exchanger or condenser tubes is the actual outside diameter in inches within a very strict tolerance. Heat exchanger tubes are available in a variety of metals which include steel, copper, admiralty, Muntz metal, brass, 70–30 copper-nickel, aluminum bronze, aluminum, and the stainless steels. They are obtainable in a number of different wall thicknesses defined by the Birmingham Wire Guage, which is usually referred to as the BWG [3] or *gage* of the tube. The sizes of tubes which are generally available are listed in the Appendix of which the ¾-in. outside diameter and 1-in. outside diameter are most common in heat-exchanger design.

Tube Pitch. Tube holes cannot be drilled very close together, since too small a width of metal between adjacent tubes structurally weakens the tube sheet. The shortest distance between two adjacent tube holes is the *clearance* or *ligament*. Tubes are laid out on either square or triangular patterns as shown in Figure 7.5*a* and *b*. The advantage of square pitch is that the tubes are accessible for external cleaning and cause a lower pressure drop when fluid flows in the direction indicated in Figure 7.5*a*. The *tube pitch*, P_T is the shortest center-to-center distance between adjacent tubes. The common pitches for square layouts are ¾-in. OD on 1-in. square pitch and 1-in. OD on 1¼-in. square pitch. For triangular layouts, the tube pitches are ¾-in. OD on 15/16-in. triangular pitch, ¾-in. OD on 1-in. triangular pitch, and 1-in. OD on 1¼-in. triangular pitch. In Figure 7.5*c* the square-pitch layout has been rotated 45°, yet it is essentially the same as in Figure 7.5*a*. In Figure 7.5*d* a mechanically cleanable modification of

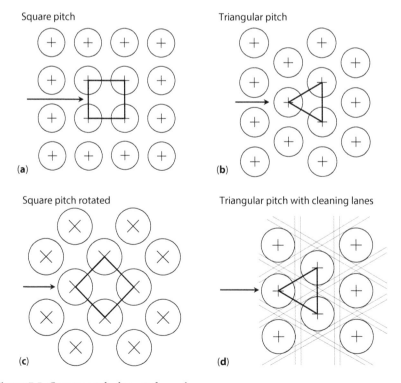

Figure 7.5 Common tube layouts for exchangers.

triangular pitch is shown. If the tubes are spread wide enough it is possible to allow the cleaning lanes indicated.

Shells. Shells are fabricated from steel pipe with nominal IPS diameters up to 12-in. Above 12 inches and including 24 inches, the actual outside diameter and the nominal pipe diameter are the same. The standard wall thickness for shells with inside diameters from 12 to 21 in. inclusive is 3/8-in., which is satisfactory for shell-side operating pressures up to 300 psi. A greater wall thickness may be required for greater pressures. Shells above 24 in. in diameter are fabricated by rolling steel plate.

Stationary Tube-sheet Exchangers. The simplest type of exchanger is the *fixed* or *stationary tube-sheet* exchanger of which the one shown in Figure 7.6 is an example. The essential parts are a shell (1), equipped with two nozzles and having tube sheets (2) at both ends, which also serve as flanges for the attachment of the two channels (3) and their respective channel covers (4). The tubes are expanded into both tube sheets and are equipped with traverse baffles (5) on the shell-side. The calculation of the effect heat-transfer surface is frequently based on the distance between the inside faces of the tube sheets instead of the overall tube length.

Baffles. It is apparent that higher heat transfer coefficients result when a liquid is maintained in a state of turbulence. To induce turbulence outside the tubes, it is customary to employ baffles which cause the liquid to flow through the shell at right angles to the axes of the tubes. This causes considerable turbulence even when a small quantity of liquid flows through the shell. The center-to-center distance between baffles is called the *baffle pitch* of *baffle spacing*. Since the baffles may be spaced close together or far apart, the mass velocity is not entirely dependent upon the diameter of the shell. The baffle spacing is usually not greater than a distance equal to the inside diameter of the shell or closer than a distance equal to one-fifth

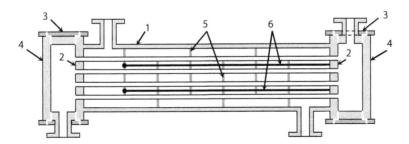

Figure 7.6 Fixed-head tubular exchanger.

the inside diameter of the shell. The baffles are held securely by means of baffle spacers (6) as shown in Figure 7.6, which consist of through-bolts screwed into the tube sheet and a number of smaller lengths of pipe which form shoulders between adjacent baffles. An enlarged detail is shown in Figure 7.7.

There are several types of baffles which are employed in heat exchangers, but by far the most common are segmental baffles as shown in Figure 7.8. Segmental baffles are drilled plates with heights which are generally 75 percent of the inside diameter of the shell. These are known as *25 percent cut baffles* and will be used through this book although other fractional baffle cuts are also employed in industry. They may be arranged, as shown, for "up-and-down" flow or may be rotated 90° to provide "side-to-side" flow, the latter being desirable when a mixture of liquid and gas flows through the shell. The baffle pitch rather than the 25 percent cut of the baffles, as shown later, determines the effective velocity of the shell fluid.

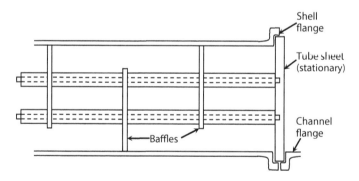

Figure 7.7 Baffle spacer detail (enlarged).

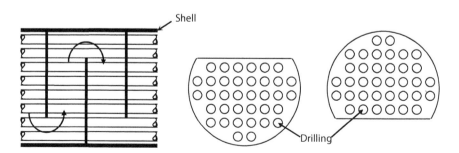

Figure 7.8 Segmental baffle detail.

Other types of baffles are the *disc* and *doughnut* of Figure 7.9 and the *orifice* baffle in Figure 7.10. Although additional types are sometimes employed, they are not of general importance.

Fixed-tube-sheet Exchanger with Integral Channels. Another type of several variations of the fixed-tube-sheet exchanger is shown in Figure 7.11, in which the tube sheets are inserted into the shell, forming

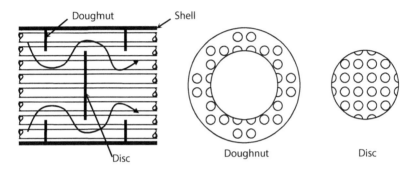

Figure 7.9 Disc and doughnut baffle.

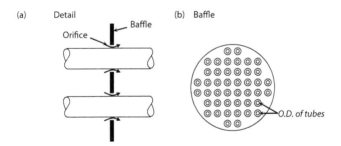

Figure 7.10 Orifice baffle.

Figure 7.11 Fixed-tube-sheet exchanger with integral channels.

channels which are integral parts of the shell. In using stationary tube-sheet exchangers, it is often necessary to provide for differential thermal expansion between the tubes and the shell during operation or thermal stresses will develop across the tube sheet. This can be accomplished by the use of an *expansion joint* on the shell, of which a number of types of flexible joints are available.

Fixed-tube-sheet 1–2 Exchangers. Exchangers of the type shown in Figure 7.12 may be considered to operate in countercurrent flow, notwithstanding the fact that the shell fluid flows across the outside of the tubes. From a practical standpoint it is very difficult to obtain a high velocity when one of the fluid flows through all the tubes in a single pass. This can be circumvented, however, by modifying the design so that the tube fluid is carried through fractions of tubes consecutively. An example of a *two-pass* fixed-tube-sheet exchanger is shown in Figure 7.13, in which all the tube fluid flows through the two halves of the tubes successively.

The exchanger in which the shell-side fluid flows in one shell pass and the tube fluid in two or more passes is the *1–2 exchanger*. A single channel is employed with a *partition* to permit the entry and exit of the tube fluid from the same channel. A bonnet is provided at the opposite end of

Figure 7.12 Fixed-head 1–2 exchanger.

Figure 7.13 Pull-through floating-head 1–2 exchanger.

the exchanger to permit the tube fluid to cross from the first to the second pass. As with all fixed-tube-sheet exchangers, the outside of the tubes are inaccessible for inspection or mechanical cleaning. The insides of the tubes are cleaned in placed by removing only the channel cover and using a rotary cleaner or a wire brush. Expansion problems are extremely critical in 1–2 fixed-tube-sheet exchangers, since both passes, as well as the shell itself, tend to expand differently and cause stress on the stationary tube sheets.

Removable-bundle Exchangers. A counterpart of the 1–2 exchanger having a tube bundle which is removable from the shell is shown in Figure 7.13. It consists of a stationary tube sheet which is clamped between the single channel flange and a shell flange. The tubes are expanded into a freely riding *floating tube sheet* or *floating head* at the opposite end of the bundle. A *floating-head cover* is bolted to the tube sheet, and the entire bundle can be withdrawn from the channel end. The shell is closed by a shell bonnet. The floating head eliminates the differential expansion problem in most cases and is called a *pull-through floating head*.

The disadvantage in the use of a pull-through floating head is one of simple geometry. To secure the floating-head cover it is necessary to bolt it to the tube sheet, and the bolt circle requires the use of space where it would be possible to insert a great number of tubes. The bolting not only reduces the number of tubes which might be placed in the tube bundle but also provides an undesirable flow channel between the bundle and the shell. These objections are overcome in the more conventional split-ring floating head 1–2 exchanger shown in Figure 7.14. Although it is relatively expensive to manufacture, it does have a great number of mechanical advantages. It differs from the pull-through type by the use of a split-ring assembly at the floating tube sheet and an oversized shell cover which accommodates it. The detail of a split ring is shown in Figure 7.15.

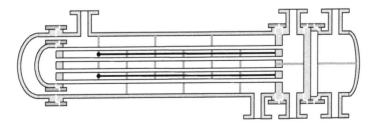

Figure 7.14 Floating-head 1–2 exchanger.

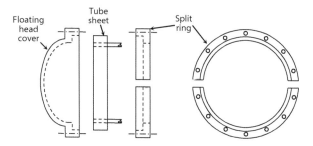

Figure 7.15 Split-ring assembly.

The floating tube sheet is clamped between the floating-head cover and a clamp ring placed in back of the tube sheet which is split in half to permit dismantling. Different manufacturers have different modifications of the design shown, but they all accomplish the purpose of providing increased surface area over the pull-through floating head in the same size shell.

A typical example of the layout of tubes for an exchanger with a split-ring floating head is shown in Figure 7.15. The actual layout is for a 13 1/4 in. ID shell with 1 in. OD tubes on 1 1/4 in. triangular pitch arranged for six tube passes. The partition arrangement is also shown for the channel and floating-head cover along with the orientation of the passes. Tubes are not usually laid out symmetrically in the tube sheet. Extra entry space is usually allowed in the shell by omitting tubes directly under the inlet nozzle so as to minimize the contraction effect of the fluid entering the shell. When tubes are laid out with minimum space allowances between partitions and adjoining tubes and within a diameter free of obstruction called the *outer tube limit*, the number of tubes in the layout is called the *tube count*. It is not always possible to have an equal number of tubes in each pass, although in large exchangers the inbalance should not be more than about 5 percent. The tube counts for ¾ and 1 in. OD tubes are given in Table 7.1 for one pass shells one, two, four, six, and eight tube pass arrangements.

These tube counts include a free entrance path below the inlet nozzle equal to the cross-sectional area of the nozzles shown in Table 7.1. When a larger inlet nozzle is used, extra entry space can be obtained by flaring the inlet nozzle at its base, removing the tubes which ordinarily lie close to the inlet nozzle.

Another modification of the floating-head 1–2 exchanger is the packed floating-head exchanger shown in Figure 7.17. This exchanger has an extension on the floating tube sheet which is confined by means of a packed gland. Although entirely satisfactory for shells up to 36 in. ID, the larger packing glands are not recommended for higher pressures or services causing vibration.

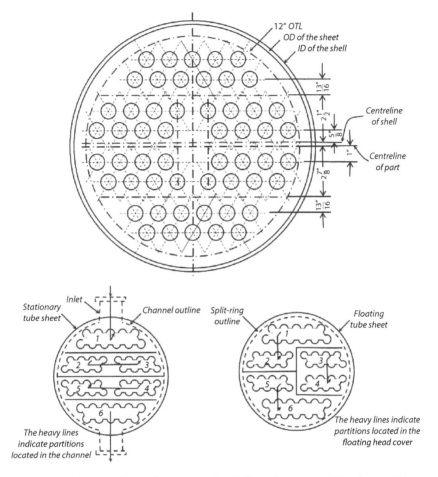

Figure 7.16 Tube-sheet layout for a 13¼-in. ID shell employing 1-in. OD tubes on 1¼-in. triangular pitch with six tube passes [2].

Table 7.1 Tube count entry allowances.

Shell ID, in.	Nozzle, in.
Less than 12	2
12–17¼	3
19¼–21¼	4
23¼–29	6
31–37	8
Over 39	10

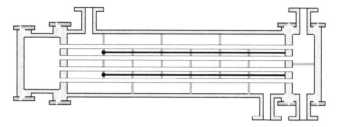

Figure 7.17   Packed floating head 1–2 exchanger.

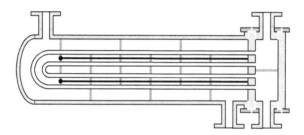

Figure 7.18   U-bend 1–2 exchanger.

The 1–2 exchanger shown in Figure 7.18 consists of tubes which are bent in the form of a U and rolled on the tube sheet. The tubes can expand freely, eliminating the need for a floating tube sheet, floating-head cover, shell flange, and removable shell cover. Baffles may be installed in a conventional manner on square or triangular pitch. The smallest diameter U-bend which can be turned without deforming the outside diameter of the tube at the bend has a diameter of three to four times the outside diameter of the tube. This means that it will usually be necessary to omit some tubes at the center of the bundle, depending upon the layout.

An interesting modification of the U-bend exchanger is shown in Figure 7.19. It employs a double stationary tube sheet and is used when leakage of one fluid stream into the other at the tube roll can cause serious corrosion damage. By using two tube sheets with an air gap between them, either fluid leaking through its adjoining tube sheets will escape to the atmosphere. In this way, neither of the streams can contaminate the other as a result of leakage except when a tube itself corrodes. Even tube failures can be prevented by applying a pressure shock test to the tubes periodically.

Design Layout of Inner Tube Bundles. When designing a shell-and-tube heat exchanger, the layout of the inner tube bundle is the main

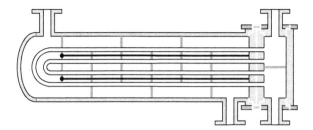

Figure 7.19 U-bend double-tube-sheet exchanger.

physical design criterion and includes some of the following aforementioned variables.

1. Tube diameter
2. Tube wall thickness
3. Tube length
4. Tube layout
5. Tube corrugation
6. Baffle design

Each are detailed below.

1. **Tube diameter.** The fouling nature of the tube-side fluid determines the correct tube diameter. Small diameter tubes allow for compact and economical design; however, mechanical cleaning of smaller tubes can be difficult. For cleaner fluids that do not foul, tubes may have an outside diameter as small as ½ in., while shell-and-tube heat exchangers that handle tars may have tubes that have outside diameters of 3 in. Most shell-and-tube heat exchangers have tubes with an outside diameter that range from ¾ to 1 in. [1].

2. **Tube Wall Thickness.** The tube wall thickness is governed by six different variables. First and foremost, the tube must have enough thickness to handle the external pressure exerted by the shell-side fluid and the internal pressure exerted by the tube-side fluid. In most cases, pressure is not a factor. Other factors that the tube wall thickness must provide are listed below:

   1. Allow enough margin for corrosion of the tube from use, or if any of the process fluids have corrosive properties.
   2. Provide enough assurances against flow-induced vibrations.

3. Provide enough axial strength to reduce sagging or bending throughout the length of the tube.
4. Meet standards for ease in ordering replacement tubes.
5. Ordering thinner or thicker tubes must justify the associated extra costs [1].

3. **Tube Length.** Generally, the most economical shell-and-tube heat exchangers have a small shell diameter and long tube length. Therefore, manufacturers strive to produce longer shell-and-tube heat exchangers. Long tubular bundles may prove difficult to replace. In practice, the maximum tube length for a fixed-bundle of tubes can be up to 9 m (30 ft) with a total weight of approximately 20 tons. However, it has been shown that the individual length of a fixed tube sheet should be limited to approximately 15 m (50 ft)—although tubes 22 m (73 ft) in length have sometimes been employed to meet process demands in industrial plants [1].

4. **Tube Layout.** When laying out a bundle of tubes, the tube pitch, i.e., the distance from the center of one tube to the next, should not exceed 1.25 times the outside diameter of the tubes. There are generally four types of bundle arrangements, as shown earlier in Figure 7.5. The fouling properties of the shell-side fluid must also be considered when determining the tube pitch. A tube pitch angle of 30° or 60° can accommodate 15% more tubes than their 45° and 90° counterparts; however, 30° triangular pitch and 60° triangular pitch do not allow for external cleaning of the outer tubes. When the shell-side fluid is in laminar region, a 45° square pitch tube layout has better heat transfer rates as well as lower pressure drops across the exchanger. The 90° square pitch provides for better heat transfer rates and a lower pressure drop across the exchanger when the shell-side fluid is in the turbulent region.

5. **Tube Corrugation.** Roped, fluted, and finned (see next chapter) tubes can be used in the design of a shell-and-tube heat exchangers to either increase the area available for heat transfer or ensure the shell-side fluid remains in the turbulent regime. Finned tubes increase the area available for heat transfer in a shell-and-tube exchanger, while fluted and roped tubes induce turbulent flow in the shell-side fluid [1].

6. **Baffle Design.** Baffles in a shell-and-tube heat exchanger serve two purposes. First, they allow the shell-side fluid to flow across all of the tubes, preventing dead spaces occurring within the heat exchanger. Secondly, they provide support for longer tubes to prevent damage from sagging or flow-induced vibrations. Baffles can either

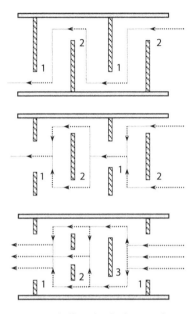

Figure 7.20 Baffle arrangements in shell-and-tube heat exchangers [1].

be designed in a single-, double-, or triple-segmental pattern, as shown in order from top to bottom in Figure 7.20 [1–3]. Single segmental baffles usually have a total length of 15–40% of the shell's inner diameter while the total length of double segmental baffles is 20–30% of the shell's inner diameter. Baffle spacing affects pressure drop; therefore, calculations of a pressure drop with certain baffle spacing must be performed to ensure that the pressure drop is reasonable. A general rule of thumb in industry is that baffles should not be spaced closer than one-fifth of the shell's inner diameter or two in. apart—whichever is greater [1–3].

Finally, the choice of which fluid should flow in the tubes is an important design decision. If one of the fluids is particularly corrosive, it is typically introduced on the tube side since more expensive resistant tubes can be purchased. If one of the fluids is more likely to form scale and/or deposits, it should flow on the tube side since the inside of the tubes are much easier to clean than the outside of the tubes. In addition, viscous fluids normally flow on the shell side because of pressure drop considerations and in order to help induce turbulence. The impact on the overall heat transfer coefficient and pressure drop should also be considered when determining the flow locations and directions of the fluids.

Example 7.1 – Number of Passes. (1) Determine the number of passes on the shell-side and the number of passes on the tube-side for a 4–6 shell-and-tube heat exchanger. (2) Provide a line diagram of a 3–6 shell-and-tube heat exchanger.

SOLUTION:

(1) First consider how many passes there are on the shell side. There are four passes on the shell-side. Finally, there are six passes on tube-side. The number of shell passes depends on the inner design of the shell while the number of tube passes depends on how many hairpins are formed by the bundle of tubes within the shell.

(2) See Figure 7.21 for line diagram of a 3–6 shell-and-tube heat exchanger.

Example 7.2 – Effect of Pressure Drop. Discuss the effect of pressure drop in the design of shell-and-tube exchangers.

SOLUTION: Shell-and-tube heat exchangers are often designed to meet upper limit specifications on shell-and-tube-side pressure drops. Many exchangers are designed for pressure drops less than 10 psi. For flow of low viscosity fluids (1 – 35 cP) through tubes, a good trial velocity is 6 fps. Equations are available in the literature [1–3] that estimates pressure drops for both the tube and shell-side. Some of these equations were presented earlier and are discussed in the next section. However, the reader should note that pressure drop (for both the tube and the shell-side) is often not a concern.

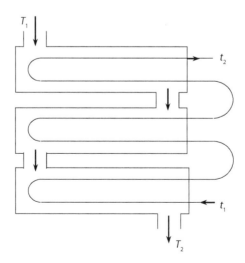

Figure 7.21 3–6 Exchanger.

7.2   Key Describing Equations

The reader should note that much of the material that follows in this section was presented in the last chapter. However, as each chapter is written on a stand-alone basis, the applicable equations for shell-and-tube exchangers receive treatment.

When two process fluids at different temperatures pass one another as in a shell-and-tube exchanger, heat transfer occurs due to the temperature difference between the two streams. The energy required to accomplish this heat transfer, i.e., the heat exchanger *duty*, can be determined for both process streams. For a well-insulated exchanger, the two duties should be equal to one another. The describing equations for the two duties are similar to those presented in Part I and for the double pipe heat exchanger (see previous chapter). However, the following equation gives the heat exchanger duty of the hot process stream when saturated steam is introduced on the shell-side (enthalpy balances between steam and condensate may also be used):

$$Q_h = \dot{m}_s \lambda + \dot{m}_s c_{hw}\left(T_{s,i} - T_{s,o}\right) = \dot{m}_s \lambda + \dot{m}_s c_{hw}\left(T_1 - T_2\right) \tag{7.1}$$

where,

c_{hw} = average heat capacity of condensate (hot water), Btu/lb - °F
h_{vap} = λ = rate of energy lost by steam upon condensing, Btu/hr
$\dot{m}_s$ = total mass flow rate of steam condensate, lb/hr
Q_h = Q_H = rate of energy lost by steam, Btu/hr
$T_{s,i}$ = T_1 = steam inlet temperature, °F
$T_{s,o}$ = T_2 = condensate outlet temperature, °F

The cold process stream duty is similarly calculated:

$$Q_c = \dot{m}_c c_{cw}\left(t_{c,o} - t_{c,i}\right) = \dot{m}_c c_{cw}\left(t_2 - t_1\right) \tag{7.2}$$

where,

c_{cw} = average heat capacity of the cold process stream (cold water), Btu/lb· °F
h_{vap} = λ = rate of latent energy lost by steam, Btu/hr
$\dot{m}_c$ = total mass flow rate of cold water (all tubes), lb/hr
Q_c = Q_C = rate of energy gained by cooling water, Btu/hr
$t_{c,i}$ = t_1 = cold water inlet temperature, °F
$t_{c,o}$ = t_2 = cold water outlet temperature, °F

If the heat exchanger duty is known, the overall heat transfer coefficient is calculated from either of the two equations provided below:

$$U_o = \frac{Q_h}{A_o \Delta T_{LM}}$$

(7.3)

or,

$$U_i = \frac{Q_c}{A_i \Delta T_{LM}}$$

(7.4)

where,

A_i = exchanger transfer area based on the inside tube surface, ft²
A_o = exchanger transfer area based on the outside tube surface, ft²
ΔT_{LM} = *log-mean temperature difference driving force*, °F
U_i = overall heat transfer coefficient based on the cold (c) process stream, Btu/ft²·hr·°F
U_o = overall heat transfer coefficient based on the hot (h) process stream, Btu/ft²·hr·°F

and,

$$\frac{1}{UA} = \frac{1}{U_i A_i} = \frac{1}{U_o A_o}$$

(7.5)

Typical values for the *overall* heat transfer coefficient in steam condensers with water in the tubes range from 1000–6000 W/m²·K [4]. *Perry's Chemical Engineers' Handbook* reports coefficients ranging from 400–1000 Btu/hr·ft²·°R [5]. These values are approximately the same after conversion of units.

For tubular heat exchangers, as with double pipe heat exchangers, the overall heat transfer coefficient is related to the individual coefficients by the following equation:

$$\frac{1}{UA} = \frac{1}{h_i A_i} + \frac{R_{d,i}}{A_i} + \frac{\ln(D_o / D_i)}{2\pi kL} + \frac{R_{d,o}}{A_o} + \frac{1}{h_o A_o}$$

(7.6)

where,

A = general notation for heat transfer surface area, ft²
A_i, A_o = inside and outside surface areas ($\pi D_i L$ or $\pi D_o L$), ft²
D_i, D_o = inside and outside pipe diameters, ft
h_i, h_o = inside and outside heat transfer coefficients, Btu/ft²·hr·°F
k = thermal conductivity, Btu/ft·hr·°F
L = tube length, ft
$R_{d,i}, R_{d,o}$ = fouling factors based on inner and outer tube surfaces, ft²·hr·°F/Btu

U =   overall heat transfer coefficient, Btu/ft²·hr·°F
U_i, U_o =   coefficients based on inner and outer tube surfaces, Btu/ft²·hr·°F

If the fouling factors, $R_{d,i}$ and $R_{d,o}$, and the tube wall resistance, i.e., the middle terms on the right-hand side of Equation (7.6) are negligible, then the relationship between the overall heat transfer coefficient and the individual coefficients simplifies once again to:

$$\frac{1}{U_o} = \frac{D_o}{h_i D_i} + \frac{1}{h_o} \tag{7.7}$$

The overall heat transfer coefficient, U, is an average value based on the average duty or $Q = (Q_h + Q_c)/2$ and an average heat transfer area or $A = (A_i + A_o)/2$. The individual coefficients above, h_i and h_o, are almost always calculated using empirical equations. First, the Reynolds number, Re, for the cold process fluid must be found in order to determine whether the flow is in the laminar, turbulent, or transition region (Refer to Chapter 1, Part I and Chapter 6 for additional details). Except for the viscosity term at the wall temperature, all of the physical properties of the tube-side fluid are usually evaluated at the average bulk temperature, $t_{c,\text{bulk}} = \bar{t} = (t_{c,i} + t_{c,o})/2$.

$$\text{Re} = \frac{4\dot{m}_c}{\pi D_i \mu_c} \tag{7.8}$$

where,

D_i =   inside diameter of one tube, ft
$\dot{m}_c$ =   mass flow rate of tube-side fluid through one tube, lb/hr
Re   =   Reynolds number
μ_c =   viscosity of tube-side fluid at the average bulk temperature, lb/ft·hr

The Nusselt number, Nu, can be determined from empirical equations. For *laminar flow* in a circular tube (Re < 2100), the Nusselt number equals 4.36 for uniform surface heat flux (Q/A) or 3.66 for constant surface temperature [2, 6]. This value should only be used for Graetz numbers, Gz, less than 10 [6]. For laminar flow with Graetz numbers from 10 to 1000, the following equation applies [6]:

$$\text{Nu} = 2\text{Gz}^{1/3}\left(\frac{\mu_c}{\mu_{c,\text{wall}}}\right)^{0.14}; \quad \text{Nu} = \frac{h_i D_i}{k_c}, \text{Gz} = \frac{\dot{m}_c c_c}{k_c L} \tag{7.9}$$

where,

Gz   =   Graetz number
k_c =   thermal conductivity of tube-side fluid at bulk temperature, Btu/ft·hr·°F
L =   length of tubes, ft
Nu   =   Nusselt number
$\mu_{c,wall}$ =   viscosity of tube-side fluid at wall temperature, lb/ft·hr

For *turbulent flow* (Re > 10,000), the Nusselt number may be calculated from the Dittus-Boelter equation if $0.7 \leq \Pr \leq 6700$, where Pr is the Prandtl number. Both the Dittus-Boelter and Sieder-Tate equations below are valid for L/D greater than 10 [2, 6].

$$\textit{Dittus-Boelter equation: } \mathrm{Nu} = 0.023\,\mathrm{Re}^{4/5}\Pr^{n} \qquad (7.10)$$

where,

$$\Pr = \frac{c_c \mu_c}{k_c}, n = 0.4 \text{ for heating or } 0.3 \text{ for cooling}$$

$$\textit{Sieder-Tate equation: } \mathrm{Nu} = 0.023\,\mathrm{Re}^{4/5}\Pr^{1/3}\left(\frac{\mu_c}{\mu_{c,wall}}\right)^{0.14} \qquad (7.11)$$

The Dittus-Boelter equation should only be used for small to moderate temperature differences. The Sieder-Tate equation applies for larger temperature differences [2, 5, 6]. Errors as large as 25% are associated with both equations. Other empirical equations with more complicated formulas and less error are available in the literature [2, 5, 6].

The above empirical equations do not apply for flows in the *transition region* (2100 < Re < 10,000). For the transition region, one should review the literature to determine the heat transfer coefficient. The reader is cautioned to use the correct mass flow rate; i.e., flow through a *single* tube rather than the total flow rate.

The outside heat transfer coefficient, h_o, is calculated using empirical correlations for condensation of a saturated vapor on a cold surface if steam is employed in the shell. Under normal conditions, a continuous flow of liquid is formed over the surface (film condensation) and condensate flows downward due to gravity. In most cases, the motion of the condensate is laminar and heat is transferred from the vapor-liquid interface to the surface by conduction through the film. Heat transfer coefficients for laminar film condensation on a single horizontal tube or on a vertical tier of N horizontal tubes can be calculated using appropriate equations that are available in the literature [2, 5, 6].

As noted earlier, the shell-and-tube heat exchanger usually consists of vertical tiers of horizontal tubes. In a vertical tier of N horizontal tubes, the average heat transfer coefficient for the stack of tubes is less than for a single tube. The equation for the average heat transfer coefficient for steam condensing on a vertical stack of horizontal tubes (laminar film condensation) is given by:

$$h_o = 0.725 \left[\frac{k_f^3 \rho_f^2 g \lambda}{N(T_s - T_w) D_o \mu_f} \right]^{1/4} \tag{7.12}$$

where,

D_o = outside tube diameter, ft
k_f = thermal conductivity of condensate at reference temperature, Btu/ft·hr·°F
N = average number of tubes in a vertical tier or stack
ρ_f = density of condensate at reference temperature, lb/ft³
T_f = reference temperature, °F
T_s = temperature of condensing vapor, °F
T_w = temperature of outside surface of wall, °F
μ_f = viscosity of condensate at reference temperature, lb/ft·hr

Figure 7.2 provided earlier is a drawing of the header of the shell-and-tube heat exchanger consisting of 56 tubes, N_{total}. The average number of tubes in a vertical tier (stack) can be determined by drawing a series of vertical lines down through the diagram of the header, then counting the tubes in each column and averaging the results. This procedure results in 11 vertical tiers of horizontal tubes with $N \approx 5$ tubes per tier. This is the correct value for N and must be used in Equation (7.12). Equation (7.12) is limited to cases where the steam-side Reynolds number, Re_s, is less than 2100 (laminar flow of shell-side fluid) [2]. Other correlations are provided in the literature for condensation heat transfer in the turbulent region [2–4]. Here,

$$Re_s = \frac{4\Gamma_b}{\mu_L}; \Gamma_b = \frac{\dot{m}_s / N_{total}}{L} \tag{7.13}$$

where,

Γ_b = condensate loading per unit length of tube, lb/hr·ft
N_{total} = total number of tubes, 56 tubes in the discussion above
L = length of tubes in Equation (7.13)

The physical properties are evaluated at a reference temperature, T_f. A reasonable value for the reference temperature in the range between $t_{c,i}$ and $T_{s,i}$ can be assumed, or an average of the four known temperatures, $(t_{c,i} + t_{c,o} + T_{s,i} + T_{s,o})/4$, can be used for a first estimate. Note once again that $t_{c,i}$ and $t_{c,o}$ represent the inlet and outlet cold liquid temperatures, respectively. A trial-and-error calculation may be required because the wall temperature and the reference temperature are both unknown and the wall temperature, T_w, depends on h_o and h_i. Details are provided in the previous chapter.

Some typical heat transfer coefficients for several specific operations are provided in Table 7.2 below. These values should be employed in lieu of any analytical calculation.

The following notation (consistent units) is employed in the shell-and-tube development to follow [1–4]:

A = heat transfer area
D_t = inside diameter of tube
L = tube length
$\dot{m}_{1,tube}$ = mass flow rate of tube-side fluid per tube
$\dot{m}_1$ = mass flow rate of tube-side fluid
N_{ttp} = number of tubes per pass on tube-side
N_p = number of passes on the tube-side
N_t = total number of tubes
U = overall heat transfer coefficient
$\bar{v}$ = average velocity of fluid in the tube

The mass flow rate per tube is given by,

$$\dot{m}_{1,tube} = \frac{\rho \bar{v} \pi D_t^2}{4} \qquad (7.14)$$

Table 7.2 Approximate coefficient values.

Heat transfer operation	Usual range for h	Suggested value
	Btu/ft²·hr·°F	
Gases flowing across tube banks	2.0–10	50
Liquids flowing across tube banks	10–100	50
Organic liquids inside tubes	50–500	250
Water flowing inside tubes	100–1,000	500
Boiling liquids	200–2,000	750
Condensing vapors	300–3,000	1,200

The number of tubes per pass is obtained from the total mass flow of the fluid on the tube-side.

$$N_{ttp} = \frac{\dot{m}_1}{\dot{m}_{1,tube}} \tag{7.15}$$

The total number of tubes and heat transfer area in the heat exchanger are:

$$N_t = N_{ttp} N_p \tag{7.16}$$

with,

$$A = N_t \pi D L \tag{7.17}$$

Examples involving the equations presented in this section are provided later in this section.

The F Factor. Some heat exchangers use true countercurrent flow. However, these heat exchangers are not as economical as multipass and crossflow units. In multipass exchangers, the flow alternates between co-current and countercurrent between the different sections of the exchanger. As a result, the driving force is not the same as a true countercurrent or a true co-current exchanger. The aforementioned ΔT_{LM} (log mean temperature difference driving force or LMTD) for these exchangers is almost always less than that of countercurrent flow and greater than that of co-current flow.

The analysis of these units requires including a *geometric F* factor in the describing equation (i.e., the heat transfer equation). For those multipass and crossflow exchangers, the log mean temperature difference (LMTD) method defined earlier is applicable, i.e.,

$$\Delta T_{LM} = \frac{\Delta T_1 - \Delta T_2}{\ln\left(\Delta T_1 / \Delta T_2\right)} \tag{7.18}$$

or,

$$\Delta T_{LM} = \frac{\Delta T_2 - \Delta T_1}{\ln\left(\Delta T_2 / \Delta T_1\right)} \tag{7.19}$$

And for the special case of $\Delta T_1 = \Delta T_2$,

$$\Delta T_{LM} = \Delta T_1 = \Delta T \tag{7.20}$$

The heat transfer rate is still given by,

$$Q = UA\Delta T_{LM} \qquad (7.21)$$

However, ΔT_{LM} must be corrected with the aforementioned geometry factor, F. This correction factor accounts for portions of multipass heat exchangers where the flow is not countercurrent (e.g., hairpin turns). Equation (7.21) is now written as,

$$Q = UA\Delta T_{LM}^{*} \qquad (7.22)$$

where,

$$\Delta T_{LM}^{*} = F\Delta T_{LM} \qquad (7.23)$$

where ΔT_{LM} (LMTD) is the default method of calculation for ΔT based on "ideal" countercurrent operation, and F is a correction factor applied to a different flow arrangement with the same hot and cold fluid temperature. The reader should note that later in this chapter, ΔT_{LM}^{*} can be replaced by ΔT^{*} when the topic of "true" temperature difference is addressed. More on this will be discussed in a later subsection.

Bowman et $al.$ [7] coordinated the results of earlier studies in his day [8–10] regarding the ΔT_{LM}^{*} for exchangers that experience neither co-current nor countercurrent flow to provide as complete a picture as possible of the various arrangements of surface conditions and flow directions. Shell-and-tube heat exchangers that experience any number of passes on the shell-and-tube-side were covered in his work as well as crossflow exchangers with different pass arrangements. Theodore [3] summarized the results for the various classifications below, of geometry factor, F, by presenting the summarized result in both graphical and equation form. Graphical results are given in Figure 7.22 thru Figure 7.25. Information on additional shell-and-tube configurations can be found in Kern's 1st Edition, Figures 18 thru 23 [2]. The parameters P and R in the graphs are:

$$R = \frac{T_1 - T_2}{t_2 - t_1} \qquad (7.24)$$

$$P = \frac{t_2 - t_1}{T_1 - t_1} \qquad (7.25)$$

Fluids Mixed. From graphical analysis [7], the correction factor approaches unity as the number of shell-side passes is increased for any value of P and

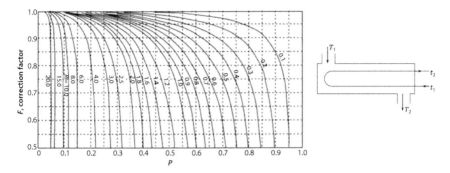

Figure 7.22 One shell pass; two or more tube passes.

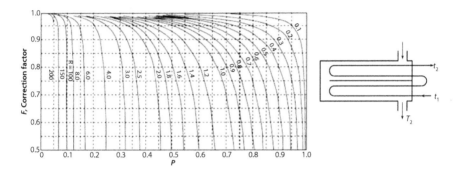

Figure 7.23 Two shell passes; four or more tube passes.

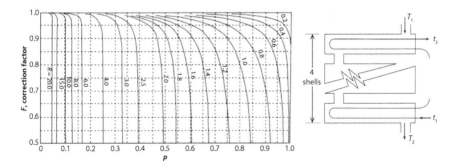

Figure 7.24 Four shell passes; eight or more tube passes.

R. This is to be expected since a multipass exchanger with several shell-side passes approaches the ideal countercurrent heat exchanger more closely than one shell-side and two or more tube passes. The following observation of the F factors' relationships can be made [3]:

$$F_{1-2} \approx F_{1-4} \approx F_{1-6} \qquad (7.26)$$

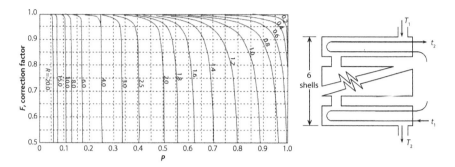

Figure 7.25 Six shell passes; twelve or more tube passes.

$$F_{1-3} \approx F_{1-6} \approx F_{1-9} \tag{7.27}$$

$$F_{2-4} \approx F_{2-8} \approx F_{2-12} \tag{7.28}$$

$$F_{3-6} \approx F_{3-12} \approx F_{3-18} \tag{7.29}$$

Kern [2] provides additional information on important design parameters associated with shell-and-tube heat exchangers. These five topics are addressed below:

1. Shell-side Film Coefficients
2. Shell-side Mass Velocity
3. Shell-side Equivalent Diameter
4. Shell-side Pressure Drop
5. Tube-side Pressure Drop

Shell-side Film Coefficients. The heat-transfer coefficients outside tube bundles are referred to as shell-side coefficients. When the tube bundle employs baffles directing the shell-side fluid across the tubes from top to bottom or side to side, the heat-transfer coefficient is higher than for undisturbed flow along the axes of the tubes. The higher transfer coefficients result from increased turbulence. In a square pitch, as seen in Figure 7.26, the velocity of the fluid undergoes continuous fluctuation because of the constricted area between adjacent tubes compared with the flow area between successive rows. In a triangular pitch even greater turbulence is encountered because the fluid flowing between adjacent tubes at high velocity impinges directly on the succeeding row. This would indicate that, when the pressure drop and cleanability are of little consequence, the

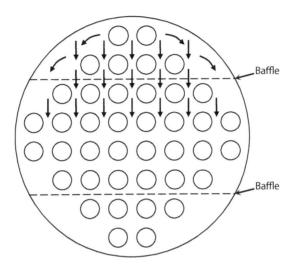

Figure 7.26 Flow across a bundle.

triangular pitch is superior for the attainment of high shell-side film coefficients. This is actually the case, and under comparable conditions of flow and tube size, the coefficients for triangular pitch are roughly 25 percent greater than for square pitch.

Several factors not treated in preceding chapters also influence the rate of heat transfer on the shell-side. Suppose the length of a bundle is divided by six baffles. All the fluid travels across the bundle seven times. If ten baffles are installed in the same length of bundle, it would require that the bundle be crossed a total of eleven times, with the closer spacing causing the greater turbulence. In addition to the effects of the baffle spacing, the shell-side coefficient is also affected by the type of pitch, tube size, clearance, and fluid flow characteristics. Furthermore, there is no true flow area by which the shell-side mass velocity can be computed, since the flow area varies across the diameter of the bundle with the different number of tube clearances in each longitudinal row of tubes. *The correlation obtained for fluids flowing in tubes is obviously not applicable to fluids flowing over tube bundles with segmental baffles*, and this is indeed borne out by experiment.

For values of Re from 2000 to 1,000,000, however, the data are closely represented by the equation:

$$\frac{h_o D_e}{k} = 0.36 \left(\frac{D_e G_s}{\mu} \right)^{0.55} \left(\frac{c\mu}{k} \right)^{1/3} \left(\frac{\mu}{\mu_w} \right)^{0.14} \tag{7.30}$$

where h_o, D_e, and G_s are as defined earlier. The different equivalent diameters used in the correlation of shell-and-tube data precludes comparison between fluids flowing in tubes and across tubes on the basis of the Reynolds number alone.

Shell-side Mass Velocity. The linear and mass velocities of the fluid change continuously across the bundle, since the width of the shell and the number of tubes vary from zero at the top and bottom to a maxima at the center of the shell. The width of the flow area is taken at the hypothetical tube row possessing the maximum flow area and corresponding to the center of the shell. The length of the flow area is taken equal to the baffle spacing B. The tube pitch is the sum of the tube diameter and the clearance C'. If the inside diameter of the shell is divided by the tube pitch, P_T, it gives a fictitious, but not necessarily integral number of tubes which may be assumed to exist at the center of the shell. Actually, in most layouts there is no row of tubes through the center but instead two equal maximum rows on either side of it having fewer tubes than computed for the center. These deviations are neglected. For each tube, there is considered to be $C' \times 1$ inch of crossflow area per inch of baffle space. The shell-side or bundle crossflow area, a_s, is then given by,

$$a_s = \frac{ID \times C'B}{P_T \times 144}[=]ft^2$$ (7.31)

and as before, the mass flux velocity is,

$$G_s = \frac{\dot{m}}{a_s}[=]lb/hr \cdot ft^2$$ (7.32)

Shell-side Equivalent Diameter [2]. By definition, the hydraulic radius corresponds to the area of a circle equivalent to the area of a noncircular flow channel and consequently in a plane perpendicular to the direction of flow. The hydraulic radius employed for correlating shell-side coefficients for bundles having baffles is not the true hydraulic radius. The direction of flow in the shell is partly along and partly at right angles to the long axes of the tubes of the bundle. The flow area at right angles to the long axes is variable from tube row to tube row. A hydraulic radius based upon the flow area across any one row could not distinguish between square and triangular pitch. In order to obtain a simple correlation combining both the size and closeness of the tubes and their type of pitch, excellent agreement is obtained if the hydraulic radius is calculated *along* instead of *across* the long axes of the tubes. The equivalent diameter for the shell is then taken as four times the hydraulic radius obtained for the pattern as laid out on the tube sheet.

Referring to Figure 7.27(a) where the crosshatch section covers the free area (the expression free area is used to avoid confusion with the free-flow area, an actual entity in the hydraulic radius) for *square pitch* is defined as:

$$D_e = \frac{4(\text{free area})}{\text{wetted perimeter}} [=] \text{ft} \tag{7.33}$$

or,

$$d_e = \frac{4\left(P_T^2 - \pi d_o^2/4\right)}{\pi d_o} [=] \text{in} \tag{7.34}$$

where P_T is the tube pitch in inches and d_o the tube outside diameter in inches. As for the triangular pitch, shown in Figure 7.27(b), the wetted perimeter of the element corresponds to half a tube:

$$d_e = \frac{4\left(0.5P_T \times 0.86P_T - 0.5\pi d_o^2/4\right)}{0.5\pi d_o} [=] \text{in} \tag{7.35}$$

It would appear that this method of evaluating the hydraulic radius and equivalent diameter does not distinguish between the relative percentage of right-angle flow to axial flow, and this is correct. It is possible, using

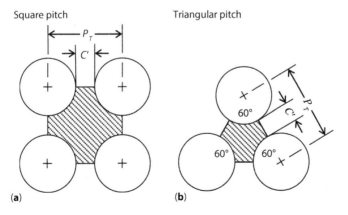

Figure 7.27 Equivalent diameter.

the same shell, to have equal mass velocities, equivalent diameters, and Reynolds numbers using (i) a large quantity of fluid and a large baffle pitch or (ii) a small quantity of fluid and a small baffle pitch-although the proportions of right-angle flow to axial flow differ. Apparently, where the range of baffle pitch is restricted between the inside diameter and one-fifth the inside diameter of the shell, the significance of the error is not too great to preclude correlation.

Shell-side Pressure Drop. The pressure drop through the shell of an exchanger is proportional to the number of times the fluid crosses the bundle between baffles. It is also proportional to the distance across the bundle each time it is crossed. Using a modification of an earlier equation, an expression for the number of crosses has been obtained using the product of the distance across the bundle, taken as the inside diameter of the shell in feet D_s and the number of times the bundle is crossed $N + 1$, where N is the number of baffles. If L is the tube length in feet,

Number of crosses, N + 1 = tube length, in/baffle space, in. (7.36)

$$N + 1 = (12)(L / B)$$

If the tube length is 16 ft and the baffles are spaced 18 inches apart, there will be 100 crosses or 10 baffles. There should always be an odd number of crosses if both shell nozzles are on opposite sides of the shell and an even number if both shell nozzles are on the same side of the shell. With close baffle spacing at convenient intervals such as 6 inches and under, one baffle may be omitted if the number of crosses is not an integer. The equivalent diameter used for calculating the pressure drop is the same as for heat transfer, the additional friction provided by the shell itself being neglected.

The isothermal equation for the pressure drop of a fluid being heated or cooled and including entrance and exit losses is,

$$\Delta P_s = \frac{f G_s^2 D_s (N+1)}{2 g \rho D_e \phi_s} = \frac{f G_s^2 D_s (N+1)}{(5.22 \times 10^{10}) D_e s \phi_s} [=] psf \qquad (7.37)$$

where s is the specific gravity of the fluid and $2 g \rho_{water} = 5.22 \times 10^{10}$. Equation (7.37) gives the pressure drop in pound-force per square foot, i.e., Ib_f / ft^2.

The reader should note that in the operation of heat exchangers, the power cost associated with the shell-side is often neglected. In effect, pressure drop is usually not a concern.

Tube-side Pressure Drop. Kern [2] recommends that Equation (6.43) from the previous chapter, which is used to compute ΔF, be used to obtain the pressure drop in tubes, but it applies principally to an isothermal fluid. Sieder and Tate have correlated friction factors for fluids being heated or cooled in tubes. In the following equation,

$$\Delta P_t = \frac{f G_t^2 L n}{\left(5.22 \times 10^{10}\right) D s \phi_t}[=]\text{psi} \tag{7.38}$$

n is the number of tube passes, L is the tube length, and Ln is the total length of path in feet. In flowing from one pass into the next at the channel and floating head the fluid changes direction abruptly by 180°, although the flow area provided in the channel and floating-head cover should not be less than the combined flow area of all the tubes in a single pass. The change of direction introduces an additional pressure drop, ΔP_r, called the return loss and can be accounted for by allowing four velocity heads per pass. A velocity head, $\bar{v}^2/2g'$, based on the mass velocity for a fluid with specific gravity, s, and the return losses for any fluid, ΔP_r, is

$$\Delta P_r = \left(\frac{4n}{s}\right)\left(\frac{\bar{v}^2}{2g'}\right)\left(\frac{62.5}{144}\right)[=]\text{psi} \tag{7.39}$$

where,

$\bar{v}$ = velocity, ft/s
s = specific gravity
n = number of tube passes
g' = acceleration of gravity, ft/s²

The total tube-side pressure drop, ΔP_T, is then,

$$\Delta P_T = \frac{\Delta P_t}{144} + \Delta P_r [=]\text{psi} \tag{7.40}$$

As with the shell-side, the reader should note that in the operation of heat exchangers, the power cost associated with the tube-side is often negligible. In effect, the pressure drop is usually not a concern. However, if there is a concern, it would more likely be with the tube-side. The reader should also note that Fannings equation (see also Chapters 3 and 6) may be employed to calculate the pressure drop.

The reader should refer to the comment provided by the authors in Chapter 6 regarding Kern's employing Fanning's equation to calculate the pressure drops in tubes and pipes. The reader should now note that Equations (7.37) and (7.38), employed by Kern to calculate the pressure drop for the shell and tube-side, respectively, are based on the Darcy (f_D), not the Fanning (f) friction factor, where $f_D = 4f$. The confusion arises since Kern employed Fanning's equation for pressure drop calculations associated with double pipe exchangers.

f and $\bar{v}$ Terms in Shell-and-Tube Exchangers. The friction factor for the tube-side can be obtained by utilizing A.F.1 from the Appendix, or by employing friction factor equations. However, the same cannot be done for the shell-side friction factor because of the difference in geometry. Mavridou et al. [18] and Hayati [19] suggest that the friction factor for shell-side can be obtained using the following approximation [7], under the condition that $400 < \text{Re}_s < 1 \times 10^6$,

$$f = \exp\left(0.576 - 0.19 \ln \text{Re}_s\right) \tag{7.41}$$

Although the approximation is fairly accurate, the reader should note that there are a number of different methods (some of which may prove to be more accurate) in obtaining the friction factor and pressure drop values for the shell-side of the exchanger. One of the most notable methods is the Bell-Delaware method [12] which takes into account multiple components of pressure drop sources such as baffle spacing, segmental cuts of baffles, tube layout, shape of heads, angle of layout, and so forth. Due to the complexity and overwhelming details of this method alone, this text will not discuss the methodology behind the Bell-Delaware calculation procedure. The reader should note that Equation (7.41) in combination with Equation (7.37) can be employed for the sake of simplifying the calculations of shell-side pressure drop.

Regarding the return pressure drop for the tube-side of the shell-and-tube exchangers, Kern originally recommended use of literature data recorded on a logarithmic graph which was provided in the original edition [2]. However, because the slope of the data curve on the log-log graph was linear, one of the authors, Toshihiro Akashige, converted the graphical relationship into equation form as follows,

$$\frac{\bar{v}^2}{2g'}\left(\frac{\rho_{\text{water}}}{144 \, \text{in}^2/\text{ft}^2}\right) = \left(1.4 \times 10^{-13}\right) G_t^2 \tag{7.42}$$

where,

$\bar{v}$ = velocity, ft/s
ρ_{water} = density of water, 62.5 lb/ft^3
G_t = mass velocity, lb/hr·ft^2
g' = acceleration of gravity, ft/s^2

Substituting the density of water in lb/ft^3 and algebraically rearranging,

$$\frac{\bar{v}^2}{2g'}\left(\frac{62.5}{144}\right)=\left(1.4\times10^{-13}\right)G_t^2\,[=]\text{psi} \tag{7.43}$$

Note that the units for 1.4×10^{-13} would be $\text{hr}\cdot\text{ft}^2/\text{in.}^2$

The "True" Temperature Difference, ΔT^*, in a 1–2 Exchanger. This section provides more detail into the aforementioned F factor, as the methodology described below is slightly more accurate than determining the F factor by graphical means.

A typical plot of temperature vs. length for an exchanger having one shell pass and two tube passes, with the nozzle arrangement indicated, was described earlier in Figure 7.1. Relative to the shell fluid, one part of the tube pass is in countercurrent flow while the other is in parallel flow. Greater temperature difference driving forces have resulted (see also Chapter 5) when the process streams are in countercurrent flow compared to cocurrent flow. The 1–2 exchanger is a combination of both, and the LMTD for countercurrent flow or cocurrent/parallel flow alone cannot be the true temperature difference for a parallel flow–countercurrent flow arrangement. As noted above, it is necessary to develop a new equation for the calculation of the effective or true temperature difference, ΔT^*, to replace the countercurrent flow LMTD. The method employed here is a modification of the derivation of Underwood [8] and is presented in the final form proposed by Nagle [9] and Bowman, Mueller, and Nagle [10] in Equation (7.44).

$$F_T=\frac{\left(\sqrt{R^2+1}\right)\ln\left[(1-P)/(1-RP)\right]}{(R-1)\ln\left[\dfrac{2-P\left(R+1-\sqrt{R^2+1}\right)}{2-P\left(R+1+\sqrt{R^2+1}\right)}\right]} \tag{7.44}$$

The fractional ratio of the true temperature difference to the idealized LMTD is F_T. The heat transfer equation for a 1–2 exchanger can now be written as:

$$Q = UA\Delta T^* = UAF_T\left(\text{LMTD}\right) = UAF_T\Delta T_{LM} \qquad (7.45)$$

Example 7.3 – Shell-side Equivalent Diameter. Compute the shell-side equivalent diameter for a ¾-in OD tube on 1-in square pitch.
SOLUTION: From Equation (7.34),

$$d_e = \frac{4\left[1^2 - \left(0.75^2\, \pi/4\right)\right]}{0.75\pi} = 0.95 \text{ in}$$

$$D_e = \frac{0.95}{12} = 0.079 \text{ ft}$$

Detailed pressure drop calculations are also provided in Section 7.4, Kern's Design Methodology.

Example 7.4 – Calculation of F. Calculate F_T for the following temperature approaches for a 1–2 exchanger (one shell, 2 tube passes).
   (a) 50° approach   (b) Zero approach   (c) 20° approach

$\left(T_1\right)$ 350   200 $\left(t_2\right)$ $\left(T_1\right)$ 300   200 $\left(t_2\right)$ $\left(T_1\right)$ 280   200 $\left(t_2\right)$

$\left(T_2\right)$ $\underline{250}$ $\underline{100}$ $\left(t_1\right)$ $\left(T_2\right)$ $\underline{200}$ $\underline{100}$ $\left(t_1\right)$ $\left(T_2\right)$ $\underline{180}$ $\underline{100}$ $\left(t_1\right)$
         100   100   100   100   100   100

SOLUTION: First calculate R and P from Equations (7.24) and (7.25), respectively.
  (a)   (b)   (c)

$$R = \frac{T_1 - T_2}{t_2 - t_1} = \frac{100}{100} = 1.0 \qquad\qquad R = 1.0 \qquad\qquad R = 1.0$$

$$P = \frac{t_2 - t_1}{T_1 - t_1} = \frac{100}{350 - 100} = 0.40 \quad P = 0.50 \qquad\qquad P = 0.555$$

$$F_T = 0.925 \;(\text{from Figure 7.22}) \quad F_T = 0.80 \qquad\qquad F_T = 0.64$$

Example 7.5 – Log Mean Temperature Difference Driving Force. Consider the shell-and-tube heat exchanger in Figure 7.28. Calculate the log mean temperature driving force. Also perform the calculation for the exchanger described in Figure 7.29.

SOLUTION: The system is first treated as an ideal countercurrent system as shown in Figure 7.28. For this ideal countercurrent system,

$$\text{LMTD} = \Delta T_{LM} = \frac{\Delta T_1 - \Delta T_2}{\ln\left(\Delta T_1/\Delta T_2\right)} \tag{7.18}$$

$$\text{LMTD} = \frac{(150-80)-(100-50)}{\ln(70/50)} = 59.4°\text{F}$$

For the system in Figure 7.29,

$$P = \frac{t_2 - t_1}{T_1 - t_1} = \frac{80-50}{150-50} = 0.30$$

$$R = \frac{T_1 - T_2}{t_2 - t_1} = \frac{150-100}{80-50} = 1.67$$

Figure 7.28 Ideal countercurrent heat exchanger; Example 7.5.

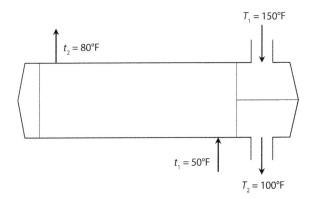

Figure 7.29 Shell-and-tube exchanger; Example 7.5.

From Figure 7.22, $F = 0.925$. Therefore,

$$\Delta T_{LM}^* = F\Delta T_{LM} = (0.925)(59.4) = 54.9°F$$

This somewhat lower value represents the LMTD for the actual (real) system shown in Figure 7.29.

Example 7.6 – A 2–4 Heat Exchanger. A shell-and-tube heat exchanger having two shell passes and four tube passes is being used for cooling. The shell-side fluid enters at 400 °F and leaves at 250 °F, and the tube-side fluid enters at 100 °F and leaves at 175 °F. What is the log mean temperature difference driving force between the hot fluid and the cold fluid?

SOLUTION: The log mean temperature difference (driving force) based ideal upon ideal (or true) countercurrent operation is first evaluated:

$$LMTD = \Delta T_{LM} = \frac{\Delta T_1 - \Delta T_2}{\ln(\Delta T_1 / \Delta T_2)} \tag{7.18}$$

$$LMTD = \frac{(400 - 250) - (175 - 100)}{\ln[(150)/(75)]} = 108°F$$

The F factor can be obtained from Figure 7.22 employing P and R:

$$P = \frac{t_2 - t_1}{T_1 - t_1} = \frac{75}{350} = 0.214$$

$$R = \frac{T_1 - T_2}{t_2 - t_1} = \frac{150}{75} = 2.0$$

From Figure 7.22,

$$F = 0.985$$

Thus,

$$\Delta T_{LM}^* = F\Delta T_{LM} = (0.985)(108)$$

$$\Delta T_{LM}^* = 106.4°F$$

Example 7.7 – Heat Exchanger Area Requirements. A shell-and-tube heat exchanger has one pass on the shell-side and two passes on the tube-side (i.e., a 1–2 shell-and-tube heat exchanger). It is being used for oil cooling. The oil flows in the tube-side. It enters at 110 °C and leaves at 75 °C. The shell-side fluid is water at a flow rate of 1.133 kg/s, entering at 35 °C and leaving at 75 °C. The heat capacity of the water is 4180 J/kg·K. The overall heat-transfer coefficient for the heat exchanger is 350 W/m²·K. The geometry factor F has been previously estimated to be 0.965. Calculate the heat-transfer area requirement for this unit in square meters [9].

SOLUTION: The heat load is:

$$Q = \dot{M}C\left(T_{out} - T_{in}\right) = \dot{m}c\left(t_2 - t_1\right) \tag{7.1}$$

$$Q = (1.133)(4180)(75 - 35)$$

$$Q = 189,400 \text{ W}$$

The countercurrent log-mean temperature difference is first calculated:

$$\Delta T_1 = 110 - 75 = 35°C$$

$$\Delta T_2 = 75 - 35 = 40°C$$

Apply Equation (7.18).

$$\Delta T_{LM} = \frac{35 - 40}{\ln(35/40)}$$

$$\Delta T_{LM} = 37.4°C$$

$$\Delta T_{LM} = 310.6 \text{ K}$$

The corrected log-mean temperature difference is calculated employed the correction factor, F, with Equation (7.23),

$$\Delta T_{LM}^* = F\Delta T_{LM}$$

$$\Delta T_{LM}^* = (0.965)(310.6)$$

$$\Delta T_{LM}^* = 299.7 \text{K}$$

The required heat-transfer area is then,

$$A = \frac{Q}{U\Delta T_{LM}^*}$$

$$A = \frac{189,000}{(350)(299.7)}$$

$$A = 1.806 \text{ m}^2$$

Example 7.8 – *F* Factor Analysis. Discuss the problem that can arise in employing the *F* factor.

SOLUTION: The LMTD method is adequate when the terminal temperatures are known. If the heat-transfer area is given and the exit temperatures are unknown, the problem often requires solution by trial and error.

Example 7.9 – Heating Water Calculation. A shell-and-tube heat exchanger is to be designed for heating 10,000 kg/hr of water from 16 °C to 84 °C by a new high heat capacity hot engine oil flowing through the shell. The oil makes a single shell pass, entering at 160 °C and leaving at 94 °C. The water flows through 11 brass tubes of 22.9 mm inside diameter and 25.4 mm outside diameter with each tube making four passes through the shell. The overall heat transfer coefficient (including the fouling resistance effect) is approximately 350 W/m² · °C and the thermal conductivity of brass is 137 W/m² · °C. Calculate:

1. the heat load, in MW,
2. the countercurrent flow log mean temperature difference, and
3. the *F* correction factor and the corrected log mean temperature difference.

SOLUTION: Calculate the average bulk temperatures of the water and oil:

$$T_{w,avg} = (16+84)/2 = 50°C$$

$$T_{oil,avg} = (160+94)/2 = 127°C$$

Obtain the properties of both fluids at the average temperature denoted above (see Appendix, AT.6), and organize this information in tabular form (see Table 7.3):

For water: $\rho = 987 \text{ kg/m}^3$, $c = 4176 \text{ J/kg}\Delta°C$

For oil: $\rho = 822 \text{ kg/m}^3$, $c = 4280 \text{ J/kg}\Delta°C$

Table 7.3 Data/Information for Illustrative Example 7.9.

Property	Stream			
	Inlet	Outlet	Inlet	Outlet
Fluid	Water (being heated)		Hot Oil	
Side/passes	Tube (four passes)		Shell (one pass)	
$t, T,$ °C	16	84	160	94
Average $t, T,$ °C	50		127	
ρ, kg/m³	987		822	
c, J/kgΔ°C	4176		4820	
$\dot{m}$, kg/s	2.778		2.480	

1. Calculate the heat load, Q, from the water-side information:

$$\dot{m}_1 = \dot{m}_w = 10{,}000 / 3600 = 2.778 \text{ kg/s}$$

$$Q = \dot{m}_1 c_w \left(t_2 - t_1\right) = \left(2.778\right)\left(4176\right)\left(84 - 16\right) = 788{,}800\text{W} = 0.788\text{MW}$$

2. One may also choose to calculate the oil flow rate, $\dot{m}_2$:

$$Q = c_{oil}\dot{m}_{oil}\left(T_1 - T_2\right);\ \dot{m}_{oil} = \dot{m}_2$$

$$788{,}800 = \left(4280\right)\left(\dot{m}_{oil}\right)\left(160 - 94\right)$$

$$\dot{m}_{oil} = \frac{788{,}800}{\left(4280\right)\left(160 - 94\right)}$$

$$\dot{m}_{oil} = \dot{m}_2 = 2.480\,\text{kg / s}$$

Calculate the log mean temperature difference based on the countercurrent flow, ΔT_{LM}:

$$\Delta T_1 = 94 - 16 = 78°\text{C}$$

$$\Delta T_2 = 160 - 84 = 76°\text{C}$$

$$\Delta T_{LM} = \frac{78 - 76}{\ln\left(78 / 76\right)} = 77°\text{C}$$

3. Calculate P and R:

$$P = \frac{t_2 - t_1}{T_1 - t_1} = \frac{84 - 16}{160 - 84} = 0.472$$

$$R = \frac{T_1 - T_2}{t_2 - t_1} = \frac{160 - 94}{84 - 16} = 0.971$$

Using Figure 7.23, for a 1–4 shell-and-tube heat exchanger, $F = 0.965$. Therefore,

$$\Delta T_{LM}^* = F\Delta T_{LM}$$

$$\Delta T_{LM}^* = (0.965)(77)$$

$$\Delta T_{LM}^* = 74.3\,°C$$

Example 7.10 – Design Calculation. Refer to the previous example. Calculate the area and the length of the tubes required for this heat exchanger.

SOLUTION: The heat transfer area may now be calculated:

$$Q = UA\Delta T_{LM}^* \qquad (7.22)$$

$$A = \frac{Q}{U\Delta T_{LM}^*} = \frac{788,800}{(350)(74.3)} = 30.33 \text{ m}^2$$

Calculate the tube length, L, by employing the following approach:

N_t = total number of tubes

$= \left(\text{number of tubes per pass, } N_{tpp}\right)\left(\text{number of passes, } N_p\right)$

$= N_{tpp}N_p \qquad (7.16)$

$= (11)(4)$

$= 44$ tubes

Therefore,

$$A = \pi D_t L N_t$$

$$L = \frac{A}{\pi D_t N_t} = \frac{30.34}{\pi (0.0229)(44)} = 31.3 \text{ ft}$$

Example 7.11 – Line Diagrams for Various Units. Steam enters the shell-side of a shell-and-tube heat exchanger at 450 °F and exits at a temperature of 300 °F. An organic material that has fluid properties approaching that of water enters the tube-side of the same heat exchanger at 200 °F and is heated by the steam to an exit temperature of 300 °F. Draw the diagram of each of the following specifications and/or heat exchangers:

    a. Countercurrent flow
    b. Co-current flow
    c. One shell-side pass and two tube-side passes
    d. One shell-side pass and four tube-side passes
    e. Two shell-side passes and four tube-side passes
    f. Six shell-side passes and twelve tube-side passes

SOLUTION: The appropriate temperature values for the heat exchanger examples are as follows:

$$T_1 = 450°F, T_2 = 300°F$$

and,

$$t_1 = 200°F, t_2 = 300°F$$

The diagrams of each specification and/or heat exchanger are shown in Figure 7.30,

Example 7.12 – F Factor Calculations. Refer to the previous example. Calculate the correction factor, F, and the log mean temperature difference, ΔT_{LM}, for each specification employing the figures provided earlier. Assume F approaches 1.0 for the exchanger provided in Figure 7.30(a).

    SOLUTION: The F and ΔT_{LM} values obtained using the figures for the given exchanger specifications are provided in Table 7.4.

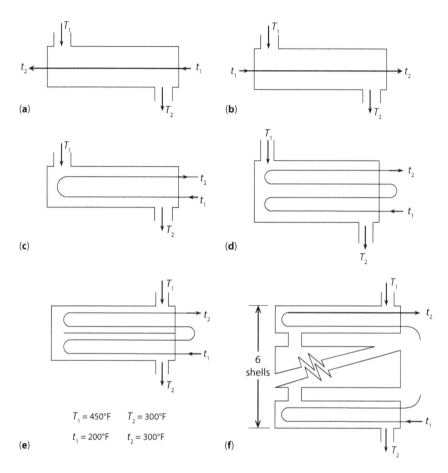

Figure 7.30 Diagrams of shell tube passes; Example 7.11.

Table 7.4 Tabulated F calculations.

		F	ΔT_{LM}
a	Countercurrent Flow	1.000	123.3
b	Co-current Flow	0	0
c	1 – 2 Multipass	0.810	99.9
d	1 – 4 Multipass	0.810	99.9
e	2 – 4 Multipass	0.960	118.4
f	6 – 12 Multipass	1.000 (est.)	123.3

Example 7.13 – Calculation of F_T for Fluids with Equal Ranges. Referring back to Example 7.4, calculate F_T using Equation (7.44):

$$F_T = \frac{\left(\sqrt{R^2+1}\right)\ln\left[(1-P)/(1-RP)\right]}{(R-1)\ln\left[\dfrac{2-P\left(R+1-\sqrt{R^2+1}\right)}{2-P\left(R+1+\sqrt{R^2+1}\right)}\right]} \qquad (7.44)$$

SOLUTION:

    (a) 50° approach   (b) Zero approach (c) 20° cross

    (T_1) 350 200 (t_2) (T_1) 300 200 (t_2) (T_1) 280 200 (t_2)

    (T_2) 250 100 (t_1) (T_2) 200 100 (t_1) (T_2) 180 100 (t_1)

Difference: 100 100   100 100   100 100

$$R = \frac{T_1-T_2}{t_2-t_1} = \frac{100}{100} = 1.0 \quad R = \frac{100}{100} = 1.0 \qquad R = \frac{100}{100} = 1.0$$

$$P = \frac{t_2-t_1}{T_1-t_1}$$

$$= \frac{100}{350-100} = 0.40 \quad P = \frac{100}{300-100} = 0.50 \quad P = \frac{100}{280-100} = 0.555$$

$$F_T = 0.92 \qquad\qquad F_T = 0.80 \qquad\qquad F_T = 0.64$$

The reader should note that the application of Equation (7.44) for $R = 1$ will give an error due to the fact that the denominator of the quotient becomes zero. Therefore, it is suggested to substitute $R = 0.99999$ (instead of 1) in order to approximate the final value for F_T.

7.3 Open-Ended Problems

Section 6.5 in Chapter 6 introduced the reader to open-ended problems where these problems involved an approach to the solution where there was not a unique solution to the problem. In addition to the earlier traditional examples in this chapter, this section extends that presentation for double-pipe exchangers to shell-and-tube units. There are eight open-ended examples in this set that attempt to provide both the beginner student and the practicing engineer with a basic understanding of heat transfer theory and principles, particularly as they apply to shell-and-tube heat exchangers.

Example 7.14 – Literature Hunt. Obtain order of magnitude values of industrial film coefficients for various boiling liquids, various condensing vapors, dropwise condensation for various vapors, film-type condensation for various vapors, heating of various liquids and vapors, cooling of various liquids and vapors, and superheated steam.

SOLUTION: This will require a rather extensive investigation. The authors suggest the reader start by reviewing the following references:

1. D. Kern, *Process Heat Transfer*, McGraw-Hill, New York City, NY, 1951 [2].
2. F. Incropera and D. De Witt, *Fundamentals of Heat and Mass Transfer*, 4th editions, Chapters 8, 11, John Wiley & Sons, Hoboken, NY, 1996 [4].
3. D. Green and R. Perry (editors), *Perry's Chemical Engineers' Handbook*, 8th edition, McGraw-Hill, New York City, NY, 2008 [5].
4. W. McCabe, J. Smith, and P. Harriott, *Unit Operations of Chemical Engineering*, 5th edition, Chapters 11–12, McGraw-Hill, New York City, NY, 1993 [6].
5. W. McAdams, *Heat Transmission*, 3rd edition, Chapters 9–10, McGraw-Hill, New York City, NY, 1954 [11].

There are also numerous other handbooks in the literature.

Example 7.15 – Troubleshooting a Tube-and-Bundle Exchanger. One option available to a plant manager when a tube within a tube-and-bundle heat exchanger fails is to simply plug the inlet of the tube. Develop an equation to describe the impact on the heat transfer performance of the exchanger as a function of both the number of tubes within the exchanger and the number of plugged tubes.

SOLUTION: This is obviously an open-ended problem. There are an infinite number of solutions since the number of tubes and the number of failed (plugged) tubes have not been specified. Solving this problem for a given number of tubes (N) and a given number of failed tubes (x) is left as an exercise for the reader.

Example 7.16 – Effect of Flow Rate on a Shell-and-Tube Exchanger. Refer to Example 6.16 in the previous chapter. Apply the same question (i.e., will doubling the flow rate in a shell-and-tube exchanger increase the heat transfer rate?).

SOLUTION: The solution presented in Example 6.16 applies to this example as well. In effect, doubling the flow rate increases both U and ΔT_{LM} (although t_2 decreases) producing an increase in Q for Condition (1). For Condition (2), U remains constant but ΔT_{LM} increases, once again leading to an increase in Q.

Example 7.17 – Effect of Plugging Tubes in a Shell-and-Tube Exchanger. During one of her heat transfer courses, Dr. Flynn decided to bypass the final exam for three students whose grades could best be classified as A+. However, as required by department regulations, she replaced the traditional final with the following take-home exam.

"Several years ago, Francesco Ricci (a junior heat transfer student) questioned Dr. Flynn if the performance of an existing shell-and-tube exchanger would be improved (i.e., increase Q) by plugging half of the tubes. Comment on Francesco's suggestion."

SOLUTION: From a qualitative prospective, once again examine the heat transfer equation:

$$Q = UA\Delta T_{LM} \tag{5.1}$$

This must also satisfy the energy (heat) conservation equation:

$$Q = \dot{m}c_p\left(t_2 - t_1\right) = \dot{m}c_p\Delta t \tag{6.46}$$

Obviously, since $\dot{m}$ is constant, t_2 — the outlet temperature — must increase for Q to increase.

Regarding Francesco's suggestion, plugging half the tubes would decrease the area for heat transfer by 50%. In addition, the area available for flow would correspondingly decrease by 50%, resulting in a doubling of the flow velocity, $\overline{v}$. For Condition (1) from Example 6.16, since $\overline{v}$ once again increases (it doubles), a decrease in $B/\overline{v}^{0.2}$ produces a corresponding decrease in t_2^* (the new t_2). Since $\dot{m}$ remains constant in Equation (6.46), Q of necessity will decrease. For Condition (2) from Example 6.16, the RHS of Equation (6.46) becomes $B/\overline{v}$; since $\overline{v}$ doubles, $B/\overline{v}$ will again decrease and t will of necessity decrease producing a corresponding decrease in Q (Δt has again decreased). Therefore, unfortunately for Frankie, his proposed recommendation will *not* do the job.

Example 7.18 – Effect of Flow Rate Change and Tube Plugging on a Shell-and-Tube Exchanger. Refer to Examples 6.16 and 7.16. Dr. Theodore, once a recognized authority in the heat transfer field, has recommended (staking his reputation on) that both doubling the flow rate *and* plugging half the tubes will increase the heat transfer rate in a shell-and-tube heat exchanger. Comment on Dr. Theodore's suggestion.

SOLUTION: The describing equations remain the same. Note that U would increase since the velocity has quadrupled since A and A_c have been halved. Since $B/\bar{v}^{0.2}$ has decreased, t_2 has decreased to t_2^* and ΔT_{LM} has increased. However, for Q to increase, Δt in Equation (6.46) must not decrease below half the original Δt (i.e., $t_2 - t_1$) to compensate for the doubling of $\dot{m}$. For this constraint,

$$\left(t_2^* - t_1\right) \le \frac{1}{2}\left(t_2 - t_1\right) \tag{7.46}$$

This equation may be solved for t_2^*:

$$t_2^* \le t_1 + \frac{t_2}{2} - \frac{t_1}{2} \tag{7.47}$$

$$t_2^* \le \frac{t_1 + t_2}{2} \tag{7.48}$$

However, the heat transfer equation — Equation (5.1) — must also be satisfied, i.e.,

$$\frac{T_c - t_1}{T_c - t_2} = e^{B/\bar{v}^{0.2}} \tag{7.49}$$

and with Dr. Theodore's changes,

$$\frac{T_c - t_1}{T_c - t_2^*} = e^{B/(4\bar{v})^{0.2}} \tag{7.50}$$

Dividing Equation (7.49) by Equation (7.50) leads to,

$$\frac{T_c - t_2^*}{T_c - t_2} = e^{(1/4)^{0.2}} \tag{7.51}$$

so that,

$$t_2^* = T_c - \left(T_c - t_2\right)e^{(1/4)^{0.2}} \tag{7.52}$$

Thus, if the calculated t_2^* in Equation (7.52) produces a value of t_2^* that is *less* than t_2, Q will *not* increase. This result applies to Condition (1). For Condition (2), Equation (7.52) becomes:

$$t_2 = T_c - \left(T_c - t_2\right)e^{1/4} \tag{7.53}$$

To test the above analysis, assume $T_c = 300°F$, $t_1 = 100°F$, and $t_2 = 250°F$. For Q to increase, t must be:

$$t \leq \frac{250 + 100}{2}$$

$$t \leq 175°F$$

For Condition (1):

$$t = 300 - \left(300 - 250\right)e^{(1/4)^{0.2}}$$

$$= 193°F$$

Thus, Q will not increase for these conditions. Condition (2) will also not produce an increase in Q, as demonstrated by applying Equation (7.53):

$$t = 300 - \left(300 - 250\right)e^{1/4}$$

$$= 235°F$$

At this point, the reader should realize — based on the result from the two previous examples — that increasing the flowrate increases Q while increasing the number of plugged tubes decreases Q. Thus, for any incremental increase in the flow rate, the increase in Q is constrained by an upper number of plugged tubes (above which Q will increase). Correspondingly, for any incremental increase in the number of plugged tubes, the decrease in Q is constrained by an upper value for the flow rate (above which Q will increase). Extreme excursions in the increase of either the flowrate or the number of plugged tubes will result in an upper value for the number of plugged tubes or the flow rate, respectively, that will produce an increase in Q. Therefore, Dr. Theodore's impeccable credentials and reputation have obviously (and justifiably) been tarnished. Too bad Louie.

Example 7.19 – Flow Rate and Tube Failure Variations: Qualitative Assessment. An existing exchanger transferring heat at a specified rate, Q, is assumed to contain a given number of tubes, N, and operate with a coolant flowrate $\dot{m}$. Qualitatively comment on the effect on performance as n (the number of tube failures) and $\dot{m}$ vary.

SOLUTION: If the flowrate is increased above $\dot{m}$ to $\dot{m}^*$, the exchanger can tolerate a number of tube failures up to n^* (where $n^* > n$) and still operate at or above the specified Q; any further increase in n^* will result in Q decreasing. Similarly, if n is increased, the exchanger will operate at a lower Q until an increased flowrate produces an increase in Q. There will also be conditions for which changes from design conditions in either the flowrate or failed number of tubes will not produce an increase in Q. In effect, there will be values of the number of failed tubes for which $\dot{m}$'s increase will not produce an increase in Q. The reader is left with the exercise of determining these excursion values for a given set of conditions.

The reader should note that the relationship between Q and the four variables, $T_c, t_1, t_2,$ and $\bar{v}$, is not only complex and nonlinear, but also a function of the other terms affecting U. No definite conclusion can be reached (at this point) regarding the merits of Dr. Theodore's recommendation in the previous example since there appears to be an infinite number of possible solutions, many of which — particularly for Condition (2) — will not increase Q.

Example 7.20 – Summary Report. Refer to Examples 7.16 thru 7.19 and provide a summary of the results.

SOLUTION: Here is a summary of the technical analysis and results of Examples 7.16, 7.17, 7.18, and 7.19. As noted earlier they are based on three key equations:

1. The conservation law for mass,

$$\dot{m} = \rho v A_c$$

2. The conservation law for energy,

$$Q = \dot{m}c_p \left(t_o - t_i\right)$$

3. The heat transfer equation,

$$Q = UA\Delta T_{LM}$$

The analysis was also performed for two conditions:

1. The coolant fluid solely contributes to the resistance to heat transfer, i.e., $h_o \to \infty$ and $k \to \infty$.
2. The condensing medium solely contributes to the resistance to heat transfer, i.e., $h_i \to \infty$ and $k \to \infty$.

These two conditions were considered for the three cases as presented in Examples 7.16 thru 7.18.

Example 7.16: Doubling the coolant flow rate. The results indicated that for Conditions (1) and (2) the heat transfer rate Q increased for both the shell-and-tube exchanger and the double pipe exchanger.

Example 7.17: Plugging half the tubes. The results indicated that Q decreased for both Conditions (1) and (2).

Example 7.18: Doubling the coolant flowrate and plugging half the tubes. The following equation resulted in:

$$2\dot{m}c_p\left(t_2-t_1\right)=\left(\frac{A}{2}\right)a\left(4\bar{v}\right)^{0.8}\frac{t_2-t_1}{\ln\left[\left(T_c-t_1\right)/\left(T_c-t_2\right)\right]}\qquad(7.54)$$

The results indicated that Q could decrease for both conditions. However, Q could increase under certain constrained situations for Condition (1). The results (for increasing Q) were more significantly constrained for Condition (2).

Example 7.19: The conclusion drawn is that the relationship for Q between N, n, and $\dot{m}$ is complex, particularly because of the four variables, T_c, t_1, t_2, and $\bar{v}$.

7.4   Kern's Design Methodology

Kern provides a design methodology for a 1–2 shell-and-tube heat exchanger [2]. His material allows for both the analysis of the performance of this unit and the design approach that should be employed. Although the material is specific to a 1–2 unit, the reader should note that Kern's methodology can be expanded relatively simply to not only other classes of tube-and-bundle exchangers but also other categories of heat exchangers.

When all the pertinent equations are used to calculate the suitability of an existing exchanger for given process conditions, it is known as *rating* an exchanger. There are three significant points in determining the suitability of an existing exchanger for a new service [2].

1. What clean coefficient, U_C, is calculated from the two fluids as a result of their flow and individual film coefficients, h_i and h_o?
2. A value of the design, or dirty coefficient, U_D is obtained from the heat balance $Q=\dot{M}C\left(T_1-T_2\right)=\dot{m}c\left(t_2-t_1\right)$, known

surface area, A, and the true temperature difference for the process temperatures. U_C must exceed U_D sufficiently so that the dirt factor, which is a measure of the excess surface required, will permit operation of the heat exchanger for a reasonable period of service.

3. The allow pressure drops for the two streams may not be exceeded.

When these requirements are fulfilled, an existing exchanger is suitable for the process conditions for which it has been rated.

The first point which arises in starting a calculation is to determine whether the hot or cold fluid should be placed in the shell. There is no hard-and-fast rule. One stream may be large and the other small, and the baffle spacing may be such that in one instance the shell-side flow area a_s will be larger. Fortunately, any selection can be checked by switching the two streams and seeing which arrangement gives the larger value of U_C without exceeding the allowable pressure drop. There is some advantage (in preparation for later calculations), however, in starting calculations with the tube-side, and it may be well to establish the habit, primarily because the tube-side pressure drop is somewhat easier to calculate. The detailed steps in the rating of an existing 1–2 exchanger are outlined below. The subscripts s and t are used to distinguish between the shell and tubes, respectively, and for the outline presented, the hot fluid has been assumed to be in the shell [2].

Calculation of an Existing 1–2 Exchanger [2]. The calculation of an existing 1–2 heat exchanger requires the following process conditions:

Process conditions required:

Hot fluid: $T_1, T_2, \dot{M}, C, S$ and $\rho, \mu, k, \Delta P, R_{d,o}$ or $R_{d,i}$

Cold fluid: $t_1, t_2, \dot{m}, c, s$ and $\rho, \mu, k, \Delta P, R_{d,o}$ or $R_{d,i}$

For the exchanger, the following data must be known:

Shell-side	Tube-side
ID	Number and length
Baffle space	OD, BWG, and pitch
Passes	Passes

(1) Heat balance, $Q = \dot{M}C(T_1 - T_2) = \dot{m}c(t_2 - t_1)$

---

(2) True temperature difference ΔT^*:

$$\text{LMTD} = \frac{\Delta t_2 - \Delta t_1}{\ln(\Delta t_2 / \Delta t_1)} \qquad R = \frac{T_1 - T_2}{t_2 - t_1} \qquad P = \frac{t_2 - t_1}{T_1 - t_1}$$

$$F_T = \left(\frac{\sqrt{R^2 + 1}}{R - 1}\right) \frac{\ln\left[(1 - P)/(1 - RP)\right]}{\ln\left[\left(2 - P\left(R + 1 - \sqrt{R^2 + 1}\right)\right) \middle/ \left(2 - P\left(R + 1 + \sqrt{R^2 + 1}\right)\right)\right]}$$

$$\Delta T^* = (\text{LMTD})(F_T)$$

---

(3) Average temperatures, T_{avg} and t_{avg}:

$$T_{avg} = \frac{T_1 + T_2}{2} \qquad\qquad t_{avg} = \frac{t_1 + t_2}{2}$$

Hot Fluid – Shell-Side (s)	*Cold Fluid – Tube-Side (t)*

(4′) Flow Area, a_s:

$$a_s = \frac{\text{ID} \times C'B}{144 P_T} [=] \text{ft}^2]$$

(4) Flow Area, a_t:

a'_t = Flow area per tube from Table AT.8

$$[=] \text{in}^2$$

$$a_t = \frac{\text{No. of tubes} \times \text{flow area/tube}}{\text{No. of passes}}$$

$$= \frac{N_t a'_t}{144 n} [=] \text{ft}^2$$

(5′) Mass Velocity, G_s:

$$G_s = \frac{\dot{M}}{a_s} [=] \text{lb} / (\text{hr} \cdot \text{ft}^2)$$

(5) Mass Velocity, G_t:

$$G_t = \frac{\dot{m}}{a_t} [=] \text{lb} / (\text{hr} \cdot \text{ft}^2)$$

(6') Compute D_e:

$$D_e = \frac{4 \times \left(P_T^2 - \pi d_o^2/4 \right)}{\pi d_o} [=] \text{ft}$$

(6) Obtain D from Table AT.8:

$$D [=] \text{ft}$$

Obtain μ at T_{avg}

Obtain μ at t_{avg}

(7') Reynolds Number, Re_s:

$$Re_s = \frac{D_e G_s}{\mu}$$

(7) Reynolds Number, Re_t:

$$Re_t = \frac{D G_t}{\mu}$$

(8') Obtain C and k at T_{avg}:

$C [=] \text{Btu/lb}\Delta$

$k [=] \text{Btu/hr}\Delta\text{ft}^2\Delta°\text{F/ft}$

(8) Obtain c and k at t_{avg}:

$c [=] \text{Btu/lb}\Delta°\text{F}$

$k [=] \text{Btu/hr}\Delta\text{ft}^2\Delta°\text{F/ft}$

(9') Solve for h_o and ϕ_s at t_w:
(trial-and-error iterative calculation)

$$t_w = t_c + \frac{h_i}{h_i + h_o}\left(T_c - t_c \right)$$

$$\phi_s = \left(\frac{\mu}{\mu_w} \right)^{0.14}$$

For Laminar Flow (viscous fluid):

$$\frac{h_o D_e}{k}$$

$$= 1.86 \times \left[\left(\frac{D_e G_s}{\mu} \right)\left(\frac{C\mu}{k} \right)\left(\frac{D_e}{L} \right) \right]^{1/3} \left(\frac{\mu}{\mu_w} \right)^{0.14}$$

For Turbulent Flow (viscous fluid):

$$\frac{h_o D_e}{k}$$

$$= 0.027 \left(\frac{D_e G_s}{\mu} \right)^{0.8} \left(\frac{C\mu}{k} \right)^{1/3} \left(\frac{\mu}{\mu_w} \right)^{0.14}$$

(9) Solve for h_i and ϕ_t at t_w:
(trial-and-error iterative calculation)

$$t_w = t_c + \frac{h_i}{h_i + h_o}\left(T_c - t_c \right)$$

$$\phi_t = \left(\frac{\mu}{\mu_w} \right)^{0.14}$$

For Laminar Flow (viscous fluid):

$$\frac{h_i D}{k} = 1.86 \left[\left(\frac{D G_t}{\mu} \right)\left(\frac{c\mu}{k} \right)\left(\frac{D}{L} \right) \right]^{1/3} \left(\frac{\mu}{\mu_w} \right)^{0.14}$$

For Turbulent Flow (viscous fluid):

$$\frac{h_i D}{k} = 0.027 \left(\frac{D G_t}{\mu} \right)^{0.8} \left(\frac{c\mu}{k} \right)^{1/3} \left(\frac{\mu}{\mu_w} \right)^{0.14}$$

For vapor or non-viscous fluids:

$$\frac{h_o D_e}{k} = 0.023\left(\frac{D_e G_s}{\mu}\right)^{0.8}\left(\frac{c\mu}{k}\right)^n$$

For vapor or non-viscous fluids:

$$\frac{h_i D}{k} = 0.023\left(\frac{DG_t}{\mu}\right)^{0.8}\left(\frac{c\mu}{k}\right)^n$$

---

(10) Clean overall coefficient, U_C:

$$U_C = \frac{h_i h_o}{h_i + h_o}[=]\text{Btu/hr} \cdot \text{ft}^2 \cdot {}^\circ\text{F}$$

---

(11) Design overall coefficient, U_D: Obtain external surface per linear ft, a'', from Appendix Table AT.8:

$$a'' [=]\text{ft}^2 \text{ per lin. ft}$$

$$A = a'' N_t L[=]\text{ft}^2$$

$$U_D = \frac{Q}{A\Delta T*}[=]\text{Btu/hr} \cdot \text{ft}^2 \cdot {}^\circ\text{F}$$

---

(12) Dirt factor R_d:

$$R_d = \frac{U_C - U_D}{U_C U_D}[=]\text{hr} \cdot \text{ft}^2 \cdot {}^\circ\text{F/Btu}$$

If R_d equals or exceeds the required dirt factor, proceed to the pressure drop.

Pressure Drop

(1′) For Re_s and T_{avg} in (6′) above:

$$f [=]\text{ft}^2 / \text{in}^2$$

$$s [=]\text{Dimensionless}$$

(2′) Number of crosses, $N + 1$

$$N + 1 = \frac{12L}{B}$$

(3′) $\Delta P_s = \dfrac{fG_s^2 D_s (N+1)}{(5.22 \times 10^{10}) D_e s\phi_s}[=]\text{psi}$

(1) For Re_t and t_{avg} in (6) above:

$$f [=]\text{ft}^2 / \text{in}^2$$

$$s [=]\text{Dimensionless}$$

(2) $\Delta P_t = \dfrac{fG_t^2 Ln}{(5.22 \times 10^{10}) D_e s\phi_t}[=]\text{psi}$

$$\Delta P_r = \left(\frac{4n}{s}\right)\left(\frac{\bar{v}^2}{2g'}\right)\left(\frac{62.5}{144}\right)[=]\text{psi}$$

(3) $\Delta P_T = \Delta P_t + \Delta P_r [=]\text{psi}$

Example 7.21 – Calculation of a Kerosene-Crude Oil Exchanger.[2]
43,800 lb/hr of a API kerosene leaves the bottom of a distilling column
at 390 °F and will be cooled 200 °F by 149,000 lb/hr of 34 °F API Mid-
continent crude coming from storage at 100 °F and heated to 170 °F. A
10 psi pressure drop is permissible on both streams, and in accordance
with Table AT.12, a combined dirt factor of 0.003 should be provided.
Available for this service is a 21¼-in ID exchanger having 158 1-in OD, 13
BWG tubes 16 ft long and laid out on a 1¼-in square pitch. The bundle is
arranged for four passes, and baffles are spaced 5 in. apart.

SOLUTION:

Shell-Side	Tube-Side
ID = 21¼ in	Number and length = 158 tubes, 16'0"
Baffle space = 5 in	OD, BWG, pitch = 1 in, 13 BWG, 1¼-in square pitch
Passes = 1	Passes = 4

(1) Heat Balance:

Hot Kerosene,

$$T_{avg} = \frac{390+200}{2} = 295°F$$

$$C = 0.61 \, \text{Btu/}(\text{lb} \cdot °\text{F}) \text{for } 42°\text{API at } 295 \, °\text{F}$$

$$Q = (43,800)(0.61)(390-200) = 5,100,000 \, \text{Btu/hr}$$

Cold Crude Oil,

$$t_{avg} = \frac{100+170}{2} = 135°F$$

$$c = 0.49 \, \text{Btu/}(\text{lb} \cdot °\text{F}) \text{for } 34°\text{API at } 135°\text{F}$$

$$Q = (149,000)(0.49)(170-100) = 5,100,000 \, \text{Btu/hr}$$

(2) ΔT^*:

$$\text{LMTD} = \frac{\Delta T_a - \Delta T_b}{\ln(\Delta T_a / \Delta T_b)} = \frac{220 - 100}{\ln(220/100)} = 152.2\,°\text{F}$$

$$R = \frac{T_1 - T_2}{t_2 - t_1} = \frac{190}{70} = 2.71$$

$$P = \frac{t_2 - t_1}{T_1 - t_1} = \frac{70}{390 - 100} = 0.241$$

$$F_T = \left(\frac{\sqrt{R^2 + 1}}{R - 1}\right) \frac{\ln\left[(1 - P)/(1 - RP)\right]}{\ln\left[\left(2 - P\left(R + 1 - \sqrt{R^2 + 1}\right)\right) \Big/ \left(2 - P\left(R + 1 + \sqrt{R^2 + 1}\right)\right)\right]}$$

$$= \left(\frac{\sqrt{2.71^2 + 1}}{2.71 - 1}\right)$$

$$\frac{\ln\left[(1 - 0.241)/(1 - (2.71)(0.241))\right]}{\ln\left[\left(2 - 0.241\left(2.71 + 1 - \sqrt{2.71^2 + 1}\right)\right) \Big/ \left(2 - 0.241\left(2.71 + 1 + \sqrt{2.71^2 + 1}\right)\right)\right]}$$

$$= 0.893$$

$$\Delta T^* = \text{LMTD} \times F_T = 152.2\,°\text{F} \times 0.893 = 136\,°\text{F}$$

(3) T_{avg} and t_{avg}: Obtained in step (1).

Since the flow areas of the shell-side and tube-side are nearly equal, assume the larger stream flows in the tubes:

Hot Fluid (Kerosene) – Shell-Side (s)

(4′) Flow Area, a_s:

$$a_s = \frac{\text{ID} \times C'B}{144 P_T}$$

$$= \frac{(21.25)(0.25)(5)}{144(1.25)} = 0.1475\,\text{ft}^2$$

Cold Fluid (Crude Oil) – Tube-Side (t)

(4) Flow Area, a_t:

$$a_t' = 0.515\,\text{in}^2 \Leftarrow \left[\text{Table AT.8}\right]$$

$$a_t = \frac{N_t a_t'}{144 n}$$

$$= \frac{(158)(0.515)}{144(4)} = 0.141\,\text{ft}^2$$

(5′) Mass Velocity, G_s:

$$G_s = \frac{\dot{M}}{a_s} = \frac{43,800}{0.1475}$$

$$= 297,000 \text{lb}/\left(\text{hr} \cdot \text{ft}^2\right)$$

(5) Mass Velocity, G_t:

$$G_t = \frac{\dot{m}}{a_t} = \frac{149,000}{0.141}$$

$$= 1,060,000 \text{lb}/\left(\text{hr} \cdot \text{ft}^2\right)$$

(6′) Compute D_e:

$$D_e = \frac{4 \times \left(P_T^2 - \pi d_o^2/4\right)}{\pi d_o}$$

$$= \frac{4 \times \left(\left(1.25^2\right) - \pi\left(1^2\right)/4\right)}{\pi(1)}$$

$$= 0.99 \text{ in} = \frac{0.99 \text{ in}}{12 \text{ in/ft}} = 0.0825 \text{ ft}$$

(6) Obtain D from Table AT.8:

$$D = 0.81 \text{ in}$$

$$= \frac{0.81 \text{ in}}{12 \text{ in/ft}} = 0.0675 \text{ ft}$$

Obtain μ at T_{avg}:

$\mu = 0.40$ cP at 280°F

$$= 0.40 \text{ cP}\left(\frac{2.42 \text{lb}/(\text{ft} \cdot \text{hr})}{1 \text{ cP}}\right)$$

$$= 0.97 \text{lb}/(\text{ft} \cdot \text{hr})$$

Obtain μ at t_{avg}:

$\mu = 3.6$ cP at 129

$$= 3.6 \text{ cP}\left(\frac{2.42 \text{lb}/(\text{ft} \cdot \text{hr})}{1 \text{ cP}}\right)$$

$$= 8.7 \text{lb}/(\text{ft} \cdot \text{hr})$$

(7′) Reynolds Number, Re_s:

$$\text{Re}_s = \frac{D_e G_s}{\mu}$$

$$= \frac{(0.0825)(297,000)}{0.97} = 25,300$$

(7) Reynolds Number, Re_t:

$$\text{Re}_t = \frac{D G_t}{\mu}$$

$$= \frac{(0.0675)(1,060,000)}{8.7} = 8,220$$

(8′) Obtain C and k at T_{avg}:

$C = 0.59$ Btu/lb · °F at 295°F

$k = 0.0765$ Btu/hr · ft² · °F/ft at 295°F

(8) Obtain c and k at t_{avg}:

$c = 0.49$ Btu/lb · °F at 135°F

$k = 0.077$ Btu/hr · ft² · °F/ft at 135°F

(9′) Solve for h_o and ϕ_s at t_w:

(9) Solve for h_i and ϕ_t at t_w:

$t_w = 202°F \Leftarrow$ from Iterative calc.

$\mu_w = 0.66\, cP$ at $202°F$

$= 0.66\, cP \left(\dfrac{2.42 lb/(ft \cdot hr)}{1\, cP} \right)$

$= 1.6 lb/(ft \cdot hr)$

$\phi_s = \left(\dfrac{\mu}{\mu_w} \right)^{0.14} = \left(\dfrac{0.97}{1.6} \right)^{0.14} = 0.93$

$h_o = 164\, Btu/hr \cdot ft^2 \cdot °F/ft$ at $202°F$

$t_w = 202°F \Leftarrow$ from Iterative calc.

$\mu_w = 1.81\, cP$ at $202°F$

$= 1.81 cP \left(\dfrac{2.42 lb/(ft \cdot hr)}{1\, cP} \right)$

$= 4.4 lb/(ft \cdot hr)$

$\phi_t = \left(\dfrac{\mu}{\mu_w} \right)^{0.14} = \left(\dfrac{8.7}{4.4} \right)^{0.14} = 1.1$

$h_i = 154\, Btu/hr \cdot ft^2 \cdot °F/ft$ at $202°F$

(10) Clean overall coefficient, U_C:

$$U_C = \frac{h_i h_o}{h_i + h_o} = \frac{(154)(164)}{154 + 164} = 79.4\, Btu/hr \cdot ft^2 \cdot °F$$

(11) Design overall coefficient, U_D:

$$a'' = 0.2618\, ft^2 \text{ per lin. ft} \Leftarrow [\text{Table AT.8}]$$

$$A = a'' N_t L = (0.2618)(158)(16) = 662\, ft^2$$

$$U_D = \frac{Q}{A \Delta T^*} = \frac{5,100,000}{(662)(136)} = 56.6\, Btu/hr \cdot ft^2 \cdot °F$$

(12) Dirt factor R_d:

$$R_d = \frac{U_C - U_D}{U_C U_D} = \frac{79.4 - 56.6}{(79.4)(56.6)} = 0.00507\, hr \cdot ft^2 \cdot °F/Btu$$

Summary

Variable	Value	Units
h_o	= 164	Btu
h_i	= 154	hr $\cdot$ ft^2 $\cdot$ °F/ft
U_C	= 79.4	Btu
U_D	= 56.6	hr $\cdot$ ft^2 $\cdot$ °F
R_d (calculated)	= 0.00507	hr $\cdot$ ft^2 $\cdot$ °F
R_d (required)	= 0.00300	Btu

Pressure Drop

(1') For Re_s and T_{avg} in (6') above:

$$f = 0.00175 \frac{ft^2}{in^2} \text{ at } Re_s = 25,300$$

$s = 0.73$ for 42° API at 280°F

$$D_s = 21.25 \text{ in} \left(\frac{1 \text{ ft}}{12 \text{ in}} \right) = 1.77 \text{ ft}$$

(1) For Re_t and t_{avg} in (6) above:

$$f = 0.000285 \frac{ft^2}{in^2} \text{ at } Re_t = 8,220$$

$s = 0.83$ for 34° API at 129°F

(2') Number of crosses, $N + 1$:

$$N + 1 = \frac{12L}{B} = \frac{12(16)}{5} = 39$$

(2) $\Delta P_t = \dfrac{fG_t^2 Ln}{\left(5.22 \times 10^{10}\right) D_e s \phi_t}$

$$= \frac{(0.000285)(1,060,000)^2 (16)(4)}{(5.22 \times 10^{10})(0.0675)(0.83)(1.1)}$$

$$= 6.3 \text{ psi}$$

$$\frac{\bar{v}^2}{2g} \left(\frac{62.5}{144} \right)$$

$$= 0.15 \text{ at } G_t = 1,060,000$$

$$\Delta P_r = \left(\frac{4n}{s} \right) \left(\frac{\bar{v}^2}{2g} \right) \left(\frac{62.5}{144} \right)$$

$$= \left(\frac{4(4)}{0.83} \right)(0.15) = 2.9 \text{ psi}$$

(3') $\Delta P_s = \dfrac{fG_s^2 D_s (N+1)}{\left(5.22 \times 10^{10}\right) D_e s \phi_s}$

$$\Delta P_s = \frac{(0.00175)(297,000)^2 (1.77)(39)}{(5.22 \times 10^{10})(0.0825)(0.73)(0.93)}$$

$\Delta P_s = 3.6 \text{ psi}$

$\Delta P_s < 10 \text{ psi as allowable } \Delta P$

(3) $\Delta P_T = \Delta P_t + \Delta P_r$

$$\Delta P_T = 6.3 + 2.9 = 9.2 \text{psi}$$

$\Delta P_T < 10 \text{ psi as allowable } \Delta P$

It is seen that a dirt factor of 0.00507 will be obtained although only 0.00300 will be required to provide a reasonable maintenance period. The pressure drops have not been exceeded and the exchanger will be satisfactory for service.

Simplifying Kern's Design Methodology. McAdams [11] simplified some of Kern's original design methodology by providing what he defined as the "General Case." For this situation, the design of most shell-and-tube exchangers involve initial conditions in which the following variables are known:

1. Process-fluid rate of flow
2. Change in temperature of process fluid
3. Inlet temperature of utility fluid (for cooling or heating)

With this information, the engineer must prepare a design for the optimum exchanger that will meet the required process conditions. Ordinarily, the following results must be determined:

1. Heat transfer area
2. Exit temperature and flow rate of utility fluid
3. Number, length, diameter, and arrangement of tubes
4. Tube-side and shell-side pressure drops

McAdams [11] also provided what he termed a "summary of procedure for the general case of optimum design." In the preceding analysis, consideration has been given to the general case in heat exchanger design in which the following conditions apply:

1. The flow rate and necessary temperature change of the process fluid are known.
2. The inlet temperature of the utility fluid is known.
3. The exchanger is a shell-and-tube type with crossflow baffling, and the flow is in the turbulent range on both the tube-side and the shell-side.
4. No partial phase changes occur.
5. Necessary safety factors are known.

The following information may be specified for the design or can be assumed as a reasonable approximation:

1. Tube diameter, wall thickness, pitch, and arrangement
2. Number of tube passes
3. Heat-transfer resistance caused by tube walls, dirt, and scale

A detailed outline and specific calculational procedures are also provided by McAdams [11].

7.5   Other Design Procedures and Applications

Kern examines other areas of interest to the practicing engineer [2]. They include:

1. Exchangers Using Water
2. Optimum Outlet Water Temperature
3. Solution Exchangers
4. Steam as a Heating Medium
5. The Optimum Use of Exhaust and Process Steam
6. 1–2 Exchanger without Baffles
7. The Efficiency of an Exchanger

An example is provided (employing Kern's [2] design methodology) for conditions 1, 3, and 6 above.

Exchangers Using Water. Cooling operations using water in tubular equipment are very common. Despite its abundance, the heat-transfer characteristics of water separate it from all other fluids. It is corrosive to steel, particularly when the tube-wall temperature is high and dissolved air is present, and many industrial plants use nonferrous tubes exclusively for heat-transfer services involving water. The commonest nonferrous tubes are admiralty, red brass, and copper, although in certain localities there is preference for Muntz metal, aluminum bronze, and aluminum. Since shells are usually fabricated of steel, water is best handled in the tubes. When water flows in the tubes, there is no serious problem of corrosion of the channel or floating-head cover since these parts are often made of cast iron or cast steel. Castings are relatively passive to water, and large corrosive allowances above structural requirements can be provided inexpensively by making the castings heavier. Tube sheets may be made of heavy steel plates with a corrosion allowance of about 1/8 in. above the required structural thickness or fabricated of naval brass or aluminum without a corrosion allowance.

When water travels slowly through a tube, dirt and slime resulting from micro-organic action adhere to the tubes which would be carried away if there were greater turbulence. As a standard practice, the use of cooling water at velocities less than 3 fps should be avoided, although in certain localities minimum velocities as high as 4 fps are required for continued operation. Still another factor of considerable importance is the deposition of mineral scale. When water of average mineral and air content is brought to a temperature in excess of 120 °F, it is found that tube action becomes excessive, and for this reason an outlet water temperature above 120 °F should be avoided.

Cooling water is rarely abundant or without cost. One of the serious problems facing the chemical and power industries today results from the gradual deficiency of surface and subsurface water in areas of industrial concentration. This can be partially overcome through the use of cooling towers, which reuse the cooling water and reduce the requirement to only 2 percent of the amount of water required in once-through use. River water may provide part of the solution to a deficiency of ground water, but it is costly and presupposes the proximity of a river. River water must usually be strained by moving screens and pumped considerable distances, and in some localities the water from rivers servicing congested industrial areas requires cooling in cooling towers before it can be used, reused, or discharged back into the river.

Many sizable municipalities have legislated against the use of public water supplies for large cooling purposes other than for make-up in cooling towers or spray-pond systems. Where available, municipal water costs may average about 1 cent per 1000 gal., although it has the advantage of being generally available from 30 to 60 psi pressure which is adequate for most process needs including the pressure drops in heat exchangers. When a cooling tower is used, the cost of the water is determined by the cost of fresh water, pumping power, fan power, and write-off on the original investment.

In a water-to-water exchanger with individual film coefficients ranging from 500 to 1500 Btu/hr $\cdot$ ft^2 $\cdot$ °F for both the shell-and-tube, the selection of the required dirt factor merits serious judgment. As an example, if film coefficients of 1000 are obtained on the shell-and-tube-sides, the combined resistance is 0.002, or $U_C = 500$. If a fouling factor of 0.004 is required, the fouling factor becomes the controlling resistance. When the fouling factor is 0.004, U_D must be less than 1/0.004 or 250. Whenever high coefficients exist on both sides of the exchanger, the use of an unnecessarily large fouling factor should be avoided.

The following heat-recovery example typically occurs in powerhouses. Although it involves a moderately sized exchanger, the heat recovery is equivalent to nearly 1500 lb/hr of steam, representing a reasonable savings in utility costs.

Example 7.22 – Calculation of a Distilled-water-Raw-water Exchanger.
175,000 lb/hr of distilled water enters an exchanger at 93 °F and leaves at 85 °F. The heat will be transferred to 280,000 lb/hr of raw water coming from supply at 75 °F and leaving the exchanger at 80 °F. A 10 psi pressure

drop may be expended on both streams while providing a fouling factor of 0.0005 for distilled water and 0.0015 for raw water when the tube velocity exceeds 6 fps.

Available for this service is a 15¼ in. ID exchanger having 160¾ in. OD, 18 BWG tubes 16'0" long and laid out on 15/16-in. triangular pitch. The bundle is arranged for two passes, and baffles are spaced 12 in. apart.

Will the exchanger be suitable?

Solution:

Shell-Side	Tube-Side
ID = 15¼ in	Number and length = 160 tubes, 16'0"
Baffle space = 12 in	OD, BWG, pitch = ¾ in, 18 BWG, 15/16 -in triangular pitch
Passes = 1	Passes = 2

Heat Balance:

$$\text{Hot Distilled Water, } T_{avg} = \frac{93+85}{2} = 89°F$$

$$C \approx 1 \, \text{Btu}/(\text{lb} \cdot °F) \text{ at } 89 \, °F$$

$$Q = (175,000)(1)(93-85) = 1,400,000 \text{ Btu/hr}$$

$$\text{Cold Raw Water, } t_{avg} = \frac{80+75}{2} = 77.5°F$$

$$c \approx 1 \, \text{Btu}/(\text{lb} \cdot °F) \text{ at } 77.5°F$$

$$Q = (280,000)(1)(80-75) = 1,400,000 \text{ Btu/hr}$$

Note: C and c are equal to each other because (i) both fluids are water, and (ii) stream temperatures are similar.

(2) ΔT^*:

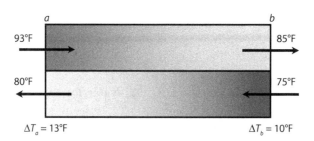

93°F   85°F

80°F   75°F

$\Delta T_a = 13°F$ $\Delta T_b = 10°F$

$$\text{LMTD} = \frac{\Delta T_a - \Delta T_b}{\ln\left(\Delta T_a / \Delta T_b\right)} = \frac{13 - 10}{\ln\left(13/10\right)} = 11.4\,^\circ\text{F}$$

$$R = \frac{T_1 - T_2}{t_2 - t_1} = \frac{8}{5} = 1.6$$

$$P = \frac{t_2 - t_1}{T_1 - t_1} = \frac{5}{93 - 75} = 0.278$$

$$F_T = \left(\frac{\sqrt{R^2 + 1}}{R - 1}\right) \frac{\ln\left[\left(1 - P\right) \Big/ \left(1 - RP\right)\right]}{\ln\left[\left(2 - P\left(R + 1 - \sqrt{R^2 + 1}\right)\right) \Big/ \left(2 - P\left(R + 1 + \sqrt{R^2 + 1}\right)\right)\right]}$$

$$= \left(\frac{\sqrt{1.6^2 + 1}}{1.6 - 1}\right)$$

$$\frac{\ln\left[\left(1 - 0.278\right) \Big/ \left(1 - \left(1.6\right)\left(0.278\right)\right)\right]}{\ln\left[\left(2 - 0.278\left(1.6 + 1 - \sqrt{1.6^2 + 1}\right)\right) \Big/ \left(2 - 0.278\left(1.6 + 1 + \sqrt{1.6^2 + 1}\right)\right)\right]}$$

$$= 0.946$$

$$\Delta T^* = \text{LMTD} \times F_T = 11.4\,^\circ\text{F} \times 0.946 = 10.8\,^\circ\text{F}$$

(3) T_{avg} and t_{avg}: Obtained in step (1).

Hot Fluid (Distilled Water) – Shell-Side (s)   *Cold Fluid (Raw Water) – Tube-side (t)*

(4′) Flow Area, a_s:

(4) Flow Area, a_t:

$$a_s = \frac{ID \times C'B}{144 P_T}$$

$$a_t' = 0.334 \text{ in}^2 \Leftarrow \left[\text{Table AT.8}\right]$$

$$a_t = \frac{N_t a_t'}{144 n}$$

$$= \frac{\left(15.25\right)\left(0.1875\right)\left(5\right)}{144\left(0.9375\right)} = 0.254 \text{ ft}^2$$

$$= \frac{\left(160\right)\left(0.334\right)}{144\left(2\right)} = 0.186 \text{ ft}^2$$

(5′) Mass Velocity, G_s:

(5) Mass Velocity, G_t:

$$G_s = \frac{\dot{M}}{a_s} = \frac{175{,}000}{0.254}$$

$$G_t = \frac{\dot{m}}{a_t} = \frac{280{,}000}{0.186}$$

$$= 690{,}000\,\text{lb}/\left(\text{hr} \cdot \text{ft}^2\right)$$

$$= 1{,}505{,}000\,\text{lb}/\left(\text{hr} \cdot \text{ft}^2\right)$$

(6') Compute D_e:

$$D_e = \frac{3.44 P_T^2 - \pi d_o^2}{\pi d_o}$$

$$= \frac{3.44\left(0.9375^2\right) - \pi\left(0.75^2\right)}{\pi\left(0.75\right)}$$

$$= 53 \text{ in} = \frac{0.53 \text{ in}}{12 \text{ in/ft}} = 0.0442 \text{ ft}$$

Obtain μ at T_{avg}:

$$\mu = 0.81 \text{ cP at } 89 \text{ °F}$$

$$= 0.81 \text{ cP}\left(\frac{2.42 \text{lb}/\left(\text{ft} \cdot \text{hr}\right)}{1 \text{cP}}\right)$$

$$= 1.96 \text{lb}/\left(\text{ft} \cdot \text{hr}\right)$$

(7') Reynolds Number, Re_s:

$$Re_s = \frac{D_e G_s}{\mu}$$

$$= \frac{\left(0.0442\right)\left(690{,}000\right)}{1.96} = 16{,}200$$

(8') Obtain C and k at T_{avg}:

$$C \approx 1 \text{ Btu/lb} \cdot \text{°F at } 89 \text{ °F}$$

$$k = 0.36 \text{ Btu/hr} \cdot \text{ft}^2 \cdot \text{°F/ft at } 89\text{°F}$$

(9') Solve for h_o and ϕ_s at t_w:
The small difference in the average temperatures eliminates the need for a tube-wall correction, and therefore ϕ_s can be approximated as 1.0 (as mentioned in step (2)).

$$\phi_s = \left(\frac{\mu}{\mu_w}\right)^{0.14} \approx 1.0$$

(6) Obtain D from Table 10:

$$D = 0.65 \text{ in} \Leftarrow \left[\text{Table AT.8}\right]$$

$$= \frac{0.65 \text{ in}}{12 \text{ in/ft}} = 0.054 \text{ ft}$$

Obtain μ at t_{avg}:

$$\mu = 0.92 \text{ cP at } 77.5 \text{ °F}$$

$$= 0.92 \text{ cP}\left(\frac{2.42 \text{lb}/\left(\text{ft} \cdot \text{hr}\right)}{1 \text{cP}}\right)$$

$$= 2.23 \text{lb}/\left(\text{ft} \cdot \text{hr}\right)$$

(7) Reynolds Number, Re_t:

$$Re_t = \frac{D G_t}{\mu}$$

$$= \frac{\left(0.054\right)\left(1{,}505{,}000\right)}{2.23} = 36{,}400$$

(Re_t is for pressure drop only)

(8) Obtain c and k at t_{avg}:

$$c \approx 1 \text{ Btu}/\text{lb} \cdot \text{°F at } 77.5 \text{ °F}$$

$$k = 0.36 \text{Btu/hr} \cdot \text{ft}^2 \cdot \text{°F/ft at } 77.5\text{°F}$$

(9) Solve for h_i and ϕ_t at t_w:
The small difference in the average temperatures eliminates the need for a tube-wall correction, and therefore ϕ_t can be approximated as 1.0 (as mentioned in step (2)).

$$\phi_t = \left(\frac{\mu}{\mu_w}\right)^{0.14} \approx 1.0$$

(9′) $h_o = 0.027 \left(\dfrac{D_e G_s}{\mu} \right)^{0.8} \left(\dfrac{C\mu}{k} \right)^{\frac{1}{3}} \left(\dfrac{\mu}{\mu_w} \right)^{0.14} \left(\dfrac{k}{D_e} \right)$

$\quad = 0.027 \left(\dfrac{G_s}{\mu} \right)^{0.8} \left(C\mu k^2 \right)^{\frac{1}{3}} \left(\dfrac{\mu}{\mu_w} \right)^{0.14} \left(\dfrac{1}{D_e} \right)^{0.2}$

$\quad = 0.027 \left(\dfrac{690{,}000}{1.96} \right)^{0.8} \left((1)(1.96)(0.36)^2 \right)^{\frac{1}{3}} (1.0) \left(\dfrac{1}{0.0442} \right)^{0.2}$

$\quad \approx 873 \dfrac{\text{Btu}}{\text{hr} \cdot \text{ft}^2 \cdot {}^\circ\text{F}}$

(9) $h_i = 0.027 \left(\dfrac{D G_t}{\mu} \right)^{0.8} \left(\dfrac{c\mu}{k} \right)^{\frac{1}{3}} \left(\dfrac{\mu}{\mu_w} \right)^{0.14} \left(\dfrac{k}{D} \right)$

$\quad = 0.027 \left(\dfrac{G_t}{\mu} \right)^{0.8} \left(c\mu k^2 \right)^{\frac{1}{3}} \left(\dfrac{\mu}{\mu_w} \right)^{0.14} \left(\dfrac{1}{D} \right)^{0.2}$

$\quad = 0.027 \left(\dfrac{1{,}505{,}000}{2.23} \right)^{0.8} \left((1)(2.23)(0.36)^2 \right)^{\frac{1}{3}} (1.0) \left(\dfrac{1}{0.054} \right)^{0.2}$

$\quad \approx 1474 \ \text{Btu/hr} \cdot \text{ft}^2 \cdot {}^\circ\text{F}$

(10) Clean overall coefficient, U_C:

$$U_C = \frac{h_i h_o}{h_i + h_o} = \frac{(873)(1474)}{873 + 1474} = 548 \ \text{Btu/hr} \cdot \text{ft}^2 \cdot {}^\circ\text{F}$$

When both film coefficients are high, the thermal resistance of the tube metal is not necessarily insignificant as assumed in the derivation of Eq. (6.38). For a steel 18 BWG tube $R_m = 0.00017$ and for copper $R_m = 0.000017$.

(11) Design overall coefficient, U_D:

$$a'' = 0.1963 \ \text{ft}^2 \ \text{per lin. ft} \Leftarrow \left[\text{Table AT.8} \right]$$

$$A = a'' N_t L = (0.1963)(160)(16) = 502 \ \text{ft}^2$$

$$U_D = \frac{Q}{A \Delta T^*} = \frac{1{,}400{,}000}{(502)(10.8)} = 258 \ \text{Btu/hr} \cdot \text{ft}^2 \cdot {}^\circ\text{F}$$

(12) Dirt factor R_d:

$$R_d = \frac{U_C - U_D}{U_C U_D} = \frac{548 - 258}{(548)(258)} = 0.00205 \, \text{hr} \cdot \text{ft}^2 \cdot {}^\circ\text{F} / \text{Btu} \qquad (6.13)$$

Summary

Variable	Value	Units
h_o	= 873	Btu
h_i	= 1474	$\overline{\text{hr} \cdot \text{ft}^2 \cdot {}^\circ\text{F}/\text{ft}}$
U_C	= 548	Btu
U_D	= 258	$\overline{\text{hr} \cdot \text{ft}^2 \cdot {}^\circ\text{F}}$
R_d (calculated)	= 0.00205	$\text{hr} \cdot \text{ft}^2 \cdot {}^\circ\text{F}$
R_d (required)	= 0.00200	$\overline{\text{Btu}}$

Pressure Drop

(1) Using Re_s and T_{avg} in (6'), obtain f:

$$f = 0.0019 \frac{ft^2}{in^2} \text{ at } Re_s = 16,200$$

$s = 1.0 \Leftarrow$ water

$$D_s = 15.25 \, \text{in} \left(\frac{1 \, \text{ft}}{12 \, \text{in}} \right) = 1.27 \, \text{ft}$$

(1') Using Re_t and t_{avg} in (6) above:

$$f = 0.00019 \frac{ft^2}{in^2} \text{ at } Re_t = 36,400$$

$s = 1.0 \Leftarrow$ water

(2) Number of crosses, $N + 1$:

$$N + 1 = \frac{12L}{B} = \frac{12(16)}{12} = 16$$

(2') $\Delta P_t = \dfrac{fG_t^2 Ln}{\left(5.22 \times 10^{10}\right) D_e s \phi_t}$

$$= \frac{(0.00019)(1,505,000)^2 (16)(2)}{\left(5.22 \times 10^{10}\right)(0.054)(1.0)(1.0)}$$

$$= 4.9 \, \text{psi}$$

$$\frac{\bar{v}^2}{2g} \left(\frac{62.5}{144} \right) = 0.33 \text{ at } G_t = 1,505,000$$

$$\Delta P_r = \left(\frac{4n}{s} \right) \left(\frac{\bar{v}^2}{2g} \right) \left(\frac{62.5}{144} \right)$$

$$= \left(\frac{4(2)}{1.0} \right) (0.33) = 2.6 \, \text{psi}$$

$$(3') \quad \Delta P_s = \frac{f G_s^2 D_s (N+1)}{(5.22 \times 10^{10}) D_e s \phi_s}$$

$$\Delta P_s = \frac{(0.0019)(690,000)^2 (1.27)(16)}{(5.22 \times 10^{10})(0.0458)(1.0)(1.0)}$$

$$\Delta P_s = 7.7 \text{ psi}$$

$\Delta P_s < 10$ psi as allowable ΔP

$$(3') \quad \Delta P_T = \Delta P_t + \Delta P_r$$

$$\Delta P_T = 4.9 + 2.6 = 7.5 \text{ psi}$$

$\Delta P_T < 10$ psi as allowable ΔP

It is seen that the overall design coefficient, U_D, for Example 7.22 is five times that of the oil-to-oil exchanger in Example 7.21. The principal reason for the difference is the excellent thermal properties of water. This exchanger is satisfactory for this service.

Optimum Outlet-water Temperature. In using water as the cooling medium for a given duty it is possible to circulate a large quantity with a small temperature range or a small quantity with a large temperature range. The temperature range of the water naturally affects the LMTD. If a large quantity is used, t_2 will be farther from T_1 and less surface is required as a result of the larger LMTD. Although this will reduce the original investment and fixed charges, since depreciation and maintenance will also be smaller, the operating cost will be increased owing to the greater quantity of water. It is apparent that there must be an optimum between the two conditions: much water and small surface or little water and large surface.

When the value of U is high or there is a large hot-fluid range, the optimum outlet-water temperature may be considerably above the upper limit of 120 °F. This is not completely correct, since the maintenance cost will probably rise considerably above 20 percent of the initial cost when the temperature rises above 120 °F. Information is usually not available on the increase in maintenance cost with increased water-outlet temperature, since such data entail not only destructive tests but also records kept over a long period of time.

Solution Exchangers. One of the more common classes of exchangers embraces the cooling or heating of solutions for which there is a minimal amount of physical data. This is understandable, since property vs. temperature plots are required not only for each combination of solute and solvent but for different concentrations as well. Some of the data available in the literature and other studies permit the formulation of rules for estimating the heat-transfer properties of solutions when the rules are used with considerable caution. They are given as follows:

Thermal conductivity (Tables AT.3 - AT.6):

- Solutions of *organic liquids*: use the weighted conductivity.
- Solutions of organic liquids and water: use 0.9 times the weighted conductivity.
- Solutions of salts and water circulated through the shell: use 0.9 times the conductivity of water up to concentrations of 30 percent.
- Solutions of salts and water circulating through the tubes and not exceeding 30 percent: use Table AT.3 with a conductivity equal to 0.8 that of water.
- Colloidal dispersions: use 0.9 times the conductivity of the dispersion liquid.
- Emulsions: use 0.9 times the conductivity of the liquid surrounding the droplets.

Specific heat (or heat capacity):

- Organic solutions: use the weighted specific heat.
- Organic solutions in water: use the weighted specific heat.
- Fusable salts in water: use the weighted specific heat where the specific heat of the salt is for the crystalline state.

Viscosity:

- Organic liquids in organics: use the reciprocal of the sum of the terms, (weight fraction/viscosity) for each component.
- Organic liquids in water: use the reciprocal of the sum of the terms, (weight fraction/viscosity) for each component.
- Salts in water where the concentration does not exceed 30 percent and where it is known that a syrup-type of solution does not result: use a viscosity twice that of water. A solution of sodium hydroxide in water under even very low concentrations should be considered syrupy and cannot be estimated.

Wherever laboratory tests are available or data can be obtained, they will be preferable to any of the foregoing rules. The following demonstrates the above via a problem involving an aqueous solution.

Example 7.23 – Calculation of a Phosphate Solution Cooler. 20,160 lb/hr of a 30% K_3PO_4 solution, specific gravity at 120 °F = 1.30, is to be cooled from 150 to 90 °F using well water from 68 to 90 °F. Pressure drops of 10 psi are allowable on both streams, and a total dirt factor of 0.002 is required.

Available for this service is a 10.02 in. ID 1–2 exchanger having 52¾ in. OD, 16 BWG tubes 16'0" long laid out on 1-in. square pitch. The bundle is arranged for two passes, and the baffles are spaced 2 in. apart. Will the exchanger be suitable?

SOLUTION:

Shell-Side	Tube-Side
ID = 10.02 in	Number and length = 52 tubes, 16'0"
Baffle space = 2 in	OD, BWG, pitch = ¾ in, 18 BWG, 1-in square pitch
Passes = 1	Passes = 2

(1) Heat Balance:

Hot K_3PO_4 solution, $\quad T_{mean} = \dfrac{150+90}{2} = 120°F$

$$C_{K_3PO_4} = 0.19 \text{ Btu}/(\text{lb}\cdot°F) \text{ at } 120°F$$

$$C_{H_2O} \approx 1.0 \text{ Btu}/(\text{lb}\cdot°F) \text{ at } 120°F$$

Therefore, $\qquad C = X_{K_3PO_4} \cdot C_{K_3PO_4} + X_{H_2O} \cdot C_{H_2O}$

$$= (0.3)(0.19)+(0.7)(1.0)$$

$$= 0.757 \text{ Btu}/(\text{lb}\cdot°F)$$

$$Q = (20{,}160)(0.757)(150-90) = 915{,}000 \text{ Btu/hr}$$

Cold Well Water, $\quad t_{mean} = \dfrac{90+68}{2} = 79°F$

$$c \approx 1.0 \text{ Btu}/(\text{lb}\cdot°F) \text{ at } 79°F$$

$$Q = (41{,}600)(1.0)(90-68) = 915{,}000 \text{ Btu/hr}$$

(2) ΔT^*:

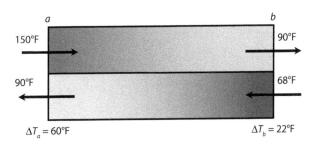

a 150°F   90°F b

90°F   68°F

$\Delta T_a = 60°F$ $\Delta T_b = 22°F$

$$\text{LMTD} = \frac{\Delta T_a - \Delta T_b}{\ln\left(\Delta T_a / \Delta T_b\right)} = \frac{60 - 22}{\ln\left(60 / 22\right)} = 37.9°\text{F}$$

$$R = \frac{T_1 - T_2}{t_2 - t_1} = \frac{60}{22} = 2.73$$

$$S = \frac{t_2 - t_1}{T_1 - t_1} = \frac{22}{150 - 68} = 0.268$$

$$F_T = \left(\frac{\sqrt{R^2 + 1}}{R - 1}\right) \frac{\ln\left[(1 - P) \big/ (1 - RP)\right]}{\ln\left[\left(2 - P\left(R + 1 - \sqrt{R^2 + 1}\right)\right) \big/ \left(2 - P\left(R + 1 + \sqrt{R^2 + 1}\right)\right)\right]}$$

$$= \left(\frac{\sqrt{2.73^2 + 1}}{2.73 - 1}\right)$$

$$\times \frac{\ln\left[(1 - 0.268) \big/ (1 - (2.73)(0.268))\right]}{\ln\left[\left(2 - 0.268\left(2.73 + 1 - \sqrt{2.73^2 + 1}\right)\right) \big/ \left(2 - 0.268\left(2.73 + 1 + \sqrt{2.73^2 + 1}\right)\right)\right]}$$

$$= 0.81$$

$$\Delta T^* = \text{LMTD} \times F_T = 37.9°\text{F} \times 0.81 = 30.7°\text{F}$$

(3) T_{avg} and t_{avg}: Obtained in step (1).

Hot Fluid (Phosphate Solution) – Shell-Side (s)	Cold Fluid (Water) – Tube-Side (t)

(4) Flow Area, a_s:

$$a_s = \frac{ID \times C'B}{144 P_T}$$

$$= \frac{(10.02)(0.25)(2)}{144(1)} = 0.0347 \text{ ft}^2$$

(4′) Flow Area, a_t:

$$a_t' = 0.302 \text{ in}^2 \Leftarrow [\text{Table AT.8}]$$

$$a_t = \frac{N_t a_t'}{144 n}$$

$$= \frac{(52)(0.302)}{144(2)} = 0.0545 \text{ ft}^2$$

(5) Mass Velocity, G_s:

$$G_s = \frac{\dot{M}}{a_s} = \frac{20,160}{0.0347}$$

$$= 578,000 \text{lb}/\left(\text{hr} \cdot \text{ft}^2\right)$$

(5′) Mass Velocity, G_t:

$$G_t = \frac{\dot{m}}{a_t} = \frac{41,600}{0.0545}$$

$$= 762,000 \text{lb}/\left(\text{hr} \cdot \text{ft}^2\right)$$

(6) Compute D_e:

$$D_e = \frac{4P_T^2 - \pi d_o^2}{\pi d_o}$$

$$= \frac{4(1^2) - \pi(0.75^2)}{\pi(0.75)}$$

$$= 0.95 \text{in} = \frac{0.95 \text{in}}{12 \text{in/ft}} = 0.079 \text{ ft}$$

Obtain μ at T_{avg}:

$$\mu = 2\mu_{water} \Leftarrow \text{for nonviscous salt sol.}$$

$$\mu_{water} = 0.60 \text{cP at } 120°\text{F}$$

$$\mu = 2(0.60) = 1.20 \text{cP}$$

$$= 1.20 \text{cP}\left(\frac{2.42 \text{lb}/(\text{ft}\cdot\text{hr})}{1 \text{cP}}\right)$$

$$= 2.90 \text{lb}/(\text{ft}\cdot\text{hr})$$

(6') Obtain D from Table AT.8:

$$D = 0.62 \text{ in}$$

$$= \frac{0.62 \text{ in}}{12 \text{in/ft}} = 0.0517 \text{ ft}$$

Obtain μ at t_{avg}:

$$\mu = 0.91 \text{cP at } 79°\text{F}$$

$$= 0.91 \text{cP}\left(\frac{2.42 \text{lb}/(\text{ft}\cdot\text{hr})}{1 \text{cP}}\right)$$

$$= 2.20 \text{lb}/(\text{ft}\cdot\text{hr})$$

---

(7) Reynolds Number, Re_s:

$$Re_s = \frac{D_e G_s}{\mu}$$

$$= \frac{(0.079)(578,000)}{2.90} = 15,750$$

(7') Reynolds Number, Re_t:

$$Re_t = \frac{DG_t}{\mu}$$

$$= \frac{(0.0517)(762,000)}{2.20} = 17,900$$

(Re_t is for pressure drop only)

---

(8) Obtain C and k at T_{avg}:

$$C = 0.757 \text{Btu/lb}\cdot°\text{F at } 120°\text{F}$$

$$k = 0.9 k_{water} \Leftarrow \text{for low conc.salt sol.}$$

$$k_{water} = 0.37 \text{ Btu/hr}\cdot\text{ft}^2\cdot°\text{F/ft at } 120°\text{F}$$

$$k = 0.9(0.37) = 0.33 \text{ Btu/hr}\cdot\text{ft}^2\cdot°\text{F/ft}$$

(8') Obtain c and k at t_{avg}:

$$c = 1 \text{Btu/lb}\cdot°\text{F at } 79°\text{F}$$

$$k = 0.35 \text{Btu/hr}\cdot\text{ft}^2\cdot°\text{F/ft at } 79°\text{F}$$

(9) Solve for h_o and ϕ_s at t_w: (9') Solve for h_i and ϕ_t at t_w:

ϕ_s and ϕ_t can be assumed approximately equal to 1.0 as the viscosity of Newtonian fluids tend to be constant over a small temperature range. Therefore, the tube-wall temperature correction factor is not necessary in this case.

$$\phi_s = \left(\frac{\mu}{\mu_w}\right)^{0.14} \approx 1.0 \qquad\qquad \phi_t = \left(\frac{\mu}{\mu_w}\right)^{0.14} \approx 1.0$$

(9) $\quad h_o = 0.027\left(\dfrac{D_e G_s}{\mu}\right)^{0.8}\left(\dfrac{C\mu}{k}\right)^{\frac{1}{3}}\left(\dfrac{\mu}{\mu_w}\right)^{0.14}\left(\dfrac{k}{D_e}\right)$

$\qquad\quad = 0.027\left(\dfrac{G_s}{\mu}\right)^{0.8}\left(C\mu k^2\right)^{\frac{1}{3}}\left(\dfrac{\mu}{\mu_w}\right)^{0.14}\left(\dfrac{1}{D_e}\right)^{0.2}$

$\qquad\quad = 0.027\left(\dfrac{578{,}000}{2.90}\right)^{0.8}\left((0.757)(2.90)(0.33)^2\right)^{\frac{1}{3}}(1.0)\left(\dfrac{1}{0.079}\right)^{0.2}$

$\qquad\quad \approx 483 \text{ Btu/hr}\cdot\text{ft}^2\cdot{}^\circ\text{F}$

(9') $\quad h_i = 0.027\left(\dfrac{D G_t}{\mu}\right)^{0.8}\left(\dfrac{c\mu}{k}\right)^{\frac{1}{3}}\left(\dfrac{\mu}{\mu_w}\right)^{0.14}\left(\dfrac{k}{D}\right)$

$\qquad\quad = 0.027\left(\dfrac{G_t}{\mu}\right)^{0.8}\left(c\mu k^2\right)^{\frac{1}{3}}\left(\dfrac{\mu}{\mu_w}\right)^{0.14}\left(\dfrac{1}{D}\right)^{0.2}$

$\qquad\quad = 0.027\left(\dfrac{762{,}000}{2.20}\right)^{0.8}\left((1)(2.20)(0.35)^2\right)^{\frac{1}{3}}(1.0)\left(\dfrac{1}{0.0517}\right)^{0.2}$

$\qquad\quad \approx 852 \text{ Btu/hr}\cdot\text{ft}^2\cdot{}^\circ\text{F}$

(13) Clean overall coefficient, U_C:

$$U_C = \frac{h_i h_o}{h_i + h_o} = \frac{(483)(852)}{483 + 852} = 308 \text{ Btu/hr}\cdot\text{ft}^2\cdot{}^\circ\text{F}$$

(14) Design overall coefficient, U_D:

$$a'' = 0.1963 \text{ ft}^2 \text{ per lin. ft} \Leftarrow [\text{Table AT.8}]$$

$$A = a'' N_t L = (0.1963)(52)(16) = 163 \text{ ft}^2$$

$$U_D = \frac{Q}{A\Delta T^*} = \frac{915{,}000}{(163)(30.7)} = 183 \text{ Btu/hr}\cdot\text{ft}^2\cdot{}^\circ\text{F}$$

(15) Dirt factor R_d:

$$R_d = \frac{U_C - U_D}{U_C U_D} = \frac{308 - 183}{(308)(183)} = 0.00222 \text{ hr} \cdot \text{ft}^2 \cdot {}^\circ\text{F/Btu}$$

Summary

Variable	Value	Units
h_o	= 483	$\dfrac{\text{Btu}}{\text{hr} \cdot \text{ft}^2 \cdot {}^\circ\text{F/ft}}$
h_i	= 852	
U_C	= 303	$\dfrac{\text{Btu}}{\text{hr} \cdot \text{ft}^2 \cdot {}^\circ\text{F}}$
U_D	= 183	
R_d (calculated)	= 0.00222	$\dfrac{\text{hr} \cdot \text{ft}^2 \cdot {}^\circ\text{F}}{\text{Btu}}$
R_d (required)	= 0.00200	

Pressure Drop

(1) For Re_s and T_{mean} in (6') obtain f:

$$f = 0.0019 \frac{\text{ft}^2}{\text{in}^2} \text{ at } Re_s = 15,750$$

$s = 1.30 \Leftarrow$ given

$$D_s = 10.02\text{in}\left(\frac{1\text{ft}}{12\text{in}}\right) = 0.833\text{ft}$$

(1) For Re_t and t_c in (6) above:

$$f = 0.00023 \frac{\text{ft}^2}{\text{in}^2} \text{ at } Re_t = 17,900$$

$s = 1.0 \Leftarrow$ water

(2) Number of crosses, $N + 1$:

$$N + 1 = \frac{12L}{B} = \frac{12(16)}{2} = 96$$

(2') $\Delta P_t = \dfrac{f G_t^2 L n}{(5.22 \times 10^{10}) D_e s \phi_t}$

$$= \frac{(0.00023)(762,000)^2 (16)(2)}{(5.22 \times 10^{10})(0.0517)(1.0)(1.0)}$$

$$= 1.6 \text{ psi}$$

$$\frac{\bar{v}^2}{2g'}\left(\frac{62.5}{144}\right) = 0.08 \text{ at } G_t = 762,000$$

$$\Delta P_r = \left(\frac{4n}{s}\right)\left(\frac{\bar{v}^2}{2g'}\right)\left(\frac{62.5}{144}\right)$$

$$= \left(\frac{4(2)}{1.0}\right)(0.08) = 0.7\text{psi}$$

(3') $\Delta P_s = \dfrac{f G_s^2 D_s (N+1)}{(5.22 \times 10^{10}) D_e s \phi_s}$

(3) $\Delta P_T = \Delta P_t + \Delta P_r$

$\Delta P_T = 1.6 + 0.7 = 2.3\text{psi}$

$\Delta P_T < 10\text{psi as allowable } \Delta P$

$\Delta P_s = \dfrac{(0.0019)(578,000)^2 (0.833)(96)}{(5.22 \times 10^{10})(0.079)(1.3)(1.0)}$

$\Delta P_s = 9.5\text{psi}$

$\Delta P_s < 10$ psi as allowable ΔP

This exchanger is satisfactory for service.

Steam as a Heating Medium [2]. Thus far, none of the heat-transfer services studied in this section has employed steam although it is by far the commonest heating medium. Steam as a heating medium introduces several difficulties: (1) Hot steam condensate is fairly corrosive, and care must be exercised to prevent condensate from accumulating within an exchanger where continuous contact with metal will cause damage. (2) The condensate line must be connected with discretion. Suppose an exhaust steam at 5 psig and 228 °F is used to heat a cold fluid entering at a temperature of 100 °F. The tube wall will be at a temperature between the two- but nearer to that of the steam, (assume 180 °F). This temperature corresponds to a saturation pressure of only 7.5 psia for the condensate at the tube wall. Although the steam entered at 5 psig, the pressure on the steam side may drop locally to a pressure below that of the atmosphere, so that the condensate will *not* run out of the heater. Instead, it will remain and build up in the exchanger until it blocks the surface available for heat transfer. Without any exposure to the cold fluid, the steam will no longer condense. Instead, the steam retains its inlet pressure, and after a period of time, the system is able to self-correct by the steam blowing out some (or all) of the accumulated condensate, thereby re-exposing the surface adjacent to the cold fluid. The heating operation will become cyclical, and to overcome this issue and attain uniform flow, it may be necessary to employ a trap or suction with suitable piping arrangements to draw off excess condensate.

The heat-transfer coefficients associated with the condensation of steam are very high compared with any which have been studied so far. It is customary to adopt a conventional and conservative value for the film coefficient since it is never the controlling film, rather than obtain one by calculation. For all heating services employing relatively air-free steam, this text will use a value of 1500 Btu/hr·ft²·°F for the condensation of steam, without regard to its location. Thus, $h_i = h_o = 1500$.

It is advantageous in heating to connect the steam to the tubes of the heater rather than the shell. In this way, since the condensate may be corrosive, the action can be confined to the tube-side alone, whereas if the steam is introduced into the shell, both may be damaged. When steam flows through the tubes of a 1–2 exchanger, there is no need for more than two tube passes. Since steam is an isothermally condensing fluid, the true temperature difference ΔT and the LMTD are identical.

When using superheated steam as a heating medium, except in desuperheaters, it is customary to disregard the temperature range of desuperheating and consider all the heat to be delivered at the saturation temperature corresponding to the operating pressure. A more intensive analysis of the condensation of steam will be undertaken in the chapters dealing with condensation.

When steam is employed in two passes on the tube-side, the allowable pressure drop should be very small, less than 1.0 psi, particularly if there is a gravity return of condensate to the boiler. In a gravity-return system, condensate flows back to the boiler because of the difference in static head between a vertical column of steam and a vertical column of condensate. The pressure drop, including entrance and end losses through an exchanger, can be calculated by taking 1/2 the pressure drop for steam when calculated in the usual manner using Equation (7.38) for the inlet vapor conditions. The mass velocity is calculated from the inlet steam rate and the flow area of the first pass (which need not be equal to that of the second pass). The Reynolds number is based on the mass velocity and viscosity of steam. The specific gravity used with Equation (7.38) is the density of the steam obtained from Table AT.2 for the inlet pressure divided by the density of water taken as 62.5 lb/ft³.

The Optimum Use of Exhaust and Process Steam. Some plants obtain power from noncondensing turbines or engines. In such places there may be an abundance of exhaust steam at low pressures from 5 to 25 psig which is considered a by-product of the power cycles in the plant. While there are arbitrary aspects to the method of estimating the cost of exhaust steam, it can be anywhere from one-quarter to one-eighth of the cost of process or live steam. Although it possesses a higher latent heat, exhaust steam is of limited process value, since the saturation temperature is usually about 215 to 230 °F. If a liquid is to be heated to 250 or 275 °F, it is necessary to use process steam at 100 to 200 psi developed at the powerhouse specifically for process heating purposes.

When a fluid is to be heated to a temperature close to or above that of exhaust steam, all the heating can be done in a single shell using only process steam. As an alternative, the heat load can be divided into two shells, one utilizing as much exhaust steam as possible and the other using as little process steam as possible. This leads to an optimum: If the outlet temperature of the cold fluid in the first exchanger is made to approach the exhaust steam temperature too closely, a small Δt and large first heater will result. On the other hand, if the approach is not close, the operating cost of the higher process steam requirement in the second heater increases so that the initial cost of two shells may not be justified. (See also Part III, Chapter 11)

1–2 Exchangers without Baffles. Not all 1–2 exchangers have 25 percent cut segmental baffles. When it is desired that a fluid pass through the shell with an extremely small pressure drop, it is possible to depart from the use of segmental baffles and use only support plates. These will usually be half-circle, 50 percent cut plates which provide rigidity and prevent the tubes from sagging. Successive support plates overlap at the shell diameter so that the entire bundle can be supported by two half circles which support one or two rows of tubes in common. These may be spaced further apart than the outside diameter of the shell. However, when they are employed, the shell fluid is said to flow along the axis, instead of across the tubes. When the shell fluid flows *along* the tubes, or the baffles are cut more than 25 percent, the flow is then analogous to the annulus of a double pipe exchanger. It can be treated in a similar manner, using an equivalent diameter based on the distribution of flow area and the wetted perimeter for the entire shell. The calculation of the shell-side pressure drop will also be similar to that of an annulus.

Example 7.24 – Calculation of a Sugar-solution Heater without Baffles. 200,000 lb/hr of a 20 percent sugar solution ($s = 1.08$) is to be heated from 100 to 122 °F using steam at 5 psi pressure. Available for this service is a 12 in ID 1–2 exchanger without baffles having 76 ¾ in OD, 16 BWG tubes 16'0" long laid out on a 1-in square pitch. The bundle is arranged for two passes. Can the exchanger provide a 0.003 dirt factor without exceeding a 10.0 psi solution pressure drop?

SOLUTION:

Shell-Side	Tube-Side
ID = 12 in	Number and length = 76 tubes, 16'0"
Baffle space = half circles	OD, BWG, pitch = ¾ in, 18 BWG, 1-in square pitch
Passes = 1	Passes = 2

Heat Balance:

Cold Sugar Solution, $\quad t_{avg} = \dfrac{100 + 122}{2} = 111°F$

$$c_{Sugar} = 0.30 \text{ Btu}/(\text{lb} \cdot °F) \text{ at } 111°F$$

$$c_{H_2O} \approx 1.0 \text{ Btu}/(\text{lb} \cdot °F) \text{ at } 111°F$$

Therefore, $\quad c = x_{Sugar} \cdot c_{Sugar} + x_{H_2O} \cdot c_{H_2O}; x = \text{mass fraction}$

$$= (0.2)(0.30) + (0.8)(1.0)$$

$$= 0.86 \text{ Btu}/(\text{lb} \cdot °F)$$

$$Q = (200,000)(0.86)(122 - 100) = 3,790,000 \text{ Btu/hr}$$

Hot Steam, $\quad H_L = 960.1 \text{ Btu/lb at } T = 228°F$

$$Q = \dot{M}H_L = (3,950)(960.1) = 3,790,000 \text{ Btu/hr}$$

(2) ΔT^*:

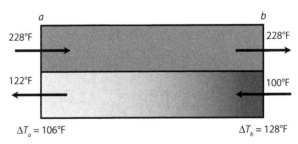

Note: There is no color gradient in darker gray stream because the steam temperature is constant.

$$\text{LMTD} = \frac{\Delta t_2 - \Delta t_1}{\ln\left(\Delta t_2 / \Delta t_1\right)} = \frac{106 - 128}{\ln\left(106 / 128\right)} = 116.7°\text{F}$$

$$R = \frac{T_1 - T_2}{t_2 - t_1} = \frac{0}{22} = 0$$

$$P = \frac{t_2 - t_1}{T_1 - t_1} = \frac{22}{228 - 100} = 0.172$$

$$F_T = \left(\frac{\sqrt{R^2 + 1}}{R - 1}\right) \frac{\ln\left[\left(1 - P\right)/\left(1 - RP\right)\right]}{\ln\left[\left(2 - P\left(R + 1 - \sqrt{R^2 + 1}\right)\right)/\left(2 - P\left(R + 1 + \sqrt{R^2 + 1}\right)\right)\right]}$$

$$= \left(\frac{\sqrt{0 + 1}}{0 - 1}\right) \frac{\ln\left[\left(1 - 0.172\right)/\left(1 - \left(0\right)\left(0.172\right)\right)\right]}{\ln\left[\left(2 - 0.172\left(0 + 1 - \sqrt{0 + 1}\right)\right)/\left(2 - 0.172\left(0 + 1 + \sqrt{0 + 1}\right)\right)\right]}$$

$$= 1$$

Whenever $R = 0$, F_T must be 1. Therefore, $\Delta T = \text{LMTD}$ at $R = 0$.

$$\Delta T^* = 116.7°\text{F}$$

(3) T_{avg} and t_{avg}: t_{avg} was obtained in step (1) and T_{avg} is simply the given T.

Hot Fluid (Steam) – Tube-Side (t)	Cold Fluid (Sugar Solution) – Shell-Side (s)
(4) Flow Area, a_t:	(4′) Flow Area, a_s:
$a_t' = 0.302 \text{ in}^2 \Leftarrow \left[\text{Table AT.8}\right]$	$a_s = \left(\text{area of shell}\right) - \left(\text{area of tubes}\right)$
$a_t = \dfrac{N_t a_t'}{144n}$	$= \left(\dfrac{1}{144}\right)\left(\dfrac{D_s^2 \pi}{4} - \dfrac{N_t d_o^2 \pi}{4}\right)$
$= \dfrac{\left(76\right)\left(0.302\right)}{144\left(2\right)} = 0.0797 \text{ft}^2$	$= \left(\dfrac{1}{144}\right)\left(\dfrac{\left(12\right)^2 \pi}{4} - \dfrac{\left(76\right)\left(0.75\right)^2 \pi}{4}\right)$
	$= 0.55 \text{ft}^2$
(5) Mass Velocity, G_t:	(5′) Mass Velocity, G_s:
$G_t = \dfrac{W}{a_t} = \dfrac{3,950}{0.0797}$	$G_s = \dfrac{w}{a_s} = \dfrac{200,000}{0.55}$
$= 49,500 \text{lb}/\left(\text{hr} \cdot \text{ft}^2\right)$	$= 364,000 \text{lb}/\left(\text{hr} \cdot \text{ft}^2\right)$

(6) Obtain D from Table At.8:

$$D = 0.62\text{in} \Leftarrow \left[\text{Table AT.8}\right]$$

$$= \frac{0.62\text{in}}{12\text{in/ft}} = 0.0517\text{ft}$$

(6′) Compute D_e:

$$D_e = \frac{4a_s}{\text{wetted perimeter}}$$

$$= \frac{4a_s}{N_t d_o \pi} = \frac{4(0.55)}{(76)(0.75/12)\pi}$$

$$= 0.148\text{ft}$$

Obtain μ at T:

$$\mu = 0.0128 \text{ cP at } 228°\text{F}$$

$$= 0.0128\text{cP}\left(\frac{2.42\text{lb/(ft}\cdot\text{hr)}}{1\text{cP}}\right)$$

$$= 0.031\text{lb/(ft}\cdot\text{hr)}$$

Obtain μ at t_{mean}:

$$\mu = 2\mu_{water} \Leftarrow \text{for nonviscous,}$$
$$\text{aqueous salt solution}$$

$$\mu_{water} = 0.65\text{cP at } 111°\text{F}$$

$$\mu = 2(0.65) = 1.30\text{cP}$$

$$= 1.30\text{cP}\left(\frac{2.42\text{lb/(ft}\cdot\text{hr)}}{1\text{cP}}\right)$$

$$= 3.14\text{lb/(ft}\cdot\text{hr)}$$

(7) Reynolds Number, Re_t:

$$\text{Re}_t = \frac{DG_t}{\mu}$$

$$= \frac{(0.0517)(49,500)}{0.031} = 82,500$$

(Re_t is for pressure drop only)

(7′) Reynolds Number, Re_s:

$$\text{Re}_s = \frac{D_e G_s}{\mu}$$

$$= \frac{(0.148)(364,000)}{3.14} = 17,100$$

(8) Skip this step since k and C are not required for a phase-changing stream.

(8′) Obtain c and k at t_{avg}:

$$c = 0.86\text{Btu/lb}\cdot°\text{F at } 111°\text{F} \Leftarrow \left[\text{Step 1}\right]$$

$$k = 0.9k_{water} \Leftarrow \text{for low concentration}$$
$$\text{salt solution}$$

$$k_{water} = 0.37 \text{ Btu/hr}\cdot\text{ft}^2\cdot°\text{F/ft at } 111°\text{F}$$

$$k = 0.9(0.37) = 0.333 \text{ Btu/hr}\cdot\text{ft}^2\cdot°\text{F/ft}$$

(9) h_i is approximately equal to 1500 Btu/hr·ft²·°F for steam condensation.

$$\phi_t = \left(\frac{\mu}{\mu_w}\right)^{0.14} \approx 1.0$$

Solve for h_o and ϕ_s at t_w:

$t_w = 212°F \Leftarrow$ from iterative calculation

$\mu_w = (0.26 \times 2)\text{cP at } 212°F$

$= 0.52\text{cP}$

$$= 0.52\text{cP}\left(\frac{2.42\text{lb}/(\text{ft}\cdot\text{hr})}{1\text{cP}}\right)$$

$= 1.26\text{lb}/(\text{ft}\cdot\text{hr})$

$$\phi_s = \left(\frac{\mu}{\mu_w}\right)^{0.14} = \left(\frac{3.14}{1.26}\right)^{0.14} = 1.14$$

$$\frac{h_o D_e}{k} = 0.023\left(\frac{D_e G_s}{\mu}\right)^{0.8}\left(\frac{C\mu}{k}\right)^{0.3}$$

$h_o = 237\text{Btu/hr}\cdot\text{ft}^2\cdot°F \text{ at } 212°F$

(10) Clean overall coefficient, U_C:

$$U_C = \frac{h_i h_o}{h_i + h_o} = \frac{(1500)(237)}{1500 + 237} = 205\text{Btu/hr}\cdot\text{ft}^2\cdot°F$$

(11) Design overall coefficient, U_D:

$a'' = 0.1963\text{ft}^2 \text{ per lin.ft} \Leftarrow [\text{Table AT.8}]$

$A = a''N_t L = (0.1963)(76)(16) = 238\text{ft}^2$

$$U_D = \frac{Q}{A\Delta T^*} = \frac{3,790,000}{(238)(116.7)} = 136\text{Btu/hr}\cdot\text{ft}^2\cdot°F$$

(12) Dirt factor R_d:

$$R_d = \frac{U_C - U_D}{U_C U_D} = \frac{205 - 136}{(205)(136)} = 0.00247\text{hr}\cdot\text{ft}^2\cdot°F/\text{Btu}$$

Summary

Variable	Value	Units
h_o	= 237	Btu
h_i	= 1500	$\overline{hr \cdot ft^2 \cdot °F/ft}$
U_C	= 205	Btu
U_D	= 136	$\overline{hr \cdot ft^2 \cdot °F}$
R_d (calculated)	= 0.00247	$hr \cdot ft^2 \cdot °F$
R_d (required)	= 0.00200	$\overline{Btu}$

Pressure Drop

(1) For Re_t and T in (6) above:

$$f = 0.000155\frac{ft^2}{in^2}\text{ at } Re_t = 82{,}500$$

$$\dot{V} = 20.0 ft^3/lb \quad \text{(specific volume)}$$

$$s = 1.0\left(\frac{\dot{V}}{\rho}\right) = 1.0\left(\frac{20.0}{62.5}\right)$$

$$= 0.00080 \Leftarrow \text{convert liq. to gas}$$

(1′) Compute D'_e and Re'_s to find f:

$$D'_e = \frac{4a_s}{\text{frictional wetted perimeter}}$$

$$= \frac{4a_s}{N_t d_o \pi + D_s \pi} = \frac{4a_s}{\left(N_t d_o + D_s\right)\pi}$$

$$= \frac{4(0.55)}{\left[(76)(0.75/12)+(12/12)\right]\pi}$$

$$= 0.122 ft \Leftarrow \left[\text{Eq.}(6.4)\right]$$

$$Re'_s = \frac{D'_e G_s}{\mu} = \frac{(0.122)(364{,}000)}{3.14}$$

$$= 14{,}100$$

$$f = 0.00025\frac{ft^2}{in^2}\text{ at } Re'_s = 14{,}100$$

$$s = 1.08 \Leftarrow \text{give}$$

(2) $\Delta P_t = \dfrac{1}{2} \cdot \dfrac{fG_t^2 Ln}{\left(5.22\times10^{10}\right)D_e s\phi_t}$

$$= \frac{(0.000155)(49{,}500)^2(16)(2)}{\left(5.22\times10^{10}\right)(0.0517)(0.0008)(1.0)}$$

$$= 2.8 psi$$

(2′) $\Delta P_s = \dfrac{fG_s^2 Ln}{\left(5.22\times10^{10}\right)D_e s\phi_s}$

$$= \frac{(0.00025)(364{,}000)^2(16)(1)}{\left(5.22\times10^{10}\right)(0.122)(1.08)(1.14)}$$

$$= 0.07 psi$$

(Note: The ½ term above arises because of the half-circle pitch)
This is a relatively high pressure drop for steam with a gravity condensate return.
However, the exchanger is satisfactory.

The Efficiency of an Exchanger. In the design of many types of exchangers it is frequently desirable to establish a standard of maximum performance. The efficiency is then defined as the fractional performance of an apparatus delivering less than the standard. Dodge [13] gives the definition of the efficiency of an exchanger, e, as the ratio of the quantity of heat actually removed from a fluid to the maximum which could have been removed. Using the usual nomenclature,

$$e = \frac{\dot{m}c\left(t_2 - t_1\right)}{\dot{m}c\left(T_1 - t_1\right)} = \frac{t_2 - t_1}{T_1 - t_1} \tag{7.54}$$

where it is presumed that $t_2 = T_1$. Depending upon whether the hot or cold terminal approaches zero, the efficiency may also be expressed by,

$$e = \frac{\dot{M}C\left(T_1 - T_2\right)}{\dot{M}C\left(T_1 - t_1\right)} \tag{7.55}$$

Although there is merit to this definition from the standpoint of thermodynamics, there is a lack of realism in an efficiency definition which involves a terminal difference and a temperature difference of zero. In effect, it is the same as defining the efficiency as the ratio of heat transferred by a real exchanger to an exchanger with infinite surface.

In process heat transfer there is another definition which is useful. The process temperatures are capable of providing a maximum temperature difference if arranged in a countercurrent manner. There appears to be some value in regarding the efficiency of an exchanger as the ratio of the temperature difference attained by any other exchanger to that for true countercurrent. This is identical with F_T, which proportionately influences the surface requirements. It will be seen in the next and later chapters that other flow arrangements besides 1–2 parallel flow-countercurrent can be attained in tubular equipment and by which the value of F_T may be increased for given process temperatures. These obviously entail flow patterns which approach true counterflow more closely than the 1–2 exchanger.

7.6   Computer Aided Heat Exchanger Design

As noted in Chapter 5, a decision was made to provide a more expansive discussion on the role of computers in the design of heat exchangers at the conclusion of this chapter. Even though the majority of heat exchangers are purchased "off-the-shelf," computer design does come into play with unique and special applications. This is particularly true in the utility (power), refineries, nuclear, electronics, and aerospace industries.

Farber [14] offers the following (employing Farber's notation) regarding the role of computers in heat exchanger calculations.

Whether an engineer is specifying a heat exchanger for design and purchase, or is actually designing the heat exchanger itself, computers play a key role in this process. The design of a heat exchanger needs to take into account both the thermal and physical requirements and the optimization of the physical design of the exchanger at the same time. Inputs to a heat exchanger design (1 = hot fluid, 2 = cold fluid) are provided in Table 7.5.

Table 7.5 Required thermal and physical property information.

	Thermal	Physical
Fluid 1	Temperature In (T_i)	Mass Flow $(\dot{M})$
	Temperature Out (T_o)	Specific Gravity (s_1)
	Condensing (Yes/No)	Viscosity In (μ_1)
	Heat Capacity In $(c_{p,1})$	Maximum Pressure Loss (ΔP_1)
		Corrosive (Yes/No)
		Erosive (Yes/No)
Fluid 2	Temperature In (t_i)	Mass Flow $(\dot{m})$
	Temperature Out (t_o)	Specific Gravity (s_2)
	Condensing (Yes/No)	Viscosity In (μ_2)
	Heat Capacity In $(c_{p,2})$	Maximum Pressure Loss (ΔP_2)
		Corrosive (Yes/No)
		Erosive (Yes/No)
Shell	Thermal Conductivity (k_s)	Shell Material Costs
	Heat Capacity $(c_{p,s})$	
	Fouling Factor (F_s)	
Tube	Thermal Conductivity (k_t)	Tube Material Costs
	Heat Capacity $(c_{p,t})$	
	Fouling Factor (F_t)	

Use of a computerized design routine can evaluate all of these variables and, with decisions made by the responsible engineer regarding design and process restrictions (e.g., pressure drop, outlet temperatures, etc.), can result in an optimized design. This computerized design procedure will generally use a nonlinear optimization routine, along with the relevant design equations and accompanying constraints to provide the heat exchanger design. See also Figure 7.31.

The design of a heat exchanger involving detailed calculation is often a complex procedure best optimized using a well-designed computer program. For example, the temperature and mass flow through either the shell-side or the tube-side may have some flexibility. Let us say that Fluid 2 is being used to cool Fluid 1, thus fixing the inlet and outlet conditions of Fluid 1. This sets the total amount of heat that needs to be exchanged, as calculated by:

$$Q_1 = \dot{M}c_{p,1}\left(T_o - T_i\right) \tag{7.56}$$

Now, in order to match the amount of heat released by Fluid 1 into Fluid 2, then either (or both) the mass of fluid 2 ($\dot{M}$) or its outlet temperature (t_o) will be varied so that,

$$Q_1 = Q_2 = \dot{m}c_{p,2}\left(t_o - t_i\right) \tag{7.57}$$

In conjunction with the above is the heat transfer equation:

$$Q = UA\Delta T_{LM}$$

Thus, as $\dot{m}$ increases, t_o will decrease and visa-versa. Minimizing $\dot{m}$ could conceivably affect the physical size of the heat exchanger required for flow but could conflict with the amount of heat transfer surface area (A) needed, the overall heat transfer coefficient (U), and the log mean temperature difference (ΔT_{LM}). Maximizing $\dot{m}$ could also affect the size of the heat exchanger but may run into restrictions for the maximum allowable pressure loss and affect the fabricated cost of the exchanger due to surface area and shell size considerations. Thus, there is a complex relationship between the heat being transferred, the maximum flow(s) possible, and potentially a maximum temperature on the outlet (caused either by other process restrictions or environmental concerns) as well as the cost of exchanger fabrication based on the size, design, and materials of construction.

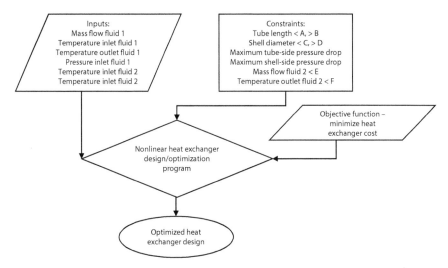

Figure 7.31 Qualitative optimization flowchart for computer aided heat exchanger design.

Nonlinear optimization [15, 16] (also referred to as nonlinear programming) is the process of solving an optimization problem defined by a system of equalities and inequalities, collectively termed constraints, over a set of unknown real variables, along with an objective function to be maximized or minimized, where some of the constraints or the objective function are nonlinear. An example of a heat exchanger design using a nonlinear optimization program is for a power plant that uses the water from a lake or reservoir to cool and condense the steam used to run the turbine in the plant. The heated water produced is then discharged back into the reservoir or the lake. For the best efficiency, the plant should run at the lowest condenser temperature possible. However, there are limits on the amount of water that can be pumped through the plant and, at the same time, there are ecological and permit constraints on how much the lake or reservoir temperature can be raised. The optimization problem is to minimize the temperature of the condensed steam, minimize the limit of the maximum rate at which water can be withdrawn from the lake, and minimize the limit on the increase in lake temperature. There are additional constraints on the physical size of the condenser, the allowable pressure drops of the condensate and cooling water through the condenser, and a need to also address/approach a minimum cost for the overall system (i.e., the capital cost of the condenser as well as the cost of the cooling water pumps and the annualized cost of the power for those pumps)."

In the end, the computer programs available to the practicing engineer — as well as the vendor — simply require the inputting of pertinent physical property data and physical specifications, as discussed in this text and earlier work of Kern [2]. There can be some "tweaking" of the output of the computer, but the process is relatively straightforward. For off-the-shelf exchangers, the design is almost always based on economics, subject to constraints and specifications provided by the user, and that difficult-to-define term referred to as sound engineering judgement [17].

7.7   Practice Problems from Kern's First Edition

1. A 1–2 exchanger is used for heating 50,000 lb/hr of methyl ethyl ketone from 100 to 200 °F using hot amyl-alcohol available at 250 °F. (*a*) What minimum quantity of amyl-alcohol is required to deliver the desired heat load in a 1–2 exchanger? (*b*) If the amyl-alcohol is available at 275 °F, how does this affect the total required quantity?

2. A 1–2 exchanger has one shell and two tube passes. The passes do not have equal surfaces. Instead X percent of the tubes are in the first pass and $(1 - X)$ percent are in the second, but if the tube-side film coefficient is not controlling, the assumption of constant U is justifiable. (*a*) Develop an expression for the true temperature difference when X percent of the tubes are in the colder of the two tube passes. (*b*) What is the true temperature difference when the hot fluid is cooled from 435 to 225 °F by a noncontrolling cooling medium in the tubes which is heated from 100 to 150 °F when 60 percent of the tubes are in the colder tube pass and (*c*) when 40 percent of the tubes are in the colder pass? How do these compare with the 1–2 true temperature difference with equal surfaces in each pass?

3. A double pipe exchanger has been designed for the nozzle arrangement shown in Figure 7.32. If the hot stream is cooled from 275 to 205 °F while the cold stream enters at 125 °F and is heated to $t_2 = 190$ °F, what is the true temperature difference? (*Hint.* Establish an equation for the temperature difference with the nozzle arrangement shown and sufficient to allow a numerical trial-and-error solution.) How does this compare with the LMTD for countercurrent?

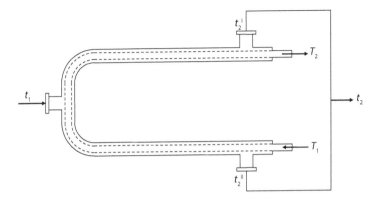

Figure 7.32 Illustration for Problem 3.

4. 43,800 lb/hr of 42°API kerosene between 390 and 200 °F is used to heat 149,000 lb/hr of 34°API Mid-continent crude from 100 to 170 °F in a 662-ft³ exchanger (Example 7.3). The clean coefficient is 69.3 Btu/hr·ft²·°F. When the 1–2 exchanger is clean, what outlet temperature will be obtained? Calculate the outlet temperatures directly from F_T. How does the total heat load compare with that which could be delivered by a true countercurrent exchanger assuming that the same U could be obtained?

5. It is necessary on a new installation to preheat 149,000 lb/hr of 34°API crude oil from 170 °F to a temperature of 285 °F, corresponding to that of the feed plate of a fractionating tower. There is a utility 33°API gas oil line running near the tower at 530 °F of relatively unlimited quantity. Because the pumping cost for cold gas oil is prohibitive, the temperature of the gas oil from the heat exchanger, returning to the line, should not be less than 300 °F.

    Available on the site is a 25 in ID 1–2 exchanger containing 252 tubes 1 in OD, 13 BWG, 16′0″ long arranged on a six-pass 1¼-in triangular pitch layout. The shell baffles are spaced at 5-in centers. A pumping head of 10 psi is allowable on the gas oil line and 15 psi on the feed line. Will the exchanger be acceptable if cleaned, and if so, what will the fouling factor be? The viscosities are 0.4 centipoise at 530 °F and 0.7 centipoise at 300 °F for the gas oil. The viscosities are 0.9 centipoise at 285 °F and 2.1 centipoise at 170 °F for the crude oil. (Interpolate by plotting °F vs. centipoise on logarithmic paper.)

6. 96,000 lb/hr of 35°API absorption oil in being cooled from 400 to 200 °F is used to heat 35°API distillate from 100 to 200 °F. Available for service is a 29 in ID 1–2 exchanger having 338 tubes 1 in OD, 14 BWG, 16'0" long on 1¼-in triangular pitch. Baffles are spaced 10 in apart, and the bundle is arranged for four tube passes. What arrangement gives the more nearly balanced pressure drops, and what is the dirt factor? The viscosity of the absorption oil is 2.6 centipoise at 100 °F and 1.15 centipoise at 210 °F. (Plot on logarithmic paper °F vs. viscosity in centipoise, and extrapolate a straight line.) The viscosity of the distillate is 3.1 centipoise at 100 °F and 1.3 centipoise at 210 °F.

7. 43,200 lb/hr of 35°API distillate is cooled from 250 to 120 °F using cooling water from 85 to 120 °F. Available for service is a 19¼ in ID 1–2 exchanger having 204 tubes ¾ in OD, 16 BWG, 16'0" long on 1-in square pitch. Baffles are spaced 5 in apart, and the bundle is arranged for four passes. What arrangement gives the more nearly balanced pressure drops, and what is the dirt factor? What is the optimum outlet-water temperature? (Viscosities of the distillate are given in Problem 7.6.)

8. 75,000 lb/hr of ethylene glycol is heated from 100 to 200 °F using steam at 250 °F. Available for the service is a 17¼ in ID 1–2 exchanger having 224 tubes ¾ in OD, 14 BWG, 16'0" long on 15/16-in triangular pitch. Baffles are spaced 7 in apart, and there are two tube passes to accommodate the steam. What are the pressure drops and what is the dirt factor?

9. 100,000 lb/hr of 20 percent potassium iodide solution is to be heated from 80 to 200 °F using steam at 15 psig. Available for the service is a 10 in ID 1–2 exchanger without baffles having 50 tubes ¾ in OD, 16 BWG, 16'0" long arranged for two passes on 15/16-in triangular pitch. What are the pressure drops and the dirt factor?

10. 78,359 lb/hr of isobutene (118°API) is cooled from 203 to 180 °F by heating butane (111.5°API) from 154 to 177 °F. Available for the service is a 17¼ in ID 1–2 exchanger having 178 tubes ¾ in OD, 14 BWG, 12'0" long on 1-in triangular pitch. Baffles are spaced 6 in apart, and the bundle is arranged for four passes. What are the pressure drops and the dirt factor?

11. A 1–2 exchanger recovers heat from 10,000 lb/hr of boiler blowdown at 135 psig by heating raw water from 70 to 96 °F. Raw water flows in the tubes. Available for the service is a 10.02 in ID 1–2 exchanger having 52 tubes ¾ in OD, 16 BWG, 8'0" long. Baffles are spaced 2 in apart, and the bundle is arranged for two tube passes. What are the pressure drops and fouling factors?

12. 60,000 lb/hr of a 25% NaCl solution is cooled from 150 to 100 °F using water with an inlet temperature of 80 °F. What outlet water temperature may be used? Available for the service is a 21¼ in ID 1–2 exchanger having 302 tubes ¾ in OD, 14 BWG, 16'0" long. Baffles are spaced 5 in apart, and the bundle is arranged for two passes. What are the pressure drops and fouling factor?

13. A liquid is cooled from 350 to 300 °F by another which is heated from 290 to 315 °F. How does the true temperature difference deviate from the LMTD if (a) the hot fluid is in series and the cold fluid flows in two parallel countercurrent paths, (b) the hot fluid is in series and the cold fluid flows in three cocurrent-countercurrent paths, (c) the cold-fluid range in (a) and (b) is changed to 275 to 300 °F.

14. A fluid is cooled from 300 to 275 °F by heating a cold fluid from 100 to 290 °F. If the hot fluid is in series, how is the true temperature difference affected by dividing the hot stream into (a) two parallel streams and (b) into three parallel streams?

References

1. I. Farag and J. Reynolds, *Heat Transfer*, A Theodore Tutorial, Theodore Tutorials, East Williston, NY, originally published by USEPA/APTI, RTP, NC, 1996.
2. D. Kern, *Process Heat Transfer*, McGraw-Hill, New York City, NY, 1950.
3. L. Theodore, *Heat Transfer Applications for the Practicing Engineer*, John Wiley & Sons, Hoboken, NJ, 2013.
4. F. Incropedia and D. De Witt, *Fundamentals of Heat and Mass Transfer*, 4th edition, Chapters 8, 10, 11. John Wiley & Sons, Hoboken, NJ, 1996.
5. D. Green and R. Perry (editors), *Perry's Chemical Engineers' Handbook*, 8th edition, McGraw-Hill, New York City, NY, 2008.
6. W. McCabe, J. Smith, and P. Harriott, *Unit Operations of Chemical Engineering*, 5th edition, McGraw-Hill, New York City, NY, 1993.

7. R. Bowman, A. Mueller, and W. Nagle, *Trans. ASME*, 62, 283, New York City, NY, 1940.

8. A. Underwood, *J. Inst. Petroleum Technol.*, New York City, NY, 20, 145–158, 1934.

9. W. Nagle, *Ind. Eng. Chem.*, 25, 604–608, New York City, NY, 1933.

10. R. Bowman, A. Mueller, W. Nagle, *Trans. ASME*, New York City, NY, 62, 283–294, 1940.

11. W. McAdams, *Heat Transmission*, McGraw-Hill, New York City, NY, 1954.

12. R.W. Serth. *Process Heat Transfer: Principles and Applications*, Elsevier Academic Press, Amsterdam, Netherlands, 2007.

13. B. Dodge, *Chemical Engineering Thermodynamics*, McGraw-Hill, Inc. New York, NY, 1944.

14. P. Farber, personal notes, private communication to L. Theodore, Chicago, IL, 2018.

15. C. Prochaska and L. Theodore, *Introduction to Mathematical Methods for Environmental Engineers and Scientists*, Scrivener-Wiley, Salem, MA, 2018.

16. L. Theodore and K. Behan, *Introduction to Optimization for Environmental and Chemical Engineers*, CRC Press/Taylor & Francis Group, Boca Raton, FL, 2018.

17. T. Akashige and L. Theodore, personal notes, Westchester County, NY, 2017.

18. S. Mavridou, G.C. Mavropoulos, D. Bouris, D.T. Hountalas, and G. Bergeles, *Applied Thermal Engineering*, "Comparative design study of a diesel exhaust gas heat exchanger for truck applications with conventional and state of the art heat transfer enhancements," 30(8–9), pp. 935–947, doi: 10.1016/j. applthermaleng.2010.01.003, London, UK, 2010.

19. M. Hayati, *Lecture Slides: Designing Shell-and-Tube Heat Exchangers Using Softwares*, University of Kashan, Kashan, I.R. Iran, 2014.

8

Extended Surface/ Finned Heat Exchangers

Introduction

Consider the case where steam flows through the inside of a metal pipe, heating the cooler air around the pipe. As previously described (Section 5.4, Controlling Resistance), the steam and metal pipe offer little resistance to heat transfer. Therefore, the total resistance to heat transfer for this system, ΣR, is the controlling resistance due to the air, or $R_{air} = 1/h_{air}A_o$.

In order to increase the heat transfer from the steam to the air, one must *decrease* the controlling resistance, R_{air}. One way to decrease R_{air} is to increase, or "extend" the surface area, A_o. In others words, as $A_o\uparrow$, $R_{air}\downarrow$.

This can be accomplished by adding pieces of metal, in various shapes and sizes, to the outside of the pipe. These add-on pieces of metal are typically called *fin*. Hence, the title of the chapter – "Extended Surface/ Finned Heat Exchangers". This process increases the overall surface area of the outside of the pipe without increasing the overall length of the pipe. If the fins are properly secured to the pipe, the temperature of the fin will be approximately equal to the temperature of the heating fluid (due to the high thermal conductivity of most metals used in practice). The fin material and the pipe material do not have to be the same. The only requirement is that the fins have a sufficiently high thermal conductivity and a low resistance to heat transfer.

The remaining sections of this chapter include:

8.1   Fin Details
8.2   Equipment Description
8.3   Key Describing Equations
8.4   Fin Effectiveness and Performance
8.5   Kern's Design Methodology
8.6   Other Fin Considerations

8.1   Fin Details

In order to accurately calculate the new heat transfer area that includes the area of attached fins, it is first necessary to become aware of the geometric shapes of fins available for use. A number of industrial types of fins are shown in Figure 8.1. Pipes and tubes with longitudinal fins are marketed by several manufacturers and consist of long metal strips or channels attached to the outside of the pipe. The strips are attached either by grooving and peening the tube as in Figure 8.2*a* or by welding continuously along the base. When channels are attached, they are integrally welded to the tube as in Figure 8.2*b*. Longitudinal fins of this type are commonly used in double pipe exchangers or in unbaffled shell-and-tube exchangers when the flow proceeds along the axis of the tube. Longitudinal fins are most commonly employed in problems involving gases and viscous liquids, or when the flowrate of one of the heat transfer streams is laminar.

Transverse or circumferential fins are employed primarily for the cooling and heating of gases in *crossflow*. The helical fins in Figure 8.3*a* are classified as circumferential fins and are attached in a variety of ways, such as by grooving and peening, expanding the tube metal itself to form the fin, or welding ribbon to the tube continuously. Disc-type fins are also circumferential fins and are usually welded to the tube or shrunken to it as shown in Figures 8.3*b* and *c*. In order to shrink a fin into a tube, a disc with an inside diameter slightly less than the outside diameter of the tube is heated until the inside diameter exceeds the outside diameter of the tube. It is slipped onto the tube, and upon cooling, the disc shrinks onto the tube and forms a bond with it. Another variation of the shrunk-on fin in Figure

Longitudinal Fins

Transverse Fins

Pegs or Studs

Discontinuous Fins

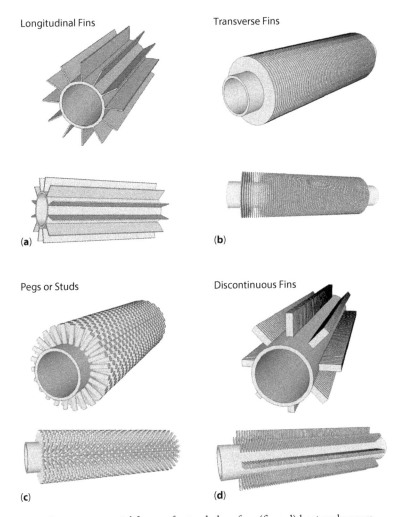

Figure 8.1 Some commercial forms of extended surface (finned) heat exchangers.

Peened

Welded

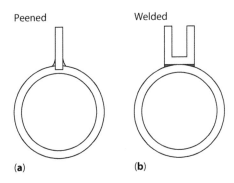

Figure 8.2 Fin attachment.

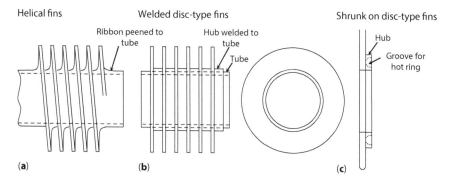

Figure 8.3 Circumferential fins.

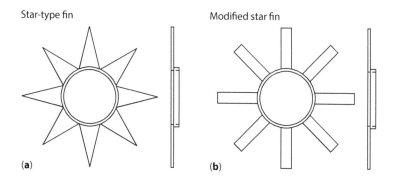

Figure 8.4 Discontinuous fins.

8.3c employs a hollow ring in its hub into which a hot metal ring is driven. Other types of transverse fins are known as discontinuous fins, and several shapes such as the star fin are shown in Figure 8.4. Spine or peg type fins employ cones, pyramids, or cylinders which extend from the pipe surface so that they are usable for either longitudinal flow or crossflow. Each type of finned tube has its own characteristics and effectiveness for the transfer of heat between the fin and the fluid.

Summarizing, and as noted above, extended surfaces, or fins, are classified into longitudinal fins, transverse fins, and spine fins. *Longitudinal fins* (also termed straight fins) are attached continuously along the length of the surface (see Figure 8.5), and they are employed in cases involving gases or viscous liquids. As one might suppose, they are primarily applied with double pipe heat exchangers. *Transverse* or *circumferential fins* are positioned approximately perpendicular to the pipe or tube axis and are usually used in the cooling of gases (see Figure 8.6). These fins find their major application with shell and tube exchangers. Transverse fins may be

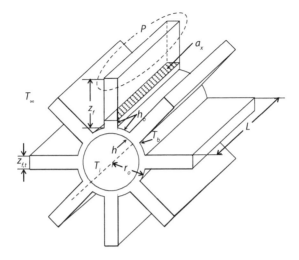

Figure 8.5 Longitudinal fins.

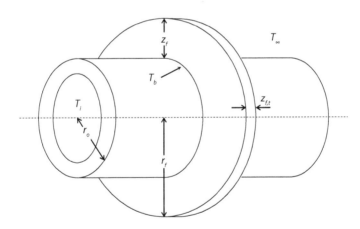

Figure 8.6 Transverse or circumferential fins.

continuous or discontinuous (segmented). *Annular fins* are examples of continuous transverse fins. *Spine* or *peg fins* employ cones or cylinders, which extend from the heat transfer surface, and are used for either longitudinal flow or cross flow.

Fins are constructed of highly conductive materials. The optimum fin design is usually one that gives the highest heat transfer for the minimum amount of metal. The metal used in their manufacture has a strong influence on fin efficiency. Table 8.1 compares the volume and mass of three different metals required to give the same amount of heat transfer for

Table 8.1 Fin metal data.

Metal	Thermal conductivity	Specific gravity	Relative volume	Relative mass
	W / m · K			
Copper	400	8.9	1.00	1.00
Aluminum	210	2.7	1.83	0.556
Steel	55	7.8	7.33	0.43

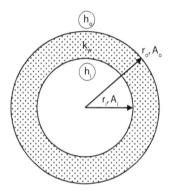

Figure 8.7 Pipe axial view.

fins with identical shapes. The values in the volume and mass columns are relative to the volume and mass of a copper fin, respectively.

In processes where convection coefficients are very different on opposite sides of a tube wall, *it is more often advantageous to increase the surface area on the side with the smaller coefficient.* This is illustrated in the following development.

The total resistance to heat transfer for the pipe in Figure 8.7 is calculated as follows:

$$\sum R = \frac{1}{h_i A_i} + \frac{\Delta x_w}{k_w \bar{A}_{LM}} + \frac{1}{h_o A_o} \tag{8.1}$$

where (once again),

$$A_{LM} = \frac{A_o - A_i}{\ln\left(A_o / A_i\right)} \tag{8.2}$$

If $h_i >>> h_o$ and $k_w >>> h_o$, then

$$\sum R = \cancelto{\approx 0}{\frac{1}{h_i A_i}} + \cancelto{\approx 0}{\frac{\Delta x_w}{k_w \bar{A}_{LM}}} + \frac{1}{h_o A_o} \qquad (8.3)$$

One may now write,

$$R = \sum R \approx \frac{1}{h_o A_o} \qquad (8.4)$$

and,

$$Q = \frac{\Delta T}{\sum R} = \frac{\Delta T}{R} \qquad (8.5)$$

Therefore, in order to increase heat transfer, it is necessary to minimize the total resistance, R. *In this particular case*, that can be accomplished by increasing A_o (the area associated with the low h) by mounting fins to the outside surface of the tube. Increasing A_o, decreases R, which in turn increases Q.

Example 8.1 – Face Area and Total Area of Fins. Consider the longitudinal (rectangular) fin pictured in Figure 8.5. Estimate the fin face area, neglecting the (top) area contribution associated with the fin thickness if $L = 1$ ft and $z_f = 1.5$ in The thickness is 0.1 in. Also calculate the total area of the fin.
SOLUTION: The face area of the fin is given as

$$A_f = 2Lz_f = (2)(12)(1.5) = 36 \,\text{in}^2 = 0.25 \,\text{ft}^2$$

The total area is given by,

$$A_{f,\text{total}} = 2Lz_f + z_{f,t}L = 36 + (0.1)(12) = 37.2 \,\text{in}^2 = 0.258 \,\text{ft}^2$$

Note that $z_{f,t}$ represents fin thickness. The notation of t has been used by some to represent thickness but this text will avoid calling the fin thickness t to avoid confusion with earlier temperature notations.

Example 8.2 – Comparison of Face Area and Total Area of Fins. Comment on the results of the previous example.
SOLUTION: Note that the external "rim" area, $z_{f,t}L$ (1.2 in²) does not contribute significantly to the total area (37.2 in²) and is normally neglected in fin calculations.

Example 8.3 – Area of Circumferential Fins. Refer to Figure 8.6. Estimate the fin area, neglecting the area contributed by fin thickness when $r_o = 4.0$ in, $r_f = 6.0$ in, and $z_{f,t} = 0.1$ in.

SOLUTION: This is an example of a circumferential (annular) fin. For a circumferential fin, the face area is given by,

$$A_f = 2\pi\left(r_f^2 - r_o^2\right) = 2\pi\left(6^2 - 4^2\right) = 125.7\,\text{in}^2 = 0.873\,\text{ft}^2$$

The total area is given by,

$$A_{f,\text{total}} = A_f + 2\pi r_f z_{f,t} = 125.7 + 2\pi\left(6\right)\left(0.1\right)$$
$$= 125.7 + 3.8 = 129.5\,\text{in}^2 = 0.899\,\text{ft}^2$$

Once again, the external "rim" area does not contribute significantly to the total area.

8.2　Equipment Description

Kern [1, 2] provides extensive details regarding heat transfer equipment mounted with fins and/or extended surfaces. The four classes of equipment most commonly employed in industry are listed below:

1. Longitudinal Fins on Double Pipe Exchangers
2. Transverse Fins on Double Pipe Exchangers
3. Extended Surface Shell-and-Tube Exchangers
4. Double Extended Surfaces

Details on the above units are presented below.

Longitudinal Fins on Double Pipe Exchangers. Manufacturers supply completely assembled double pipe exchangers with fins on the inner pipe. Because the finned pipe or tube cannot be inserted through a packed gland, the method of confining the inner pipe is somewhat more complicated. A typical example of a standard preassembled double pipe finned exchanger is called a *hairpin* (note that a bare double pipe exchanger is also called a hairpin). This type of exchanger is preferable when one of the fluids is a gas or a viscous liquid. Operated in series or in parallel, they are often superior to shell and tube exchangers, notwithstanding the increased number of leakable joints introduced. When operated with cooling water in the inner pipe, a nonferrous pipe may be selected with either steel or nonferrous fins attached.

Transverse Fins on Double Pipe Exchangers. Transverse fin exchangers in crossflow are used only when the film coefficients of the fluids passing over them are low. This applies particularly to gases and air at low and moderate pressures. Tubes are also available which have many very small fins internally shaped from the tube metal itself and which are usable in conventional 1–2 exchangers with baffled side to side flow. Calculations can be accomplished by using a suitable shell side heat transfer curve and an appropriate efficiency curve.

Perhaps the most interesting applications of transverse fins are found in the larger gas cooling and heating services such as on furnaces (economizers), tempering coils for air conditioning, air cooled steam condensers for turbine and engine work, and miscellaneous special services. An application which is growing in popularity is the air-cooled steam condenser for localities with inadequate cooling water supplies. The steam enters the tubes, and an induced draft fan circulates air over the transverse finned tubes. In this way it is possible to attain a closer approach to the atmospheric temperature than could be accomplished with a reasonable amount of surface composed entirely of bare tubes.

Extended-Surface Shell-and-Tube Exchangers. The use of extended surfaces in double pipe exchangers permits the transfer of a great deal of heat in a compact unit. The same advantages may be obtained from the use of longitudinally finned tubes in shell-and-tube arrangements equivalent to the 1–1, 1–2, or 2–4 exchanger. Because they are relatively uncleanable, extended-surface tubes are usually laid out on triangular pitch and are never spaced so closely that the fins of adjacent tubes intermesh. To prevent sagging and the possibility of tube vibration which might result from intermeshing, each tube in a bundle is supported individually. This cannot be done with conventional supported plates because they introduce a certain amount of flow across the bundle which cannot be achieved with longitudinal finned tubes. Support is accomplished, however, by welding or shrinking small circumferential rings about each tube which enclose the fins but at different points along the length of each tube. The rings prevent any tubes from intermeshing and at the same time afford a positive elimination of vibration damage. The entire bundle is then bound with circumferential bands at several points along its length which keep all of the finned tubes firmly pressed against the rings of adjacent tubes.

Longitudinal fin exchangers are relatively expensive and, since they are not cleanable, can only be used for fluids which ordinarily have very low film coefficients and which are clean or form dirt that can be boiled out. This makes them ideal for gases at low pressure where the density is low and the allowable pressure drop is accordingly small.

Double Extended Surfaces. Another type of heating element, which at first inspection appears to offer unlimited possibilities, is the double extended surface. Suppose two flowing fluids are separated by a metal wall and the surface is extended into both fluids by means of spines or pegs whose bases are superimposed. Per square foot of wall area, it should be possible to add as much surface as desired as long as there are no restrictions along the axes of the spines. Such an arrangement will be called for only when the coefficients from both fluids to their respective fins is small. Under these conditions it is usually observed that an unpractical length of fin is required for even a moderate rate of heat transfer. The fin efficiencies and total transfer can be determined by the methods already discussed. The heat transfer coefficients generally must be approximated from more conventional arrangements.

8.3 Key Describing Equations

There are two problems in calculating the heat transfer coefficient for smooth finned tubes.

1. The mean surface temperature of the fin is lower than the surface temperature of a smooth tube under the same conditions (due to the flow of heat through the metal of the fin).
2. There is a question as to whether or not the flow of the fluid outside the tube is as great at the bottom of the space between the fins as in the unobstructed space. Both these factors depend on the size and thickness of the fins as well as their spacing and the conditions of flow [1].

To analyze the heat transfer in extended surfaces, the following assumptions are usually made [3, 4].

1. Steady-state operation
2. Constant properties
3. Constant surrounding air temperature of T_∞
4. Homogeneous isotropic material, with thermal conductivity, k

5. One-dimensional heat transfer by conduction in the radial direction
6. No internal heat generation
7. Heat transfer coefficient, h, is uniform along the fin surface
8. Negligible thermal radiation
9. The fin perimeter at any cross section is P
10. The fin cross-sectional area is a_x
11. The temperature of the heat transfer surface (exposed and unexposed) at the base of the fin is constant, T_b
12. The maximum temperature driving force for convection is $T_b - T_\infty$

The maximum rate of heat transfer, $Q_{f,\max}$, from a fin will occur when the entire fin surface is isothermal at $T = T_b$. In the case of a fin of total surface area A_f, $Q_{f,\max}$ is written as,

$$Q_{f,\max} = hA_f\left(T_b - T_\infty\right) = hA_f\theta_b \tag{8.6}$$

where θ, termed the *excess temperature*, is defined as $\left(T_b - T_\infty\right)$, where T_∞ is the fluid temperature [2, 3].

Since the fin has a finite thermal conductivity, a temperature gradient will exist along the fin. Thus, the actual heat transfer rate from the fin to the outside fluid, Q_f, will be less than $Q_{f,\max}$. The *fin efficiency*, η_f, is a measure of how close Q_f approaches $Q_{f,\max}$ and is defined as:

$$\eta_f = \frac{Q_f}{Q_{f,\max}} = \frac{Q_f}{hA_f\theta_b} \tag{8.7}$$

From Equations (8.6) and (8.7),

$$Q_f = \eta_f hA_f\theta_b = \frac{T_b - T_\infty}{\left(1/\eta_f hA_f\right)} = \frac{T_b - T_\infty}{R_f} \tag{8.8}$$

where,

$$R_f = \text{fin thermal resistance} = \frac{1}{\eta_f hA_f} \tag{8.9}$$

Figure 8.8 [2, 6] is a plot of the efficiency of straight (longitudinal) fins ($\eta_f\%$) versus the following dimensionless group,

$$z_{f,c}^{3/2} = \sqrt{\frac{h}{kA_p}} \tag{8.10}$$

where $z_{f,c}$ is the corrected fin length and A_p is the profile area of the fin. Table 8.2 provides expressions to calculate the corrected length, $z_{f,c}$, the projected or profile area, A_p, and the surface area, A_f, in terms of the fin length, L, fin thickness at the base, $z_{f,t}$, and the fin width (or fin height), z_f,

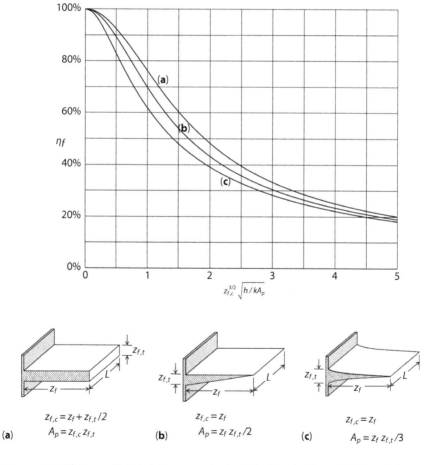

FIGURE 8.8 Efficiency of (a) straight rectangular fins, (b) triangular fins, and (c) parabolic profile fins. (Adapted from Incroprera and De Witt, *Fundamentals of Heat and Mass Transfer*, John Wiley & Sons, 1981 [6] and from Kern, *Process Heat Transfer*, McGraw-Hill, 1950 [2].

for various fin types. *Note*: the parameter L is used to define the pipe length *and* the length of a longitudinal fin (as they are both the same length). The fin efficiency figures are valid for Biot numbers ≤ 0.25, i.e.,

$$\mathrm{Bi}_f = \frac{h\left(z_{f,t}/2\right)}{k} \leq 0.25 \tag{8.11}$$

Barkwill *et al.*[5] recently converted the graphical results presented in Figure 8.8 [5, 6] into equation form. His results are provided in Table 8.3.

Table 8.2 Fin data.

Variable	Rectangular Fin	Triangular Fin	Parabolic Fin
$z_{f,c}$ = corrected height	$z_f + \dfrac{z_{f,t}}{2}$	z_f	z_f
A_p = profile area	$z_{f,c}z_{f,t}$	$\dfrac{z_f \cdot z_{f,t}}{2}$	$\dfrac{z_f \cdot z_{f,t}}{3}$
A_f = fin surface area (for 2 sides)	$2Lz_{f,c}$	$2L\sqrt{z_f^2 - \left(\dfrac{z_{f,t}}{2}\right)^2}$	$2.05L\sqrt{z_f^2 - \left(\dfrac{z_{f,t}}{2}\right)^2}$

Table 8.3 Efficiency of straight rectangular fins, triangular fins, and parabolic profile fins.

For rectangular fins:	
$Y = (4.5128)X^3 - (10.079)X^2 - (31.413)X + 101.47$	(a)
For triangular and parabolic profile fins:	
$Y = (3.1453)X^3 - (7.5664)X^2 - (25.536)X + 101.18$	(b)
Note:	
$Y = \eta_f, \%$	
$X = z_{f,c}^{3/2}\left(\dfrac{h}{kA_p}\right)^{1/2} = \sqrt{\dfrac{z_{f,c}^3 h}{kA_p}}$	

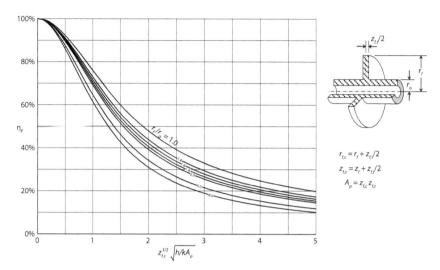

Figure 8.9 Efficiency of annulus fins of rectangular profiles. (Adapted from Incropera and DeWitt, *Fundamentals of Heat and Mass Transfer*, John Wiley & Sons, 1981 [6] and from Kern, *Process Heat Transfer*, McGraw-Hill, 1950 [2].)

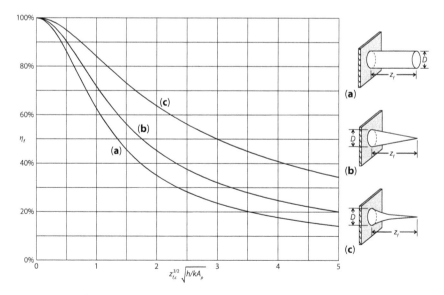

Figure 8.10 Efficiency of pegs and studs fins. (Adapted from Incropera and DeWitt, *Fundamentals of Heat and Mass Transfer*, John Wiley & Sons, 1981 [6] and from Kern, *Process Heat Transfer*, McGraw-Hill , 1950 [2].)

Figure 8.9 [2, 6] shows the fin efficiency (η_f%) of annular fins of rectangular profile. The variables L_c and A_p used in the abscissa of the graph are related to the fin height, L, pipe radius, r_o, and fin radius, r_f, in the following manner,

$$z_f = \text{fin height} = r_f - r_o$$

$$r_{f,c} = \text{corrected fin outside radius} = r_f + \left(z_{ft}/2\right)$$

$$z_{f,c} = \text{corrected fin height} = L + \left(z_{ft}/2\right)$$

$$A_p = \text{profile (cross-sectional area)} = z_{fc}z_{ft}$$

$$A_f = \text{fin surface area} = 2\pi\left(r_{f,c}^2 - r_o^2\right)$$

Figure 8.10 [2, 6] provides efficiency information on pegs and studs. The parameter of the curves is the ratio, $r_{f,c}/r_o$. Barkwill et al.[5] also converted the results of Figure 8.9 into equation form.

For a straight fin (one of uniform cross-section as opposed to one that, for example, tapers down to a point), the heat transfer from the fin may be represented mathematically by,

$$Q = \left(\sqrt{hPkA_c}\right)\theta_c \tanh\left(mz_{f,c}\right) \tag{8.12}$$

where P is the fin cross-section perimeter, A_c is the (cross-sectional) area of the fin, and $m = \sqrt{hP/kA_c}$.

Calculating pressure drops in finned double-pipe exchangers must be viewed as a state of the art; past experience is usually heavily relied on. In lieu of this, the authors [7] recommend estimating the pressure drop by employing equation (8.13):

$$\Delta P = \frac{fG_s^2 L_p}{\left(5.22\times10^{10}\right)Ds\phi_i}\left(\frac{D}{S_T}\right)^{0.4}\left(\frac{S_L}{S_T}\right)^{0.6} \tag{8.13}$$

For longitudinal forms, assume $\left(D/S_T\right)^{0.4}$ and $\left(S_L/S_T\right)$ are unity. The factors S_T is the pitch in a transverse bank and S_L is the center to center distance to the neatest tube in the next bank. For the annular region, $D = D_e$ and $\phi = \phi_a$. For the tube, $D = D$ and $\phi_i = \phi_t$, and the term L_p represents the total tube length. The friction factor f may be estimated by [7],

$$f = \left(5.8 \times 10^{-4}\right) - \left(Re\right)\left(6.3 \times 10^{-8}\right) \tag{8.14}$$

For transverse fins, $D = D'_{e,v}$ where $D'_{e,v}$ is the volumetric equivalent diameter is defined by,

$$D'_{e,v} = \frac{(4)(\text{new free volume})}{A_f + A_o} \tag{8.15}$$

where A_f = surface of fins (both sides), ft² and A_o = bare tube surface in between fins on the outside of the tube, ft². The net free volume is the volume between the center line of two vertical banks of tubes, less the volumes of the half-tubes and fins within the center lines.

Almost all the heat transfer coefficient data available on commercial scale applications have been made with air or flue gas. It has been found that the heat-transfer coefficient is not influenced by the spacing of succeeding rows although nearly all the data were obtained for triangular pitch arrangements. The recommended equivalent diameter to employ in the available heat transfer correlations is,

$$D_e = \frac{2\left(A_f + A_o\right)}{\pi\left(\text{projected perimeter}\right)} \tag{8.16}$$

The projected perimeter is the sum of all the external distances in the plane view of a transverse finned tube. The mass velocity is computed from the free flow area in a single bank of tubes at right angles to the gas flow.

Example 8.4 – Estimating Fin Efficiency. The following information is provided for a straight rectangular fin: $h = 15\,\text{W/m}^2 \cdot \text{K}$, $k = 300\,\text{W/m} \cdot \text{K}$, $z_f = 3\,\text{in.}$, and $z_{f,t} = 1\,\text{in.}$ Estimate the fin efficiency.

SOLUTION: Refer to Figure 8.8. For a rectangular fin:

$$z_{f,c} = z_f + \frac{z_{f,t}}{2} = 3 + \frac{1}{2} = 3.5\,\text{in} = 0.0889\,\text{m}$$

and,

$$A_p = \left(z_{f,c}\right)\left(z_{f,t}\right) = (3.5)(1) = 3.5\,\text{in}^2 = 0.00226\,\text{m}^2$$

Generate the X-coordinate of Figure 8.8:

$$\sqrt{\frac{z_{f,c}^3 h}{kA_p}} = z_{f,c}^{3/2}\left(\frac{h}{kA_p}\right)^{1/2} = (0.0899)^{3/2}\left[\frac{15}{(300)(0.00226)}\right]^{1/2}$$

$$= (0.0265)(22.124)^{1/2}$$

$$= 0.1246$$

The following value is read from Figure 8.8,

$$\eta_f \approx 98\%$$

Example 8.5 – Estimating Fin Efficiency – Barkwill's Equation. Estimate the fin efficiency in the previous example using the equation developed by Barkwill et al.[4].

SOLUTION: Apply Equation (a) from Table 8.3:

$$Y = (4.5128)X^3 - (10.079)X^2 - (31.413)X + 101.47$$

with,

$$X = 0.1246$$

and,

$$Y = \eta_f$$

Substituting,

$$\eta_f = (4.5128)(0.1246)^3 - (10.079)(0.1246)^2$$
$$- (31.413)(0.1246) + 101.47$$
$$= 0.008730 - 0.15648 - 3.91406 + 101.47$$
$$= 97.4\%$$
$$= 0.974$$

The two results are in agreement with each other.

Example 8.6 – Heat Transfer Calculation for a Single Fin. A set of micro-fins is designed to cool an electronic circuit. Each micro-fin has a square cross-section of 0.2 cm by 0.2 cm and a length of 1 cm. The conductivity of the fin material is $400\,W/m\cdot K$ and the air heat transfer coefficient is

16 W/m$^2 \cdot$K. The circuit temperature is 100 °C and the air temperature is 25 °C. Calculate the heat transfer from each micro-fin in W.

SOLUTION: Write the appropriate equation for this heat transfer application.

$$Q = \left(\sqrt{hPkA_c} \right) \theta_c \tanh\left(mz_{f,c} \right) \tag{8.12}$$

Substitute known values and compute the heat transfer:

$$P = (4)(0.2) = 0.8 \, \text{cm}$$

$$z_{f,c} = z_f + \frac{A_c}{P} = 1.0 + \frac{(0.2)(0.2)}{(4.0)(0.2)} = 1.05 \, \text{cm} = 0.0105 \, \text{m}$$

$$m = \sqrt{\frac{hP}{kA_c}} = \sqrt{\frac{(16)(4)(0.002)}{(400)(0.002)(0.002)}} = 8.95 \, \text{m}^{-1}$$

Substitution,

$$Q = \left(\sqrt{(16)(4)(0.002)(400)(0.002)(0.002)} \right)(100 - 25)$$
$$\tanh\left[(8.95)(0.0105) \right]$$
$$= 0.10 \, \text{W}$$

Example 8.7 – Heat Transfer Calculations for Fins. Calculate the total heat transfer from the set of micro-fins in the previous example.

SOLUTION: Although the heat transfer from one micro-fin is known, the total number of fins in the set is *not* known. Therefore, the total heat transfer cannot be calculated.

8.4   Fin Effectiveness and Performance

Another dimensionless quantity used to assess the benefit of adding fins is the *fin effectiveness*, ε_f, or its accompanying definition, *fin performance coefficient*, FPC [2,3]. It is defined as,

$$\varepsilon_f = \text{FPC} \tag{8.17}$$

where,

$$\varepsilon_f = \text{FPC} = \frac{Q_f}{Q_{w/o,f}} \qquad (8.18)$$

and $Q_{w/o,f}$ is the rate of heat transfer without fins, i.e.,

$$Q_{w/o,f} = hA_b\theta_b \qquad (8.19)$$

Fins are not usually justified unless ε_f or FPC ≥ 2.

The fin efficiency (see also Equation 8.7), η_f, and the aforementioned fin effectiveness, ε_f, characterize the performance of a single fin. As indicated above, arrays of fins are attached to the base surface in many applications. The distance from the center of one fin to the next one along the same tube surface is termed the *fin pitch, S*. In this case, the total heat transfer area, A_t, includes contributions due to the fin surfaces and the exposed (unfinned) base surface, i.e.,

$$A_t = A_{be} + N_f A_f \qquad (8.20)$$

The total heat transfer area without fins, $A_{t,w/o,f}$, is,

$$A_{t,w/o,f} = A_{be} + N_f A_b \qquad (8.21)$$

where N_f = number of fins, A_f = surface area per fin, A_{be} = total exposed (unfinned) base area of the surface, and A_b = the base area of one fin. Equation (8.20) is often used to determine A_{be} from knowledge of the surface geometry, fin base area, and number of fins. The total heat transfer rate, without fins, $Q_{t,w/o,f}$, is,

$$Q_{t,w/o,f} = hA_{t,w/o,f}\theta_b = h\left(A_{be} + N_f A_b\right)\theta_b \qquad (8.22)$$

The total heat transfer rate from the finned surface, Q_t, is,

$$\begin{aligned}
Q_t &= Q_{be} + Q_{f,t} \\
&= Q_{be} + N_f Q_f \\
&= hA_{be}\theta_b + N_f hA_f \eta_f \theta_b \\
&= h\left(A_{be} + N_f A_f \eta_f\right)\theta_b
\end{aligned} \qquad (8.23)$$

where $Q_{f,t}$ is the heat transfer rate due to all (total) fins and Q_f is that due to a single fin.

The maximum heat transfer rate, $Q_{t,max}$, of the surface occurs when $\eta_f = 1.0$, i.e., when the temperature of the base and all the fins is T_b, and is given by

$$Q_{t,max} - h\left(A_{be} + N_f A_f\right)\theta_b = hA_t\theta_b \tag{8.24}$$

The corresponding *overall fin efficiency*, $\eta_{o,f}$, is defined as,

$$\eta_{o,f} = \frac{Q_t}{Q_{t,max}} = \frac{A_{be} + N_f A_f \eta_f}{A_t} \tag{8.25}$$

Substituting from Equation (8.20) into (8.25) yields,

$$\eta_{o,f} = 1 - \left(\frac{N_f A_f}{A_t}\right)\left(1 - \eta_f\right) \tag{8.26}$$

Finally, the *overall surface effectiveness*, $\varepsilon_{o,f}$, is defined as,

$$\varepsilon_{o,f} = \frac{Q_t}{Q_{t,w/o,f}} \tag{8.27}$$

The optimum fin may be considered the one which has a constant heat flux at any cross section between the outer edge and the base. These would correspond to tall narrow shapes with small base areas for the common longitudinal and transverse fins. Furthermore, the sides of the fins should have parabolic curvatures. Although the thermal efficiency for ideal fins may be high, the cost of their manufacture is usually excessive and they are rarely structurally adaptable to industrial applications. The calculations of the optimum shapes have led, however, to the present type of manufacture using 20 BWG and lighter metal ribbons, except where the conditions of heat transfer require a more rugged construction. Jacob (8) earlier presented an excellent survey of optimum fins defined by other criteria.

Example 8.8 – Annular Fins. A circular tube has an outside diameter of 2.5 cm and a surface temperature, T_b, of 170 °C. An annular aluminum fin of rectangular profile is attached to the tube. The fin has an outside radius, r_f, of 2.75 cm, a thickness, $z_{f,t}$, of 1 mm, and a thermal conductivity, k, of

200 W/m·K. The surrounding fluid is at a temperature $T_\infty = 25°C$ and the associated heat transfer coefficient, h, is 130 W/m²·K. Calculate the heat transfer rate without the fin, $Q_{w/o, f}$, the corrected length, $z_{f,c}$, the outer radius, $r_{o,c}$, the maximum heat transfer rate from the fin, $Q_{f,max}$, the fin efficiency, η_f, the fin heat transfer rate, Q_f, and the fin thermal resistance, $R_{t,f}$.

SOLUTION: Determine the area of the base of the fin:

$$r_o = \frac{D_o}{2} = \frac{0.025}{2} = 0.0125\,\text{m}$$

$$A_b = 2\pi r_o z_{f,t} = 2\pi (0.0125)(0.001) = 7.854 \times 10^{-5}\,\text{m}^2$$

Calculate the excess temperature at the base of the fin:

$$\theta_b = T_b - T = 170 - 25 = 145\,\text{K}$$

The total heat transfer rate without the fin, $Q_{w/o, f}$, is then (see Equation (8.19)),

$$Q_{t,w/o,f} = Q_{w/o,f} = hA_b\theta_b = (130)(7.854 \times 10^{-5})(145)$$
$$= 1.48\,\text{W} = 5.0\,\text{Btu/hr}$$

Calculate the Biot number. Employ Equation (8.11). Determine if it is valid to use the fin efficiency figures provided in this section since the Biot number must be less than 0.25,

$$\text{Bi}_f = \frac{h(z_{f,t}/2)}{k_f} = \frac{(130)(0.001/2)}{200} = 3.25 \times 10^{-4} < 0.25$$

The fin efficiency figures may be used. Calculate the fin height, z_f. Since,

$$r_f = 0.0275\,\text{m}$$
$$z_f = r_f - r_o = 0.0275 - 0.0125 = 0.015\,\text{m}$$

Calculate the corrected radius and height:

$$r_{f,c} = r_f + \frac{z_{f,t}}{2} = 0.0275 + \frac{0.001}{2} = 0.028\,\text{m}$$

$$z_{f,c} = z_f + \frac{z_{f,t}}{2} = 0.015 + \frac{0.001}{2} = 0.0155\,\text{m}$$

Calculate the profile area and the fin surface area:

$$A_p = z_{f,c} z_{f,t} = (0.0155)(0.001) = 1.55 \times 10^{-5} \, \text{m}^2$$
$$A_p = 2\pi \left(r_{f,c}^2 - r_o^2 \right) = 2\pi \left(0.028^2 - 0.0125^2 \right) = 3.94 \times 10^{-3} \, \text{m}^2$$

The maximum fin heat transfer rate, $Q_{f,\text{max}}$, is therefore:

$$Q_{f,\text{max}} = hA_f \theta_b \qquad\qquad (8.19)$$
$$= (130)(3.94 \times 10^{-3})(145) = 74.35 \, \text{W}$$

Calculate the curve parameter and the abscissa for the fin efficiency in Figure 8.8. Employ Equation (8.10) and the information provided in Figure 8.8:

$$\text{Abscissa} = z_{f,c}^{3/2} \left(\frac{h}{kA_p} \right)^{1/2} = \sqrt{\frac{z_{f,c}^3 h}{kA_p}} \qquad\qquad (8.10)$$

$$= \sqrt{\frac{(0.0155)^3 (130)}{(200)(1.55 \times 10^{-5})}} = 0.395$$

From Figure 8.8 (interpolating),

$$\eta_f = 86\% = 0.86$$

Therefore, the fin heat transfer rate, Q_f, as per Equation (8.7) is,

$$Q_f = \eta_f Q_{f,\text{max}} \qquad\qquad (8.17)$$
$$= (0.86)(74.35) = 64 \, \text{W} = 218 \, \text{Btu/hr}$$

The corresponding fin resistance, from Equations (8.7) – (8.9), is,

$$R_{t,f} = \frac{\theta_b}{Q_f}; \quad \theta_b = T - T_\infty$$

$$R_{t,f} = \frac{145}{64} = 218 \, \text{K/W}$$

Example 8.9 – Calculating Fin Effectiveness. Refer to Example 8.8. Calculate the fin effectiveness, ε_f, and whether the use of the fin is justified.

SOLUTION: Calculate the fin effectiveness or performance coefficient using Equation (8.18):

$$\varepsilon_f = \frac{Q_f}{Q_{w/o,f}} = \frac{64}{1.48} = 43.2 \tag{8.18}$$

Since,

$$\varepsilon_f = 43.2 > 2.0$$

the use of the fin is justified (refer to Equation (8.19)).

Example 8.10 – Total Heat Rate and Effectiveness Calculation. If the tube described in Example 8.8 had a length of one meter and fin pitch (S) of 10 mm, what would be the total surface area for heat transfer, the exposed tube base total heat transfer rate, the overall efficiency of the surface, and the overall surface effectiveness?

SOLUTION: Calculate the number of fins in the tube length:

$$N_f = \frac{L}{S} = \frac{1}{0.01} = 100 \text{ fins}; \quad S = 10\text{mm} = 0.01\text{m}$$

Calculate the unfinned base area:

$$L_{be} = \text{unfinned (exposed) base length} = L - N_f z_{f,t}$$
$$= 1 - (100)(0.001) = 0.9\,\text{m}$$
$$A_{be} = \text{unfinned base area} = 2\pi r_o L_{be}$$
$$= 2\pi (0.0125)(0.9) = 0.0707\,\text{m}^2$$

The total transfer surface area, A_t, may now be calculated:

$$A_t = A_{be} + N_f A_f \tag{8.20}$$
$$= 0.0707 + (100)(3.94 \times 10^{-3}) = 0.465\,\text{m}^2$$

Equation (8.22) is used to obtain the total heat rate without fins:

$$Q_{t,w/o,f} = h(2\pi r_o X)\theta_b \tag{8.22}$$

$$= (130)(2\pi)(0.0125)(1)(145) = 1480\,\text{W}$$

Calculate the heat flow rate from the exposed tube base:

$$Q_{be} = hA_{be}\theta_b \tag{8.23}$$

$$= (130)(0.0707)(145) = 1332.7\,\text{W} = 4548\,\text{Btu/hr}$$

Calculate the heat flow rate from all the fins:

$$Q_{f,\text{total}} = N_f Q_f \tag{8.23}$$

$$= (100)(64) = 6400\,\text{W}$$

The total heat flow rate may now be calculated:

$$Q_t = Q_{be} + Q_{f,\text{total}} \tag{8.23}$$

$$= 1332.7 + 6400 = 7732.7\,\text{W}$$

Calculate the maximum heat transfer rate from Equation (8.24):

$$Q_{t,\text{max}} = hA_t\theta_b \tag{8.24}$$

$$= (130)(0.465)(145) = 8765.3\,\text{W}$$

By definition, the overall fin efficiency is given in Equation (8.25):

$$\eta_{o,f} = \frac{Q_t}{Q_{t,\text{max}}} \tag{8.25}$$

$$= \frac{7732.7}{8765.3} = 0.882 = 88.2\%$$

The corresponding overall effectiveness is therefore given by Equation (8.27):

$$\varepsilon_{o,f} = \frac{Q_t}{Q_{t,w/o,f}} \tag{8.27}$$

Substitution,

$$= \frac{7732.7}{1480} = 5.22$$

Example 8.11 – Single Fin Calculations. Consider the case of aluminum fins of triangular profile that are attached to a plane wall with a surface temperature of 250°C. The fin base thickness is 2 mm and its length is 6 mm. The system is in ambient air at a temperature of 20°C and the surface convection coefficient is 40 W/m² · K. Consider a 1 m width of a single fin. Determine:

1. the heat transfer rate without the fin,
2. the maximum heat transfer rate from the fin, and
3. the fin efficiency, thermal resistance, and effectiveness.

Properties of the aluminum may be evaluated at the average temperature, $\overline{T} = (T_b + T_\infty)/2 = (250 + 20)/2 = 135°C = 408\,K$, where $k \approx 240\,W/m \cdot K$.
SOLUTION: Determine the base area of the fin:

$$A_b = z_{f,t} w = (0.002)(1) = 0.002\,m^2$$

Calculate the excess temperature at the base of the fin:

$$\theta_b = T_b - T_\infty = 250 - 20 = 230°C = 230\,K$$

Calculate the heat transfer rate without a fin, $Q_{w/o,f}$, employing Equation (8.19).

$$Q_{w/o,f} = hA_b\theta_b \qquad (8.19)$$

Substitution,

$$Q_{w/o,f} = (40)(0.002)(230) = 18.4\,W$$

Also, calculate the maximum heat transfer rate, $Q_{f,max}$:

$$Q_{f,max} = hA_f\theta_b$$
$$= (40)(0.012)(230) = 110\,W$$

where,

$$A_f = 0.0118 \approx 0.012\,m^2$$

Check the Biot number criterion to determine if the use of the fin efficiency figure is valid,

$$Bi_f = \frac{h(z_{f,t}/2)}{k} \qquad (8.11)$$

$$= \frac{(40)(0.002/2)}{240} = 1.67 \times 10^{-4} < 0.25$$

Since $Bi_f < 0.25$, the use of the figure is permitted. Calculate the fin $z_{f,c}$, A_p, and P, noting that this is a triangular fin:

$$z_{f,c} = \text{corrected length} = z_f = 0.006\,\text{m}$$

$$A_p = \text{profile area} = L\left(z_{f,t}/2\right) = (0.006)(0.002/2) = 6 \times 10^{-6}\,\text{m}^2$$

$$A_f = \text{fin surface area} = 2L\sqrt{z_f^2 - \left(z_{f,t}/2\right)^2} \approx 0.012\,\text{m}^2$$

Determine the abscissa for the fin efficiency figure (see Figure 8.8):

$$\text{Abscissa} = \sqrt{\frac{z_{f,c}^3 h}{kA_p}} \tag{8.10}$$

$$= \sqrt{\frac{(0.006)^3 (40)}{(240)(6 \times 10^{-6})}} \approx 0.0775$$

Obtain the fin efficiency from Figure 8.8 for triangular profile straight fins,

$$\eta_f = 0.99$$

Calculate the fin heat transfer rate, Q_f, employing Equation (8.7),

$$Q_f = \eta_f Q_{f,\text{max}} = (0.99)(110) = 108.9\,\text{W}$$

The fin thermal resistance, $R_{t,f}$, is therefore,

$$R_{t,f} = \frac{\theta_b}{Q_f} \tag{8.9}$$

$$= \frac{230}{108.9} = 2.1°\text{C/W}$$

Example 8.12 – Estimating Conductive Resistance. Air and water are separated by a 1.5 mm plane wall made of steel ($k = 38\,\text{W/m·K}$; density, $\rho = 7753\,\text{kg/m}^3$; heat capacity, $c = 486\,\text{J/kg·K}$). The air temperature, T_1, is 19°C, and the water temperature, T_4, is 83°C. Denote the temperature at the air-wall interface T_2 and let T_3 be the temperature at the wall-water interface. The air-side heat transfer coefficient, h_1, is $13\,\text{W/m}^2 \cdot \text{K}$ and the water side heat transfer coefficient, h_3, is $260\,\text{W/m}^2 \cdot \text{K}$. Assume an area of the wall that is 1 m high and 1 m wide as a basis.

1. Show whether the conduction resistance may be neglected.
2. What is the rate of heat transfer from water to air?

To increase the rate of heat transfer, it is proposed to add steel fins to the wall. These straight rectangular steel fins will be 2.5 cm long, 1.3 mm thick, and will be spaced such that the fin pitch, S, is 1.3 cm between centers.

3. Calculate the percent increase in the steady-state heat transfer rate that can be realized by adding fins to the air side of the plane wall.
4. Calculate the percent increase in the steady-state heat transfer rate that can be realized by adding fins to the water side of the plane wall.

SOLUTION: The base wall area, A, is $1.0\,\text{m}^2$.

1. R_1 = air resistance $= \dfrac{1}{h_1 A} = \dfrac{1}{13} = 0.0769°\text{C/W}$

   R_2 = conduction resistance $= \dfrac{L_2}{k_1 A} = \dfrac{0.0015}{38} = 3.95 \times 10^{-5}°\text{C/W}$

   R_3 = water resistance $= \dfrac{1}{h_4 A} = \dfrac{1}{260} = 0.00385°\text{C/W}$

   Clearly, $R_2 \lll R_1$ or R_3. Therefore, R_2 may be neglected.

2. $R_{\text{total}} = R_1 + R_3 = 0.0769 + 0.00385 = 0.0807°\text{C/W}$

   Therefore,

   $$Q = \frac{T_4 - T_1}{R_{\text{total}}} = \frac{83 - 19}{0.0807} = 793.1\,\text{W}$$

   This represents the total heat transfer from the base pipe without fins.

   i.e., $Q_{t,w/o,f}$.

3. For the case where fins have been added on the air side, write an expression for the total heat transfer rate, Q_t:

$$Q_t = Q_{be} + Q_{f,total} = Q_{be} + N_f Q_f$$
$$= Q_{be} + N_f \eta_f Q_{f,max}$$
$$= h\theta_b \left(A_{be} + N_f \eta_f Q_f \right)$$

See Equations (8.20) and (8.21). For a unit area, 1 m in length and 1 m in width (w), calculate the number of fins, N_f, and the exposed base surface area, A_{be}:

$$N_f = \text{number of fins} = \frac{1}{0.013} = 77 \text{ fins; } S = 1.3 \text{ cm}$$

L_{be} = unfinned exposed base surface net length
$$= L - N_f z_{f,t}; \ z_{f,t} = 1.3 \text{ mm}$$
$$= 1 - (77)(0.0013) = 1 - 0.1 = 0.9 \text{ m}$$

$$A_{be} = w L_{be} = (1)(0.9) = 0.9 \text{ m}^2$$

For a rectangular fin, calculate the corrected length, $z_{f,c}$, the profile area, A_p, and the fin surface area, A_f:

$$z_{f,c} = \text{corrected length} = L + \frac{z_{f,t}}{2} = 0.025 + \frac{0.0013}{2} = 0.02565 \text{ m}$$

$$A_p = \text{profile area} = z_{f,c} z_{f,t} = (0.02565)(0.0013) = 3.334 \times 10^{-5} \text{ m}^2$$

$$A_f = \text{fin surface area (per one fin)} = 2Lz_{f,c} = 2(1)(0.02565)$$
$$= 0.0513 \text{ m}^2$$

Check the Biot number to verify that the use of the fin efficiency figure is valid.

$$Bi_f = \frac{h(z_{f,t}/2)}{k_f} = \frac{(13)(0.001312/2)}{38} = 2.2 \times 10^{-4}$$

Since $Bi_f < 0.25$, the use of the figure is valid. Calculate the abscissa of the fin efficiency diagram and obtain the fin efficiency:

$$\sqrt{\frac{z_{f,c}^3 h}{kA_p}} = \sqrt{\frac{(0.02565)^3 (13)}{(13)(3.334 \times 10^{-5})}} = \sqrt{0.1731} = 0.416$$

From Figure 8.8, $\eta_f \approx 0.88$. Calculate the air thermal resistances:

$$R_{base} = \frac{1}{hA_{be}} = \frac{1}{(13)(0.9)} = 0.0855°C/W$$

$$R_{f,total} = \frac{1}{hN_f A_f \eta_f} = \frac{1}{(13)(77)(0.0513)(0.88)} = 0.0221°C/W$$

The total resistance of the fin array is therefore given by,

$$\frac{1}{R_{total}} = \frac{1}{R_{base}} + \frac{1}{R_{fins}} = \frac{1}{0.0855} + \frac{1}{0.0221} = 11.69 + 45.2 = 56.9 \, W/°C$$

$$R_{total} = \frac{1}{56.9} = 0.0176°C/W \; (\text{due to finned surface})$$

This is the *outside* resistance; the water side resistance remains the same. Determine the total resistance to heat transfer and the rate of heat transfer:

$$R_{total} = 0.0176 + 0.00385 = 0.0214°C/W$$

$$Q_t = \frac{T_1 - T_4}{R_{total}} = \frac{83 - 19}{0.0214} = 2867 \, W$$

Calculate the percent increase in Q due to the fins on the air side:

$$\% \text{ increase} = 100 \left(\frac{2867}{793.1} - 1 \right) = 261.5\%$$

Therefore, Q will increase by 261.5% when fins are added on the air side.
4. For the case where fins have been added on the water side, determine the new fin efficiency:

$$\text{Abscissa} = \sqrt{\frac{z_{f,c}^3 h}{kA_p}} = \sqrt{\frac{(0.02565)^3 (260)}{(38)(3.334 \times 10^{-5})}} = \sqrt{3.46} = 1.86$$

From Figure 8.8, $\eta_f \approx 0.38$. Calculate the thermal resistance of the base, fins, and the total resistance of the finned surface:

$$R_{base} = \frac{1}{hA_{be}} = \frac{1}{(260)(0.9)} = 0.00427°C/W$$

$$R_{f,total} = \frac{1}{hN_f A_f \eta_f} = \frac{1}{(260)(77)(0.0513)(0.38)} = 0.00256°C/W$$

$$R_{total} = \frac{1}{1/R_{base} + 1/R_{f,total}} = \frac{1}{1/0.00427 + 1/0.00256} = \frac{1}{624.6}$$
$$= 0.0016°C/W$$

This resistance is on the water side; the air resistance remains the same (i.e., as it was before fins were added to the air side):

$$R_{total} = 0.0769 + 0.0016 = 0.0785°C/W$$

$$Q_t = (83 - 19)/0.0785 = 815.3\,W$$

Finally, calculate the percent increase in Q due to water-side fins:

$$\%increase = 100\left(\frac{815.3 - 793.1}{793.1}\right) = 2.8\%$$

Example 8.13 – Fin Advantage. Comment on the results of the calculations from the previous example.

SOLUTION: It is concluded that fins should be added on the air side; the fins on the water side are practically useless.

Example 8.14 – Fin Economic Consideration. Determine whether the use of the fin is justified in the previous example.

SOLUTION: Calculate the fin effectiveness, ε_f, or performance coefficient, FPC, employing Equation (8.18).

$$\varepsilon_f = FPC = \frac{Q_f}{Q_{w/o,f}} \tag{8.18}$$

$$= \frac{108.9}{18.4} = 5.92$$

Is the use of the fin justified? Since $\varepsilon_f = 5.94 > 2$, the use of the fin is justified. Note that the triangular fin is known to provide the maximum heat transfer per unit mass.

Example 8.15 – Justifying Fin Use. Annular aluminum fins of rectangular profile are attached to a circular tube. The outside diameter of the tube is 50 mm and the temperature of its outer surface is 200 °C. The fins are 4 mm thick and have a length of 15 mm. The system is in ambient air at a temperature of 20 °C and the surface convection coefficient is 40 W/m² · K. The thermal conductivity of aluminum is 240 W/m · K. What are the efficiency, thermal resistance, effectiveness, and heat transfer rate of a single fin? Is the use of the fin justified?

SOLUTION: Determine the base area of the fin:

$$A_b = 2\pi r_o z_{f,t} = 2\pi(0.025)(0.004) = 6.283 \times 10^{-4} \, \text{m}^2$$

Calculate the excess temperature at the fin base:

$$\theta_b = T_b - T_\infty = 200 - 20 = 180°C = 180\,\text{K} = 324°\text{R} = 324°\text{F}$$

Calculate the heat transfer rate without a fin, $Q_{w/o,f}$, using Equation (8.19).

$$Q_{w/o,f} = hA_b\theta_b \tag{8.19}$$

Substitution,

$$Q_{w/o,f} = (40)(6.283 \times 10^{-4})(180) = 4.52\,\text{W}$$

Check the Biot number criterion to determine if the use of the fin efficiency figure is valid:

$$Bi_f = \frac{h(z_{f,t}/2)}{k} \tag{8.11}$$

$$= \frac{(40)(0.004/2)}{240} = 3.33 \times 10^{-4} < 0.25$$

Since $Bi_f < 0.25$, the use of the figure is permitted.
Calculate the fin corrected radius and length, profile area, and surface area:

$$r_f = r_o + z_f = 0.025 + 0.015 = 0.04\,\mathrm{m}$$

$$r_{f,c} = r_f + \left(z_{f,t}/2\right) = 0.04 + 0.002 = 0.042\,\mathrm{m}$$

$$z_{f,c} = z_f + \left(z_{f,t}/2\right) = 0.015 + 0.002 = 0.017\,\mathrm{m}$$

$$A_p = \left(z_{f,c}\right)\left(z_{f,t}\right) = (0.017)(0.004) = 6.8 \times 10^{-5}\,\mathrm{m^2}$$

$$A_f = 2\pi\left(r_{f,c}^2 - r_o^2\right) = 2\pi\left(0.042^2 - 0.025^2\right) = 7.157 \times 10^{-3}\,\mathrm{m^2}$$

Calculate the maximum heat transfer rate from a *single* fin using Equation (8.19).

$$Q_{f,\max} = hA_f\theta_b \tag{8.19}$$

Substitution,

$$Q_{f,\max} = (40)\left(7.157 \times 10^{-3}\right)(180) = 51.53\,\mathrm{W}$$

Determine the abscissa and curve parameter for the efficiency figure (for annular rectangular fins):

$$\frac{r_{f,c}}{r_o} = \frac{0.042}{0.025} = 1.68$$

$$\text{Abscissa} = \sqrt{\frac{z_{f,c}^3\,h}{kA_p}} \tag{8.10}$$

$$= \sqrt{\frac{(0.0017)^3\,(40)}{(240)\left(6.8 \times 10^{-5}\right)}} = 0.11$$

Read the fin efficiency from Figure 8.9 for annular rectangular straight fins:

$$\eta_f = 0.97$$

The fin heat transfer rate, Q_f, is therefore,

$$Q_f = \eta_f Q_{f,\text{max}} \qquad (8.7)$$

Substituting,

$$Q_f = (0.97)(51.53) = 50\,\text{W}$$

Calculate the fin resistance and fin effectiveness:

$$R_{t,f} = \frac{\theta_b}{Q_f} \qquad (8.9)$$

$$= \frac{180}{50} = 3.6°\text{C/W}$$

$$\varepsilon_f = \frac{Q_f}{Q_{w/o,f}} \qquad (8.18)$$

$$= \frac{50}{4.52} = 11.06$$

Since $\varepsilon_f = 11.06 > 2$ the use of the fin is justified.

Example 8.16 – Total Fin Efficiency and Effectiveness. What is the rate of heat transfer per unit length of tube in the previous illustrative example if there are 125 such fins per meter of tube length? Also calculate the total efficiency and effectiveness.

SOLUTION: Calculate the unfinned base area for an array of 125 fins per meter where $N_f = 125$.

$$L_{be} = \text{unfinned (exposed) base length} = L - N_f z_{f,t}$$
$$= 1 - (125)(0.004) = 0.5\,\text{m}$$

$$A_{be} = \text{unfinned base area} = 2\pi r_o L_{be}$$
$$= 2\pi (0.025)(0.5) = 0.0785\,\text{m}^2$$

Calculate the total heat transfer surface area, A_t:

$$A_t = A_{be} + N_f A_f \qquad (8.20)$$
$$= 0.0785 + (125)(7.157\times10^{-3}) = 0.973\,\text{m}^2$$

Calculate the heat rate without fins:

$$Q_{w/o,f} = h\left(2\pi r_o L\right)\theta_b$$
$$= (40)(2\pi)(0.025)(1)(180) = 1131\,\text{W}$$

Calculate the heat rate from the base and fins, and the total heat rate, employing Equation (8.23).

$$Q_{be} = hA_{be}\theta_b$$
$$= (40)(0.0785)(180) = 565.2\,\text{W}$$

$$Q_{f,total} = N_f Q_f$$
$$= (125)(50) = 6250\,\text{W}$$

$$Q_t = Q_{be} + Q_{f,total}$$
$$= 565.2 + 6250 = 6815.2\,\text{W} \approx 6.82\,\text{kW}$$

Calculate the maximum heat transfer rate:

$$Q_{max} = hA_t\theta_b \tag{8.24}$$
$$= (40)(0.973)(180) = 7005.6\,\text{W}$$

The overall fin efficiency is therefore,

$$\eta_{o,f} = \frac{Q_t}{Q_{t,max}} \tag{8.25}$$

$$= \frac{6815.2}{7008.6} = 0.973$$

The corresponding overall fin effectiveness is,

$$\varepsilon_{o,f} = \frac{Q_t}{Q_{t,w/o,f}} \tag{8.27}$$

$$= \frac{6815.2}{1131} = 6.03$$

Finally, the thermal resistances are,

$$R_{base} = \frac{1}{hA_{be}} \tag{8.9}$$

$$= \frac{1}{(40)(0.0785)} = 0.318°C/W$$

$$R_f = \frac{1}{hN_f A_f \eta_f} \qquad (8.9)$$

$$= \frac{1}{(40)(125)(7.157 \times 10^{-3})(0.97)} = 0.0288°C/W$$

Example 8.17 – Fin Tip Calculation. A metal fin 1-inch high and 1/8-inch thick has a thermal conductivity, k, of 25 Btu/hr·ft·°F and a uniform base temperature of 250 °F. It is exposed to an air stream at 70 °F with a velocity past the fin such that the convection coefficient of heat transfer, h, is 15 Btu/hr·ft·°F. Calculate the temperature at the tip of the fin and the heat transfer from the fin per foot of fin. Solve the problem analytically.

SOLUTION: Once again, set the fin height, thickness, and length equal to z_f, $z_{f,t}$, and L, respectively. Bennet and Meyers [8] have shown that the equation describing the temperature profile of this system is:

$$\frac{d^2 T}{dx^2} - \frac{2h}{kz_{f,t}}(T - 70) = 0; \quad x = \text{height coordinate}$$

Let,

$$\alpha^2 = \frac{2h}{kz_{f,t}} = \frac{(2)(15)}{(25)(1/96)} = 115$$

Therefore,

$$\frac{d^2 (T - 70)}{dx^2} - \alpha^2 (T - 70) = 0$$

The solution to this ordinary differential equation [10] is:

$$T - 70 = c_1 e^{\alpha x} + c_2 e^{-\alpha x}$$

or,

$$T = c_1 e^{\alpha x} + c_2 e^{-\alpha x} + 70$$

In addition,

$$\frac{dT}{dx} = \alpha c_1 e^{\alpha x} - \alpha c_2 e^{-\alpha x}$$

and,

$$\frac{d^2T}{dx^2} = \alpha^2 c_1 e^{\alpha x} - \alpha^2 c_2 e^{-\alpha x}$$

The reader is left the exercise of showing that this solution satisfies the above second order ordinary differential equation.

The solution must also satisfy the two boundary conditions (BC) below that are employed to evaluate the above two integration constants c_1 and c_2 [10].

$$BC(1):\ T = 250; \quad \text{at } x = 0 \text{ (fin base)}$$

$$BC(2):\ -kA\frac{dT}{dx} = hA(T - 70); \quad A = z_{f,t}L$$

or,

$$\frac{dT}{dx} = -\frac{h}{k}(T - 70); \quad \text{at } x = z_f = \frac{1}{12}\text{ft}$$

From BC(1),

$$180 = c_1 + c_2$$

From BC(2),

$$-k\left(\alpha c_1 e^{\alpha x} - \alpha c_2 e^{-\alpha x}\right) = h\left(c_1 e^{\alpha x} + c_2 e^{-\alpha x}\right)$$

Solving these two equations simultaneously gives,

$$c_1 = 25.9 \quad c_2 = 154.1$$

The fin top $\left(x = z_f = \dfrac{1}{12}\text{ft}\right)$ temperature is therefore,

$$T - 70 = (25.9)(2.4392) + (154.1)(0.40997)$$

$$T - 70 = 63.2 + 63.2$$

$$T - 70 = 126.4$$

$$T = 126.4 + 70$$

$$T = 196.4°F$$

The heat transfer rate is given by,

$$Q = -kA_{base}\frac{dT}{dx}\bigg|_{x=0} \quad ; \quad \frac{dT}{dx}\bigg|_{x=0} = a(c_1 - c_2)$$

$$= -(25)\left(\frac{1}{8}\right)\left(\frac{1}{12}\right)(w)(10.7)(25.9 - 154.1); \quad L = 1.0$$

$$= 357.2 \text{ Btu/hr} \cdot \text{ft of fin length}$$

Example 8.18 – Modified Fin Efficiency. Estimate the fin efficiency in the previous example if the efficiency is defined (as noted earlier) as the actual heat transfer divided by the heat rate of the entire fin with the same temperature as its base.

SOLUTION: For the entire fin at $T = 250°F$,

$$Q = hA(T - 70)$$
$$= (15)\left[\frac{(1)(w) + (1)(w) + (1/8)(w)}{12}\right](T - 70)$$

Assume a one-foot length of fin, i.e., $L = 1$ ft.

$$Q = (15)\left(\frac{2.125}{12}\right)(250 - 70)$$
$$= 478 \text{ Btu/hr} \cdot \text{ft of fin length}$$

The fin efficiency from Equation (8.25) is therefore,

$$\eta_f = \left(\frac{357.2}{478}\right)(100) \tag{8.25}$$

$$= 74.7\%$$

Example 8.19 – Temperature Driving Force for Crossflow Transverse Fin Exchanger. A crossflow traverse fin heat exchanger is represented diagrammatically in Figure 8.11. Assuming the temperature change is not too large for either fluid, estimate the LMTD. Apply a correction factor either suggested by McAdams [11] or based on experience.

SOLUTION: For the system pictured in Figures 8.11, one can adopt the same procedure provided earlier by applying a correction factor to the LMTD. However, if the temperature of each fluid stream is small, it is sufficiently accurate for most applications to assume that an arithmetic mean temperature difference for a pure cross-flow heat exchanger applies. Therefore, based on experience:

$$\overline{\Delta T} = \frac{(T_1 + T_2) - (t_1 + t_2)}{2} \tag{8.28}$$

Thus, if the inlet and outlet temperatures for a cross-flow unit are,

$$T_1 = 300\,°\text{F} \quad t_1 = 70\,°\text{F}$$

$$T_2 = 240\,°\text{F} \quad t_2 = 92\,°\text{F}$$

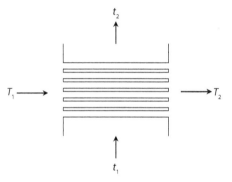

Figure 8.11 Crossflow schematic.

The arithmetic mean temperature difference driving force is given by

$$\overline{\Delta T} = \frac{(300 + 240) - (70 + 92)}{2}$$

$$= \frac{540 - 162}{2} = \frac{378}{2} = 189\,°\text{F}$$

Note that applications involving large temperature differences for either fluid should not use this approach.

8.5   Kern's Design Methodology

Kern's design methodology employed here closely follows that of the outline presented in Chapters 6 and 7. Therefore, no design procedure specific to finned exchangers is provided in this section. The true temperature difference for a double pipe exchanger is the LMTD calculated for counterflow. The effective tube length does not include the return bend between legs of a hairpin or any of the unfinned portion of the inner tube. The inside of the tube is usually employed as the reference surface. However, most manufacturers prefer to use the outside area as the reference surface, since it has a larger numerical value. The solutions of problems can be converted to manufacturers' data by multiplying the inside surface by the ratio $(A_f + A_o)/A_i$ and dividing the overall coefficient $U_{D,i}$ by this ratio. Tube lengths of 12, 15, 20, and 24 ft are considered reasonable for extended-surface hairpins. Large tube lengths are permissible since the fins on the inner pipe rest snugly on the outer pipe and there generally is no sagging. As explained previously, fouling factors cannot be combined as in ordinary double pipe exchangers because impact each type of surface in a different way and each must be treated separately. The next example will demonstrate the design methodology of finned exchangers. Two additional examples compliment the presentation in this section. The main difference from the examples provided in Chapters 6 and 7 is that the finned exchanger calculational procedure will require a fin efficiency calculation.

Example 8.20 – Calculation of a Double Pipe Extended-surface Gas Oil Cooler [2]. It is desired to cool 18,000 lb/hr of 28°API gas oil from 250 to 200°F in double pipe exchangers consisting of 3 in. IPS shells with 1½ in. IPS inner pipes on which are mounted 24 fins ½ in. high by 0.035 in. wide (20 BWG). Water from 80 to 120°F will serve as the cooling medium. Pressure drops of 10.0 psi are allowable on both streams, and fouling factors of 0.002 for the gas oil and 0.003 for the water are required. Assuming the pipe and fins are cast iron, how many 20-ft hairpins will be required?

SOLUTION:

(1) Heat Balance:

Hot Gas Oil,

$$T_{avg} = \frac{250 + 200}{2} = 225°F$$

$$C = 0.53 \, \text{Btu}/(\text{lb} \cdot °F) \text{ for } 28°\text{API at } 225°F$$

$$Q = (18,000)(0.53)(250 - 200) = 477,000 \, \text{Btu/hr}$$

Cold Water,

$$t_{avg} = \frac{80 + 120}{2} = 100°F$$

$$c = 1.0 \, \text{Btu}/(\text{lb} \cdot °F) \text{ at } 100°F$$

$$Q = (11,925)(1.0)(120 - 80) = 477,000 \, \text{Btu/hr}$$

(2) ΔT^{*}:

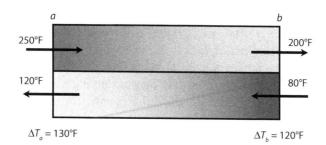

$\Delta T_a = 130°F$ $\Delta T_b = 120°F$

$$\Delta T^{*} = \text{LMTD} = \frac{\Delta T_a - \Delta T_b}{\ln\left(\Delta T_a / \Delta T_b\right)} = \frac{130 - 120}{\ln\left(130/120\right)} = 125°F$$

(3) T_{avg} and t_{avg}: obtained in step (1).

Hot Fluid (Gas Oil) – Annulus (a)	Cold Fluid (Water) – Inner Pipe (p)
(4') Flow Area, a_a	(4) Flow Area, a_p

(4')   Flow Area, a_a

$d_2 = 3.068\,\text{in} \Leftarrow [\text{Table AT.8}]$

$d_1 = 1.90\,\text{in} \Leftarrow [\text{Table AT.8}]$

$z_{f,t} = 0.035\,\text{in} \Leftarrow [\text{Table AT.7}]$

$a_f = N_f z_{f,t} L \Leftarrow$ for longitudinal

        fins

$= (24)(0.035)(0.5) = 0.42\,\text{in}^2$

$a_a = \frac{\pi}{4}\left(d_2^2 - d_1^2\right) - a_f$

$= \left[\frac{\pi}{4}\left(3.068^2 - 1.90^2\right) - 0.42\right]$

$= 4.14\,\text{in}^2 \left(\frac{1\text{ft}^2}{144\text{in}^2}\right)$

$= 0.0287\,\text{ft}^2$

Wetted Perimeter*, z_{wp}

(4)   Flow Area, a_p

$D = 1.61\,\text{in}\left(\frac{1\text{ft}}{12\text{in}}\right) \Leftarrow [\text{Table AT.8}]$

$= 0.134\,\text{ft}$

$a_p = \frac{D^2 \pi}{4}$

$= \frac{(0.134)^2 \pi}{4}$

$= 0.0141\,\text{ft}^2$

$$z_{wp} = d_1\pi + 2N_f z_f$$

$$= 1.90\pi + 2(24)(0.5)$$

$$= 29.97\,\text{in} \cdot \frac{1\,\text{ft}}{12\,\text{in}} = 2.498\,\text{ft}$$

Equivalent Diameter, D_e

$$D_e = \frac{4a_a}{z_{wp}} = \frac{4(0.0287)}{2.498} = 0.0460\,\text{ft}$$

(5') Mass Velocity, G_a	(5) Mass Velocity, G_p
$$G_a = \frac{\dot{M}}{a_a} = \frac{18,000}{0.0287}$$ $$= 627,000\,\text{lb}/\left(\text{hr}\cdot\text{ft}^2\right)$$	$$G_p = \frac{\dot{m}}{a_p} = \frac{11,925}{0.0141}$$ $$= 846,000\,\text{lb}/\left(\text{hr}\cdot\text{ft}^2\right)$$
(6') Obtain μ at T_{avg}	(6) Obtain μ at t_{avg}
$\mu = 2.50\,\text{cP at } 225°\text{F}$	$\mu = 0.72\,\text{cP at } 100°\text{F}$
$$= 2.50\,\text{cP}\left(\frac{2.42\text{lb}/(\text{ft}\cdot\text{hr})}{1\text{cP}}\right)$$ $$= 6.05\,\text{lb}/(\text{ft}\cdot\text{hr})$$	$$= 0.72\,\text{cP}\left(\frac{2.42\text{lb}/(\text{ft}\cdot\text{hr})}{1\text{cP}}\right)$$ $$= 1.74\,\text{lb}/(\text{ft}\cdot\text{hr})$$
(7') Reynolds Number, Re_a	(7) Reynolds Number, Re_p
$$Re_a = \frac{D_e G_a}{\mu}$$ $$= \frac{(0.0460)(627,000)}{6.05}$$ $$= 4,770$$	$$Re_p = \frac{D G_p}{\mu}$$ $$= \frac{(0.134)(846,000)}{1.74}$$ $$= 65,200$$
(8') Obtain k at T_{avg}	(8) Obtain k at t_{avg}
$k = 0.072\,\text{Btu/hr}\cdot\text{ft}^2\cdot°\text{F/ft at } 225°\text{F}$	$k = 0.36\,\text{Btu/hr}\cdot\text{ft}^2\cdot°\text{F/ft at } 100°\text{F}$

*In the derivation, the outermost edge of the fins was assumed to have zero heat transfer.

(9') Solve for h_o, ϕ_a, and t_w of turbulent viscous stream flowing through annulus:

$$\frac{h_o D_e}{k} = 0.027 \left(\frac{D_e G_a}{\mu}\right)^{0.8} \left(\frac{C\mu}{k}\right)^{1/3} \left(\frac{\mu}{\mu_w}\right)^{0.14} \Leftarrow \text{Sieder-Tate Equation}$$

$$h_o = 0.027 \left(\frac{D_e G_a}{\mu}\right)^{0.8} \left(\frac{C\mu}{k}\right)^{1/3} \left(\frac{\mu}{\mu_w}\right)^{0.14} \left(\frac{k}{D_e}\right)$$

where,

$\mu_w = \mu$ at t_w

$$t_w = t_c + \left(\frac{h_i}{h_i + h_o}\right)(T_c - t_c)$$

$h_o = 129\,\text{Btu/hr} \cdot \text{ft}^2 \cdot {}^\circ\text{F}$ at $t_w = 204.7\,{}^\circ\text{F} \Leftarrow$ from iterative calculation

$$\phi_a = \left(\frac{\mu}{\mu_w}\right)^{0.14} = 0.862$$

---

(9) Solve for h_i of non-viscous water stream flowing through inner pipe:

$$\frac{h_i D}{k} = 0.023 \left(\frac{D G_p}{\mu}\right)^{0.8} \left(\frac{c\mu}{k}\right)^{0.3} \Leftarrow \text{Dittus-Boelter eq. for cooling fluid}$$

$$h_i = 0.023 \left(\frac{D G_p}{\mu}\right)^{0.8} \left(\frac{c\mu}{k}\right)^{0.3} \left(\frac{k}{D}\right)$$

$$= 0.023 \left[\frac{(0.134)(846,000)}{1.74}\right]^{0.8} \left[\frac{(1.0)(1.74)}{0.36}\right]^{0.3} \left(\frac{0.36}{0.134}\right)$$

$$= 704\,\text{Btu/hr} \cdot \text{ft}^2 \cdot {}^\circ\text{F}$$

$\phi_p \approx 1$. Assume non-viscous fluid

---

(10) Surface area calculation of inner pipe and fins (per 1 lin. ft length of exchanger):

Fin total surface, $$A_{f,t} = \frac{2 N_f z_{f,c} L}{L} = \frac{2 N_f \left(z_f + 0.5 z_{f,t}\right) L}{L}$$
$$= 2 N_f \left(z_f + 0.5 z_{f,t}\right)$$

$$= (2)(24) \big[0.5 + (0.5)(0.035) \big] \cdot \frac{1 \, \text{ft}}{12 \, \text{in}}$$

$$= 2.07 \, \text{ft}^2 / \text{lin.ft}$$

Bare exposed pipe surface,
$$A_{be} = \frac{d_1 \pi L - N_f z_{f,t} L}{L} = d_1 \pi - N_f z_{f,t}$$

$$= \big[1.90\pi - (24)(0.035) \big] \cdot \frac{1 \, \text{ft}}{12 \, \text{in}}$$

$$= 0.567 \, \text{ft}^2 / \text{lin.ft}$$

Total outer surface,
$$A_t = A_{f,\text{total}} + A_{be} = 2.07 + 0.567 = 2.64 \, \text{ft}^2 / \text{lin.ft}$$

Total inside surface,
$$A_i = 0.422 \, \text{ft}^2 / \text{lin.ft} \Leftarrow [\text{Table AT.8}]$$

Pipe outer area,
$$A_{i(o)} = 0.498 \, \text{ft}^2 / \text{lin.ft} \Leftarrow [\text{Table AT.7}]$$

Log mean average of Pipe surface area,
$$A_{\text{LM}} = \frac{A_{i(o)} - A_i}{\ln \big(A_{i(o)} / A_i \big)} = \frac{0.498 - 0.422}{\ln (0.498/0.422)}$$

$$= 0.459 \, \text{ft}^2 / \text{lin.ft}$$

---

(11) Fin efficiency, η_f and $\eta_{o,f}$ (for a rectangular fin):

Refer to Table 8.3.

$$z_{f,c} = z_f + \frac{z_{f,t}}{2} = 0.5 + \frac{0.035}{2} = 0.5175 \, \text{in} = 0.0431 \, \text{ft}$$

$$A_p = (z_{f,c})(z_{f,t}) = (0.5175)(0.0175) = 0.00906 \, \text{in}^2 = 6.29 \times 10^{-5} \, \text{ft}^2$$

$$k_w = 25 \, \text{Btu/hr} \cdot \text{ft}^2 \cdot {}^\circ\text{F/ft}$$

Generate the X-coordinate (abscissa).

$$X = \sqrt{\frac{z_{f,c}^3 h_o}{k_w A_p}} = \sqrt{\frac{(0.0431)^3 (129)}{(25)(6.29 \times 10^{-5})}} = 2.56$$

Calculate η_f and $\eta_{o,f}$ (employ Equation (a) in Table 8.3).

$$\eta_f = (4.5128) X^3 - (10.079) X^2 - (31.413) X + 101.47$$

$$= (4.5128)(2.56)^3 - (10.079)(2.56)^2 - (31.413)(2.56) + 101.47$$

$$= 30.7\% = 0.307$$

$$\eta_{o,f} = 1 - \left(\frac{N_f A_f}{A_t}\right)(1 - \eta_f) = 1 - \left(\frac{A_{f,t}}{A_t}\right)(1 - \eta_f)$$

$$= 1 - \frac{2.07}{2.64}(0.307 - 1)$$

$$= 1.54$$

(12)   Clean overall coefficient, U_C:

$$\frac{1}{U_i A_i} = \frac{1}{h_i A_i} + \frac{z_{f,t}}{k_w A_{LM}} + \frac{1}{\eta_o h_o A}$$

$$U_C = \left[\left(\frac{1}{h_i A_i} + \frac{z_{f,t}}{k_w A_{LM}} + \frac{1}{\eta_o h_o A_t}\right) A_i\right]^{-1}$$

$$= \left[\left(\frac{1}{(704)(0.422)} + \frac{0.035}{(25)(0.459)} + \frac{1}{(1.54)(129)(2.64)}\right)(0.422)\right]^{-1}$$

$$= 285 \, \text{Btu/hr} \cdot \text{ft}^2 \cdot {}^\circ\text{F}$$

(13)   Design overall coefficient, U_D:

$$\frac{1}{U_D} = \frac{1}{U_C} + R_d = \frac{1}{U_C} + R_{d,i} + R_{d,o}$$

$$U_D = \left[\frac{1}{U_C} + R_{d,i} + R_{d,o}\right]^{-1}$$

$$= \left[\frac{1}{285} + 0.002 + 0.003\right]^{-1}$$

$$= 118 \, \text{Btu/hr} \cdot \text{ft}^2 \cdot {}^\circ\text{F}$$

(14)   Length of exchanger, L:

$$Q = U_D(A_i L)\Delta t$$

Note that L is the length of the whole exchanger. A_i is in terms of ft² per ft, so by multiplying it by L which is in terms of ft, provides the total inside surface area of exchanger.

$$L = \frac{Q}{U_D A_i \Delta t} = \frac{477,000}{(118)(0.422)(125)} = 76.6\,\text{ft}$$

In order to cover 76.6 ft of piping, a total of <u>four</u> 20-ft hairpins will be necessary. The new length will be 80 ft total.

Pressure Drop

(1') Obtain Re'_a:	(1) $Re_p = 65,200$ from (7) above.

(1') Obtain Re'_a:

$$D'_e = \frac{4a_a}{z_{wp} + D_2 \pi}$$

$$= \frac{4(0.0287)}{2.498 + \dfrac{3.068\pi}{12\,\text{in.}/\text{ft}}}$$

$$= 0.0348\,\text{ft}$$

$$Re'_a = \frac{D'_e\, G_a}{\mu}$$

$$= \frac{(0.0348)(627,000)}{6.05}$$

$$= 3610$$

$f = 0.00037\,\dfrac{\text{ft}^2}{\text{in.}^2}$ at $Re'_a = 3610$

$S = 0.82$ for 28°API at 225°F

---

(1) $Re_p = 65,200$ from (7) above.

$f = 0.000192\,\dfrac{\text{ft}^2}{\text{in.}^2}$ at $Re_p = 65,200$

$s \approx 1.0 \Leftarrow$ water

(2') Calculate ΔP_a:

$$\Delta P_a = \frac{f G_a^2 L}{(5.22 \times 10^{10}) D'_e\, S\phi_a}$$

$$= \frac{(0.00037)(627,000)^2 (80)}{(5.22 \times 10^{10})(0.0348)(0.82)(0.86)}$$

$$= 9.08\,\text{psi}$$

$\Delta P_a < 10\,\text{psi}$ as the allowable ΔP

---

(2) Calculate ΔP_p:

$$\Delta P_p = \frac{f G_p^2 L}{(5.22 \times 10^{10}) D s\phi_p}$$

$$= \frac{(0.000192)(846,000)^2 (80)}{(5.22 \times 10^{10})(0.134)(1.0)(1.0)}$$

$$= 1.57\,\text{psi}$$

$\Delta P_p < 10\,\text{psi}$ as the allowable ΔP

Summary

Parameter	Condition #1	Condition #2	Units
Dirt Factor	0.002	0.003	hr · ft · °F/Btu
h	704	129	Btu/hr · ft² · °F
U_C	282		Btu/hr ·ft² · °F
U_D	117		Btu/hr · ft² · °F
Calculated ΔP	9.08	1.57	psi
Allowable ΔP	10.0	10.0	psi

Example 8.21 – Calculation of a Longitudinal Fin Shell and tube Exchanger [2]. 30,000 lb/hr of oxygen at 3 psig pressure and 250 °F is to be cooled to 100 °F using water from 80 to 100 °F. The maximum allowable pressure drop for the gas is 2.0 psi and for the water 10.0 psi. Fouling factors of not less than 0.0060 should be provided. Available for the service is a 19¼ in. ID 1–2 exchanger equipped with 70 16-ft tubes each with 20 fins ½ in. high of 20 BWG (0.035 in.) steel. The tubes are 1 in. OD, 12 BWG and are laid out on 2 in. triangular pitch for four passes. Will the exchanger fulfill the service? What is the final dirt factor?

In this derivation, heat transfer from the edge/tip of the fin is neglected.

SOLUTION:

Shell Side	Tube Side
ID = 19¼ in	Number and length = 70 tubes, 16'0", 20 fins, 20 BWG, in
Baffle space = ring supports	OD, BWG, pitch = 1 in, 12 BWG, 2-in triangular pitch
Passes = 1	Passes = 4

(1)   Heat Balance:

Hot Oxygen
(at 17.7psia),

$$T_{avg} = \frac{250 + 100}{2} = 175°F$$

$$C \approx 0.225\,Btu/(lb \cdot °F) \text{ at } 175°F$$

$$Q = (30,000)(0.225)(250 - 100) = 1,010,000\,Btu/hr$$

Cold Water,

$$t_{avg} = \frac{80 + 100}{2} = 90°F$$

$$c \approx 1\,Btu/(lb \cdot °F) \text{ at } 90°F$$

$$Q = (50,500)(1)(100 - 80) = 1,010,000\,Btu/hr$$

(2) ΔT^{*}:

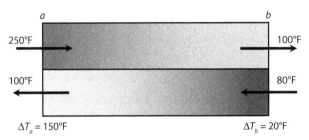

$$\Delta T = \text{LMTD} = \frac{\Delta T_{a} - \Delta T_{b}}{\ln(\Delta T_{a}/\Delta T_{b})} = \frac{150 - 20}{\ln(150/20)} = 64.5\,^{\circ}\text{F}$$

$$R = \frac{T_{1} - T_{2}}{t_{2} - t_{1}} = \frac{150}{20} = 7.5$$

$$P = \frac{t_{2} - t_{1}}{T_{1} - t_{1}} = \frac{20}{250 - 80} = 0.1176$$

$$F_{T} = \left(\frac{\sqrt{R^{2} + 1}}{R - 1}\right) \frac{\ln\left[(1 - S)/(1 - RP)\right]}{\ln\left[\left(2 - P\left(R + 1 - \sqrt{R^{2} + 1}\right)\right)/\left(2 - P\left(R + 1 + \sqrt{R^{2} + 1}\right)\right)\right]}$$

$$= \left(\frac{\sqrt{7.5^{2} + 1}}{7.5 - 1}\right)$$

$$\times \frac{\ln\left[(1 - 0.1176)/(1 - (7.5)(0.1176))\right]}{\ln\left[\left(2 - 0.1176\left(7.5 + 1 - \sqrt{7.5^{2} + 1}\right)\right)/\left(2 - 0.1176\left(7.5 + 1 + \sqrt{7.5^{2} + 1}\right)\right)\right]}$$

$$= 0.825$$

$$\Delta T^{*} = \text{LMTD} \times F_{T} = (64.5\,^{\circ}\text{F})(0.825) = 53.2\,^{\circ}\text{F}$$

(3) T_{avg} and t_{avg}: Obtained in step (1).

Hot Fluid (Oxygen Gas) – Shell Side (s)	Cold Fluid (Water) – Tube Side (t)
(4') Flow Area, a_{s} $$a_{s} = \frac{(\text{ID})^{2}\,\pi}{4}$$ $$-N_{t}\left(\frac{d_{2}^{2}\,\pi}{4} + N_{f}z_{f,t}z_{f}\right)$$	(4) Flow Area, a_{t} $$a_{t}' = 0.479\,\text{in}^{2} \Leftarrow [\text{Table AT.8}]$$

$$= \frac{(19.25)^2 \pi}{4}$$

$$-70\left[\frac{(1)^2 \pi}{4} + (20)(0.035)(0.5)\right]$$

$$= 211.6 \, \text{in}^2 \left(\frac{1 \, \text{ft}^2}{144 \, \text{in}^2}\right) = 1.47 \, \text{ft}^2$$

Wetted Perimeter*, z_{wp}

$$z_{wp} = N_t\left(d_2\pi + 2N_f z_f\right)$$
$$= (70)\left[(1)\pi + 2(20)(0.5)\right]$$
$$= 1620 \, \text{in} \cdot \frac{1 \, \text{ft}}{12 \, \text{in}} = 135 \, \text{ft}$$

Equivalent Diameter, D_e

$$D_e = \frac{4a_a}{z_{wp}} = \frac{4(1.47)}{135} = 0.0436 \, \text{ft}$$

$$a_t = \frac{N_t a_t'}{144n}$$

$$= \frac{(70)(0.479)}{144(4)} = 0.0582 \, \text{ft}^2$$

$$D = 0.782 \, \text{in}\left(\frac{1 \, \text{ft}}{12 \, \text{in}}\right) \Leftarrow [\text{Table AT.8}]$$
$$= 0.0652 \, \text{ft}$$

(5') Mass Velocity, G_s

$$G_s = \frac{\dot{M}}{a_s} = \frac{30,000}{1.47}$$
$$= 20,400 \, \text{lb}/\left(\text{hr} \cdot \text{ft}^2\right)$$

(5) Mass Velocity, G_t

$$G_t = \frac{\dot{m}}{a_t} = \frac{50,500}{0.0582}$$
$$= 868,000 \, \text{lb}/\left(\text{hr} \cdot \text{ft}^2\right)$$

(6') Obtain μ at T_{avg}

$$\mu = 0.0225 \, \text{cP at } 175 \, °\text{F}$$

$$= 0.0225 \, \text{cP}\left(\frac{2.42 \, \text{lb}/\left(\text{ft} \cdot \text{hr}\right)}{1 \, \text{cP}}\right)$$

$$= 0.0545 \, \text{lb}/\left(\text{ft} \cdot \text{hr}\right)$$

(6) Obtain μ at t_{avg}

$$\mu = 0.80 \, \text{cP at } 90 \, °\text{F}$$

$$= 0.80 \, \text{cP}\left(\frac{2.42 \, \text{lb}/\left(\text{ft} \cdot \text{hr}\right)}{1 \, \text{cP}}\right)$$

$$= 1.94 \, \text{lb}/\left(\text{ft} \cdot \text{hr}\right)$$

(7') Reynolds Number, Re_s

$$Re_s = \frac{D_e G_s}{\mu}$$

$$= \frac{(0.0436)(20,400)}{0.0545}$$

$$= 16,300$$

(7) Reynolds Number, Re_t

$$Re_t = \frac{D G_t}{\mu}$$

$$= \frac{(0.0652)(868,000)}{1.94}$$

$$= 29,200$$

(8') Obtain k at T_{avg}	(8) Obtain and k at t_{avg}
$k = 0.0175 \, \text{Btu/hr} \cdot \text{ft}^2 \cdot °\text{F/ft}$ at 175°F	$k = 0.36 \, \text{Btu/hr} \cdot \text{ft}^2 \cdot °\text{F/ft}$ at 90°F

(9') Solve for h_o of gaseous stream flowing through shell:

$$\frac{h_o D_e}{k} = 0.023 \left(\frac{D_e G_s}{\mu} \right)^{0.8} \left(\frac{C\mu}{k} \right)^{0.4} \impliedby \text{Dittus-Boelter equation (heating fluid)}$$

$$h_o = 0.023 \left(\frac{D_e G_s}{\mu} \right)^{0.8} \left(\frac{C\mu}{k} \right)^{0.4} \left(\frac{k}{D_e} \right)$$

$$= 0.023 \left[\frac{(0.0436)(20,400)}{0.0545} \right]^{0.8} \left[\frac{(0.225)(0.0545)}{0.0175} \right]^{0.4} \left(\frac{0.0175}{0.0436} \right)$$

$$= 18.8 \, \text{Btu/hr} \cdot \text{ft}^2 \cdot °\text{F}$$

$$\phi_s \approx 1 \impliedby \text{For gases}$$

(9) Solve for h_i of non-viscous water stream flowing through tubes:

$$\frac{h_i D}{k} = 0.023 \left(\frac{DG_t}{\mu} \right)^{0.8} \left(\frac{c\mu}{k} \right)^{0.3} \impliedby \text{Dittus-Boelter equation (cooling fluid)}$$

$$h_i = 0.023 \left(\frac{DG_t}{\mu} \right)^{0.8} \left(\frac{c\mu}{k} \right)^{0.3} \left(\frac{k}{D} \right)$$

$$= 0.023 \left[\frac{(0.0652)(868,000)}{1.94} \right]^{0.8} \left[\frac{(1.0)(1.94)}{0.36} \right]^{0.3} \left(\frac{0.36}{0.0652} \right)$$

$$= 786 \, \text{Btu/hr} \cdot \text{ft}^2 \cdot °\text{F}$$

$$\phi_t \approx 1 \impliedby \text{Assume non-viscous fluid}$$

(10) Surface area calculation (per tube):

Total fin surface, $\quad A_{f,t} = \dfrac{2N_f z_{f,c} L}{L} = 2N_f \left(z_f + 0.5 z_{f,t} \right)$

$$= (2)(20) \left[0.5 + (0.5)(0.035) \right] \cdot \left(\frac{1 \, \text{ft}}{12 \, \text{in}} \right)$$

$$= 1.725 \, \text{ft}^2 / \text{lin.ft}$$

Bare exposed tube surface,

$$A_{be} = \frac{d_2 \pi L - N_f z_{f,t} L}{L} = d_2 \pi - N_f z_{f,t}$$

$$= \left[(1)\pi - (20)(0.035)\right] \cdot \left(\frac{1\,ft}{12\,in}\right)$$

$$= 0.203\,ft^2/lin.\,ft$$

Total outside surface,

$$A_t = A_{f,t} + A_{be} = 1.725 + 0.203 = 1.928\,ft^2/lin.\,ft$$

Total inside surface, $A_i = 0.205\,ft^2/lin.\,ft \Leftarrow [\text{Table AT.8}]$

---

(11) Fin efficiency, η_f and $\eta_{o,f}$ (for a rectangular fin):

Refer to Table 8.3.

$$z_{f,c} = z_f + \frac{z_{f,t}}{2} = 0.5 + \frac{0.035}{2} = 0.5175\,in = 0.0431\,ft$$

$$A_p = (z_{f,c})(z_{f,t}) = (0.5175)(0.0175) = 0.00906\,in^2 = 6.29 \times 10^{-5}\,ft^2$$

$$k_w = 26\,Btu/hr \cdot ft^2 \cdot °F/ft$$

Generate the X-coordinate.

$$X = \sqrt{\frac{z_{f,c}^3 h_o}{k_w A_p}} = \sqrt{\frac{(0.0431)^3 (18.8)}{(26)(6.29 \times 10^{-5})}} = 0.959$$

Calculate η_f and $\eta_{o,f}$.

$$\eta_f = (4.5128) X^3 - (10.079) X^2 - (31.413) X + 101.47$$

$$= (4.5128)(0.959)^3 - (10.079)(0.959)^2 - (31.413)(0.959) + 101.47$$

$$= 66.1\% = 0.661$$

$$\eta_{o,f} = 1 - \left(\frac{N_f A_f}{A_t}\right)(1 - \eta_f) = 1 - \left(\frac{A_{f,t}}{A_t}\right)(1 - \eta_f)$$

$$= 1 - \frac{1.725}{1.928}(0.661 - 1)$$

$$= 1.30$$

(12) Clean overall coefficient, U_C:

$$\frac{1}{U_i A_i} = \frac{1}{h_i A_i} + \frac{1}{\eta_{o,f} h_o A} \Leftarrow \text{Assume} \frac{z_{f,t}}{k_w A_{LM}} \text{ is negligible}$$

$$U_C = \left[\left(\frac{1}{h_i A_i} + \frac{1}{\eta_{o,f} h_o A_t} \right) A_i \right]^{-1}$$

$$= \left[\left(\frac{1}{(786)(0.205)} + \frac{1}{(1.30)(18.8)(1.928)} \right) (0.205) \right]^{-1}$$

$$= 177.8 \, \text{Btu/hr} \cdot \text{ft}^2 \cdot °\text{F}$$

(13) Design overall coefficient, $U_{D,i}$:

$a'' = 0.2048 \, \text{ft}^2$ per lin. ft $\Leftarrow$ [Table AT.7]

(Note that instead of obtaining the exterior surface area, use the interior surface area where the fins are not present.)

$$A_{t(i)} = a'' N_t L = (0.205)(70)(16) = 230 \, \text{ft}^2$$

$$U_{D,i} = \frac{Q}{A_{t(i)} \Delta T_{LM}} = \frac{1,010,000}{(230)(53.2)} = 82.5 \, \text{Btu/hr} \cdot \text{ft}^2 \cdot °\text{F}$$

(14) Dirt factor R_d :

$$R_d = \frac{U_C - U_D}{U_C U_D} = \frac{177.8 - 82.5}{(177.8)(82.5)} = 0.00650 \, \text{hr} \cdot \text{ft}^2 \cdot \text{F/Btu}$$

Summary

Parameter	Value	Units
h_i	786	$\text{Btu/hr} \cdot \text{ft}^2 \cdot °\text{F}$
h_o	18.8	$\text{Btu/hr} \cdot \text{ft}^2 \cdot °\text{F}$
U_C	177.8	$\text{Btu/hr} \cdot \text{ft}^2 \cdot °\text{F}$
U_D	82.5	$\text{Btu/hr} \cdot \text{ft}^2 \cdot °\text{F}$
Calculated R_d	0.00650	$\text{hr} \cdot \text{ft} \cdot °\text{F/Btu}$
Required R_d	0.00650	$\text{hr} \cdot \text{ft} \cdot °\text{F/Btu}$

Pressure Drop

(1') Obtain Re'_s	(1) For $Re_t = 29,200$ in (7) above

(1') Obtain Re'_s

$$D'_e = \frac{4a_u}{z_{wp} + (ID)\pi}$$

$$= \frac{4(1.47)}{135 + \dfrac{19.25\pi}{12\,\text{in./ft}}}$$

$$= 0.0420\,\text{ft}$$

$$Re'_s = \frac{D'_e G_s}{\mu}$$

$$= \frac{(0.0420)(20,400)}{0.0545}$$

$$= 15,700$$

$$f = 0.00025\frac{\text{ft}^2}{\text{in}^2}\ \text{at}\ Re'_s = 15,700$$

At 14.7 psia and 32 °F (492 °R),

$$\rho_{std} = \frac{1\,\text{lbmol}}{359\,\text{ft}^3} \times \text{Molecular weight}$$

Molecular weight of O_2 is 32 lb/lbmol

Therefore, at 17.7 psia and 175 °F (635 °R),

$$\rho = \frac{1\,\text{lbmol}}{359\ \text{ft}^3}\left(\frac{17.7}{14.7}\right)\left(\frac{492}{635}\right)(32)$$

$$= 0.083\,\text{lb/ft}^3$$

$$S = \frac{\rho}{\rho_{water}} = \frac{0.083}{62.5} = 0.00133$$

(1) For $Re_t = 29,200$ in (7) above

$$f = 0.00021\frac{\text{ft}^2}{\text{in}^2}\ \text{at}\ Re_t = 29,200$$

$$s \approx 1.0 \Leftarrow \text{water}$$

(2') Calculate ΔP_s:	(2) Calculate ΔP_p:

(2') Calculate ΔP_s:

$$\Delta P_s = \frac{fG_s^2 Ln}{(5.22\times10^{10})D'_e\, S\phi_s}$$

(2) Calculate ΔP_p:

$$\Delta P_t = \frac{fG_t^2 Ln}{(5.22\times10^{10})Ds\phi_t}$$

$$= \frac{(0.00025)(20,400)^2(16)(1)}{(5.22\times10^{10})(0.0420)(0.00133)(1)}$$

$$= 0.6\,\text{psi}$$

$\Delta P_s < 2$ psi as allowable ΔP

$$= \frac{(0.00021)(868,000)^2(16)(4)}{(5.22\times10^{10})(0.0652)(1.0)(1.0)}$$

$$= 3.0\,\text{psi}$$

$\Delta P_t < 10$ psi as allowable ΔP

This exchanger is satisfactory and will fulfill its service.

8.6   Other Fin Considerations

The selection of fin provides an overall comparison of:

1. economics,
2. mass of fin,
3. space (if available),
4. pressure drop (if applicable), and
5. the heat transfer characteristics of the fin.

Generally, fins are effective with gases; are less effective with liquids in forced (or natural) convection; are very poor with boiling liquids; and, are extremely poor with condensing vapors.

As discussed earlier, fins should be placed on the side of the heat exchanger surface where h is the lowest. Thin, closely spaced fins (subject to economic constraints) are generally superior to fewer thicker fins. Their thermal conductivity, k, should obviously be high.

In summary, the same basic heat transfer equation applies for fins:

$$Q = UA\Delta T \tag{8.28}$$

with

$$\frac{1}{U} = \frac{1}{h_o} + \frac{\Delta x}{k} + \frac{1}{h_i} \tag{8.29}$$

For many applications, $h_i \gg h_o$ and $\Delta x/k \ll h_o$ so that,

$$\frac{1}{U} \approx \frac{1}{h_o} \tag{8.30}$$

The outside coefficient is generally the controlling resistance and it may be large. As noted earlier, one way to increase Q is to increase A and this may be accomplished by the addition of fins to the appropriate heat transfer surface. These can be mounted longitudinally or circumferentially.

8.7   Practice Problems from Kern's First Edition

1. 4620 lb/hr of a 28 °API gas oil will be used to preheat 5700 lb/hr of a 110 °API butane reactor feed at an elevated pressure from 260 to 400 °F. The gas oil will enter at 575 °F and leave at 350 °F. A pressure drop of 10 psi is permissible on the gas oil, but it should not exceed 2 to 3 psi on the reactor feed. Dirt factors of 0.002 should be provided on each side. The gas oil is the controlling fluid and should flow in the annulus.

2. In a regenerative gas absorption process, 10,300 lb/hr of 15 °Bé caustic soda ($s = 1.115$) leaves the regenerator at 240 °F and is cooled to 170 °F. The heat is absorbed by 10,300 lb/hr of 15 °Bé caustic soda at 100 °F being sent to the regenerator. There are a number of 20 ft hairpins consisting of 3 in. IPS shells and a 1½ in. IPS inner tubes with 24 fins 20 BWG, ½ in. high is available for the service. Pressure drops of 10 psi are allowable. Dirt factors of 0.002 should be provided on each side. How many sections are required, and how shall they be arranged?

3. 20,000 lb/hr of nitrogen at 0 gage pressure is to be heated from 100 to 175 °F using exhaust steam at 212 °F. Fouling factors of 0.002 should be provided for both. A 19¼ in. ID 1–2 exchanger containing 15 1¼ in. OD tubes, 12′0″ long and having 24 fins 20 BWG, ½ in. high arranged for two passes is available for service. Is the exchanger satisfactory? What are the pressure drops?

4. A textile impregnating room measures 50 by 100 by 12 ft. Because of the possibility of developing an explosive and toxic concentration, it is necessary to change the air eight times an hour. The room should be kept at 75 °F to provide comfortable conditions, although no provision will be made for humidity control. The lowest winter temperature anticipated is 30 °F (Middle Atlantic). The air intake will be through an existing 4 by 4 ft duct in which a tempering coil is to be provided using exhaust steam at 212 °F.

References

1. W. Badger and J. Banchero, adapted from, *Introduction to Chemical Engineering*, McGraw-Hill, New York City, NY, 1955.
2. D. Kern, *Process Heat Transfer*, McGraw-Hill, New York City, NY, 1950.
3. I. Farag and J. Reynolds, *Heat Transfer*, A Theodore Tutorial, Theodore Tutorials, East Williston, NY, originally published by USEPA/APTI, RTP, NC, 1996.
4. L. Theodore, *Heat Transfer Applications for the Practicing Engineer*, John Wiley & Sons, Hoboken, NJ, 2013.
5. B. Barkwill, M. Spinelli, and K. Valentine, project submitted to L. Theodore, Manhattan College, Bronx, NY, 2009.
6. F. Incropera and D. De Witt, *Fundamentals of Heat and Mass Transfer*, John Wiley & Sons, Hoboken, NJ, 1981.
7. A.M. Flynn and L. Theodore, personal notes, Floral Park, NY, 2015.
8. M. Jakob, *Heat Transfer*, John Wiley & Sons, Hoboken, NJ, 1949.
9. C. Bennet and J. Meyers, *Momentum, Heat, and Mass Transfer*, McGraw-Hill, New York City, NY, 1962.
10. C. Prochaska and L. Theodore, *Mathematical Methods for Environmental Engineers and Scientists*, Scrivener-Wiley, Salem, MA, 2018.
11. W. McAdams, *Heat Transfer*, McGraw-Hill, New York City, NY, 1954.

9

Other Heat Exchangers

Contributing author: Kleant Daci

Introduction

The purpose of this last chapter in Part II is to extend the material presented in the three previous chapters to the design, operation, and predicative calculations of several other types of heat exchangers. There are a number of collateral uses for these heat transfer equipment which have not appeared in any of the preceding chapters. Some of these include the commonest and least expensive forms of heat transfer surfaces such as coils, submerged pipes in boxes, and trombone coolers. For example, the following were treated earlier by Kern [1]:

1. Jacketed vessels
2. Coils
3. Submerged-pipe coils
4. Trombone coolers

5. Atmospheric coolers
6. Evaporative condensers
7. Bayonet units
8. Falling-film exchangers
9. Granular materials in tubes
10. Electric resistance heaters

As described earlier, heat exchangers are devices used to transfer heat from a hot fluid to a cold fluid. They can be classified by their functions. Faraq and Reynolds [2] provide an abbreviated summary of these units and their functions. The four general classifications of direct-contact gas–liquid heat transfer operations are [2, 3]:

1. simple gas cooling,
2. gas cooling with vaporization of coolant,
3. gas cooling with partial condensation, and
4. gas cooling with total condensation.

Most of the direct-exchange applications listed earlier are accomplished with the following devices:

1. Baffle-tray columns
2. Spray chambers
3. Packed columns
4. Crossflow-tray columns
5. Pipeline contractors

Example 9.1 – Exchanger Description. Which of the following is NOT a heat exchanger?

a. Reboiler
b. Condenser
c. Absorber
d. Superheater

SOLUTION: A reboiler is connected to the bottom of a fractionating tower and provides the reboil heat necessary for distillation. The heating medium, which may be steam or a hot process fluid, transfers heat to the bottoms product of the column. Therefore, a reboiler is classified as a heat exchanger.

A condenser (see later section) condenses a vapor or mixture of vapors either alone or in the presence of a non-condensable gas. The cooling medium, which may be cooling water or air, absorbs heat from the hot process vapor. Therefore, a condenser is classified as a heat exchanger.

An absorber [4] is a mass transfer device, used for separating components from either a liquid or gaseous stream. Therefore, an absorber is NOT a heat exchanger.

A superheater heats a vapor above its saturation temperature. Therefore, a superheater is classified as a heat exchanger, and (c) is the correct answer.

The following eight topics are treated in this chapter:

9.1. Condensers
9.2. Evaporators
9.3. Boilers and Furnaces
9.4. Waste Heat Boilers
9.5. Cogeneration/Combined Heat and Power (CHP)
9.6. Quenchers
9.7. Cooling Towers
9.8. Heat Pipes

Cooling towers receive superficial treatment; however, additional material is available in the literature [4]. Equipment associated with unsteady state operations are briefly reviewed Part III, Chapter 10.

9.1   Condensers

It should be noted that phase-change processes involve changes (sometimes significantly) in density, viscosity, heat capacity, and thermal conductivity of the fluid in question. The heat transfer process and the applicable heat transfer coefficients for condensation and boiling is more involved and complicated than that for a single-phase process. It is therefore not surprising that most real-world applications involving condensation and boiling require the use of empirical correlations.

The transfer of heat, which accompanies a change of phase, is often characterized by high rates. Heat fluxes as high as 50 million Btu/(hr·ft^2) have been obtained in some boiling systems. This mechanism of transferring heat has become important in rocket technology and nuclear-reactor design where large quantities of heat are usually produced in confined spaces. Although condensation rates have not reached a similar magnitude, heat transfer coefficients for condensation as high as 20,000 Btu/(hr·ft²·°F) have been reported in the literature [3, 4]. Due to the somewhat complex nature of these two phenomena, simple pragmatic calculations and numerical details are available in the literature [2, 3].

Phase Changes. Several sections in this chapter addresses phenomena associated with the change in phase of a fluid. The processes almost always occur at a solid–liquid interface and are referred to as *condensation* and *boiling*. The change from liquid to vapor due to boiling occurs because of heat transfer from the solid surface; alternatively, condensation of vapor to liquid occurs due to heat transfer to the solid surface.

Phase changes of substances can only occur if heat transfer is involved in the process. The phase change processes that arise in practice include:

1. Boiling (or evaporation)
2. Condensation
3. Melting (or thawing)
4. Freezing (or fusion)
5. Sublimation

The corresponding heat of transformation arising during these processes are:

1. Enthalpy of vaporization (or condensation)
2. Enthalpy of fusion (or melting)
3. Enthalpy of sublimation

These common phase change operations are listed in Table 9.1. Other phase changes besides these are possible. During any phase change, it is usual (but not necessary) to have heat transfer without an accompanying change in temperature.

Table 9.1 Phase change operations.

Process	Description
Solidification	Change from liquid to solid
Melting	Change from solid to liquid
Boiling	Change from liquid to vapor
Condensation	Change from vapor to liquid
Sublimation	Change from solid to vapor

The applications of phase change involving heat transfer are numerous and include utility units where water is boiled, evaporators in refrigeration systems where a refrigerant maybe be either vaporized or boiled, or both, and condensers that are used to cool vapors to liquids. For example, in a power cycle, pressurized liquid is converted to vapor in a *boiler*. After expansion in a turbine, the vapor is restored to its liquid state in a *condenser*; it is then pumped to the boiler to repeat the cycle. Evaporators, in which the boiling process occurs, and condensers, are also essential components in vapor-compression refrigeration cycles. Thus, the practicing engineer needs to be familiar with phase change processes [3].

Applications involving the solidification or melting of materials are also important. Typical examples include the making of ice, freezing of foods, freeze-drying processes, solidification and melting of metals, and so on. The freezing of food and other biological matter usually involves the removal of energy in the form of both sensible heat (enthalpy) and latent heat of freezing. A large part of biological matter is liquid water, which has a latent enthalpy of freezing, h_{sf}, of approximately 335 kJ/kg (144 Btu/lb or 80 cal/g). When meat is frozen from room temperature, it is typically placed in a freezer at approximately –30 °C, which is considerably lower than the freezing point of water. The sensible heat to cool any liquids from the initial temperature to the freezing point is first removed, followed by the latent heat, h_{sf}, to accomplish the actual freezing. Once frozen, the substance is often cooled further by removing some sensible heat of the solid. Additional details are provided in Part III, Chapter 10.

Dimensionless parameters arise in condensation and boiling. The Nusselt and Prandtl numbers, presented earlier in convection analyses appear once again. The new dimensionless parameters are the Jakob number, Ja, the Bond (no relation to James) number, Bo, and the condensation number, Co. The Jakob number is the ratio of the maximum sensible energy absorbed by the liquid (vapor) to the latent energy absorbed by the liquid (vapor). However, in many applications, the sensible energy is much less than the latent energy and Ja has a small numerical value. The Bond number is the ratio of the buoyancy force to the surface tension force and finds applications in some nucleate boiling equations. The condensation number, as one might suppose, is employed in condensation calculations. A heat transfer coefficient correlation is reviewed in Illustrative Example 9.7.

Gibb's Phase Rule (GPR) comes into play when phase changes occur at equilibrium. The *degrees of freedom*, F, or the *variance* of a system is defined as the *smallest number* of independent variables (such as pressure, temperature, concentration) that must be specified in order to completely define (the remaining variables of) the system. The significance of the degrees of freedom of a system may be drawn from the following examples. In order to specify the density of gaseous (vapor) steam, it is necessary to state both the temperature and pressure to which this density corresponds. Thus, the density of the steam has a particular value at 100 °C and 1 atm pressure. A statement of this density at 100 °C without mention of pressure does not completely define the state of the steam, for at 100 °C the steam may exist in many other possible pressures. Similarly, mention of the pressure without the temperature leaves ambiguity. Therefore, for the complete description of the state of the steam, two variables must be given, and this phase, when present alone in a system, possesses two degrees of freedom, or the system is said to be *bivariant*. When liquid water and steam exist in equilibrium, however, the temperature and the densities of the phases are determined only by pressure, and a statement of some arbitrary value of the latter is sufficient to define all the other variables. The same applies to the choice of temperature as the independent variable. At each arbitrarily chosen temperature (within the range of existence of the two phases), equilibrium is possible only at a given pressure, and once again the system is defined in terms of one variable. Under these conditions, the system possesses only one degree of freedom or it is *monovariant*.

There is a definite relation in a system between the number of degrees of freedom, the number of components, and the number of phases present. This relationship was first established by J. Willard Gibbs in 1876 [5]. This relation, known as the *Gibbs Phase Rule*, is a principle of the widest generality. It is one of the most often used rules in thermodynamic analyses, particularly in the representation of equilibrium conditions existing in heterogeneous systems [6].

A mathematical description of the phase rule, for a system of C components in which P phases are present is given by:

$$F = \text{number of variables} - \text{number of equations} \quad (9.1)$$

$$= \left[P(C-1) + 2 \right] - \left[C(P-1) \right]$$

$$= C - P + 2$$

Equation (9.1) is the celebrated Gibbs Phase Rule (GPR) [5].

The simplest applicable case of GPR is one in which only a single component in a single phase is present, as with ice or steam. When more than one component and/or phase is present in a system, the number of degrees of freedom correspondingly increases in accordance with Equation (9.1).

From a thermodynamic point-of-view, condensation of a condensable vapor in a condensable vapor-noncondensable gas mixture can be induced by either increasing the pressure or decreasing the temperature, or both. Condensation most often occurs when a vapor mixture contacts a surface at a temperature lower than saturation (*dew point*) temperature of the mixture. The dew point [6] of a vapor mixture is the temperature at which the vapor pressure exerted by the condensable component(s) is equal to the(ir) partial pressure in the vapor. For example, consider an air–water mixture at 75 °F that is 80% saturated with water (or has a relatively humidity, RH, of 80%). The describing equation for RH is,

$$\% \, Sat = \% \, RH = \frac{P_{H_2O}}{P'}(100) \tag{9.2}$$

where,

P_{H_2O} = partial pressure of water

P' = vapor pressure of water = 0.43 psia at 75 °F (see Steam Table in AT.2)

For this condition, $P_{H_2O} = (0.8)(0.43) = 0.344 \, psia$.

Pure vapors condense (or vaporize) at their vapor pressure at a given temperature. For example, water at 212 °F and 1 atm will vaporize to steam (or steam will condense to water). A vapor or a mixture of vapors in a non-condensable gas is more difficult to analyze. A typical example is steam (water) in air or a high molecular weight organic in air.

There are two key vapor-liquid mixtures of interest to the practicing engineer: air-water and steam-water (liquid). Information on the former is available on a psychometric chart (see Appendix AF.2 and AF.3) while steam tables provide information on the later. A discussion on both follows.

A vapor-liquid phase equilibrium example involving raw data is the psychometric or humidity chart [6]. A humidity chart is used to determine the properties of moist air and to calculate moisture content in air. The ordinate of the chart is the absolute humidity $\mathcal{H}$, which is defined as the mass of water vapor per mass of *bone-dry* air. (Some charts base the ordinate on moles instead of mass.) Based on this definition, Equation (9.3) gives $\mathcal{H}$ in terms of moles and also in terms of partial pressure:

$$\mathcal{H} = \frac{18n_{H_2O}}{29\left(n_T - n_{H_2O}\right)} = \frac{18P_{H_2O}}{29\left(P - P_{H_2O}\right)} \tag{9.3}$$

where,

n_{H_2O} = number of moles of water vapor

n_T = total number of moles in gas

P_{H_2O} = partial pressure of water vapor

P = total system pressure

Curves showing the *relative humidity* (ratio of the mass of the water vapor in the air to the maximum mass of water vapor that the air could hold at that temperature, i.e., if the air were saturated) of humid air, also appear on the charts. The curve for 100% relative humidity is also referred to as the *saturation curve*. The abscissa of the humidity chart is the air temperature, also known as the *dry-bulb* temperature (T_{DB}). The *wet-bulb* temperature (T_{WB}) is another measure of humidity; it is the temperature at which a thermometer with a wet wick wrapped around the bulb stabilizes. As water evaporates from the wick to the ambient air, the bulb is cooled; the rate of cooling depends on how humid the air is. No evaporation occurs if the air is saturated with water where T_{WB} and T_{DB} are the same. The lower the humidity, the greater the difference between these two temperatures. On the psychometric chart, constant wet-bulb temperature lines are straight with negative slopes. The value of T_{WB} corresponds to the value of the abscissa at the point of intersection of this line with the saturation curve.

Given the dry bulb and wet bulb temperatures, the relative humidity (along with any other quantity on the chart) may be determined by finding the point of intersection between the dry bulb abscissa and wet bulb ordinate. The point of intersection describes all "humidity" properties of the system. As noted, psychometric charts are provided in the Appendix A.F.2 and A.F.3)

The steam tables comprise a tabular representation of the thermodynamic properties of water. These tables are divided into three separate categories:

1. The *saturated*-steam tables provide the value of enthalpy, specific volume, and entropy of saturated steam and saturated water as functions of pressures and/or temperatures (condensation or boiling points). Changes in these extensive properties during the evaporation of 1 lb of the saturated liquid are also tabulated. The tables normally extend from 32 to 705 °F, the temperature range where saturated liquid and vapor can coexist.

2. The *superheat* tables list the same properties in the superheated-vapor region. Degrees superheat or number of degrees above the boiling point (at the pressure in question) are also listed.

3. The Mollier chart (enthalpy-entropy diagram) for water is frequently used in engineering practice that it also deserves mention. This diagram is useful since the *entropy function stays constant during any reversible adiabatic expansion or compression* [6].

The steam tables are located in Table AT.2 of the Appendix.

Regarding solid-vapor equilibrium, a solid, like a liquid, has a definite vapor pressure at each temperature (some solids cannot exist in contact with vapor). Examples are helium below the critical temperature and various forms of ice; this pressure may be extremely small but it is nevertheless finite. The vapor pressure of a solid increases with temperature and the variation can be represented by a curve similar to that of a liquid; it is generally referred to as a *sublimation curve*. The term *sublimation* indicates the direct conversion of solid to vapor by absorption of heat. This property is referred to as the latent enthalpy of sublimation.

Example 9.2 – Degrees of Freedom. Calculate the number of degrees of freedom for a one-component, one-phase system.

SOLUTION: Refer to Equation (9.1)

$$F = C - P + 2$$

Since $C = 1$ and $P = 1$,

$$F = 1 - 1 + 2 = 2$$

Thus, two independent variable must be satisfied to completely define the system.

Example 9.3 – Steam Calculation. A large *enclosed* main contains steam at 500 psia and 233 °F superheat. Find the temperature and pressure in the main after half the steam is condensed.

SOLUTION: From the saturated steam tables in the Appendix, the condensation temperature is 467 °F at 500 psia. Therefore, the steam temperature is $467 + 233 = 700°F$. The corresponding initial specific volume is approximately 1.30 ft³/lb. After condensation, only half of the steam remains. Therefore, the final specific volume will be $1.30/0.5 = 2.60\,\text{ft}^3/\text{lb}$. Since two phases exist at equilibrium, Gibbs phase rule indicate that,

$$F = (C - P) + 2 = (1 - 2) + 2 = 1$$

Employing the saturated tables, one notes (by interpolation)[4] that $T = 370°F$ and $P = 175\,\text{psia}$ as the final state. The specific volume of liquid is approximately 0.018 ft³/lb.

Example 9.4 – Steam Heat Removal. Refer to the previous illustrative example. How much heat is removed from the system during this process?

SOLUTION: This is a constant volume process since the main is enclosed. From the first law [2, 3],

$$Q = \Delta U = U_2 - U_1$$

where,

$$U = \text{internal energy}$$

From the steam tables (see Appendix Table AT.2),

$$U_1 = 1237.1 \text{ Btu/lb}$$
$$U_{2,g} = 1112.2 \text{ Btu/lb}$$
$$U_{2,l} = 343.15 \text{ Btu/lb}$$

Substituting,

$$Q = (0.5)(1112.2 + 343.15) - (1)(1267.1)$$
$$= 727.1 - 1237.1$$
$$= -509.4 \text{ Btu / lb } (\text{heat is removed})$$

Condensation Fundamentals [2, 3]. Although several earlier chapters dealt with situations in which the fluid medium remained in a single phase, a significant number of real-world engineering applications involve a phase change that occurs simultaneously with the heat transfer process. As discussed earlier, the process of condensation of a vapor is usually accomplished by allowing it to come into contact with a surface where the temperature is maintained at a value lower than the saturation temperature of the vapor for the pressure at which it exists. The removal of thermal energy from the vapor causes it to lose its latent heat of vaporization and, hence, to condense onto the surface.

The appearance of the liquid phase on the cooling surface, either in the form of individual drops or in the form of a continuous film, offers resistance to the continued removal of heat from the vapor. In most applications, the condensate is removed by the action of *gravity*. As one would expect, the rate of removal of condensate (and the rate of heat removal from the vapor) is greater for vertical surfaces than for horizontal surfaces. Most condensing equipment consists of an assembly of tubes around which the vapor to be condensed is allowed to flow. The cool temperature of the outer tube surface is maintained by circulating a colder medium, often water, through the inside of the tube.

There are primarily three types of condensation processes:

1. *Surface Condensation.* This type of condensation occurs when vapor is in contact with a cool surface. This process is common in industrial applications and is discussed below.

2. *Homogenous Condensation.* Homogenous condensation occurs when the vapor condenses out as droplets in the gas phase.
3. *Direct Contact Condensation.* This process occurs when vapor is in contact with a cold liquid.

Surface condensation (1) — the primary mode of condensation — may occur in one of two modes depending upon the conditions of the surface.

1. *Film Condensation.* When the surface is clean and uncontaminated, the condensed vapor forms a liquid film that covers the entire condensing surface; this film contributes a finite additional resistance to heat transfer.
2. *Dropwise Condensation.* When the surface is coated with a substance that inhibits wetting, the condensed vapor forms drops in cracks and cavities on the surface. The drops often grow and coalesce and "drop" from the surface available for heat transfer. Thus, the additional resistance to heat transfer can be either reduced or eliminated. Up to 90% of the surface can be covered by drops.

The aforementioned liquid condensate provides resistance to heat transfer between the vapor and the surface. This resistance naturally increases as the thickness of the condensate later increases. For design purposes, it is usually desirable to have the condensation occur on surfaces that discourage the formation of thick liquid layers (e.g., short vertical surfaces or horizontal cylinders, or tube bundles, through which a coolant liquid flows). Dropwise condensation has a lower thermal resistance (and therefore a higher heat transfer rate) than filmwise condensation. In industrial practice, surface coatings that inhibit wetting are often used. Examples of such coating include Teflon, silicones, waxes, and fatty acids. The drawback is that these coatings lose their effectiveness over time, which results in the condensation mode, eventually changing from dropwise to filmwise. Obviously, all other things being equal, dropwise condensation is preferred to film condensation. In fact, steps are often put in place to induce this "dropwise" effect on heat transfer surfaces.

Finally, it should be noted that the local heat transfer coefficient varies along a flat surface and along the length of a vertical tube and around the perimeter of a horizontal tube (i.e., the local coefficient of heat transfer for a vapor condensing in a horizontal tube is a function of position, just like a vertical tube). As one would suppose, the highest coefficient is located at the top of a tube where the condensate film is thinnest. The condensing temperature in the application of many of the describing equations involving film coefficients is taken as the temperature at the vapor-liquid interface.

Example 9.5 – Describing Condensation Mechanisms. Describe dropwise and filmwise condensation in laymen terms.

SOLUTION: When a saturated pure vapor comes into contact with a cold surface such as a tube, it condenses and may form liquid droplets on the surface of the tube [1]. These droplets may not exhibit an affinity for the surface and instead of coating the tube fall from it, leaving bare metal on which successive droplets of condensate may form. When condensation occurs by this mechanism, it is called *dropwise* condensation. Usually, however, a distinct film may appear as the vapor condenses and coats the tube. Additional vapor is then required to condense into the liquid film rather than form directly on the bare surface. This is *film* or *filmwise* condensation. The two mechanisms are distinct and independent of the quantity of vapor condensing per square foot of surface. *Filmwise* condensation is therefore not a transition from *dropwise* condensation because of the rapidity at which condensate forms on the tube. Due to the resistance of the condensate film to the heat passing through it, the heat-transfer coefficients for dropwise condensation are four to eight times greater than those for filmwise condensation.

Example 9.6 – Condenser Separation Techniques. Describe a common separation technique in industry that involves a condenser.

SOLUTION: In the chemical industry it is a common practice to separate a liquid mixture by *distilling* [4] off the compounds which have lower boiling points in the pure condition from those having higher boiling points. By successively boiling off only part of a liquid mixture, condensing the vapor formed, and boiling off only a part of the condensate, it is possible to obtain a nearly pure quantity of the most volatile compound by numerous repetitions of the procedure. Thus, the separation by distillation is accomplished by partial vaporization and subsequent condensation via a condenser located at the top of the distillation column [7].

Example 9.7 – Average Heat Transfer Coefficient. A horizontal 4-inch OD tube is surrounded by saturated steam at 2.0 psia. The tube is maintained at 64 °F. What is the average heat-transfer coefficient? Assume laminar flow.

SOLUTION: Assuming laminar flow, the average heat-transfer coefficient is given by Equation (9.4) below. The liquid properties are evaluated at the mean film temperature, $\overline{T} = (T_{sat} + T)/2 = (126 + 64)/2 = 95°F$. The approximate values (at 100 °F) of the key steam properties are:

$$h_{vap} = 1022 \, \text{Btu} / \text{lb} \left(\text{at} \, T_{sat} \right)$$

$$\rho_v = 0.00576 \, \text{lb} / \text{ft}^3 \left(\text{at} \, T_{sat} \right)$$

$$\rho_L = 62.03 \, \text{lb} / \text{ft}^3$$

$$k_L = 0.364 \, \text{Btu} / \left(\text{hr} \cdot \text{ft} \cdot {}^\circ\text{F} \right)$$

$$\mu_L = 4.26 \times 10^{-4} \, \text{lb} / \text{ft} \cdot \text{s}$$

Substitute into Equation (9.4) below and solve for $\bar{h}$.

$$\bar{h} = 0.725 \left[\frac{\rho_L \left(\rho_L - \rho_v \right) g h_{vap} k_L^3}{\mu_L D \left(T_{sat} - T_S \right)} \right]^{1/4} \tag{9.4}$$

$$= 0.725 \left[\frac{(62.03)(62.03 - 0.00576)(32.2)(1022)(0.364)^3}{\left(4.26 \times 10^{-4} / 3600 \right)(4/12)(126-64)} \right]$$

$$= 911.6 \, \text{Btu} / \left(\text{hr} \cdot \text{ft}^2 \cdot {}^\circ\text{F} \right)$$

Equipment [8–10]. Condensation can be accomplished by increasing the pressure or the decreasing temperature (removing heat), or both. In practice, condensers operate through extraction of heat. Condensers differ in the means of removing heat and the type of device used. The two different means of condensing are direct contact (or contact), where the cooling medium with vapors and condensate are intimately mixed and combined, and indirect (or surface), where the cooling medium and vapors/condensate are separated by surface area of some type. The reader is referred to Chapters 6–8 for additional details.

Contact condensers are simpler, less expensive to install, and require less auxiliary equipment and maintenance. The condensate/coolant from a contact condenser has a volume flow 10 to 20 times that of a surface condenser. This condensate cannot be reused and may pose a waste disposal problem unless the dilution of any pollutant present is sufficient to meet regulatory requirements. Some typical contact condensers are shown in Figure 9.1.

Surface condensers form the bulk of the condensers used in industry. Some of the applicable types of surface condensers are: shell and tube, double pipe, spiral plate, flat plate, air-cooled, and various extended surface tubular units. Condensing can be accomplished in either the shell or tubes. The economics, maintenance, and operational ramifications of the allocation of fluids are extremely important, especially if extended surface tubing

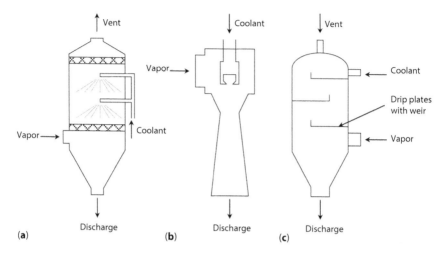

Figure 9.1 Contact condensers: (a)spray; (b) jet; (c) barometric.

is being considered. The designer should be given as much latitude as possible in specifying the condenser details.

The design of contact condensers involves calculating the quantity of coolant required to condense and subcool the vapor, and proper sizing of the discharge piping and hotwell (a feedwater tank that is non-pressurized and vented to the atmosphere). The calculation of coolant flow rate is available in the literature [9]. Use of a contact condenser requires consideration of coolant availability, any liquid waste disposal, or treatment facilities. Contact condensers are relatively efficient scrubbers as well as condensers and have the lowest equipment cost. Reputable manufacturers of the various types of contact condensers should be consulted for sizing and layout recommendations.

Example 9.8 – Typical Heat Transfer Coefficient. Provide some typical condensation heat transfer coefficients.

SOLUTION: Some typical condensing coefficients are presented in Table 9.2 [10].

Example 9.9 – Selecting Coolants. Discuss condenser/coolant selection.

SOLUTION: The choice of a coolant will depend upon the particular plant and the efficiency required of the condenser. The most common coolant is the primary plant coolant, usually cooling tower or river water.

9.2   Evaporators

The process of converting a liquid into a vapor is also of importance to practicing engineers. The production of steam for electrical power generation

Table 9.2 Condensation heat transfer coefficients.

Fluid	h, Btu $/ \left(\text{hr} \cdot \text{ft}^2 \cdot {}^\circ\text{F} \right)$
Steam	1500
Steam, 10% gas	600
Steam, 20% gas	400
Steam, 40% gas	220
Pure light (molecular weight) hydrocarbons	250
Mixed light hydrocarbons	175
Medium hydrocarbons	100
Medium hydrocarbons with steam	125
Pure organic solvents	250

is a prime example. Many other processes, particularly in the refining of petroleum, require the vaporization of a liquid. The vaporization of a liquid for the purpose of concentrating a solution is also a common operation in the chemical process industry. The simplest device is an open pan or kettle that receives heat from a coil or jacket or by direct firing underneath the pan. Perhaps the traditional unit is the horizontal-tube evaporator in which a liquid (to be concentrated) in the shell side of a closed, vertical cylindrical vessel is evaporated by passing steam or another hot gas through a bundle of horizontal tubes contained in the lower part of the vessel; empty space above the unit permits disengagement of entrained liquid from the vapor passing overhead [7]. Additional information on evaporator equipment is provided in a later subsection.

When steam flows through a tube which is submerged in a pool of liquid, minute bubbles of vapor form at random points on the surface of the tube. The heat passing through the tube surface where no bubbles form enters the surrounding liquid by convection. Some of the heat in the liquid then flows toward the bubble, causing evaporation from its inner surface onto itself. When sufficient buoyancy has been developed between the bubble and the liquid, the bubble breaks loose from the forces holding it to the tube and rises to the surface of the liquid pool. In order for this behavior to prevail, the liquid must be hotter than the saturation temperature in the incipient bubble. This is possible, since the spherical nature of the bubble establishes liquid surface (tension) forces on it so that the saturation pressure inside the bubble is less than that of the surrounding liquid. Since the saturation temperature of the bubble is lower than that of the liquid surrounding it, heat flows into the bubble. The number of points at which bubbles originate is dependent upon the texture of the tube surface with the number increasing with roughness [1].

Heat transfer by vaporization without mechanical agitation is obviously a combination of ordinary liquid free convection and the additional convection produced by the rising stream of bubbles. Under very small temperature differences between the tube wall and boiling liquid, the formation of bubbles proceeds slowly and the rate of heat transfer is essentially that of free convection. The aforementioned surface tension has an influence on bubble formation and growth is another variable. If the surface tension of the liquid is low, it tends to wet surfaces so that the bubble is blocked by the liquid and rises [1].

Boiling Fundamentals [2, 3]. Evaporation involves boiling. Boiling is the opposite of condensation. Boiling occurs when a liquid at its saturation temperature, T_{sat}, is in contact with a solid surface at a temperature, T_S, which is above T_{sat}. The excess temperature, ΔT_e, is defined as $\Delta T_e = T_S - T_{sat}$. The ΔT_e driving force causes heat to flow from the surface into the liquid, which results in the formation of vapor bubbles that move up through the liquid. Because of bubble formation, the surface tension of the liquid has an impact on the rate of heat transfer.

Boiling may be classified as *pool boiling* or *forced convection boiling.* In pool boiling, the liquid forms a "pool" in a container while submerged surfaces supply the heat. The liquid motion is induced by the formation of bubbles as well as density variations. In forced convection, the liquid motion is induced by external means (e.g., pumping of liquids through a heated tube.)

Another common classification for types of boiling is based on the relationship of the liquid temperature to its saturated temperature. Boiling is *subcooled* (also known as *local*) when the liquid temperature, T, is below the saturation temperature (i.e., $T < T_{sat}$). When the liquid is at the saturation temperature, the boiling is referred to as *saturated* boiling.

The explanation of the strange behavior of boiling systems lies in the fact that boiling heat transfer occurs by several different mechanisms and the mechanism is often more important in determining the heat rate than the temperature difference driving force. Consider, for example, the heating of water in an open pot. Heat is initially transferred within the water by natural convection. As the heat rate is further increased, the surface temperature at the base of the pot increases to and above 212 °F. Bubbles begin to form and then rise in columns from the heating surface, creating a condition favorable to heat transfer. As the temperature increases further, more sites become available until the liquid can no longer reach the heating surface at a sufficient rate to form the required amount of vapor. This ends the *nucleate boiling stage*. At this point, the mechanism changes to *film boiling* since the heating surface is now covered with a film of vapor. The temperature of the base of the pot rises (even though the heat rate is constant). Vaporization takes place at the liquid-vapor interface and the vapors disengage from the film in irregularly-shaped bubbles at random locations.

Although the phenomena of superheating occurs in most boiling systems, the temperature of the boiling liquid, measured some distance from the heated surface, is higher than the temperature of the vapor above the liquid (which is at the saturation temperature). The liquid superheat adjacent to the heated surface may be as high as 25 to 50 °F. Most of the temperature change occurs in a narrow thin film from the surface. This superheat occurs because the internal pressure in the vapor bubble is higher due to surface tension effects.

The formation, growth, and release of bubbles referred to earlier, is an extremely rapid sequence of events. The rapid growth and departure of vapor bubbles causes turbulence in the liquid, especially in the aforementioned zone of superheat near the heat surface. This turbulence assists the transport of heat from the heated surface to the liquid evaporating at the bubble surface. The rapid growth of bubbles and the turbulence in the liquid complement each other, resulting in high heat transfer coefficients.

As the superheat of a boiling liquid is further increased, the concentration of active centers on the heating surface increases and the heat rate correspondingly increases. The mass rate of the vapor rising from the surface must be equal to the mass rate of liquid proceeding toward the surface if steady-state conditions prevail. As the boiling rate increases, the rate of liquid influx must increase since the area available for flow decreases with the increasing number of bubble columns; in addition, the liquid velocity must also increase. At this limiting condition, the liquid flow toward the heated surface cannot increase and the surface becomes largely blanketed with vapor. If the heat rate to the surface is held constant, the surface temperature will rise to a high value at which point heat is transmitted to the fluid by the mechanism of film boiling. Film boiling occurs when the superheat is sufficiently high to keep the heated surface completely blanketed with vapor. Heat may then be transmitted through the gas film by conduction, convection, and radiation.

Finally, the *Leidenfrost* phenomenon is one of the many complex properties of boiling. If a liquid is dropped on a surface that has a temperature significantly higher than the liquids boiling point, the liquid will skitter across the surface and evaporate at a slower rate than expected. This is due to an instantaneous layer of vapor, suspending the liquid, which acts as insulation between the hot surface and the drop of liquid. At temperatures closer to boiling, the liquid will evaporate at a much faster rate.

Boiling point elevation is another interesting phenomenon. When a solute (e.g., NaOH) is dissolved in water, the vapor pressure of the aqueous solution is less than that of water at the same temperature. Thus, at a given pressure, the solution boils at a higher temperature than water. This increase is termed the boiling point elevation (BPE) of the solution. Dilute solutions of organic compounds exhibit small (negligible) BPEs. For aqueous solutions of inorganic compounds, the BPE may be as high as 150 °F. This BPE represents a loss in the thermal driving force and has to be accounted for in some calculations.

The Effect of Noncondensables on Heat Transfer. This effect is often a concern. Most of the heat transfer in steam evaporators does not occur from pure steam but from vapor evolved in a preceding effect. This vapor usually contains inert gases from air leakage if the preceding effect was under vacuum, from air entrained or dissolved in the feed, or from gases liberated by decomposition reactions. To prevent these inerts from seriously impeding heat transfer, the gases should be channeled past the heating surface and vented from the system while the gas concentration is still quite low. The influence of inert gases on heat transfer is due partially to the effect on ΔT of lowering the partial pressure and hence the condensing temperature of the steam. The primary effect, however, results from the formation of an insulating blanket of gas at the heating surface through which the steam must diffuse before it can condense. The latter effect can be treated as an added resistance or fouling factor (in $s \cdot m^2 \cdot K / J$) equal to approximately 6.5×10^{-5} times the local mole percent inert gas [4].

Example 9.10 – Describing Heat Flow Rates. Which of the following 6 choices adequately completes this sentence?

The rate of heat flow per unit area of heat transfer surface to an evaporating liquid…

    a. increases with the temperature of the heating medium.
    b. decreases with the temperature of the heating medium.
    c. increases with the temperature of the boiling liquid.
    d. decreases with the temperature of the boiling liquid.
    e. may increase or decrease with the temperature of the heating medium.
    f. none of the above.

Solution: This answer is (e) due to the aforementioned "bubble" effect.

Evaporator Equipment Classification [1]. There are two principal types of tubular vaporizing equipment used in industry: *boilers* and *vaporizing exchangers*. Boilers are directly fired tubular apparatus which primarily convert fuel energy into latent heat of vaporization. Vaporizing exchangers are unfired and convert the latent or sensible heat of one fluid into the latent heat of vaporization of another. If a vaporizing exchanger is used for the evaporation of water or an aqueous solution, it is conventional to call it an *evaporator*. If used to supply the heat requirements at the bottom of a distilling column, whether the vapor formed be steam or not, it is a *reboiler*. When not used for the formation of steam and not a part of a distillation process, a vaporizing exchanger is simply called an *evaporator* or a *vaporizer*.

When an evaporator is used in connection with a power-generating system for the production of pure water or for any of the evaporative processes associated with power generation, it is a *power-plant evaporator*. When an evaporator is used to concentrate a chemical solution by the evaporation of solvent water, it is a *chemical evaporator*. Both classes differ in design. The primary object of *reboilers* to supply part of the heat required for distillation [2].

One of the main purpose of power-plant evaporators is to provide relatively pure water for boiler feed. The principal features incorporated in power-plant evaporators are a tubular heating element, a space in which the liquid droplets which are carried up by the bursting bubbles may be *disengaged* and a means of removing the scale from the outside of the tubes. The impurities are continuously withdrawn from the system as *blowdown*.

The manufacture of heavy chemicals such as caustic soda, table salt, and sugar in the chemical industry starts with dilute aqueous solutions from which large quantities of water must be removed before final crystallization [7] can take place in suitable equipment. In the power-plant evaporator, the unevaporated portion of the feed is residue, whereas it is the product in the chemical evaporator. This leads to the first of several differences between power-plant and chemical evaporation.

Chemical evaporators do not operate with blowdown, and instead of liquid being fed in parallel to each body, it is usually fed to multiple-effect systems in series. The feed to the first effect is partially evaporated in it and partially in each of the succeeding effects. When the liquid feed flows in the same direction as the vapor, it is *forward feed*, and when fed in the reverse direction, it is *backward feed*. From the standpoint of effectively using the temperature potentials, forward feed is preferable. If the liquid is very viscous, there is an advantage to the use of a backward feed since the temperature of the first effect is always the greatest and the corresponding viscosity will be less. The absence of blowdown enables greater heat recovery in a chemical evaporator. Additional details are provided in the next section.

Chemical evaporators fall into two classes: natural circulation and forced circulation. Natural-circulation evaporators are used singly or in multiple effects for simpler evaporation requirements. Forced-circulation evaporators are used for viscous, salting and scale-forming solutions. Natural-circulation evaporators fall into four main classes:

1. Horizontal tube
2. Calandria vertical tube
3. Basket vertical tube
4. Long tube vertical

Product losses in evaporation may result from foaming, splashing, or entrainment. Vapor-liquid separation is normally accomplished by providing sufficient horizontal area so that any entrained liquid (droplets) can settle over (under the influence of gravity) against the rising flow of vapor. Entrainment separations can be employed to enhance the separation process. The most common separator employed by industry is a low pressure drop system, but knitted wire mesh has also been used when it cannot be plugged by solid in the entrained liquid. Theodore [11] provides extensive details on the various entrainment separators and associated calculations.

Additional information on evaporator equipment and applications is provided in *Perry's Chemical Engineers' Handbook* [4].

Evaporative Coolers — Cooling Towers. The two major evaporative cooling processes are concerned with either cooling a liquid (usually water) or cooling a gas (usually air). Both processes involve the evaporation of water with the medium to be cooled. The latent heat required to vaporize the water is obtained from the sensible heat of the medium to be cooled; the sensible heat decrease produces a temperature drop.

A cooling tower is usually employed if water is to be cooled; this operation is considered in this subsection. A spray chamber or the equivalent is employed if air is to be cooled; this evaporative cooling operation is addressed in the next subsection. The overlapping principles involved in both processes — the enthalpy of vaporization of water and the sensible enthalpy reduction of the medium to be cooled — is highlighted in both sections.

For example, refineries use large quantities of water for cooling purposes. The cooling water normally absorbs heat from process streams in indirect (noncontact) heat exchangers. However, before the water can be reused, it must be cooled. Cooling towers are used to transfer heat from the cooling water to the atmosphere by allowing the water to cascade through a series of decks and slat-type grids. Water that enters the tower may unfortunately contain hydrocarbons from leaking heat exchangers [12]. Atmospheric emissions from the cooling tower may thus consist of volatile organic compounds (VOCs) and gases stripped from the cooling water as the air and water come into direct contact. The choice of a coolant will depend on the particular plant and the efficiency required of the condenser; however, the most common coolant is usually river water.

Evaporative Cooler and Air Cooling. No discussion on evaporative coolers and air cooling would be complete without a discussion of the term, *air conditioning*, particularly as most have come to understand this term (note that this topic will be revisited in Section 9.7 – Cooling Towers). Air conditioning is a process of treating air so as to control temperature, humidity, cleanliness, and (at times) distribution to meet the requirements of space. Industrial buildings have to be designed according to their intended use from an air-conditioning perspective. In the design of domestic air-conditioning systems, odors arising from occupants, cooking, or other sources are usually controlled by introducing fresh air to reduce odor concentrations to an acceptable level by dilution. However, in industrial air-conditioning systems, harmful environmental gases, vapors, dusts, and fumes are controlled either by exhaust systems at the source or by dilution, ventilation, or by a combination of the two methods. These air-conditioning systems consist of a fan which forces a mixture of fresh outdoor air and room air through a series of units which act upon the air to clean it, to increase or decrease its temperature, and to increase or decrease its water-vapor content or humidity.

Industrial cooling of air is accomplished by employing either a traditional heat exchanger or an evaporative cooler. Air cooling in an exchanger is usually achieved with air as the coolant while evaporative coolers employ water (that is vaporized) to accomplish the cooling. This section will primarily be concerned with the evaporative cooling process; it has come to be defined by some as wet surface air cooling. The end result with this operation is that process streams can be economically cooled to temperatures approaching the wet-bulb temperature.

The theory and principles for the design of these units are a combination of those known for evaporative coolers and heat exchanger design. However, it should be noted that the design practices for evaporative coolers remain a largely proprietary, technical state-of-the-art of which details are not presented in this text. The evaluation of any particular industrial application requires direct input from a reputable vendor and/or consultant.

Evaporative cooling operations in the past have traditionally been concerned with employing a liquid (almost always water) to remove undesirable heat in liquid process streams in cooling towers [7] and transferring it to the environment. More recently, evaporatively coolers — often referred to by some as swamp coolers — have been employed to cool air; they have also been used to increase the efficiency of air conditioning operations where the cooler basically serves as a pretreatment device. The development in this section is concerned with the application of the above principle to cooling air.

The basic operation involved in evaporative cooling is that heat is required to vaporize water. This heat can be "extracted" from the medium surrounding the water, resulting in a decrease in temperature of the surrounding medium. Technically, if water is sprayed into relatively dry air, a mass transfer concentration difference driving force exist which leads to the evaporation of the water. The required heat of vaporization of the water is provided at the expense of the sensible heat of the air [6].

Evaporative cooling for air conditioning employs the evaporation of water to cool the air. See Figure 9.2. A prime mover (e.g. a pump) either sprays water or deposits water on to a cooling surface. A fan draws hot air to be cooled through the unit. It is cooled by evaporation as it passes through the water. Evaporative coolers thus reduce energy consumption and equipment requirements when compared to compressor-based air conditioners. The net effect is that energy is conserved, pollution prevention [13] has occurred, and the overall cooling operation is more economical.

Calculation details for the evaporative cooler are essentially identical to that of water-quenching.

As noted earlier, as the temperature is lowered, the temperature at which a vapor begins to condense is defined as the dewpoint. It is determined by calculating the temperature at which a given vapor mixture is saturated. Referring to the psychrometric chart, if moist air is humidified (adiabatically) in an evaporative cooler by bringing it in contact with

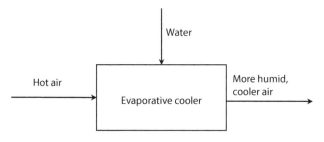

Figure 9.2 Evaporative cooler line diagram.

water at a temperature approximately equal to the wet-bulb temperature of the air, the operation is represented by a straight line drawn along the wet-bulb temperature line of the air between the limits of the process. In this process, the total enthalpy of the air remains unchanged because the sensible heat extracted from the air is returned as latent heat by an increase in moisture content. This process is distinguished by a change in dry bulb temperature, relative humidity, specific volume, moisture content, dewpoint temperature, vapor pressure, and with no change in wet-bulb temperature.

Various efficiency equations have been proposed to describe the performance of an evaporative cooler. Theodore [13, 14] has proposed the following equation for the fractional efficiency of the cooler, E.

$$E = \frac{T_1 - T_2}{T_1 - T^*}$$

where,

T_1 = inlet dry-bulb temperature
T_2 = outlet dry-bulb temperature
T^* = inlet wet-bulb temperature

Note that $T_1 - T^*$ represents the maximum temperature cooling change that can occur prior to condensation occurring, while $T_1 - T_2$ represents the actual temperature reduction in the process.

Domestic examples of evaporative cooling include:

1. The application of alcohol to skin surfaces that require cooling.
2. Cooling experienced by one of the authors [14] at Jones Beach, Long Island when the wind out of the south off the Atlantic Ocean experiences a cooling phenomenon due to the vaporization of ocean water sprays in the air.
3. Large open canopies in Dubai [15] are cooled by spraying a fine mist of air above patrons.

Advantages of evaporative coolers include:

1. It is less expensive to install and operate relative to traditional air conditioners.
2. The coolant is water.
3. Ease of installation and maintenance.

4. The cooler increases the humidity of the discharged air.
5. These coolers serve as pollution prevention measures and enhanced sustainability [13].

Disadvantages include:

1. Lower performance relative to traditional air conditioners.
2. High inlet dewpoints (or humidity) decrease efficiency of the cooler.
3. Evaporative coolers add moisture and may decrease comfort.
4. High air humidity increases the potential for corrosion.
5. The coolers require a constant supply of water.
6. Evaporative coolers are a potential health problem due to the presence of molds, bacteria, and mosquito breeding.

As one might suppose, evaporative cooling is partially well suited when the humidity of the air is low, and at elevated temperatures. This constraint led to the employment of indirect evaporative cooling in some applications. Traditional evaporative cooling processes — the direction addition of water to the air — were expanded to include indirect cooling processes. These latter processes involved the employment of conventional heat exchangers. The indirect cooling process can involve direct cooling followed by indirect cooling or vise-versa. In the former process, the cold humified air from a direct cooler is used to cool external drier air, resulting in a cool drier air stream. In the latter process, cool dry air from the heat exchanger is passed onto a direct evaporative cooling unit. The end result is a cool drier air stream. Additional details follow.

One of the pollution prevention/sustainability [13] heat transfer application for the future thus involves evaporative cooling. It has been proposed to replace traditional air conditioning processes with a two-step pretreatment operation described above prior to entering an air conditioner, as seen in Figure 9.3. The unique aspect of the proposed process eliminates the problem that arises with the higher humidity air stream discharged from the cooler. As can be seen in Figure 9.3, a heat exchanger is employed to cool the inlet air using the discharge from the cool air prior to entering the air conditioner. The net effect is to reduce the quantity of water in the air stream entering the air conditioner, resulting in an improved overall efficiency of the process.

Example 9.11 – Dry-bulb temperature. An air stream in Phoenix, Arizona, with a dry-bulb and wet-bulb temperature of 100 °F and 70 °F,

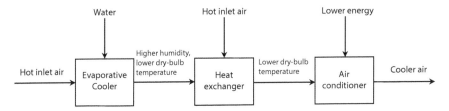

Figure 9.3 Evaporative cooler hybrid process.

respectively, passes through an evaporative cooler, resulting in an increase in humidity of 0.0043 lb/lb dry air (BDA). Calculate the resulting dry-bulb temperature and the subsequent reduction (cooling) in temperature.

SOLUTION: Refer to the psychrometric chart in Figure AF.3. The initial humidity is approximately 0.0090 lb/lb BDA. Thus, the final humidity is $0.0090 + 0.0043 = 0.0133$ lb/lb BDA. As noted above, adiabatic evaporative cooling follows the negative sloped wet-bulb temperature line upwards and to the left (toward the saturation curve). The intersection of this line with a humidity of 0.0133 occurs at a dry bulb temperature of 80 °F. Thus, the evaporative cooling process has reduced the air temperature from 100 °F to 80 °F, a 20 °F difference.

Interestingly, Theodore [14] has removed the need for these types of calculations by noting that (of typical conditions encountered in most industrial and residential application) every pound of water added to 1000 lb of air (or 13,000 ft³ of air) produces an approximate 4.5 °F decrease in temperature.

As noted above, the general subject of cooling will be revisited later in Section 9.7 – Cooling Towers.

Describing Equations. The describing equation of an evaporator, like that of any heat exchanger, is given by,

$$Q = UA\Delta T_{LM} \tag{9.6}$$

where the term $(UA)^{-1}$ is equal to the sum of the individual resistances of the heating medium, the walls of the tubes, the boiling liquid, and any fouling that may be present.

Consider Figure 9.4 [6, 11]. Assume F lb of feed to the evaporator per hour, whose solid content is x_F. (The symbol x is employed for weight fraction.) Also, assume the enthalpy of the feed per lb to be h_F. L lb of thick liquor, whose composition in weight fraction of solute is x_L and whose enthalpy is h_L leaves from the bottom of the evaporator. V lb of vapor, having a solute concentration of y_V and an enthalpy of h_V Btu/lb, leaves the unit. In most evaporators, the vapor is pure water, and therefore y_V is zero.

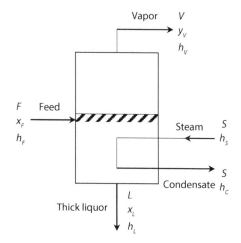

Figure 9.4 Material and enthalpy balance for a single-effect evaporator.

(Note that F, L, and V notations for this section are not related to pressure drop, length, or volume, respectively, from earlier chapters of the book.)

The material balance equations for this evaporator are relatively simple. A total material balance gives,

$$F = L + V \qquad (9.7)$$

A componential balance leads to,

$$Fx_F = Lx_L + Vy_V \qquad (9.8)$$

In order to furnish the heat necessary for evaporation, S lb of steam is supplied to the heating surface with an enthalpy of h_S Btu/lb; S lb of condensate with an enthalpy of h_C Btu/lb leaves as condensate. One simplifying assumption usually made is that in an evaporator there is very little cooling of the condensate. This is usually less than a few degrees in practice; the sensible heat recovered from cooling the condensate is so small compared to the latent heat of the steam supplied to the heating surface that the condensate will leave at the aforementioned condensing temperature of the steam. The enthalpy balance equation is therefore,

$$Fh_F + Sh_S = Vh_V + Lh_L + Sh_C \qquad (9.9)$$

Both Equations (9.6) and (9.9) are applied in tandem when designing and/or predicting the performance of an evaporator.

The calculations for an evaporator can be complicated because of an adiabatic solution temperature change. When two or more pure substances are mixed to form a solution, a heat effect usually results. Many have experienced this effect on mixing concentrated sulfuric acid with water. This heat of mixing is defined as the enthalpy change that occurs when two or more pure substances are mixed at constant temperature and pressure to form a solution.

Enthalpy-concentration diagrams offer a convenient way to calculate/obtain enthalpy of mixing effects and temperature changes associated with this type of process. These diagrams, for a two-component mixture, are graphs of the enthalpy of a binary solution plotted as a function of composition (mole fraction or weight fraction of one component), with the temperature as a parameter. For an ideal solution, isotherms on an enthalpy-concentration diagram would be straight lines. For a real solution, the actual isotherm is displaced vertically from the ideal solution isotherm at a given point by the value of Δh at that point, where Δh is the enthalpy of mixing. With reference to the enthalpy concentration diagram in Figure 9.5, Δh is negative over the entire composition range. This means that heat must be evolved whenever the pure components at the given temperature are mixed to form a solution at the same temperature. Such a system is said to be *exothermic*. An *endothermic* system is one for which the heats of solution are positive, i.e., solution at constant temperature is accompanied by the absorption of heat. Organic mixtures often fit this description.

One of the principal operating expenses of evaporators is the cost of steam for heating. A considerable reduction in those costs can be achieved by operating a battery of evaporators in which the overhead vapor from one evaporator (or "effect") becomes the heating medium in the steam chest of the next evaporator, thus saving both the cost on condensing the vapor from the first unit and supplying heat for the second. Several evaporators may operate in a battery in this fashion.

Basically, a multiple-effect evaporator may be thought of as a number of resistances, in series, to the flow of heat. The main resistances are those to heat transfer across the heating surface of each effect in the evaporator and across the final condenser if a surface condenser is used. The resistances of the heating surfaces are equivalent to the reciprocal of the product of area and overall heat transfer coefficient ($1/UA$) for each effect. (Recall that $\Sigma R = 1/UA$, and $Q = \Delta T/\Sigma R$).

Neglecting all resistances but those noted above and assuming they are equal, it can be seen that if ΣR is doubled, the flow of heat, Q, (steam consumption) will be cut in half. With half as much heat, about half as much water will evaporate. But since there are twice as many effects, the total evaporation will be the same. Thus, under these simplifying assumptions, the same output would be obtained regardless of the number of effects (each of equal resistances), and the steam consumption would be

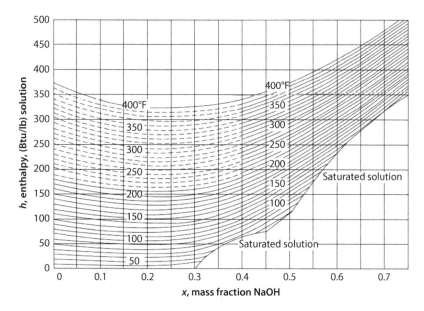

Figure 9.5 Enthalpy-concentration diagram for NaOH-H$_2$O [13, 14]. (Adapted from W.L. McCabe, *Trans. AIChE*, 31, 129, New York City, NY, 1935; R.H. Wilson and W.L. McCabe, *Ind. Eng. Chem*, 34, 558, New York City, NY, 1942.)

inversely proportional to the number of effects or the total heating surface installed.

This resistance concept is also useful in understanding the design and operation of such units. The designer places as many resistances (effects) in series as can be afforded in order to reduce the steam consumption, a positive feature from a sustainability perspective [13]. One can also show that the lowest total area required arises when the ratio of temperature drop to area is the same for each effect.

Example 9.12 – Evaporator Functions. Qualitatively describe what an evaporator does.

SOLUTION: This is an open-ended question of which Standiford [16] provides an answer to this example. The requirements for the correct functioning of any evaporator are:

1. It must transfer a great deal of heat–on the order of 1000 Btu/lb of water evaporated. This, more than anything else, determines the type, size, and cost of each effect of the evaporator.
2. It must efficiently separate the vapor from the residual liquid. What is efficient in one evaporator may be many orders

of magnitude different from what is efficient in another (e.g., from loss of salt value at only a few dollars a ton, to lithium chemicals valued at a dollar a pound, to radioactive waste). Separation may be important solely because of the value of the product lost, pollution problems, or because of fouling or corrosion in the equipment in which the vapor is condensed or in which the condensate is subsequently used.

3. It must make as efficient use of the available heat or mechanical energy as is economically feasible. This means using the vapor evaporated in one part (effect) of the evaporator as the heating steam in another effect that is operating at a lower temperature (as in a multiple-effect evaporator), or compressing the vapor evolved so that it can be used as the heating medium in the same evaporator, or by employing a combination of these. As in most other cases, efficiency is usually gained only as capital cost is increased and, for evaporators, the designer has a wide range to choose from.

4. It must meet conditions dictated by the characteristics of either the liquid being evaporated or the product. In a crystallizing evaporator, crystal size, shape, and purity may be of the utmost importance [7]. If a salting or scaling liquid is to be handled, the evaporator type selected must be capable of dealing with it. Product quality characteristics that may also be important are corrosiveness and degradation at high temperature, long holdup time, or contact with certain metals.

Other considerations are the size of the operation, the foaming characteristics of the liquor, the need for special types of materials of construction such as polished stainless steel (required for many food products), and easy access for cleaning.

Example 9.13 – NaOH Evaporator Application. A single-effect evaporator is to concentrate 10,000 lb/hr of a 10% NaOH solution to 75%. The feed enters at 120 °F and the evaporator is to operate at an absolute pressure of 14.7 psi. The 75% NaOH solution leaves at the evaporator equilibrium temperature. For what heat transfer rate (Btu/hr) should the evaporator be designed? Also calculate the area requirement in the evaporator if the overall heat transfer coefficient is 500 Btu/(hr · ft² · °F) and 103 psig (340 °F) saturated steam is employed in the steam chest. The NaOH-H$_2$O enthalpy-concentration diagram is provided in Figure 9.5. The enthalpy of saturated steam of 14.7 psia is approximately 1150 Btu/lb.

SOLUTION: Assume a basis of one hour of operation. Calculate the flow rate of steam, $\dot{m}_s$, and the 75% NaOH-H$_2$O solution, $\dot{m}_{sol.}$, leaving the evaporator. From a NaOH balance,

$$(10,000)(0.1) = (0.75)\dot{m}_{sol.}$$

$$\dot{m}_{sol.} = 1333.3 \, lb/hr$$

From an overall material balance,

$$F = L + V$$

$$V = 10,000 - 1333 = 8667 \, lb/hr$$

Estimate the enthalpy of solution entering the unit, h_F, in Btu/lb, (see Figure 9.3),

$$h_F = 81 \, Btu/lb \text{ solution}$$

Estimate the enthalpy of the 75% NaOH solution, h_L, leaving the unit.

$$h_L = 395 \, Btu/lb \text{ solution}$$

The evaporator heat required, Q, in Btu/hr is then,

$$Q = \sum \dot{m}_i h_i$$

$$= (8667)(1150) + (1333)(395) - (10,000)(81)$$

$$= 9,683,600 \, Btu/hr$$

Finally, the area of the evaporator in ft^2 may be calculated from,

$$Q = UA\Delta T_{LM} \qquad (9.5)$$

Rearranging and substituting,

$$A = \frac{9,683,600}{(500)(340-212)} = 151.3 \, ft^2$$

Example 9.14 – Evaporator Application (adapted from Badger and Banchero [17]). An evaporator is to be fed with 5000 lb of solution containing 2% solids weight. The feed, F, is at a temperature of 100 °F. It is to be concentrated to a solution of 5% solute by weight in an evaporator operating at a pressure of 1 atm in the vapor space. In order to carry out the evaporation, the heating surface is supplied with steam at 5 psig (227 °F) and the overall heat transfer coefficient of the evaporator, U, is 280 Btu/(hr · ft² · °F). What is the mass of vapor produced, the total mass of steam required, and the surface area required? *Neglect* enthalpy of solution effects.

SOLUTION: In order to simply the solution, it will be assumed that the solution is so dilute that its boiling point is the same as the boiling point of water and that its heat capacity and latent enthalpy are the same as that of water. Under these circumstances, the thermal properties of the solution (both feed and product) and of the steam can be taken from the steam tables. This results in the following values for pertinent quantities on a per hour basis:

$F = 5000\,\text{lb}$

$x_F = 0.02$

$x_F = 0.05$

$\text{Total solids in feed} = (5000)(0.02) = 200\,\text{lb}\,(\text{componential balance})$

$\text{Total water in feed} = 5000 - 100 = 4900\,\text{lb}\,(\text{componential balance})$

$\text{Total solids in liquor} = (5000)(0.02) = 100\,\text{lb}$

In addition,

$T_F = 100°\text{F}$

$h_F = 68\,\text{Btu}\,/\,\text{lb}\,\left(\text{estimated from the steam tables at }100°\text{F}\right)$

$L = 100\,/\,0.05 = 2000\,\text{lb}$

$T_L = 212°\text{F}\left(\text{at 1 atm}\right)$

$h_L = 180\,\text{Btu}\,/\,\text{lb}\,\left(\text{estimated from the steam tables}\right)$

$V = 5000 - 2000 = 3000\,\left(\text{overall balance}\right)$

$T_V = 212°\text{F}$

$h_V = 1156\,\text{Btu}\,/\,\text{lb}\,\left(\text{estimated from the steam tables at 1 atm}\right)$

$T_S = 227°F$

$h_S = 1156$ Btu / lb (estimated from the steam tables

at 227°F and 5psig)

$T_C = 227°F$ (condensate)

$h_C = 195$Btu / lb (estimated from the steam tables)

The following enthalpy (energy) balance results (S representing steam):

$$Fh_F + Sh_S = Vh_V + Lh_C + Sh_C \qquad (9.9)$$

Substituting and solving yields,

$$(5000)(68)+(1156)S = (3000)(1150)+(2000)(180)+(195)S$$

$$S = 3611 lb$$

The total heat requirement is,

$$Q = (3611)(1156 - 195)$$

$$= 3,470,000 \text{ Btu}$$

The required area is (assuming all of the above is based on one hour):

$$A = \frac{Q}{U\Delta T_{LM}} \qquad (9.5)$$

$$= \frac{3,470,000}{(280)(227 - 212)}$$

$$= 826 ft^2$$

Example 9.15 – Required Evaporator Area (adapted from Badger and Banchero [17]). An evaporator is fed with 5000 lb/hr of a 20% solution of sodium hydroxide at 100 °F. This is to be concentrated to a 40% solution. The evaporator is supplied with saturated steam at 5 psig. Although the unit operates with the vapor space at a pressure of 4 in. Hg absolute, the boiling temperature of the solution in the evaporator is 198 °F (due to the superheat created by the exposed heating element). The overall heat transfer coefficient is 400 Btu/(hr · ft² · °F). Calculate the steam rate and the required heat transfer area.

SOLUTION: Assume a basis of one hour. The following data is known:

$$F = 5000 \qquad\qquad x_F = 0.20$$

$$T_F = 100°F \qquad\qquad x_L = 0.40$$

$$x_V = y_V = 0.00$$

From the steam tables (see Appendix AT.2) at 228°F and 5 psig,

$$h_S = 1156 \text{ Btu / lb}$$

$$h_C = 196 \text{ Btu / lb}$$

$$h_{vap} = 1156 - 196 = 960 \text{ Btu / lb}$$

In addition,

The boiling point of water at 4 in. Hg (absolute) = 125.4 °F

Enthalpy of saturated team at 125°F

$$= 1116 \text{ Btu / lb (estimated}$$

from the steam tables)

$$h_F (100°F) \approx 55 \text{ Btu / lb}$$

$$h_L (198°F) \approx 177 \text{ Btu / lb}$$

A componential material balance yields,

$$Fx_F = Lx_L + Vy_V \qquad\qquad (9.8)$$

$$(5000)(0.20) = (0.40)L + (0.00)V$$

$$1000 = (0.40)L$$

$$L = 2500 \text{ lb}$$

$$V = 2500 \text{ lb}$$

An enthalpy balances yields,

$$Fh_F + S(h_S - h_C) = Vh_V + Lh_L \qquad\qquad (9.9)$$

In calculating h_V, the enthalpy of vapor leaving the solution, it should be remembered that this vapor is in equilibrium with the boiling solution at a pressure of 4 in. Hg absolute and therefore, is superheated in comparison with vapor in equilibrium with water at the same pressure. Since the heat capacity of superheated steam in this range may be approximated as 0.46 Btu/lb· °F, then,

$$h_V = 1116 + (0.46)(198.0 - 125)$$

$$= 1150 \text{ Btu / lb}$$

Substituting into the enthalpy balance, Equation (9.9), gives,

$$(5000)(55.0) + S(1156 - 196) = (2500)(1150) + (2500)(117)$$

Solving for S, the steam rate yields,

$$S = 3170 \text{ lb / hr}$$

9.3   Boilers and Furnaces

The two most important commercial applications of radiant-heat-transfer calculations are encountered in the design of steam-generating boilers and petroleum-refinery furnaces. Since the art of the construction of these units developed before the theory, empirical methods were evolved for the calculation of radiant-heat transfer in such furnaces. (The general subject of waste heat boilers is addressed separately in the next section.) Various contributions to the literature on general and specific radiant-heat transfer problems have made possible a more fundamental approach to furnace design. Several semi-theoretical calculations for furnace radiant-section heat-transfer are now available. These methods can be adapted to the solution of problems in kilns, ovens, heat-treating and chemical furnaces, and miscellaneous equipment in which radiant-heat transfer is of importance. Two other important commercial and industrial applications of boilers (and furnaces) are condensing boilers and cogeneration. These receive an abbreviated review later in this section.

The purpose of this section is to present some of the empirical and semi-theoretical methods of boilers and furnace calculations, data for their use, and examples of their application. The limitations of these methods are pointed out, and their adaptability to miscellaneous heat transfer problems indicated. A brief description of several types of boilers and oil heaters currently in use is included. The section concludes with three illustrative examples.

Steam-generating Boilers [1]. There are two types of steam generating boilers: the *fire-tube* boiler and the *water-tube* boiler. Both classes of boilers are revisited in the next section. The former consists of a cylindrical vessel having tubes passing through it which are rolled into the heads at each end of the vessel. The tube bundle is generally horizontal, and the upper section of the vessel is not tubed. Combustion gases pass through the tubes, and a water level is carried in the vessel to immerse the tubes completely but at the same time allowing disengaging space between the water level and top portion of the vessel. In some vertical-tube boilers of this type, the tubes must be immersed in water for that portion of their length required to reduce the temperature of the gases sufficiently to avoid overheating the uncooled upper portion of the tubes. Some of the water-cooled parts such as shell or tube sheets can be subjected to radiation from the combustion gases, since these parts may form a portion of the enclosure of the combustion chamber.

The major mechanism of heat transfer from the gases to the tubes is convection. Fire-tube boilers seldom exceed 8 ft in diameter, and the steam pressure is generally limited to 100 to 150 psig. These boilers are used for low-capacity services up to 15,000 to 20,000 lb/hr of steam production for domestic, industrial, and process heating and for small-scale power generation as in locomotives, etc. Fuels employed may be coal, oil, or gas, and in some cases, local combustibles such as wood, sludge, etc., are used.

Water-tube boilers, as their name implies, have water within their tubes. Combustion of stoker or pulverized coal and coke or gas and oil fuel provides radiation to the boiler tubes, and further heat transfer is accomplished by arranging the flow of hot gases over the tubes to provide convection heat-transfer. There are three important classifications of water-tube boilers: longitudinal drum, cross-drum straight tube, and cross-drum bent tube.

A typical low-pressure boiler is designed to generate approximately 200,000 lb/hr of steam at 235 psig and 500 °F. Since the saturation temperature at this pressure is only 401 °F, nearly 100 °F of superheat is required. Only a small superheater is necessary, since the superheat duty is about 5 percent of the total boiler duty. The radiant boiler tubes cover the entire wall and roof surface, forming a "water wall" by means of which the temperature of the refractory walls is kept down, thus decreasing their maintenance. Often the water tubes are partially embedded in the walls. The radiant-section furnace walls are sometimes protected from overheating by circulating cooling air outside them.

In a typical furnace, water is fed by gravity from the upper drums to headers at the bottom end of the water wall tubes and all four radiant walls. Circulation is upward through these tubes, and the steam is disengaged from water in the upper drums, then passes through a steam separator before being superheated. In a low-pressure boiler, the convection tubes reduce the flue-gas temperature sufficiently that they proceed directly to the air preheater, obviating the need of an economizer (feed-water preheater). These convection tubes are of bent tube variety running from the upper drums to the lower drum. Circulation in these tubes is, in general, downward in the cooler bank and upward through the hotter bank.

A typical power-generating steam boiler might have a capacity of 450,000 lb/hr of 900 psig steam delivered at 875 °F. Since the saturation temperature at 900 psig is 532 °F, considerable superheat duty is required. Very little boiler convection surface can be placed between the radiant boiler and the superheater since high-temperature combustion gases must be used to attain the required superheat temperature level with a reasonable amount of superheater tube surface. Since feed water must be brought essentially to the saturation temperature before it is admitted to the boiler drum, considerable heat is absorbed in the economizer section wherein the feed water is preheated, and the thermal efficiency of the unit is further increased by preheating the combustion air with the flue gas before they are sent to the discharge stack.

Steam generators are generally designed to produce steam for process requirements, for process needs along with electric power generation, or solely for electric power generation. In each case, the goal is the most efficient and reliable boiler design for the least cost. Many factors influence the selection of the type of steam generator and its design; Perry's [4] provides some of the chief operating characteristics of a range of boilers from small-scale heating systems to large-scale utility boilers.

Boiler design involves the interaction of many variables: water-steam circulation, fuel characteristics, firing system, heat input, and heat transfer. The (furnace) enclosure is one of the most critical components of a steam generator and must be conservatively designed to assure high boiler availability. The furnace configuration and its size are determined by combustion requirements, fuel characteristics, emission standards for gaseous effluents and particulate matter, and the need to provide a uniform gas flow and temperature entering the convection zone to minimize ash deposits and excessive superheater metal temperatures [4].

Petroleum-refinery Furnaces [1]. In atmospheric and vacuum crude distillation, thermal cracking, and modern high-temperature gas processing, the direct-fired tubular furnace is the primary unit in the refinery equipment. Furnaces also are widely employed in various heating, treating, and vaporizing services. Refinery furnaces of various types are required for handling fluids at temperatures as high as 1500 °F and at a combination of temperature and pressure as severe as 1100 °F and 1600 psig.

Oil or gas fuels are used exclusively in these furnaces, although in the near future the need may develop for firing with by-product petroleum coke. In general, the thermal efficiency of refinery furnaces is considerably less than that of large boilers, since in many cases fuel has little worth in the refinery. With the trend toward utilization of a greater percentage of the crude oil produced, fuel is becoming scarcer and more valuable and refiners are recognizing the need for higher thermal efficiencies. It is expected that the range of thermal efficiencies will rise from 65 to 70 percent employed in the past to 75 to 80 percent in the future.

As in boilers, refinery surfaces usually contain both radiant- and convection-heat-transfer surface. Occasionally only a radiant surface is employed for very low-capacity furnaces with duties up to 5,000,000 Btu/hr. Air preheaters have been used to a very limited extent because of the relative unimportance of fuel efficiency in the past; however, even at moderate fuel prices their use can be shown to be economical. Furnaces of this type can have capacities ranging from 25,000,000 to 100,000,000 Btu/hr heat input to the oil. Radiant tubes cover the side wall, roof, and bridge-wall (partition between radiant and convection section) surfaces. Oil is preheated in the bottom and top rows of the convection bank, then passed through the radiant tubes. After reaching an elevated temperature (900 to 1000 °F) it is passed through a large number of convection-section tubes wherein it is maintained at a high temperature for sufficient time to accomplish the desired degree of cracking.

The common methods for calculating heat absorption in furnace radiant sections have been surveyed by Kern [1]. Four of the earlier proposed calculations included:

1. Method of Lobo and Evans [18]
2. Method of Wilson, Lobo, and Hottel [19]
3. The Orrok-Hudson [20] Equation
4. Wohlenberg [21] Simplified Method

Boiler/furnace calculations involve the application of combustion processes under controlled conditions for fuels to liberate energy. For fossil fuels, the required three T's of combustion must be present along with sufficient oxygen for the reaction.

1. The most important T, *temperature*, of the combustion zone, must be maintained: exothermic reactions of combustion of the fuel (if required) must provide enough heat to raise the burning mixture (air and fuel) to a sufficient temperature.

2. *Turbulence*, the second most important T, the constant mixing of fuel and oxygen, must exist.

3. Elapsed time of exposure to combustion temperatures (*residence time*) must be adequately long in duration to ensure that even the slowest combustion fuel component has gone essentially to completion. In other words, transport of the burning mixture through the high-temperature region must occur over a sufficient period of *time, T*.

4. Adequate free *oxygen* must always be available in the combustion zone.

Thus, four parameters influence the mechanism associated with providing an energy/heat source: temperature, turbulence, residence time, and oxygen [22].

Condensing Boilers. Condensing boilers are water heaters fueled by gas or oil. They achieve high efficiency (typically greater than 90% on the higher heating value) by using waste heat in flue gases to pre-heat cold water entering the boiler. Water vapor produced during combustion is condensed into liquid form, which leaves the system via a drain.

In a conventional boiler, fuel is burned and the hot gases produced pass through a heat exchanger where much of their heat is transferred to water, thus raising the water's temperature. One of the hot gases produced in the combustion process is water vapor (steam), which arises from burning the hydrogen content of the fuel. A condensing boiler extracts additional heat from the waste gases by condensing this water vapor to liquid water, thus recovering its latent heat (enthalpy) of vaporization. When latent energy is extracted from water vapor, acidic condensate will usually be left behind on the surface of heat exchangers. Unless built from the highest-quality materials and designed to drain freely, a heat exchanger will corrode over time.

The typical increase in efficiency associated with condensing boilers can be as much as 10–12%. Moreover, a condensing boiler will almost always have a better operating efficiency than a conventional non-condensing one, due to its larger and more efficient heat exchanger.

Today's high-efficiency boilers are engineered to condense water. As noted, the condensate produced is often slightly acidic (3–5 pH), so suitable materials must be used in areas where liquid is present. The heat exchangers are made of high-quality materials and designed to drain freely, which allows them to withstand the condensing operation with no significant corrosion. Aluminum alloys and stainless steel are the materials most commonly used. The production of condensate also requires the installation of a heat exchanger condensate drainage system. To economically manufacture a condensing boiler's heat exchanger (and for the unit to be manageable at installation), the smallest practical size for its output is preferred. This approach has resulted in heat exchangers with high combustion side resistance, often requiring the use of a combustion fan to move the products through the narrow passageways. This has also had the benefit of providing energy for the flue system as the expelled combustion gases are usually below 100 °F (212 °F) and as such, have a density close to air, with little buoyancy.

Today's high-efficiency boilers offer a variety of venting options. Units can be common-vented through a ceiling or individually vented through a sidewall, with the forced-draft design of some equipment able to dramatically reduce the length and diameter of flue runs. This translates to reduced project costs and more "usable" building space. Even within the confines of a mechanical room, most equipment is doorway sized, with a small installed footprint. The ability to support variable flows (and withstand no-flow conditions) eliminates the need for dedicated pumping equipment, saving both money and space.

Control of a condensing boiler is crucial to ensuring that it operates in the most economical, environmental, and fuel-efficient way. Almost all have modulating burners. The burners are usually controlled by an embedded system with built-in logic to control the output of the burner to match the load and provide optimum performance. The highest-efficiency boilers combine accurate temperature control with high burner turndown to precisely and cost-effectively match plant output to heating demand. These boilers may use a PID (proportional integral derivative) controller to maintain supply-water temperature to within ±2 °F. A PID controller is a type of feedback control that can make sure that industrial process systems are running under appropriate conditions. It can ensure the systems are operating within certain range of temperature, pressures, flow rate, and so forth. PID controllers will be revisited in Part III, Chapter 10. After load and temperature requirements are determined, these PID controllers modulate high-turndown burners in 1-percent increments. This allows boilers to change input/output to match load exactly. There is no temperature overshoot, and these boilers can operate over their entire range in a matter of seconds. Such high turndown, coupled with latent energy recaptured during condensing, can generate as much as a 30-to-40 percent increase in efficiency when compared with conventional hydronic systems.

Condensing boiler manufacturers claim that up to 98% thermal efficiency can be achieved, compared to 70%-80% with conventional designs (based on the higher heating value of fuels). Typical models offer efficiencies in the 90%-95% range, which brings most brands of condensing gas boilers into the highest available categories for energy efficiency. It should be noted that condensing boilers can be up to 50% more expensive to buy and install than conventional types. However, the extra cost of installing a condensing instead of a conventional boiler should be recovered in around 2–5 years through lower fuel use. Exact figures will depend on the efficiency of the original boiler installation, boiler utilization patterns, the costs associated with the new boiler installation, and how frequently the system is used.

Heat Sources [1]. The heat to a boiler or furnace is provide primarily in the combustion reaction and in the sensible heat of the combustion air if it has been preheated. Gas fuels generally provide nonluminous flames. Oil fuels can be fired to provide flames of varying degrees of luminosity, depending upon burner design, extent of atomization, and percentage of excess air used. Pulverized coal burners produce a flame containing incandescent solid particles and of a greater degree of luminosity than the minimum obtainable with oil burners. Stoker firing provides an incandescent fuel bed.

The differences in the characteristics of the flames or heat patterns produced in the conventional firing of the various fuels have resulted in the development of methods of calculating radiant-heat transmission which apply on the one hand to oil- or gas-fired refinery furnaces and on the other hand to coal (either stoker or pulverized) -fired units. There is no simple, universally applicable method of calculating heat absorption in all types of furnaces. At the outset, then, the calculations must differentiate between furnaces fired with gas or oil and furnaces fired with solid fuels. Details on both energy sources and calculations follow.

Skipka and Theodore [23] have classified all the major heat/energy resources in the following 13 categories:

1. Natural gas
2. Liquid fuels (oil)
3. Coal
4. Shale oil
5. Tar sands
6. Solar
7. Nuclear (fission)
8. Hydroelectric
9. Wind
10. Geothermic
11. Hydrogen
12. Bioenergy
13. Waste

Extensive details on each of these heat/energy resources are available on the Department of Energy (DOE) website and the Internet, as well as the aforementioned work of Skipka and Theodore [23].

Example 9.16 – Rankine Cycle Application. A Rankine cycle has a T-S (Temperature-Entropy) diagram, as shown in Figure 9.6, and rejects 2043 kJ/kg of heat during the constant pressure process between points 3 and 4. The data provided in Table 9.3 is also associated with this cycle. Calculate the enthalpy change across the boiler. Properties associated with the condenser are as follows:

$$P = 0.1235 \text{ bar}$$

$$h_f = 209 \text{ kJ / kg}$$

$$h_g = 2592 \text{ kJ / kg}$$

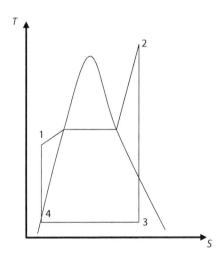

Figure 9.6 Rankine cycle T-S diagram.

Table 9.3 T-S Data for Example 9.16.

Point	x	h, kJ/kg	s, kJ/kg·K
1		548	
2		3989	7.5190
3	0.9575	2491	7.7630
4			1.4410

Note: x = mass fraction vapor

$$s_f = 0.7308 \text{ kJ / kg} \cdot \text{K}$$

$$s_g = 8.0763 \text{ kJ / kg} \cdot \text{K}$$

SOLUTION: Determine the steam enthalpy at the entry and exit to the boiler (points 1 and 2), turbine (points 2 and 3), and pump (points 4 and 1) from the problem statement and data provided in Table 9.3:

$$h_1 = 548 \text{ kJ / kg}$$

$$h_2 = 3989 \text{ kJ / kg}$$

$$h_3 = 2491 \text{ kJ / kg}$$

In addition,

$$h_4 = h_3 - Q_H$$

where,

$$Q_H = \text{heat rejected by the condenser}$$

Thus,

$$h_4 = 2491 - 2043$$

$$= 448 \text{ kJ / kg}$$

Calculate the heat added to the boiler (which is equal to the enthalpy change across the boiler):

$$Q_{\text{boiler}} = h_2 - h_1$$

$$= 3989 - 548$$

$$= 3441 \text{ kJ / kg}$$

Example 9.17 – Rankine Cycle Thermal Efficiency. What is the thermal efficiency of the cycle in Example 9.16?

SOLUTION: Calculate the work produced by the turbine by determining the enthalpy change across the turbine:

$$W_{\text{turbine}} = h_2 - h_3$$

$$= 3989 - 2491$$

$$= 1498 \text{ kJ / kg}$$

Calculate the work by the pump, which is equal to the enthalpy change across the pump:

$$W_{pump} = h_1 - h_4$$

$$= 548 - 448$$

$$= 100 \text{ kJ} / \text{kg}$$

Calculate the net work by subtracting the pump work from the turbine work:

$$W_{net} = W_{turbine} - W_{pump}$$

$$= 1498 - 100$$

$$= 1398 \text{ kJ} / \text{kg}$$

Calculate the thermal efficiency, η_{th}. By definition,

$$\eta_{th} = \frac{W_{net}}{Q_{in}} = \frac{W_{net}}{Q_{boiler}} \qquad (9.10)$$

Substituting,

$$\eta_{th} = \frac{1398}{3441}$$

$$= 0.406 = 40.6\%$$

Comment: Unless otherwise specified, it is generally assumed that the turbine and pump operate adiabatically.

Example 9.18 – Rankine Cycle Temperature Calculations. Using the Rankine cycle T-S diagram in Figure 9.6, and the data from Example 9.16, calculate the temperature at point 3.

SOLUTION: Identify the known properties at point 3, which includes x, h, and s from Table 9.3:

$$x_3 = 0.9575$$

$$h_3 = 2491 \text{ kJ} / \text{kg}$$

$$s_3 = 7.7630 \text{ kJ} / \text{kg} \cdot \text{K}$$

Identify a process that is associated with point 3. Since it is inside the vapor dome, it will therefore be a constant pressure process. The heat rejection of,

$$h_3 - h_4 = Q_{out} = 2043 kJ / kg \approx Q_{rev}$$

is therefore associated with a constant temperature, constant pressure condensation process.

Determine the relationship between the process and the point in question. Since the entropy and the heat transfer associated with the condenser are known, a relationship therefore exists that allows the temperature to be determined by rearranging [2, 3, 6],

$$T_3 = \frac{Q_{out}}{\Delta s} = \frac{Q_{out}}{s_3 - s_4} \tag{9.11}$$

Substitute the known values,

$$T_3 = \frac{2043}{7.7630 - 1.4410}$$

$$= 323K = 50°C$$

9.4   Waste Heat Boilers

Energy has become too valuable to discard. As a result, waste heat and/or heat-recovery boilers are now common in many process plants [24]. As the chemical processing industries become more competitive, no company can afford to waste or dump thermal energy. This increased awareness of economic, environmental, and sustainability considerations has made waste-heat boilers another product of the boiler industry. The term "waste heat boiler" includes units in which steam is generated primarily from the sensible heat of an available hot flue or hot gas stream rather than solely by firing fuel.

An obvious by-product of incineration processes is thermal energy–in many cases, a large amount of thermal energy. The total heat load generated by a typical hazardous waste incinerator, for example, is in the range of 10 to 150 million Btu/hr. While waste heat boilers are capable of recovering 60–70% of this energy, the effort may or may not be justified economically. In assessing the feasibility of recovery, a number of factors must be taken into account; among these are the amount of heat wasted, the fraction of that heat that is realistically recoverable, the irregularity in availability of heat, the cost of equipment to recover the heat, the cost of energy, and the aforementioned environmental/sustainability concerns. The last factor is particularly critical and may be the most important consideration in a decision involving whether or not to harness the energy generated by a particular incinerator. Other important considerations are the incinerator capacity and nature of the waste/fuel being handled. Generally, heat recovery on incinerators of less than 5 million Btu/hr may not be economical because of capital cost considerations. Larger capacity incinerators may also be poor candidates for heat recovery if steam is not needed at the plant site or if the combustion gases are highly corrosive; in the latter case, the maintenance cost of the heat recovery equipment may be prohibitive.

Process Description. The main purpose of a boiler is to convert a liquid, usually water, into a vapor. In most industrial boilers — and as noted earlier in this chapter — the energy required to vaporize the liquid is provided by the direct firing of a fuel in the combustion chamber. The energy is transferred from the burning fuel in the combustion chamber by convection and radiation to the metal wall separating the liquid from the combustion chamber. Conduction then takes place through the metal wall and conduction/convection into the body of the vaporizing liquid. In a waste heat boiler, no combustion occurs in the boiler itself; the energy for vaporizing the liquids is provided by the sensible heat of hot gases which are usually product (flue) gases generated by a combustion process occurring elsewhere in the system. The waste heat boilers found at many facilities make use of the flue gases for this purpose.

In a typical waste heat boiler installation, the water enters the unit after it has passed through a water treatment plant or the equivalent. This boiler feed water is sent to heaters/economizers and then into a steam drum. Steam is generated in the boiler by indirectly contacting the water with hot combustion (flue) gases. These hot gases may be around 2000 °F. The steam, which is separated from the water in the steam drum, may pass through a superheater, and is then available for internal use or export. The required steam rate for the process or facility plus the steam temperature and pressure are the key design and operating variables on the water side. The inlet and outlet flue gas temperatures also play a role, but it is the chemical properties of the flue gas that can significantly impact boiler performance. For example, acid gases can arise due to the presence of any chlorine or sulfur in the fuel or waste. The principal combustion product of chlorine is hydrogen chloride, which is extremely corrosive to most metal heat transfer surfaces. This problem is particularly aggravated if the temperature of the flue gas is below the dewpoint temperature of HCl (i.e., the temperature at which the HCl condenses). This usually occurs at temperatures of about 300 °F. In addition to acid gases, problems may also rise from the ash of incineration processes; some can contain a fairly high concentration of alkali metal salts that have melting points below 1500 °F. The lower melting point salts can slag and ultimately foul (and, in some cases, corrode) boiler tubes and/or heat transfer surfaces.

Equipment. As noted in the previous section, boilers may be either *fire tube* or *water tube* (water-wall). Both are commonly used in practice; the fire-tube variety is generally employed for smaller applications (<15 × 10⁶ Btu/hr). In the fire-tube waste heat boiler, the hot gases from the process are passed through the boiler tubes. The bundle of tubes is immersed in the water to be vaporized; the vaporizing water and tube bundles are encased in a large insulated container called a shell. The steam generated is stored in a surge drum, usually located above the shell and connected to the shell through vertical tubes called risers. Because of construction constraints, the steam pressure in fire-tube boilers is usually limited to around 1000 psia.

Fire-tube boilers are compact, low initial cost, and easy to modularize based on plant requirements. However, they are also slow to respond to changes in demand for steam (load) compared to water-tube boilers, and the circulation is slower. Also, stresses are greater in fire-tube boilers because of their rigid design and subsequent inability to expand and contract easily. Fire-tube boilers are usually directly fired with fuels, either liquid or gaseous. They also serve to recover heat from incinerators fired on waster fuels.

In the water-tube waste heat boiler, the water is contained *inside* the tubes and the hot flue gases flow through the tube bundle, usually in a direction perpendicular to the tubes (cross-flow). Because of the increased turbulence that accompanies cross-flow, the overall heat transfer coefficient for water-tube boilers is higher than that for fire-tube boilers. This advantage is somewhat offset, however, because it is more difficult to clean the outside surfaces of the tubes than the inside surfaces. As a result, heat transfer losses and maintenance problems due to flue gas fouling tend to be greater in water-tube boilers.

Traditional water-tube boilers can be physically divided into two sections, the furnace and the convection pass. Furnaces (fireboxes, combustion chambers, etc.) will vary in configuration and size, but their function is to contain the flaming combustion gases and transfer the heat energy to the water-cooled walls. The convection pass contains the superheaters, reheater, economizer, and air preheater heat exchangers, where the heat of the combustion flue gases is used to increase the temperature of the steam, water, and combustion air. The *superheaters* and *reheaters* are designed to increase the temperature of the steam generated within the tubes of the furnace walls. Steam flows inside the tubes and flue gas passes along the outside surface of the tubes. The *economizer* is normally a countercurrent heat exchanger designed to recover energy from the flue gas after the superheater and the reheater. The boiler economizer is a tube bank type, hot-gas-to-water heat exchanger. It increases the temperature of the water entering the steam drum. The *air heater* is not a portion of the steam-water circuit, but serves a key role in the steam generator system to provide additional heat transfer and efficiency. In many cases, especially in a high-pressure boiler, the temperature of the flue gas leaving the economizer is still quite high. The air heater recovers much of this energy and adds it to the combustion air. Heating the combustion air prior to its entrance to the furnace reduces fuel usage and reduces environmental and sustainability concerns.

Water-tube boilers respond quickly to changes in demand for steam due to improved water circulation. They can withstand much higher operating pressures and temperatures than fire-tube boilers. In addition, the water-tube boiler design is safer [12]. They can also burn a wide variety of fuels as well as wastes and have the ability to expand and contract more easily than fire-tube boilers. The major drawback is that water-tube boilers are more expensive to install. They also require more complicated furnaces and repair techniques.

Describing Equations. The design of waste heat boilers involve calculations that are based on energy balances and estimates of the rates of heat transfer. Although some units operate in an unsteady-state or cyclical mode, the calculation procedures are invariably based on steady-state conditions.

In heat transfer equipment, there is no shaft work, and potential and kinetic energy effects are small in comparison with the other terms in an energy balance equation [6]. Heat flow to or from the surroundings is not usually desired in practice and is usually reduced to a small magnitude by suitable insulation. It is therefore customary to consider this heat loss or gain negligible in comparison to the heat transfer through the walls of the tubes from the hot combustion gases to the water in the boilers. Thus, all the sensible heat lost by the hot gases may be assumed transferred to the steam.

The heart of the heat transfer calculation is the heat flux, which is based on the area of the heating surface and is a function of the temperature difference driving force and the overall heat transfer coefficient, as discussed earlier (see also Chapters 5 thru 8). This relatively simple equation can be used to estimate heat transfer rates, area requirements, and temperature changes in a number of heat transfer equipment. However, problems develop if more exact calculations are required. The properties of the fluid (viscosity, thermal conductivity, heat capacity, and density) are more important parameters in these calculations. Each of these properties, especially viscosity, is temperature dependent. Since a temperature profile, in which the temperature varies from point to point, exists in a flowing stream undergoing heat transfer, a problem arises in the choice of temperature at which the properties should be evaluated. When temperature changes within the stream become large, the difficulty of calculating heat transfer to fluids with phase change (as in a waste heat boiler) is complex and, in practice, is treated empirically rather than theoretically [4, 22, 24].

The rigorous design and/or performance evaluation of a waste heat boiler is an involved procedure. Fortunately, several less rigorous methods are available in the literature. One approach that is fairly simple and yet reasonably accurate has been devised by Ganapathy [25]. This method provides a technique for sizing waste heat boilers for the fire-tube type and involves the use of performance evaluation chart (see Figure 9.7) that is based on fundamental heat transfer equations plus some simplifying assumptions.

Details regarding Ganapathy's methods are now discussed, employing this author's notation. Earlier equations may be combined and applied to the boiler where the temperature outside the tubes (T_c) can be assumed constant due to the water-to-steam phase change. The result is once again,

$$Q = \dot{m}_h c_h \left(T_{h,1} - T_{h,2}\right) = U_i A_i \frac{\left(T_{h,1} - T_c\right) - \left(T_{h,2} - T_c\right)}{\ln\left[\left(T_{h,1} - T_c\right)/\left(T_{h,2} - T_c\right)\right]} \qquad (9.12)$$

where (employing Ganapathy's notation) T_c is the boiling temperature of water. Note that $U_i A_i$ is used in Equation (9.17) instead of $U_o A_o$. The two expressions are equal. In this procedure, it is slightly more convenient to use the inside heat transfer area rather than the outside as the basis for U. Since the outside film coefficient, h_o, associated with a boiling liquid is much greater than the inside coefficient for the flue gas, the inside resistance dominates and is responsible for about 95% of the total resistance. The inside overall heat transfer coefficient may therefore be simplified by the approximate relationship.

$$U_i = 0.95 h_i \qquad (9.13)$$

The total inside heat transfer area for the boiler is given by,

$$A_i = \pi D_i N L \qquad (9.14)$$

where N is the number of tubes. The Dittus-Boelter equation may be used to solve explicitly for h_i (the inside film coefficient):

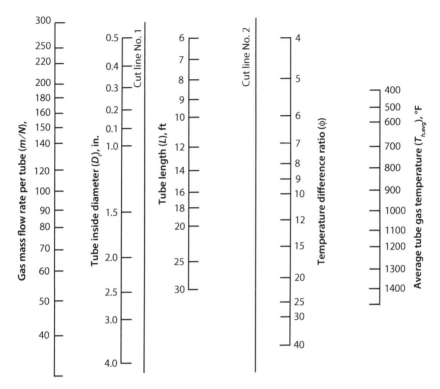

Figure 9.7 Waste heat boiler performance evaluation chart [25].

$$h_i = 0.023\frac{k^{0.6}G^{0.8}c_h^{0.4}}{D_i^{0.2}\mu^{0.4}} \tag{9.15}$$

and the gas mass flux, G, may be represented in terms of $\dot{m}_h$,

$$G = \frac{4\dot{m}_h}{D_i^2 N} \tag{9.16}$$

Equations (9.12) through (9.16) may be combined to give,

$$\ln\left(\frac{T_{h,1}-T_c}{T_{h,2}-T_c}\right) = \frac{U_iA_i}{\dot{m}_hc_h} = \frac{(C)(L)\left[F(T)\right]}{D_i^{0.8}\left(\dot{m}_h/N\right)^{0.2}} \tag{9.17}$$

where $F(T) = k^{0.6}/c_h^{0.6}\mu^{0.4}$ and C is a constant equal to 0.0833 when D_i is in feet, or 0.608 when D_i is in inches.

Keeping in mind that this procedure is used for engineering design purposes and is not intended for rigorous analytical calculations, the value of $F(T)$ does not, in practice, vary over a very large range for most gas streams.

The following describes a procedure for the use of the above chart as a design tool. It is assumed in this procedure that the inside tube diameter and the number of tubes have been chosen and the tube length is to be determined. Two later examples further demonstrate the use of Figure 9.7.

1. From the inlet and desired outlet gas temperature, and the water saturation (boiling) temperature, calculate the arithmetic average gas temperature, $T_{h,\text{avg}}$, and the value of the temperature difference ratio, ϕ, where,

$$\phi = \frac{T_{h,1} - T_c}{T_{h,2} - T_c} \qquad (9.18)$$

   Mark these points on the appropriate axes.
2. Draw a straight line connecting these points and extend the line to the left to cut line No. 2. Mark this point B.
3. Mark the value of the flue gas mass flow rate per tube and the value of the inside diameter on the appropriate axes. (For design purposes, a good starting value of $\dot{m}_h / N$ is 80–150 lb/hr.)
4. Draw a straight line connecting these points and extend the line to the right to cut line No. 1. Mark this point A.
5. Connect points A and B by a straight line. The intersection of this line with the L axis yields the appropriate tube length.

The chart could be used in similar fashion as a performance evaluation tool. In this case, the outlet gas temperature would be unknown and hence would have to be estimated in order to determine the average gas temperature. In order to avoid a time-consuming trail-and-error procedure involving the gas outlet temperature and the average gas temperature, Equation (9.19) may be used to estimate $T_{h,\text{avg}}$ without too much loss of accuracy:

$$T_{h,\text{avg}} = 0.5\left(T_{h,1} + T_c\right) \qquad (9.19)$$

Example 9.19 – Water Heater Variables. What factors should be considered when assessing whether it is feasible to implement energy recovery using a water heat boiler?

SOLUTION: As discussed above, important factors include:

1. the amount of heat wasted,
2. the fraction of heat that could be realistically recovered,
3. irregularity in availability of heat (scheduling),
4. cost of equipment,
5. cost of energy,
6. environmental considerations,
7. incinerator capacity, and
8. the nature of the waste handled.

Example 9.20 – Waste Heat Boiler Fouling. Discuss the following with respect to waste heat boilers:

1. The effect of fouling on the water/steam side
2. The effect of fouling on the gas side
3. Monitoring procedures to follow to account for fouling and/ or scaling

SOLUTION:

1. The exit gas temperature will increase, thus resulting in the loss of energy recovery and reduced steam production. Tube wall temperatures can increase significantly, leading to tube failures due to scale formation.
2. This leads to a loss of steam output and can increase the gas side pressure drop. However, it does not significantly increase the tube wall temperatures.
3. In monitoring boiler performance, one should be aware of increases (excursions) in exit gas temperatures, tube wall temperatures or loss of steam production for the same gas inlet conditions.

Example 9.21 – Ganapathy's Design Methodology. Using Ganapathy's method, determine the required "length" of a waste heat boiler to be used to cool hot gases (average heat capacity is 0.279 Btu/lb·°F) from 2000 to

550 °F and generate 30,000 lb/hr of steam at 330 °F from water at 140 °F. The boiler contains 800 1.5 in. ID tubes.

SOLUTION: Calculate the temperature difference ratio, ϕ:

$$\phi = \frac{2000 - 330}{550 - 330} = 7.59 \qquad (9.18)$$

The average gas temperature is,

$$T_{h,avg} = 0.5\left(T_{h,1} + T_{h,2}\right) \qquad (9.19)$$

$$= 0.5\left(2000 + 550\right)$$

$$= 1275°F$$

From the steam tables, see Appendix:

$$h_{steam} = 1187.7\,\text{Btu / lb}$$

$$h_{water} = 107.89\,\text{Btu / lb}$$

Therefore,

$$Q = \dot{m}\left(h_{steam} - h_{water}\right) = \left(30{,}000\right)\left(1187.7 - 107.89\right)$$

$$= 32.39 \times 10^6\,\text{Btu / hr}$$

$$\dot{m}_h = \frac{Q}{c\left(T_{h,1} - T_{h,2}\right)} = \frac{32.39 \times 10^6}{\left(0.279\right)\left(2000 - 550\right)} = 80{,}070\,\text{lb / hr}$$

$$\frac{\dot{m}_h}{N} = \frac{80{,}070}{800} = 100\,\text{lb / hr per tube}$$

From Figure 9.7, the tube length is $L = 15\,\text{ft}$.

9.5   Cogeneration/Combined Heat and Power (CHP)

Cogeneration is an energy conversion process wherein heat from a fuel is simultaneously converted to useful thermal energy (e.g., process steam) *and* electrical energy. The need for either form can be the primary incentive for cogeneration, but there must be an opportunity for economic captive use or sale of the other. In a chemical plant the need for process and other heating steam is likely to be the primary; in a public utility plant, electricity is usually the primary product.

Thus, a cogeneration system is designed from one of two perspectives: it may be sized to meet the process heat and other steam needs of a plant or a community of industrial and institutional users so that the electric power is treated as a by-product which it must be either used on site or sold; it may be sized to meet electric power demand, and the rejected heat used as supply, at or near the site. The latter approach is the likely one if a utility owns the system; the former if a chemical plant is the owner [4].

Industrial use of cogeneration leads to small, dispersed electric-power-generation installations–an alternative to complete reliance on large central power plants. Because of the relatively short distances over which thermal energy can be transported, process-heat generation is characteristically an on-site process with or without cogeneration.

Cogeneration systems will generally not match the varying power and heat demands at all times for most applications. Thus, an industrial cogeneration system's output frequently must be supplemented by the separate on-site generation of heat or the purchase of utility-supplied electric power. If the on-site electric power demand is relatively low, an alternative option is to match the cogeneration system to the heat load and contract for the sale for excess electricity to the local utility grid.

Fuel saving is the major incentive for cogeneration. Since all heat-engine-based electric power systems reject heat to the environment, that rejected heat can frequently be used to meet all or part of the local thermal energy needs. Using rejected heat usually has no effect on the amount of primary fuel used, yet it leads to a saving of all or part of the fuel that would otherwise be used for the thermal-energy process. Heat engines [4] also require a high-temperature thermal input, usually receiving the working fluid directly from a heating source; but in some situations, they can obtain the input thermal energy as the rejected heat from a higher-temperature process. In the former case, the cogeneration process employs a heat-engine topping cycle; in the latter case, a bottoming cycle is used.

The choice of fuel for a cogeneration system is determined by the primary heat-engine cycle. Closed-cycle power systems which are externally fired–the steam turbine, the indirectly fired open-cycle gas turbine, and closed-cycle gas turbine systems–can use virtually any fuel that can be burned in a safe and environmentally acceptable manner: coal, municipal solid waste, biomass, and industrial wastes are burnable with closed power systems. Internal combustion engines, on the other hand, including open-cycle gas turbines, are restricted to fuels that have combustion characteristics compatible with the engine type and that yield combustion products clean enough to pass through the engine without damaging it. In addition to natural gas, butane, and gaseous fuels derived from shale, coal, or biomass are in this category. Direct-coal-fired internal combustion engines have been an experimental reality for decades but are not yet a practical reality technologically or economically.

There are at least three broad classes of applications for topping-cycle cogeneration systems:

1. Utilities or municipal power systems supplying electric power and low-grade heat (e.g., 422 K [300 °F]) for local district heating systems.
2. Large residential, commercial, or institutional complexes requiring space heat, hot water, and electricity.
3. Large industrial operations with on-site needs for electricity and heat in the form of process steam, direct heat, and/or space heat.

Typical Systems. All cogeneration systems involve the operation of a heat engine for the production of mechanical work which, in nearly all cases, is used to drive an electric generator. The four common heat-engine types appropriate for topping-cycle cogeneration systems are:

1. Steam turbines (backpressure and extraction configurations)
2. Open-cycle (combustion) gas turbines
3. Indirectly fired gas turbines: open cycles and closed cycles
4. Diesel engines

CHP is not a technology, but an approach to applying various technologies. This approach generates electricity and useful thermal energy in a single integrated system. As noted, heat that is normally wasted in conventional power generation is recovered as useful energy, avoiding the energy losses that are incurred when heat and power are generated separately.

While the conventional method of producing heat and power separately has a typical combined efficiency of 45%, CHP systems often have total efficiencies of 70 to 80%. This increased efficiency reduces or eliminates the purchase of energy from offsite generating sources such as an electric utility and so saves on utility costs. Transmission losses associated with traditional grid-sourced power are also avoided.

A highly efficient CHP system uses less fuel input to produce the same energy output as conventional methods that generate heat and power. The conventional system has two parts: the boiler that generates heat and the power station that generates electricity. The majority of the energy lost in a power station is lost in the conversion from heat to electricity. Although the boiler may be able to convert up to 85% of the energy in its fuel into steam or another form of thermal energy, this efficiency cannot make up for the inefficiency of the power station generator. In contrast, the CHP approach minimizes overall conversion losses by using a *single* system to generate both electricity and heat.

CHP is a versatile and flexible energy resource. As a system, it can be powered by engines, turbines, fuel cells, or other prime movers. A variety of fuels can be used to feed the system, including natural gas, biomass, diesel, and waste energy streams. As a scalable power source, a CHP system can be sized to meet energy demand loads ranging from that of a single family home to that of a large industrial facility. Such systems can be deployed in either existing or new buildings in the industrial, commercial, residential, and agricultural sectors. They can be good investments and provide environmental/sustainability benefits when appropriately designed to meet local needs.

CHP can comprise of a number of different equipment types and use a number of different fuels. Each installed project is a combination of many different project elements. These elements are chosen based upon the unique constraints and challenges presented by each potential CHP host site, such as cost limitations, space availability, local emissions profiles, and noise concerns.

Combined heat and power systems are most often classified by the type of machine or engine that creates energy from the fuel or fuels. This piece of equipment is called the prime mover [21] and generally falls into one of five categories.

1. *Reciprocating engines,* which can be sized up to about 7 MW, are the most commonly used type of prime mover. A facility may choose a reciprocating engine for the starting speed, or ability to reach the optimal level the fastest, as well as for it reliability rating.
2. *Steam turbines* use boiler-generated steam to generate electricity or other types of energy. Steam turbines can run on a variety of fuels and can range in size from 50 kW to 250 MW.
3. *Combustion turbines,* which can burn a variety of different gases, are used in applications that require high-pressure and high-temperature steam. These turbines can be sized up to well over 100 MW and are often used by utilities interested in deploying CHP within their generating fleets.
4. *Microturbines* are small, often modular combustion turbines with availabilities nearing 500 kW. They are able to run on

a variety of fuels and resemble small jet engines. These turbines typically offer lower emissions than other types of CHP equipment and contain few moving parts, which can be advantageous in some applications.

5. *Fuel cells* create electricity using an electrochemical reaction process, concurrently producing heat byproducts such as hot water and hot air. Fuel cells have recently been used in the hospitality and residential sectors, utilizing heat for space conditioning and for general potable hot water needs in homes and hotel rooms. Fuel cells can be an expensive investment but are known for their clean operating characteristics, high efficiency, and relatively quiet operation.

Fuel availability and long-term price forecasts often dictate which prime mover technologies are considered and ultimately chosen at a given site. CHP systems are often operated on fuels such as natural gas purchased from a local utility or purchased as commodities from vendors. Some prime movers can also run on waste heat, waste pressure, or waste fuel streams from industrial processes. Biomass or biogas fuels can also be used when municipal waste management activities, forest management activities, or local agricultural waste streams are located in close proximity. One of the benefits of CHP solutions is that they can take advantage of multiple fuel sources to create useful outputs from what would have otherwise have become waste or pollution. These systems offer considerable flexibility and can meet a wide variety of energy needs.

9.6   Quenchers

Hot gases must often be cooled before being discharged to the atmosphere or entering another device(s) which normally is not designed for very high temperature operation (>50 °F). These gases are usually cooled either by recovering the energy in a waste heat boiler, as discussed in an earlier section of this chapter, or by quenching [24]. Both methods may be used in tandem. For example, a waste heat boiler can reduce an exit gas temperature down to about 500 °F; a water quench can then be used to further reduce the gas temperature to around 200 °F, as well as saturate the gas with water. This secondary cooling and saturation can later eliminate the problem of water evaporation and can also alleviate other potential problems, including environmental ones.

Although quenching and the use of a waste heat boiler are commonly used methods for gas cooling applications, there are several other techniques for cooling hot gases. All methods may be divided into two categories: *direct-contact* and *indirect-contact* cooling. The direct-contact cooling methods include (a) dilution with ambient air, (b) quenching with water, and (c) contact with high heat capacity solids. Among the indirect contact methods are (d) natural convection and radiation from ductwork, (e) forced-draft heat exchangers, and (f) the aforementioned waste heat boilers.

With the *dilution* method, the hot gaseous effluent is cooled by adding sufficient ambient air that results in a mixture of gases at the desired temperature. The *water quench* method uses the heat of vaporization of water to cool the gases. As discussed earlier, when water is sprayed into the hot gases under conditions conducive to evaporation, the energy contained in the gases evaporates the water, and this results in a cooling of the gases. The hot exhaust gases may also be quenched using submerged exhaust quenching. This is another technique employed in some applications. In the *solids contact* method, the hot gases are cooled by incoming air to be used elsewhere in the process. *Natural convection and radiation* occur whenever there is a temperature difference between the gases inside a duct and the atmosphere surrounding it. Cooling hot gases by this method requires only the provision of enough heat transfer area to obtain the desired amount of cooling. In *forced-draft heat exchangers*, the hot gases are cooled by forcing cooling fluid past the barrier separating the fluid from the hot gases. These methods are further discussed below.

Dilution with Ambient Air [24]. The cooling of gases by dilution with ambient air is the simplest method that can be employed. Essentially, it involves the mixing of ambient air with a gas of known volume and temperature to produce a low-temperature mixture that can be admitted to another device. In designing such a system, the amount of ambient air required to provide a gaseous mixture of the desired temperature is first determined.

The quantity of air required for cooling may be calculated directly from an enthalpy balance, i.e., the heat "lost" by the gas is equal to that "gained" by the dilution air. The design of the vessel or duct to accomplish this mixing process is based on a residence time of approximately 0.8–1.5 secs (based on the combined flow rate) at the average temperature.

Quenching with Liquids [24]. When a large volume of hot gas is to be cooled, a method other than dilution with ambient air should be used.

This is usually the case and the cooling method most often used is liquid quenching.

Cooling by liquid quenching is essentially accomplished by introducing the hot gases into a liquid (usually water) contacting device. When the water evaporates, the energy necessary to vaporize the water is obtained at the expense of the hot gas, resulting in a reduction in the gas temperature. The temperature of the gases discharged from the quencher is at the adiabatic saturation temperature of the gas if the operation is adiabatic and the gas leaves the quencher saturated with water vapor. (As noted in an earlier section, a saturated gas contains the maximum water vapor possible at that temperature; any increase in water content will result in condensation.) Simple calculational and graphical procedures are available for estimating the adiabatic saturation temperature of a gas.

There are four types of liquid quenchers that may be employed: spray towers, venturi scrubbers, submerged exhaust tanks, and packed towers (generally a poor choice). The venturi scrubber is actually an air pollution control device but can also be used simultaneously for both particulate removal from, and quenching of, the hot gases [11].

Contact with High Heat Capacity Solids [24]. Inert ceramics with high heat capacities have, in a few cases, been used to cool hot effluent gases from one unit to temperatures suitable for another process unit. Besides cooling the gases by absorbing heat, this method also allows a certain amount of energy recovery by using the heat absorbed by the solids to subsequently preheat an airstream being introduced to another process unit. In one such system, the solid elements are packed into beds called *regenerative thermal oxidizers* (RTO's) [25]. These chambers operate on a *regenerative* principle– alternatively absorbing, storing, and recycling heat energy. The chambers are thus used in a cyclic fashion. In the first half-cycle, the process air is passed through the previously heated ceramic elements in a particular chamber, simultaneously cooling the bed and transferring heat into the gas that is then moved on to another unit. At any point during the cycle, the temperature at the bed varies over 1000 °F from the cold side (i.e., the inlet side for the ambient air) to the hot side. In the second half-cycle, hot gases are introduced to the hot side of the bed, flow through the bed in the opposite direction of the cold flow in the first half-cycle, and the heat lost by the gases is absorbed by the bed that is brought to its maximum average temperature for the start of the next cycle. Typical cycle times range from 2–10 min. Manufacturers claim over 90% thermal energy recovery with this device.

Forced-Draft Cooling. As discussed in, Part I, Chapter 3, heat transfer by convection is due to fluid motion. Cold fluid adjacent to a hot surface receives heat, which is imparted to the bulk of the fluid by mixing. With *natural convection*, the heated fluid adjacent to the hot surface rises and is replaced by colder fluid. By agitating the fluid, mixing occurs at a much higher rate than with natural currents, and heat is taken away from the hot surface at a much higher rate. In most process applications, this agitation is induced by circulating the fluid at a rapid rate past the hot surface. This method of heat transfer is called *forced convection*. Since forced convection transfers heat much faster than natural convection, most process applications used forced-convection heat exchangers. Whenever possible, heat is exchanged between the hot and cold streams to reduce the heat input to the process. There are, however, many industrial applications where it is not feasible to exchange heat in this fashion, and a cooling fluid such as water or air is used, and the heat removed from the stream by the coolant is transferred to the environment. When water is used, the heat is often taken from the process stream in a shell-and-tube cooler, and the heat picked up by the water is transferred to the atmosphere in a cooling tower (see next section). When air is used as the cooling medium in either shell-and-tube or fin-tube coolers, the heated air is discharged to the atmosphere and is not recirculated through the cooler. Because there are so many different types of forced-convection heat exchangers, it is impossible to present a specific design procedure unless a specific type is chosen for discussion.

Natural Convection and Radiation. When a hot gas flows through a duct, the duct becomes hot and heats the surrounding air. As the air becomes heated, natural drafts are formed, carrying the heat away from the duct. This phenomenon is *natural convection*. Heat is also discharged from the hot duct to its surroundings by *radiant energy*. Both of these modes of heat transfer have already been discussed in Part I, Chapters 3 and 4, respectively.

The rate of heat transfer is determined by the amount of heat to be removed from the hot gaseous effluent entering the system. For any particular basic process, the mass flow rate of gaseous effluent and its maximum temperature are fixed. The cooling system must therefore be designed to recover sufficient heat to lower the effluent temperature to the operating temperature of the device to be used. The rate of heat transfer can be calculated by the enthalpy difference of the gas at the inlet and outlet of the cooling system.

Example 9.22 – Quencher Systems. Discuss the equipment employed for most convection-radiation cooling systems.

SOLUTION: For most convection-radiation systems, the only equipment used is sufficient ductwork to provide the required heat transfer area. Unless the temperature of the gases discharged is exceptionally high, or if there are corrosive gases or fumes present, black iron ductwork is generally satisfactory. The temperature of the duct wall, T_w, can usually be determined for any portion of the ductwork. If T_w proves to be greater than black iron can withstand, either a more heat-resistant material should be used for that portion of the system or a portion of the cooled gas should be recirculated to lower the gas temperature at the cooling system inlet.

Example 9.23 – Air Quenching. It is proposed to cool 144,206 lb/hr of a hot flue combustion gas by using ambient air at 70 °F. Calculate the quantity (mass, mole, and volume basis) of air required to cool the gases from 2050 °F to an acceptable temperature of 560 °F. Assume an average flue gas heat capacity of 0.3 Btu/lb·°F. Also design the air quench tank if a 1.5-second residence time is required.

SOLUTION: Under adiabatic conditions,

$$Q_{flue} = Q_{air}$$

$$Q_{flue} = \dot{m}_{flue}\bar{c}\Delta T$$

$$= (144,206)(0.3)(2050-560)$$

$$= 64.5\times10^6 \text{ Btu / hr}$$

Since $Q_{flue} = Q_{air}$,

$$64.5\times10^6 = \dot{m}_{air}(0.243)(560-70)$$

so that,

$$\dot{m}_{air} = 541,700 \text{ lb / hr}$$

Since the molecular weight (MW) of air is 29 lb/lbmol,

$$\dot{m}_{air} = 18,680 \text{ lbmol / hr}$$

$$= 7,215,000 \text{ ft}^3 \text{ / hr} (\text{at } 70°F)$$

Note: The combined mass (and volume) flow of gas is now significantly higher. This can adversely impact on the economics for the downstream equipment.

The total mass flow rate of the gas is now,

$$\dot{m} = 144,200+541,700$$

$$= 685,900 \text{ lb / hr}$$

The volumetric flow at 560 °F (assuming air) is calculated employing the ideal gas law.

$$q = \frac{\dot{m}_t RT}{P(\text{MW})}$$

Substituting,

$$q = \frac{(685,900)(0.73)(1020)}{(1.0)(29)}$$

$$= 1.76 \times 10^7 \text{ ft}^3 / \text{hr}$$

$$= 4890 \text{ ft}^3 / \text{s}$$

The volume of the tank is, therefore,

$$V_t = (4890)(1.5)$$

$$= 7335 \text{ ft}^3$$

Example 9.24 – Air Quencher Design. Calculate the physical dimensions of the tank in the previous example.

SOLUTION: The physical dimensions of the tank are usually set by minimizing surface (materials) cost [12]. The total surface is given by,

$$S = 2\left(\frac{\pi D^2}{4}\right) + \pi DH; \quad H = \text{tank height}$$

To minimize the total surface, the derivative of S with respect to D is set equal to zero, i.e.,

$$\frac{dS}{dD} = \pi D + \pi H = 0$$

or,

$$D = H$$

The dimensions of the tank may now be calculated:

$$V_t = \left(\frac{\pi D^2}{4}\right) H = \frac{\pi D^3}{4}$$

$$D = 21.06 \text{ ft}$$

$$H = 21.06 \text{ ft}$$

Note: The size of the tank is excessive. This is another reason why air quenching is rarely used.

9.7   Cooling Towers

This section is an expansion of Section 9.2 – Evaporators. The evaporation of water into air for the purpose of increasing the air humidity is known as *humidification*. Closely allied to this is the evaporation of water into air for the purpose of cooling the water. *Dehumidification* consists of condensing water from air to decrease the air humidity. All these processes are of considerable industrial importance and involve the contacting of air and water accompanied by heat and mass transfer [7, 26]. These processes are discussed in this section.

Direct contact of a condensable vapor-noncondensable gas mixture with a liquid can produce any of several results, including humidifying of the gas or cooling of the liquid. The direction of liquid transfer (either humidification or dehumidification) depends on the difference in humidity of the bulk of the gas and that at the liquid surface. If the liquid is normally a pure substance, no concentration gradient exists within it and the resistance to mass transfer lies entirely within the gas. Since evaporation and condensation of a vapor simultaneously involves a latent enthalpy of vaporization or condensation, there will always be a transfer of latent heat in the direction of mass transfer. The temperature differences existing within the system additionally control the direction of any sensible heat transfer that may occur. Furthermore, since temperature gradients may reside within the liquid, within the gas, or within both, the sensible heat transfer resistance may include effects in either or both phases. The effects of latent and sensible heat transfer may be simultaneously considered in terms of the enthalpy changes which occur.

The operation of an adiabatic humidifier normally involves some make-up water entering the unit at the adiabatic saturation temperature. Under these conditions, the temperature of the water in the system is assumed constant at the adiabatic saturation temperature, and both the air temperature and humidity remain constant. In addition, all the heat required to vaporize the water is supplied from the sensible heat of the air.

When water is present in air, it is possible to extract the water by cooling the air-water mixture below the mixture's dew point temperature. The dew point was defined earlier as the temperature at which the air can no longer absorb more water (i.e., it is the 100% relative humidity point) and if the temperature is then reduced further, water is forced out of the air-water mixture as condensation (dew) is formed. This process is described as *dehumidification*. The amount of water that is removed from a mixture as a result of cooling can be determined by drawing a line on a psychrometric chart from the mixture's initial conditions (e.g., dry-bulb temperature and relative humidity), horizontally to the left (i.e., the cooling direction) until the 100% relative humidity line is encountered (see also Appendix AF.2).

The dehumidification process can also be accomplished by bringing moist air into contact with a spray of water — the temperature of which is lower than the dew point of the entering air. An example is passing the air through sprays. Furthermore, dehumidification of air may also be accomplished by passing a cold fluid through the inside of (finned) tubes arranged in banks through which the air is blown. The outside surface of the metal tubes must be below the dew point of the air so that water will condense out of the air.

Cooling towers — the title of this section — also find application in industry. The same operation that is used to humidify air may also be used to cool water. There are many cases in practice in which warm water is discharged from condensers or other equipment and where the value of this water is such that it is more economical to cool it and reuse it than to discard it. Water shortages and thermal pollution have made the cooling tower a vital part of many plants in the chemical process industry. Cooling towers are normally employed for this purpose and they may be destined to have an increasingly important role in almost all phases of industry. Modern (newer) power-generating stations remain under construction or in the planning stage, and both water shortages and thermal pollution are serious problems that must be dealt with.

The cooling of water in a cooling tower is accomplished by bringing the water into contact with unsaturated air under such conditions that the air is humidified and the water brought approximately to the wet-bulb temperature. This method is applicable only in those cases where the wet-bulb temperature of the air is below the desired temperature of the exit water.

As noted earlier, water is cooled in cooling towers by the exchange of sensible heat, latent heat, and water vapor with a stream of relatively cool dry air. The basic relationship developed for dehumidifiers also apply to cooling towers although the transfer is in the opposite direction since the unit acts as a humidifier rather than as a dehumidifier of air.

In all the preceding chapters in Part II, the hot and cold fluids were separated by impervious surfaces. In tubular equipment, the tube does not permit the intimacy of contact between the hot and cold fluids and also serves as a surface upon which resistances accumulate as fouling and scale films. In order that a turbulent fluid in a tube may receive heat, "particles" in the eddying fluid body must contact a warm film at the tube wall, take in heat by conduction, and then mix with the eddying fluid body. A similar process occurs on the shell side, and the net heat exchange may occur — as discussed earlier — through as many as seven individual resistances.

One of the principal reasons for employing tubes is to prevent the potential contamination of the hot fluid by the cold fluid. When one of the fluids is a gas and the other a liquid, an impervious surface is often unnecessary, since there may be no problem of mutual contamination, the gas and liquid being readily separable after mixing and exchanging heat. Fouling resistances are automatically eliminated by the absence of a surface on which they can collect and permit direct-contact equipment to operate indefinitely with uniform thermal performance. The greater intimacy of the direct contact approach generally permits the attainment of greater heat-transfer coefficients than are usual in tubular equipment.

Perhaps the key application of a heat exchanger unit with a direct contact between a gas and a liquid is the aforementioned cooling tower. It is usually a boxlike redwood structure with redwood internals. Cooling towers are employed to contact hot water coming from process cooling systems with air for the purpose of cooling the water and allowing its reuse in a process. The function of the wooden internals, or *fill*, is to increase the contact surface between the air and the water. A cooling tower ordinarily reduces the fresh cooling-water requirement by about 98 percent, although there is some mutual contamination caused by the saturation of the air with water vapor.

The prospective use for direct-contact equipment in other services requiring rapid rates of heat transfer is perhaps greater than for any other type of heat-transfer equipment. Although applied now almost exclusively to the humidification of air or the cooling of water, the principles of direct-contact heat transfer may be applied to the cooling or heating of other insoluble gases or liquids. This is especially true in the cooling of gases over wide temperature ranges. The condenser referred to earlier in this chapter is also an example of direct contact as applied to condensation in which a large heat load may be condensed in an apparatus of small volume. A modification of the same principle might readily be applied to the condensation of organic vapors by a spray of water and particularly to the problem of condensing an oil vapor in the presence of a non-condensable gas.

The Diffusion Process. If dry air at constant temperature is saturated by water at the same temperature in a direct-contact apparatus, the water vapor entering the air carries with it its latent heat of vaporization. The humidity of the air-water-vapor mixture increases during saturation because the vapor pressure of water in the liquid is greater than it is in the unsaturated air and vaporization is the result. When the vapor pressure of water in the air equals that of the liquid, the air is saturated and vaporization ceases. The temperature of the water can be kept constant during air saturation if heat is supplied to it to replace that lost from it to the gas as latent heat (enthalpy) of vaporization. Clearly, then, the heat transfer during the saturation of a gas with a liquid can be made to proceed without a temperature difference, although such a limitation is rarely encountered. It is seen, however, that there is a fundamental difference between this type of heat transfer and traditional conduction, convection, or radiation transfer addressed in Part I.

When a movement of material is promoted between two phases by a vapor pressure (or concentration) difference, it is *diffusion* and is characterized by the fact that material is transferred from one phase to another or between both phases. This behavior is called *mass* or *material transfer* to set it apart from the ordinary concepts of heat transfer. While the earlier phase-rule definitions apply to systems at equilibrium, if a phase is not homogenous, *self-diffusion* may occur as the phase approaches homogeneity.

In the condensation of a vapor from a noncondensable gas an inert film near the tube wall retards the condensable vapor from reaching the condensate film at the tube wall. The rate at which the vapor passes through the inert film is a function of the pressure of the vapor in the gas body an in the condensate film adjoining the tube walls and follows the aforementioned mechanism of diffusion.

Diffusion involves the passage of one fluid through another. Consider a gas, such as air, containing a small amount of acetone vapor which is soluble in water while the air may be considered insoluble in water. Suppose the air-acetone mixture is fed to a tower which has fresh water flowing continuously down its walls so that any acetone molecules which might be bound into the water are removed by it from the gas body. How fast will acetone molecules be removed from the gas body?

An idealized picture of the problem is shown in Figure 9.8. It may be assumed that a relatively stagnant air film forms at the liquid surface owing to the loss of momentum of the air molecules striking the liquid film and being dragged along by it. This is represented by 1-1′ and 2-2′. The liquid film may also be considered relatively stagnant compared with the air body. This is the basis of the "two-film" theory. Because of the mutual solubility of acetone in water, the rate at which acetone molecules can pass through

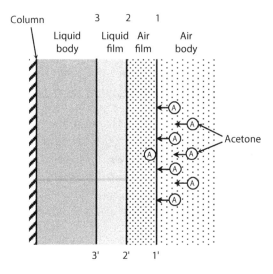

Figure 9.8 Film theory showing principal resistances.

the liquid film is exceedingly great. Thus, the acetone molecules in the air film which arrive at the liquid film are depleted so rapidly by solution in the liquid film that the concentration of acetone molecules in the air film is less than it is in the gas body. This establishes a pressure or concentration gradient for both heat transfer and mass transfer.

Humidifiers: The Cooling Tower [1]. The largest present-day uses of diffusional heat transfer are found in the cooling tower, the air-conditioning spray chamber, the spray drier, the spray tower, and the spray pond. The use of cooling towers has grown tremendously in the past owing to an increasing necessity. In many industrial localities cold fresh water is too scarce to permit its unlimited use as a cooling medium. The problem of supplying sufficient surface and subsurface cooling water has grown to such an extent that new plants are often required to develop the continual reuse of the limited water they may obtain from public or private sources. In some communities, even river water, which may be present in abundance, requires precooling. This is especially true of some of the rivers in the southern part of the Unites States having river sources starting in the North and which are heated to the dry-bulb temperature by the time water in the river reaches the South.

The available temperature of cooling water has been shown to be an important economic factor in the design of modern chemical and power plants. In the chemical plant it fixes the operating pressure on the condensers of distillation and evaporation processes and consequently on the equipment preceding them. In the power plant it fixes the turbine- or engine-discharge pressure and the ultimate recovery of heat. For these vital reasons, the study of the cooling tower and the temperature of the water made available by it is of great importance in the planning of a process.

Modern cooling towers are classified according to the means by which air is supplied to the tower. All employ stacked horizontal rows of the aforementioned fill to provide increased contact surface between the air and water. In *mechanical-draft* towers the air is supplied on either of two ways. If the air is sucked into the tower by a fan at the top through louvers at the bottom, it is *induced draft*. If the air is forced in by a fan at the bottom and discharges through the top, it is *forced draft*. *Natural-circulation* towers are of two types, *atmospheric* and *natural draft* as in Figure 9.9 (a) and (b). Details on these two towers follow.

Mechanical-draft Towers. Towers of this class are the commonest erected in the United States at present, and of these the vast majority are now induced-draft towers. The trend toward the induced-draft tower has been pronounced in recent years, but it represents a logical transition, since there are advantages to its use which exceed all others except under very special conditions. In the forced-draft type, the air enters through a circular fan opening and a relatively large ineffective height and volume of tower must be provided as air-inlet space. The air distribution is relatively poor

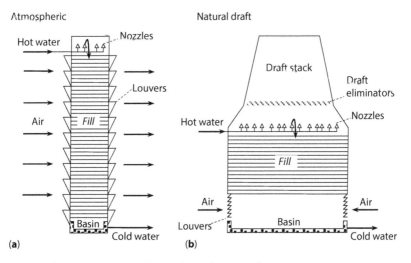

Figure 9.9 The common types of natural circulation cooling towers.

since the air must make a 90° turn while at a high velocity. In the induced-draft tower, on the other hand, the air can enter along one or more entire lengths of wall, and as a result, the height of tower required for air entry is very small.

In the forced-draft tower the air is discharged at a low velocity from a large opening at the top of the tower. Under these conditions the air possesses a low velocity head and tends to settle into the path of the fan intake stream. This means that the fresh intake air is contaminated by partially saturated air which has already passed through the tower before. When this occurs, it is known as *recirculation* and reduces the performance capacity of the cooling tower. In the induced-draft tower the air discharges through the fan at a high velocity so that it is driven up into the natural air currents which prevent it from settling at the air intake. In induced-draft towers, however, the pressure drop is on the intake side of the fan, which increase the total fan-power requirements. The higher velocity of discharge of the induced-draft towers also causes a somewhat greater entrainment or *drift* loss of water droplets carried from the system by the air stream.

Natural-circulation Towers. The atmospheric tower avails itself of atmospheric wind currents. Air blows through the louvered sides in one direction at a time, shifting with the time of year and other atmospheric conditions. In exposed places having average wind velocities of 5 or 6 mph the atmospheric tower may prove to be the most economical type, and where power costs are high, it may even be preferable with wind velocities as low as 2½ to 3 mph. Since the atmospheric currents must penetrate the entire width of the tower, the towers are made very narrow by comparison with other types and must be very long to afford equal capacity. Towers of this type have been built which are over 200 ft long. Drift losses occur over the entire side and are greater than for other types. These towers make less efficient use of available cooling potential, since they operate in crossflow whereas the most effective cooling occurs in countercurrent arrangement. When cooling water is desired at a temperature close to the wet bulb, this type is incapable of producing it. Atmospheric towers are consequently extremely large and have high initial costs, and when the air is becalmed, they may cease to operate. They have one great advantage, however, in that they may eliminate the principal operating cost of mechanical draft towers, i.e., the cost of fan power. In areas with low average wind velocities, the fixed charges and pumping costs offset the advantage.

Natural-draft towers operate in the same way as a furnace chimney. Air is heated in the tower by the hot water it contacts, so that its density is lowered. The difference between the density of air in the tower and outside it causes a natural flow of cold air in at the bottom and the rejection of less dense, warm air at the top. Natural-draft towers must be tall for sufficient buoyancy and must have large cross sections because of the low rate at which the air circulates compared with mechanical draft. Although, natural-draft towers consume more pumping power, they eliminate the cost of fan power and may be more reliable in some localities than are atmospheric towers. In atmospheric towers, emphasis must be placed on wind characteristics. In natural-draft towers, primary consideration must be given to the temperature characteristics of the air. If it is customary for the air to rise to a high temperature during the day, at least relative to the hot-water temperature, the natural-draft tower will cease to operate during the hot portion of the day. The initial cost and fixed charges on these towers are rather great, and they seem to be passing out of use.

Closely allied to natural-circulation cooling towers is the spray pond consisting of a number of up-spray nozzles, which spray water into the air without inducing air currents. These do not operate with an orderly attempt at air flow and consequently are not capable of producing water approaching the wet-bulb temperature as closely as cooling towers. Where the water must be cooled over a short range and without a close approach to the wet-bulb temperature, spray ponds may provide the economical solution to a water-cooling problem. Drift losses in spray ponds are relatively high. Spray towers are also used widely. They are similar to atmospheric towers except that they employ little or no fill.

Cooling towers are occasionally equipped with bare-tube bundles, which are inserted just above the water basin at the bottom of the tower. These are referred to as *bare-tube* or *atmospheric coolers*. The primary cooling water flows inside the cooler while tower water is continuously circulated over it. The primary cooling water is thus contained in a totally closed system.

In the study of cooling towers, the impression sometimes arises that the cooling tower cannot operate when the inlet air is at its wet-bulb temperature. This, of course, is not so. When air at the wet bulb temperature enters the tower, it receives sensible heat from the hot water, and its temperature is raised thereby so that it is no longer saturated. Water then evaporates continuously into the air as it travels upward in the tower.

One of the objectionable characteristics of cooling towers is known as *fogging*. When the hot and saturated exit air discharges into a cold atmosphere, condensation will occur. It may cause a dense fog to fall over a portion of the plant with attendant safety hazards [12]. If provision is made during the initial design, condensation can be reduced by any means which reduces the outlet temperature of the air. If it is desired to maintain a fixed range for the cooling water in coolers and condensers, fogging can be reduced by the recirculation of part of the basin water back to the top of the tower where it combines with hot water from the coolers and condensers. This reduces the temperature of the water to the tower, while the heat load is unchanged. The principal expense of the operation, aside from initial investment, will be that of the pumping power for the recirculation water which does not enter the coolers and condensers.

Caution is emphasized in the application of the information in the literature to use in actual services. Many of the data have been obtained on small packed, or filled heights, or on laboratory-scale apparatus of small cross section. A problem in the design of large towers is the attainment of a uniform distribution of both air and water over the *entire* cross section and height of the tower. A tower of large cross section should be easier to control than a small one because of the lower ratio of wall perimeter to cross section, which, in a sense, is a rough index of the fraction of the liquid flowing the tower walls.

The Calculation of Cooling-tower Performances [1]. Operators usually buy cooling towers rather than erect them themselves. This is undoubtedly the wisest policy, since it makes available to the operator a great deal of "know-how" in a field in which it is of great value. The operator will specify the quantity of water and the temperature range required for the process. The fabricator will propose a tower which will fulfill the conditions furnished by the operator for the 5 percent wet bulb (pertaining to relative humidity) in the locality of the plant and guarantee the fan power with which it will be accomplished. With the first cost, fan power, and the approximate height of the pumping head the operator can compute the cost of the cooling-tower water based on a period of depreciation of about twenty years.

Assume that a cooling tower has been erected and placed in operation on this basis. An acceptance run is made to determine whether or not the cooling tower is meeting the guarantee. This consists of a wet-bulb determination in the windward side of the tower and the determination of the air rate by a velometer or other pitot arrangement. The air leaving the cooling tower is always assumed to be saturated at its outlet temperature. Tests have shown it to be anywhere from 95 to 99 percent saturated.

The performance in a heat exchanger is satisfactory if the measured overall coefficient during initial operation at the process conditions equals or exceeds the stipulated clean coefficient. In cooling towers, the basis for design is one which is very rarely present. A cooling tower designed for a given approach to the 5 percent wet bulb temperature, and erected in the fall of the year, will probably not encounter the design conditions for another 8 to 9 months. On the basis of its performance in fall season, will it operate at the design conditions when they eventually return? A series of calculations answering questions of this nature often need to be undertaken.

Brown and Associates [27] have provided empirical correlations from the literature [28, 29] for estimating (roughly) sizes and capacities of conventional towers. Theodore and Ricci [7] also provide details of cooling tower calculations.

Example 9.25 – Dehumidifier Application. A dehumidification unit in a large factory draws in moist air at 500 lb/hr. Moist air enters the unit at a temperature of 82.5 °F and wet-bulb temperature of 75 °F. The exit stream of the unit is 60.8 °F and saturated with water. Determine the amount of water in (lb/hr) condensed in the dehumidifier.

SOLUTION: This problem is meant to serve as an example of a basic material balance. As with material balance problems, it is first recommended that a picture be drawn as follows. (See also Figure 9.10.)

In order to solve this problem, one must make use of the aforementioned *relative humidity* (Y_R). As noted earlier, the relative humidity is the ratio of the partial pressure of water in the gas phase to the vapor pressure of water at temperature T. Since the outlet stream is saturated with water, its relative humidity is unity. The relative humidity of the inlet stream, however, can be determined via the psychometric chart (see also Appendix, A.F.3). Locating the appropriate intersection of T and T_{WB} for the inlet conditions,

$$Y_R = 0.73 = 73\%$$

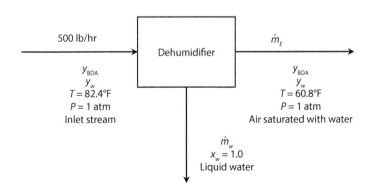

Figure 9.10 Dehumidifier in Example 9.25.

Note that,

$$Y_R = \frac{y_w P}{P_w^{'}}$$

At the inlet temperature, the vapor pressure of water is approximately 0.0373 atm (see steam tables in the Appendix). Therefore, at the inlet:

$$y_W = \frac{Y_R P_w^{'}}{P}$$

$$= \frac{(0.73)(0.0373)}{1}$$

$$= 0.027$$

Also note that $y_{BDA} = 1 - y_W$. At the exit conditions, the vapor pressure of water is approximately 0.0179 atm. Therefore, at the exit:

$$y_W = \frac{Y_R P_w^{'}}{P}$$

$$= \frac{(1)(0.0179)}{1}$$

$$= 0.0179$$

The total material balance equation for the system may not be written,

$$500 = \dot{m}_W + \dot{m}_E$$

Similarly, a componential balance for water may be written around the dehumidifier as,

$$(0.027)(500) = \dot{m}_W + (0.0179)\dot{m}_E$$

The details of solving the above two equations simultaneously is left as an exercise for the reader. The final result is,

$$\dot{m}_W = 4.6 \, \text{lb} / \text{hr}$$

9.8 Heat Pipes

This last section introduces the reader to heat pipes [30]. A heat pipe is a relatively new heat-transfer device that combines the principles of both thermal conductivity *and* phase transition to efficiently manage the transfer of heat between two interfaces. Refer to Figure 9.11. When heat is applied to the heat pipe, the liquid in the wick (chosen for compatibility with both the heat pipe material of construction and the temperatures of the hot and cold surfaces) heats and evaporates. As the evaporating fluid fills the heat pipe's hollow center, it diffuses axially along its length. Condensation of the downstream vapor occurs wherever the temperature is even slightly below that of the evaporation area. As it condenses, the vapor gives up the heat it acquired during evaporation and establishes a reverse vapor flow pattern in the pipe. The condensed liquid flows through capillary action within the wick, back to the heat source, completing the heat transfer cycle.

According to Farber [30] the heat pipe thermal cycle operates in the following manner. Refer to Figure 9.11.

1. Working fluid evaporates to vapor, absorbing thermal energy.
2. Vapor migrates along the cavity to the lower temperature end.
3. Vapor condenses back to the fluid releasing thermal energy in the cooler end and is absorbed by the wick.
4. The working fluid flows back to the higher temperature end.

Thus, a heat pipe (referred to by some as a superthermal conductor) consists of three components:

1. a containing enclosure
2. a wick (or equivalent) unit
3. the circulating fluid

The unit process itself consists of three operating sections:

1. the evaporator section
2. the adiabatic section
3. the condenser

The enthalpy of phase change is employed to achieve the desired heat transfer. The overall heat transfer process is accomplished by evaporating part of the circulating fluid in the evaporator section. A pressure difference driving force is thus created that moves the moving vapor to the condenser section where the enthalpy of condensation converts the vapor to liquid. The resultant flowing liquid is transferred into the wick (or equivalent) and back into the evaporator section. The net effect of the cyclic process is that the heat transferred in the evaporator section is rejected out in the

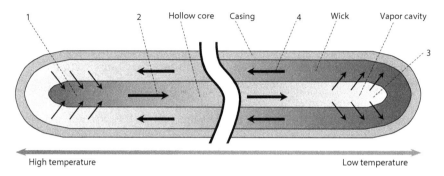

Figure 9.11 Heat pipe principles.

condenser section; in effect, the transfer of heat occurs *without* direct contact of the source and receiver.

Heat pipes are thus capable of transferring process heat to the surroundings. They may also be used to recover thermal energy from process gas streams. It may also be employed to heat up iced surfaces of ground by transferring the heat below the Earth's surface to cooler external surfaces. A host of other applications are predicted for the future, including a desalination unit developed by one of the authors [31, 32].

Calculational details regarding heat pipes are provided by Hagen [32]. Information on the pipe thermal resistances, the heat source, the cooling (sink) source, evaporator and condenser pipes, and for the wick on both the evaporator and condenser sections are provided.

Heat pipes have been employed within computer systems to transfer heat from CPU's to heat sinks so the heat can be rejected to the environment. Heat pipes are also used in HVAC systems to recover heat by transfer from an exit stream to a fresh air stream in the winter and to cool an entering air stream in the summer.

References

1. D. Kern, *Process Heat Transfer*, McGraw-Hill, New York City, NY, 1950.
2. I. Faraq and J. Reynolds, *Heat Transfer*, A Theodore Tutorial, Theodore Tutorials, East Williston, NY, originally published by the USEPA/APTI, RTP, NC, 1996.
3. L. Theodore, *Heat Transfer Applications for the Practicing Engineer*, John Wiley & Sons, Hoboken, NJ, 2011.
4. D. Green and R. Perry (editors), *Perry's Chemical Engineers' Handbook*, 8th edition, McGraw-Hill, New York City, NY, 2008.

5. J. Gibbs, mathematical physicist, publication/location/date unknown, 1839–1903.
6. L. Theodore, F. Ricci, and T. VanVliet, *Thermodynamics for the Practicing Engineer,* John Wiley & Sons, Hoboken, NJ, 2009.
7. L. Theodore and F. Ricci, *Mass Transfer Operations for the Practicing Engineer,* John Wiley & Sons, Hoboken, NJ, 2010.
8. L. Theodore and R. Allen, *Air Pollution Control Equipment*, A Theodore Tutorial, Theodore Tutorials, East Williston, NY, originally published by the USEPA/APTI, RTP, NC, 1993.
9. A. Colburn and O. Hougen, *Ind. Eng. Chem.,* 26(11), 1178–1182, New York City, NY, 1934.
10. W. Connery, *Condensers (Chapter 6), Air Pollution Control Equipment,* L. Theodore and A.J. Buonicore (editors), Prentice-Hall, Upper Saddle River, NJ, 1982.
11. L. Theodore, *Air Pollution Control Equipment Calculations,* John Wiley & Sons, Hoboken, NJ, 2008.
12. L. Theodore and R. Dupont, *Environmental Health and Hazard Risk Assessment: Principles and Calculations*, CRC Press/Taylor & Francis Group, Boca Raton, FL, 2012.
13. R. Dupont, K. Ganeson, and L. Theodore, *Pollution Prevention*, 2nd edition, CRC Press/Talyor & Francis Group, Boca Raton, FL, 2017.
14. L. Theodore, personal notes, East Williston, NY, 2018.
15. G. Theodore, personal communications to L. Theodore, Interboro Partners, Brooklyn, NY, 2018.
16. F. Staudiford, *Evaporation*, Chem. Eng., New York City, NY, December 9, 1963.
17. W. Badger and J. Banchero, *Introduction to Chemical Engineering,* McGraw-Hill, New York City, NY, 1955.
18. W. Lobo and J. Evans, *Trans. AIChE*, 35, 743, New York City, NY, 1939.
19. D. Wilson, W. Lobo, and H. Hottel, *Ind. Eng. Chem.*, 24, 486, New York City, NY, 1932.
20. G. Orrok, *Trans. ASME*, 47, 1148, New York City, NY, 1925.
21. W. Wohlenberg and H. Mullikin, *Trans. ASME*, 57, 531, New York City, NY, 1935.
22. J. Santoleri, J. Reynolds, and L. Theodore, *Introduction to Hazardous Waste Incineration,* 2nd edition, John Wiley & Sons, Hoboken, NJ, 2000.
23. K. Skipka and L. Theodore, *Energy Resources: Availability, Management, and Environmental Impacts*, CRC/Taylor & Francis Group, Boca Raton, FL, 2014.
24. L. Theodore and E. May, *Hazardous Waste Incineration,* A Theodore Tutorial, Theodore Tutorials, East Williston, NY, originally published by the USEPA/APTI, RTP, NC, 1994.
25. V. Ganapathy, "Size and Check Waste Heat Boilers Quickly," Hydrocarbon Processing, 169–170, Washington D.C., September 1984.

26. L. Theodore and J. Barden, *Mass Transfer Operations,* A Theodore Tutorial, Theodore Tutorials, East Williston, NY, originally published by the USEPA/APTI, RTP, NC, 1998.

27. G. Brown and Associates, *Unit Operations,* John Wiley & Sons, Hoboken, NJ, 1950.

28. Fluor Corp. LTD., *Bulletin T 337,* location unknown, 1939.

29. R.C. Kelly, paper published by the Fluor Corp., presented before the California Natural Gasoline Association, location unknown, Dec. 3, 1942.

30. P. Farber, personal communication to L. Theodore, Willowbrook, IL, 2016.

31. L. Theodore, personal notes, East Williston, NY, 2017.

32. K. Hagen, *Heat Transfer with Applications,* Prentice-Hall, Upper Saddle River, NJ, 1999.

Part III
PERIPHERAL TOPICS

10

Other Heat Transfer Considerations

Contributing author: Michael Pryor

Introduction

The development presented in earlier chapters on heat exchangers may be expanded to include heat transfer considerations and their effects.

This chapter contains six sections:

10.1   Insulation and Refractory
10.2   Refrigeration and Cryogenics
10.3   Instrumentation and Controls
10.4   Batch and Unsteady-State Processes
10.5   Operation, Maintenance and Inspection (OM & I)
10.6   Economics and Finance

The reader should note that recommendations related to purchasing heat exchangers can be found in the Operation, Maintenance, and Inspection (OM & I) section. Finally, the subject of accounting is treated superficially in the last section.

10.1 Insulation and Refractory

The development presented in earlier chapters may be expanded to include insulation plus refractory materials and their effects. Industrial thermal insulation usually consists of materials of low thermal conductivity combined to achieve a higher overall resistance to heat flow. Webster [1] defines insulation as: "to separate or cover with a non-conducting material in order to prevent the passage or leakage of ... heat ... etc." Insulation is defined in Perry's handbook [2] in the following manner: "Materials or combinations of material which have air- or gas-filled pockets or void spaces that retard the transfer of heat with reasonable effectiveness are thermal insulators. Such materials may be particulate and/or fibrous, with or without binders, or may be assembled, such as multiple heat-reflecting surfaces that incorporate air- or gas-filled void spaces." Refractory materials also serve the chemical process industries. In addition to withstanding heat, refractory materials also provide resistance to corrosion, erosion, abrasion, and/or deformation.

The bulk of the material presented in this section keys on insulation since it has found more application than refractories, particularly in its ability to reduce heat losses. The reader should note that there is significant overlap of common theory, equations, and applications with Part I, Chapter 2 which addresses with heat conduction.

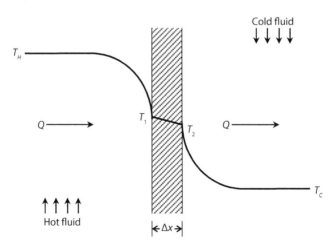

Figure 10.1 Flow past a flat plate.

Describing Equations [2–8]. When insulation is added to a surface, the heat transfer between the wall surface and the surroundings will take place by a two-step steady-state process (see Figure 10.1 for flow through a plane wall and Figure 10.2 for flow through an insulated plane wall): conduction from the wall surface at T_0 through the wall to T_1 and through the insulation from T_1 to T_2, and convection from the insulation surface at T_2 to the surrounding fluid at T_3. The temperature drop across each part of the heat flow path in Figure 10.2 is given below. The temperature drop across the wall and insulation is

$$T_0 - T_1 = QR_0 \tag{10.1}$$

and,

$$T_1 - T_2 = QR_1 \tag{10.2}$$

The temperature drop across the fluid film is,

$$T_2 - T_3 = QR_2 \tag{10.3}$$

where R_0 is the thermal resistance due to the conduction through the plane wall $(L_0 / k_0 A_0)$; R_1, the thermal resistance due to the conduction through the insulation $(L_1 / k_1 A_1)$; and R_2, the thermal resistance due to the convection through the fluid $(1 / h_2 A_2)$.

As with earlier analyses, the same heat rate, Q, flows through each thermal resistance under steady-state conditions. Therefore, Equation (10.1) through (10.3) can be combined to give,

$$T_0 - T_3 = QR \tag{10.4}$$

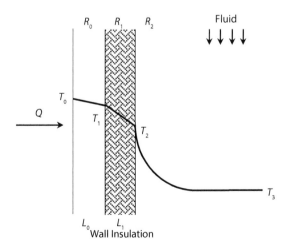

Figure 10.2 Flow past a flat insulated plate.

The heat transfer is then

$$Q = \frac{T_0 - T_3}{R} = \frac{\text{total thermal driving force}}{\text{total thermal resistance}} \tag{10.5}$$

Note that for the case of plane walls, the areas A_0, A_1, and A_2 are the same (see Figure 10.2). Dividing Equation (10.2) by Equation (10.3) yields

$$\frac{T_1 - T_2}{T_2 - T_3} = \frac{R_1}{R_2} = \frac{h_2 L_1}{k_1} = \frac{\text{conduction insulation resistance}}{\text{convection resistance}} \tag{10.6}$$

The group $h_2 L_1 / k_1$ is a dimensionless number termed earlier as the *Biot number*, Bi (refer to Table 1.1 in Part I, Chapter 1),

$$\text{Bi} = \frac{(\text{fluid convection coefficient})(\text{characteristic length})}{\text{thermal conductivity of insulation surface}} = \frac{hL}{k} \tag{10.7}$$

Insulation [5–6]. *Fiber, powder,* and *flake-type* insulation consist of finely dispersed solids throughout an air space. The ratio of the air space to the insulator volume is called the *porosity* or *void fraction*, denoted as ε. In *cellular* insulation, a material with a rigid matrix contains entrapped air pockets. An example of such rigid insulation is *foamed insulation* which is made from *plastic* and *glass* material. Another type of insulation consists of multi-layered thin sheets of *foil* of high reflectivity. The spacing between the foil sheets is intended to restrict the motion of air. This type of insulation is referred to as *reflective* insulation.

Most thermal insulation systems consist of the insulation *and* a so-called "finish." The finish provides protection against water or other liquid entry, mechanical damage, and ultraviolet degradation; it can also provide fire protection. The finish usually consists of any form of coating (e.g., polymeric paint material, etc.), a membrane (e.g., felt, plastic laminate, foil, etc.), or a sheet material (e.g., fabric, plastic, etc.). Naturally, the finish must be able to withstand any potential temperature excursion(s) in its immediate vicinity.

It would normally seem that the thicker the insulation, the less the heat loss, i.e., increasing the insulation should increase the resistance and reduce the heat loss to the surroundings. But this is not always the case with pipes. There is a *critical insulation thickness* below which the system will experience a greater heat loss due to an increase in insulation. This situation arises for "small" diameter pipes when the increase in area increases more rapidly than the resistance opposed by the thicker insulation.

Applying earlier theory to a pipe/cylinder system shown in Figure 10.3 leads to [5, 6]

$$
\begin{aligned}
Q &= \cfrac{T_i - T_o}{\cfrac{1}{2\pi r_i L}\left(\cfrac{1}{h_i}\right) + \cfrac{\Delta x_w}{k_w 2\pi L r_{LM,w}} + \cfrac{\Delta x_i}{k_i 2\pi L r_{LM,i}} + \cfrac{1}{2\pi r L}\left(\cfrac{1}{h_o}\right)} \\[2mm]
&= \cfrac{2\pi L\left(T_i - T_o\right)}{\cfrac{1}{r_i h_i} + \cfrac{\ln\left(r_o / r_i\right)}{k_w} + \cfrac{\ln\left(r / r_o\right)}{k_i} + \cfrac{1}{r h_o}} \\[2mm]
&= \cfrac{2\pi L\left(T_i - T_o\right)}{f(r)}
\end{aligned}
$$

(10.8)

Assuming that Q goes through a maximum or minimum as r is varied, L'Hôpital's rule can be applied to Equation (10.8):

$$
\frac{dQ}{dr} = 2\pi L\left(T_i - T_o\right)\left[\frac{-\left[df(r)/dr\right]}{\left(f(r)\right)^2}\right] = 0
$$

(10.9)

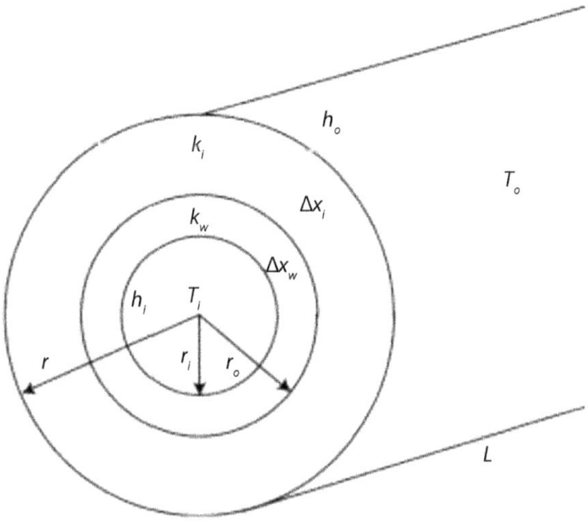

Figure 10.3 Critical insulation thickness for a pipe.

with,

$$-\frac{df(r)}{dr} = -\frac{1}{rk_i} + \frac{1}{r^2 h_o}$$

For $dQ/dr = 0$, one may therefore write,

$$\frac{dQ}{dr} = -\frac{df(r)}{dr} = -\frac{1}{rk_i} + \frac{1}{r^2 h_o} = 0 \qquad (10.10)$$

For this maximum/minimum condition, set $r = r_c$ and solve for r_c.

$$r_c = \frac{k_i}{h_o} \qquad (10.11)$$

The second derivative of dQ_r/dr of Equation (10.10) provides information as to whether Q experiences a maximum or minimum at r_c.

$$\frac{d}{dr}\left(\frac{dQ_r}{dr}\right) = \frac{1}{r^2 k_i} - \frac{2}{r^3 h_o} = \frac{h_o^2}{k^3} - \frac{2h_o^2}{k^3} = \frac{h_o^2}{k^3}(1-2) \qquad (10.12)$$

Clearly, the second derivative is a *negative* number; Q is therefore a *maximum* at $r = r_c$. Q then decreases as r is increased beyond r_c. However, one should exercise care in interpreting the implications of the above development. This result applies *only* if r_o is *less* than r_c, i.e., it generally applies to "small" diameter pipes/tubes. Thus, r_c represents the outer radius (not the thickness) of the insulation that will maximize the heat loss and at which point any further increase in insulation thickness will result in a decrease in heat loss.

A graphical plot of the resistance R versus r is provided in Figure 10.4. (The curve is inverted for the plot of Q or r.) In other words, the maximum heat loss from a pipe occurs when the critical radius equals the ratio of the thermal conductivity of the insulation to the surface coefficient of heat transfer. This ratio has the dimension of length (e.g., ft).

To reduce Q below that for a bare wall ($r = r_o$), r must be greater than r_c, i.e., $r > r_c$. The radius at which this occurs is denoted as r'. The term r' may be obtained by solving the equation,

$$Q_{bare} = Q_{r'} \qquad (10.13)$$

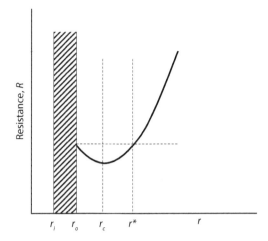

Figure 10.4 Resistance associated with the critical insulation thickness for a bare surface.

so that,

$$\frac{2\pi L(T_i - T_o)}{\dfrac{1}{r_i h_i} + \dfrac{\ln(r_o/r_i)}{k_w} + \dfrac{1}{r_o h_o}} = \frac{2\pi L(T_i - T_o)}{\dfrac{1}{r_i h_i} + \dfrac{\ln(r_o/r_i)}{k_w} + \dfrac{\ln(r'/r_o)}{k_i} + \dfrac{1}{r' h_o}} \qquad (10.14)$$

One may now solve for r' using a suitable trial-and-error procedure.

Note that the above development applies when r is less than r_c. If r_o is larger than r_c, the above analysis again applies, but only to the results presented for $r > r_c$, i.e., Q will decrease indefinitely as r increases. There is *no* maximum/minimum (inflection) for this case since values of $\ln(r_o/r_i)$ and indeterminate for $r < r_0$. Once again, Q approaches zero in the limit as r approaches infinity.

Properties of some common insulation materials is provided in Table 10.1. Perry [7] provides typical insulation thickness for various pipe sizes in Table 10.2.

Refractory. Webster [1] defines refractory as: "resistant to heat, hard to melt or work... not yielding to treatment ... a heat resistant material used in lining furnaces, etc." Refractory materials must obviously be chemically and physically stable at high temperatures. Depending on the operating environment, they must also be resistant to thermal shock, chemically inert, and resistant to wear. Refractories normally require special warm-up periods to reduce the possibility of thermal shock and/or drying stresses.

Table 10.1 Properties of Non-metals.

Substance	T	k	ρ	c	α
	°C	W / m · °C	kg / m³	kJ / kg · °C	m² / s × 10⁷
Asphalt	20–55	0.74–0.76			
Brick					
Building brick, common face	20	0.69	1600	0.84	9.2
		1.32	2000		
Carborundum brick	600	18.5			
	1400	11.1			
Chrome brick	200	2.32	3000	0.84	9.2
	550	2.47			9.8
	900	1.99			7.9
Diatomaceous earth, molded and fired	200	0.24			
	870	0.31			
Fire clay, burnt 2426 °F	500	1.04	2000	0.96	5.4
	800	1.07			
	1100	1.09			
Burnt 2642 °F	500	1.28	2300	0.96	5.8
	800	1.37			
	1100	1.40			
Fireclay brick	200	1.0	2645	0.96	3.9
	655	1.5			
	1205	1.8			
Missouri	200	1.00	2600	0.96	4.0
	600	1.47			
	1400	1.77			
Magnesite	200	3.81		1.13	
	650	2.77			
	1200	1.90			

(*Continued*)

Table 10.1 Cont.

Substance	T	k	ρ	c	α
	°C	W / m · °C	kg / m³	kJ / kg · °C	m² / s × 10⁷
Cement					
Portland		0.29	1500		
Mortar	23	1.16			
Concrete					
Cinder	23	0.76			
Stone, 1-2-4 mix	20	1.37	1900–2300	0.88	8.2–6.8
Glass					
Window	20	0.78 (avg)	2700	0.84	3.4
Borosilicate	30–75	1.09	2200		
Plaster					
Gypsum	20	0.48	1440	0.84	4.0
Metal lath	20	0.47			
Wood lath	20	0.28			
Stone					
Granite		1.73–3.98	2640	0.82	8–18
Limestone	100–300	1.26–1.33	2500	0.90	5.6–5.9
Marble		2.07–2.94	2500–2700	0.80	10.0–13.6
Sandstone	40	1.83	2160–2300	0.71	11.2–11.9
Wood (across the grain)					
Balsa, 8.8 lb/ft³	30	0.055	140		
Cypress	30	0.097	460		
Fir	23	0.11	420	2.72	0.96
Maple or oak	30	0.166	540	2.4	1.28
Yellow pine	23	0.147	640	2.8	0.82
White pine	30	0.112	430		
Asbestos					
Loosely packed	–45	0.149			
	0	0.154	470–570	0.816	3.3–4
	100	0.161			
	20	0.74			

(Continued)

Table 10.1 Cont.

Substance	T	k	ρ	c	α
	°C	W / m · °C	kg / m³	kJ / kg · °C	m² / s ×10⁷
Asbestos–Cement boards	20	0.74			
Sheets	51	0.166			
Felt					
40 laminations per inch	38	0.057			
	150	0.069			
	260	0.083			
20 laminations per inch	38	0.078			
	150	0.095			
	260	0.112			
Corrugated					
4 piles per inch	38	0.087			
	93	0.100			
	150	0.119			
Asbestos cement	–	2.08			
Balsam wood, 2.2 lb/ft³	32	0.04	35		
Cardboard					
Corrugated	–	0.064			
Celotex	32	0.048			
Corkboard, 10 lb/ft³	30	0.043	160		
Cork					
Regranulated	32	0.045	45–120	1.88	2–5.3
Ground	32	0.043	150		

(Continued)

Table 10.1 Cont.

Substance	T	k	ρ	c	α
	°C	W / m · °C	kg / m³	kJ / kg · °C	m² / s × 10⁷
Diatomaceous earth (Sil-o-cel)	0	0.061	320		
Felt					
Hair	30	0.036	130–200		
Wool	30	0.052	330		
Fiber, insulating board	20	0.048	240		
Glass wool, 1.5 lb/ft³	23	0.038	24	0.7	22.6
Insulex, dry	32	0.064			
		0.144			
Kapok	30	0.035			
Magnesia, 85%	38	0.067	270		
	93	0.071			
	150	0.074			
	204	0.080			
Rock wool, 10 lb/ft³	32	0.040	160		
Loosely packed	150	0.067	64		
	260	0.087			
Sawdust	23	0.059			
Silica aerogel	32	0.024	140		
Wood shavings	23	0.059			

Table 10.2 Thickness of piping insulation.

Nominal iron-pipe size		Outer diameter		Insulation, nominal thickness													
				1		1½		2		2½		3		3½		4	
				25		38		51		64		76		89		102	
in.	mm	in.	mm	Approximate wall thickness													
				in.	mm	in.	mm	in.	mm	in.	mm	in.	mm	in.	mm	in.	mm
½		0.84	21	1.01	26	1.57	40	2.07	53	2.88	73	3.38	86	3.88	99	4.38	111
¾		1.05	27	0.90	23	1.46	37	1.96	50	2.78	71	3.28	83	3.78	96	4.28	109
1		1.32	33	1.08	27	1.58	40	2.12	54	2.64	67	3.14	80	3.64	92	4.14	105
1¼		1.66	42	0.91	23	1.66	42	1.94	49	2.47	63	2.97	75	3.47	88	3.97	101
1½		1.90	48	1.04	26	1.54	39	2.35	60	2.85	72	3.35	85	3.85	98	4.42	112
2		2.38	60	1.04	26	1.58	40	2.10	53	2.60	66	3.10	79	3.60	91	4.17	106
2½		2.88	73	1.04	26	1.86	47	2.36	60	2.86	73	3.36	85	3.92	100	4.42	112
3		3.50	89	1.02	26	1.54	39	2.04	52	2.54	65	3.04	77	3.61	92	4.11	104
3½		4.00	102	1.30	33	1.80	46	2.30	58	2.80	71	3.36	85	3.86	98	4.36	111
4		4.50	114	1.04	26	1.54	39	2.04	52	2.54	65	3.11	79	3.61	92	4.11	104
4½		5.00	127	1.30	33	1.80	46	2.30	58	2.86	73	3.36	85	3.86	98	4.48	114

(Continued)

Table 10.2 Cont.

Nominal iron-pipe size in.	mm	Outer diameter in.	mm	Insulation, nominal thickness 1 (25) in.	mm	1½ (38) in.	mm	2 (51) in.	mm	2½ (64) in.	mm	3 (76) in.	mm	3½ (89) in.	mm	4 (102) in.	mm
				Approximate wall thickness													
5		5.56	141	0.99	25	1.49	38	1.99	51	2.56	65	3.06	78	3.56	90	4.18	106
6		6.62	168	0.96	24	1.46	37	2.02	51	2.52	64	3.02	77	3.65	93	4.15	105
7		7.62	194			1.52	39	2.02	51	2.52	64	3.15	80	3.65	93	4.15	105
8		8.62	219			1.52	39	2.02	51	2.65	67	3.15	80	3.65	93	4.15	105
9		9.62	244			1.52	39	2.15	55	2.65	67	3.15	80	3.65	93	4.15	105
10		10.75	273			1.58	40	2.08	53	2.58	66	3.08	78	3.58	91	4.08	104
11		11.75	298			1.58	40	2.08	53	2.58	66	3.08	78	3.58	91	4.08	104
12		12.75	324			1.58	40	2.08	53	2.58	66	3.08	78	3.58	91	4.08	104
14		14.00	356			1.46	37	1.96	50	2.46	62	2.96	75	3.46	88	3.96	101
Over 14, up to and including 36						1.46	37	1.96	50	2.46	62	2.96	75	3.46	88	3.96	101

The oxides of aluminum (alumina), silicon (silica), and magnesium (magnesia) are the most common materials used in the manufacture of refractories. Another oxide usually found in refractories is the oxide of calcium (lime). Fireclays are also widely used in the manufacture of refractories.

Refractories are selected based primarily on operating conditions. Some applications require special refractory materials. Zirconia is used when the material must withstand extremely high temperatures. Silicon carbide and carbon are two other refractory materials used in some very severe temperature conditions, but they cannot be used in contact with oxygen, as they will oxidize and burn.

There is no single design and selection procedure for refractories. Three general rules can be followed:

1. Design for compressive loading.
2. Allow for thermal expansion.
3. Take advantage of the full range of materials, forms, and shapes.

These apply to whether the design is essentially brickwork and masonry construction or whether the refractory is one that might have been made of specialty metal.

Example 10.1 – Heat Flux Calculation. One wall of an oven has a 3-inch insulation cover. The temperature on the inside of the wall is at 400 °F; the temperature on the outside is at 25 °C. What is the heat flux (heat flow rate per unit area) across the wall if the insulation is made of glass wool ($k = 0.022$ Btu/hr · ft · °F)?

SOLUTION: Once again, the thermal resistance associated with conduction is defined as,

$$R = \frac{L}{kA}$$

where,

R = thermal resistance
k = thermal conductivity
A = area across which heat is conducted
L = length across which heat is conducted

The rate of heat transfer, Q, is then

$$Q = \frac{\Delta T}{R}$$

Thus,

$$Q = \frac{kA\Delta T}{L}$$

Since 25 °C is approximately 77 °F and L is 0.25 ft,

$$\frac{Q}{A} = \frac{k\Delta T}{L}$$
$$= \frac{(0.022)(400-77)}{0.25}$$
$$= 28.4 \, \text{Btu/hr} \cdot \text{ft}^2$$

Example 10.2 – Required Insulation Thickness. A cold-storage room has a plane rectangular wall 8 m wide (w) and 3 m high (H). The temperature of the outside surface of the wall T_1 is −18°C. The surrounding air temperature T_3 is 26 °C. The convective heat transfer coefficient between the air and the surface is 21 W/m² · K. A layer of cork board insulation (thermal conductivity, k = 0.0433 W/m · K) is to be attached to the outside wall to reduce the cooling load by 80%.

1. Calculate the rate of heat flow through the rectangular wall without insulation. Express the answer in tons of refrigeration (1 ton of refrigeration = 12,000 Btu/hr). Which direction is the heat flowing?
2. Determine the required thickness of the insulation board.

SOLUTION:

1. Calculate the heat transfer area, A:

$$A = wH = (8)(3) = 24 \, \text{m}^2$$

Calculate the rate of heat flow in the absence of insulation. Heat is transferred by convection from the wall surface to the surroundings. Apply Newton's law of cooling,

$$Q = hA(T_1 - T_3)$$
$$= (21)(24)(-18-26)$$
$$= -22,176 \, \text{W} = -22.18 \, \text{kW}$$

$$= \left(-22{,}176 \ \text{W}\right)\left(\frac{3.4123 \ \text{Btu/hr}}{\text{W}}\right) = -75{,}671 \ \text{Btu/hr}$$

$$= -75{,}671 \ \text{Btu/hr}\left(\frac{1\text{ton of refrigeration}}{12{,}000 \ \text{Btu/hr}}\right)$$

$$= -6.3 \ \text{ton of refrigeration}$$

The negative sign indicates that heat flows from the surrounding air into the cold room.

2. Calculate the heat rate with insulation. Since the insulation is to reduce Q by 80%, then,

$$Q = \left(0.2\right)\left(-22{,}176\right) = -4435.2 \ \text{W}$$

Calculate the total thermal resistance:

$$R_t = \frac{T_1 - T_3}{Q} = \frac{-18 - 26}{-4435.2} = 0.00992°\text{C/W}$$

$$R_t = R_1 + R_2$$

Calculate the convection thermal resistance, R_2:

$$R_2 = \frac{1}{hA} = \frac{1}{\left(21\right)\left(24\right)} = 0.00198°\text{C/W}$$

Also calculate the insulation conduction resistance, R_1:

$$R_1 = R_t - R_2 = 0.00992 - 0.00198 = 0.00794°\text{C/W}$$

The required insulation thickness, L_1, is given by

$$R_1 = \frac{L_1}{k_1 A}$$

Substituting,

$$L_1 = R_1 k_1 A$$
$$L_1 = \left(0.00794\right)\left(0.0433\right)\left(24\right)$$
$$= 0.00825 \ \text{m} = 8.25 \ \text{mm}$$

Example 10.3 – Factors Affecting Insulation Diameter. It has come to the attention of a young engineer that there is a bare pipe that is releasing a significant amount of heat into the atmosphere. List factors that should be considered in selecting the optimum insulation diameter for the pipe.

SOLUTION: Two important factors include durability and maintainability. If it is determined that cost is the number one factor, and a cheaper insulation is chosen, it would be wise to investigate these two factors for the insulation. The end result might be that the cheaper insulation does not have a long life and might have to be maintained much more often than a more expensive one. This could cause the more expensive one to be more cost effective than the cheaper insulation.

Temperature differences will also play a role in determining the insulation diameter. It may be imperative to keep the temperature of the material in the pipe just above freezing, or it may be that the temperature needs to be 60 °F above freezing. These different situations call for different insulation diameters. Another factor that falls under the temperature category is the location of the pipe. It would be extremely different if a pipe is insulated in New York, Alaska, or Tahiti. All of these places have different climates and it is imperative that these be investigated in order to know how large or small the diameter of the insulation needs to be.

Finally, the other factors that need to be looked at, and may be as important as the first, are whether the materials that constitute the insulation are harmful. First and foremost, there is asbestos, and due to environmental considerations, insulation with no or very little asbestos should be used. Another insulation material is fiberglass. A good number of insulators are made with fiberglass. When the insulation is cut, fiberglass escapes into the air and workers should not breathe this harmful material.

Example 10.4 – Outer Critical Radius Calculation. Calculate the outer critical radius of insulation on a 2.0 inch OD pipe. Assume the air flow coefficient to be 1.32 Btu/hr · ft². °F and the loosely-packed insulation's thermal conductivity to be 0.44 Btu/hr·ft· °F. Comment on the effect of insulation on the heat rate lost from the pipe.

SOLUTION: Employ Equation (10.11):

$$r_c = \frac{k_i}{h_o} \tag{10.1}$$

$$= \frac{0.44}{1.32} = 0.333 \, \text{ft} = 4.0 \, \text{in}$$

The critical insulation thickness is therefore:

$$L_1 = \Delta x_1 = 4.0 - \frac{2.0}{2} = 3.0 \, \text{in}$$

Since $r_o = 1.0$, $r_o < r_c$ so that the heat loss will increase as insulation is added, but start to decrease when the radius of the insulation increases above r_c, i.e., when $r > r_c$.

Example 10.5 – Incinerator Insulationx [4, 5]. An incinerator is 30 ft long, has a 12 ft ID and is constructed of ¾ inch carbon steel. The inside of the steel shell is protected by 10 in. of firebrick ($k = 0.608$ Btu/hr·ft· °F) and 5 in. of Sil-o-cel insulation ($k = 0.035$ Btu/hr·ft· °F) cover the outside. The ambient air temperature is 85 °F and the average inside temperature is 1800 °F. The present heat loss through the furnace wall is 6% of the heat generated by the combustion of a fuel. Calculate the thickness of Sil-o-cel insulation that must be added to cut the losses to 3%.

SOLUTION: A diagram of this system is presented in Figure 10.5. Although the resistance of the steel can be neglected, the other two need to be considered. For cylindrical systems, the effect of the radius of curvature must once again be included in the resistance equations. These take the form presented below:

$$R_{\text{firebrick}} = \frac{\ln\left(r_{f,o} / r_{f,i}\right)}{2\pi L k_f}$$

$$= \frac{\ln\left(6.000 / 5.167\right)}{2\pi\left(30\right)\left(0.608\right)}$$

$$= 1.304 \times 10^{-3} \, \text{hr} \cdot {}^\circ\text{F/Btu}$$

$$R_{\text{sil-o-cel}} = \frac{\ln\left(r_{s,o} / r_{s,i}\right)}{2\pi L k_s}$$

$$= \frac{\ln\left(6.479 / 6.063\right)}{2\pi\left(30\right)\left(0.035\right)}$$

$$= 10.059 \times 10^{-3} \, \text{hr} \cdot {}^\circ\text{F/Btu}$$

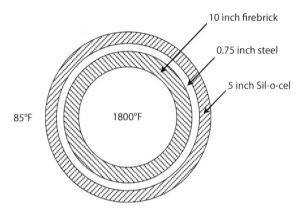

Figure 10.5 Diagram for Example 10.5.

Thus,

$$R = \sum R = R_{firebrick} + R_{sil-o-cel}$$
$$= 11.363 \times 10^{-3} \, hr \cdot °F/Btu$$

To cut the heat loss in half, R must be double. The additional Sil-o-cel resistance is therefore $11.363 \times 10^{-3} \, hr \cdot °F/Btu$. The new outside radius, r_o, is calculated from:

$$R_{added\ sil-o-cel} = 11.363 \times 10^{-3} = \frac{\ln(r_o / 6.479)}{2\pi(30)(0.035)}$$

$$r_o = 6.983 \, ft$$

The extra thickness is $6.983 - 6.479 = 0.504 \, ft = 6.05 \, in$

10.2   Refrigeration and Cryogenics

Refrigeration and cryogenics have aroused considerable interest among those in engineering and science [8]. All refrigeration processes involve work, specifically, the extraction of heat from a body of low temperature and the rejection of this heat to a body willing to accept it. Refrigeration generally refers to operations in the temperature range of 120 to 273K, while cryogenics usually deals with temperatures below 120K where gases, include methane oxygen, argon, nitrogen, hydrogen, and helium, can be liquefied.

In addition to being employed for domestic purposes (when a small "portable" refrigerator is required), refrigeration and cryogenic units have been used for the storage of materials such as antibiotics, other medical supplies, specialty foods, etc. Much larger cooling capacities than this are required in air conditioning equipment. Some of these units, both small and large, are especially useful in applications that require the accurate control of temperature. Most temperature-controlled enclosures are provided with a unit that can maintain a space below ambient temperature (or at precisely ambient temperature) as required. The implementation of such devices led to the recognition that cooling units would be well suited to the refrigeration of electronic components and to applications in the field of instrumentation. Such applications usually require small compact refrigerators, with a relatively low amount of cooling power, where economy of operation is often unimportant. In addition, air conditioning represents an extension of refrigeration.

One of the main cost considerations when dealing with refrigeration and cryogenics is the cost of building and powering the equipment. This is a costly element in the process, so it is important to efficiently transfer heat so that money is not wasted in lost heat in the refrigeration and cryogenic processes. Since the cost of equipment can be expensive, there are a number of factors to be considered when choosing equipment. Equipment details are discussed in later sections.

Cryogenics plays a major role in the chemical processing industry. Its importance lies in the recovery of valuable feedstocks from natural gas streams, upgrading the heat content of fuel gas, purifying many process and waste streams, producing ethylene, as well as other chemical processes.

Cryogenic air separation provides gases (nitrogen, oxygen, and argon) used in:

1. the manufacturing of metals such as steel,
2. chemical processing and manufacturing industries,
3. electronic industries,
4. enhanced oil recovery, and
5. partial oxidation and coal gasification processes.

Other cryogenic gases, including hydrogen and carbon monoxide, are used in the chemical and metal industries while helium is used in welding, medicine, and gas chromatography.

Cryogenic liquids have their own applications. Liquid nitrogen is commonly used to freeze food, while cryogenic cooling techniques are used to reclaim rubber tires and scrap metal from old cars. Cryogenic freezing and storage is essential in the preservation of biological materials that include blood, bone marrow, skin, tumor cells, tissue cultures, and animal semen. Magnetic resonance imaging (MRI) also employs cryogenics to cool the highly conductive magnets that are used for the types of nonintrusive body diagnostics [9].

As will become apparent throughout this section, there is a wide variety of applications, uses, and methods to produce and to utilize the systems of refrigeration and cryogenics, Multiple factors must be considered when dealing with these practices including the choice of refrigerant or cryogen, the choice of equipment and methods of insulation, and all hazards and risks must be accounted for to ensure the safest environment possible.

Insulation must certainly be considered as an integral part of any refrigeration or cryogenic unit. The extent of the problem of keeping heat out of a storage vessel containing a liquid refrigerant or a cryogenic liquid varies widely. Generally, one must decide on the permissible and / or allowable heat losses (leaks) since insulation costs money, and an economic analysis must be performed. Thus, the main purpose of insulation is to minimize radiative and convective heat transfer and to use as little material as possible in providing the optimal insulation. When choosing appropriate insulation, the following factors are normally taken into consideration:

1. ruggedness
2. convenience
3. volume and weight
4. ease of handling
5. thermal effectiveness
6. cost

The thermal conductivity (k) of a material is a major consideration in determining the thermal effectiveness of the insulation material. Different types of insulation obviously have different k values and there are five categories of insulation. These include:

1. Vacuum insulation which employs an evacuated space that reduces radiant heat transfer.
2. Multilayer insulation, referred to by some as superinsulation, which consists of alternating layers of highly reflective material and low conductivity insulation in a high vacuum.

3. Powder insulation which utilizes finely divided particulate material packed between surfaces.
4. Foam insulation which employs non-homogeneous foam whose thermal conductivity depends on the amount of insulation.
5. Special insulation which includes composite insulation that incorporates many of the advantageous qualities of the other types of insulation.

It should also be noted that multilayer insulation has revolutionized the design of cryogenic refrigerant vessels.

In double-walled vessel, typical of cryogen storage to be discussed in the next section, heat is usually transferred to the inner vessel by three methods:

1. conduction through the vessel's "jacket" by gases present in this space
2. conduction along solid materials through both the inner and outer containers
3. radiation from the outer vessel

It was discovered in 1898, that the optimum material to place inside the space created by the double-walled vessel is "nothing" (i.e., a vacuum). This Dewar vessel, named after Sir James Dewar, is still one of the most widely used insulation techniques for cryogenic purposes.

Refrigeration. The development of refrigeration systems was rapid and continuous at the turn of the 20th century, leading to a history of steady growth. The purpose of refrigeration, in a general sense, is to make materials colder by extracting heat. As described in earlier chapters, heat moves in the direction of decreasing temperature (i.e., it is transferred from a region of high temperature to one of a lower temperature). When the opposite process needs to occur, it cannot do so by itself, and a refrigeration system (or its equivalent) is required [10].

Refrigeration, in a commercial setting, usually refers to food preservation and air conditioning. When food is kept at colder temperatures, the growth of bacteria and the accompanying spoiling of food is either reduced or prevented. People learned early on that certain foods had to be keep cold to maintain freshness and humans kept these foods in ice boxes where melting ice usually absorbed the heat from the foods. Household refrigerators became popular in the early 1900s and only the wealthy could afford them at the time. Freezers did not become a staple part of the refrigerator until after World War II when frozen food became popular.

The equipment necessary in refrigeration is dependent upon many factors, including the substances and fluids working in the system. One very important part of refrigeration is the choice of refrigerant being employed and the refrigerant choice obviously depends on the system in which it will be used. The following criteria are usually considered in refrigerant selection:

1. practical evaporation and condensation pressures
2. high critical and low freezing temperatures
3. low liquid and vapor densities
4. low liquid heat capacities
5. high latent heat of evaporation (enthalpy)
6. high vapor heat capacities

Ideally, a refrigerant should also have a low viscosity and a high coefficient of performance (to be defined later in this chapter). Practically, a refrigerant should have:

1. a low cost
2. chemical and physical inertness at operating conditions
3. no corrosiveness towards materials of construction
4. low explosion hazard
5. be non-poisonous and non-irritating

Solid refrigerants are not impossible to use but liquid refrigerants are most often used in practice. These liquid refrigerants include hydrocarbon and non-hydrocarbon refrigerants. The most commonly used hydrocarbon refrigerants include:

1. propane
2. ethane
3. propylene
4. ethylene

Non-hydrocarbon liquid refrigerants include:

1. nitrogen
2. oxygen
3. neon

4. hydrogen
5. helium

A basic refrigeration cycle is shown in Figure 10.6. The cycle begins when a refrigerant enters the compressor as a low pressure gas [1]. Once compressed, it leaves as a hot, high pressure gas. Upon entering the condenser [2], the gas condenses to a liquid and releases heat to the outside environment, which may be air or water. The cool liquid then enters the expansion valve at a high pressure [3]; the flow is restricted and the pressure lowered. In the evaporator [4], heat from the source to be cooled is absorbed and the liquid becomes a gas. The refrigerant then repeats the process (i.e., the cycle continues).

In the refrigerator, the working fluid enters the evaporator in a wet condition and leaves dry and saturated (or slightly superheated). The heat absorbed, Q_C, by the evaporator can therefore be estimated by multiplying the change in the fluid's entropy [10], Δs, as it passes through the evaporator by the fluid's saturation temperature, T_S, at the evaporator pressure since the fluid's temperature will be constant while it is in a wet condition at constant pressure. Thus,

$$Q_C = T_S \Delta s \tag{10.15}$$

where,

Q_C = heat absorbed by the evaporator (e.g., kJ/kg)
T_S = fluid saturation temperature at evaporator temperature (e.g., K)
Δs = fluid entropy change (e.g., kJ/kg · K)

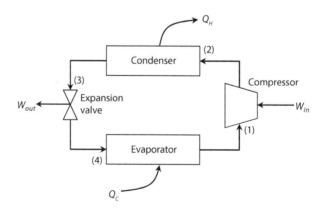

Figure 10.6 Basic components of a refrigeration system.

The performance and energy saving ability of a refrigerator is measured in terms of the system's Coefficient of Performance (COP). This is defined as the heat removed at a low temperature, i.e., the cooling effect, Q_C, divided by the work input, W_{in}, into the system:

$$COP = \frac{Q_C}{W_{in}} \tag{10.16}$$

where,

COP = coefficient of performance, dimensionless
Q_C = cooling effect (e.g., kJ/kg)
W_{in} = work input (e.g., kJ/kg)

The traditional system of units used in refrigeration are English units (i.e., Btu, etc.) but SI units are also acceptable and are employed in two later examples. The cooling effect, Q_C, is equal to the change in enthalpy of the working fluid as it passes through the evaporator, and the work input (W_{in}) is equal to the increase in the working fluid's enthalpy as it passes through the compressor.

The performance of a steam powered plant process can be measured in a manner somewhat analogous to the COP for the refrigeration system. The thermal efficiency, η_{th}, of a work-producing cycle is defined as the ratio of work produced to the heat added. Thus,

$$\eta_{th} = \frac{W_{net}}{Q_{in}} \tag{10.17}$$

where,

η_{th} = thermal efficiency, dimensionless
W_{net} = net work produced by the cycle (e.g., J/kg)
Q_{in} = heat added to the cycle (e.g., J/kg)

This can be rewritten as:

$$\eta_{th} = \frac{W_{out} - W_{in}}{Q_{in}} \tag{10.18}$$

where,

W_{out} = work produced by the cycle (e.g., J/kg)
W_{in} = work consumed by the cycle (e.g., J/kg)

For this type of cycle, the compressor, evaporator, and expansion valve in Figure 10.6 are replaced by a turbine, boiler, and pump, respectively, with both Q_C and Q_H as well as W_{in} and W_{out} reversed (see also Figure 10.7). When no velocity information is provided, velocity effects can be neglected and this equation can be expressed in terms of enthalpies [10] at points in the entry and exit to the boiler, turbine, and pump, which for a simple power cycle is:

$$\eta_{th} = \frac{\left(h_2 - h_3\right) - \left(h_1 - h_4\right)}{\left(h_2 - h_1\right)} \qquad (10.19)$$

where,

h_1 = enthalpy on entry to the boiler (e.g., J/kg)
h_2 = enthalpy on exit from the boiler, or entry to the turbine (e.g., J/kg)
h_3 = enthalpy on exit from the turbine (e.g., J/kg)
h_4 = enthalpy on entry to the pump (e.g., J/kg)

Note that the change in enthalpy across the pump is often neglected since it is close to zero relative to the other enthalpy changes. In effect, $h_1 - h_4 \approx 0$.

Air conditioning is the process of treating air so as to simultaneously control its temperature, humidity, cleanliness, and distribution to meet the requirements of a conditioned space. Applications of air conditioning include the promotion of human comfort and the maintenance of proper conditions for the manufacture, processing, and preserving of material and equipment. Also, in industrial environments where for economical or other reasons, conditions cannot be made entirely comfortable, air conditioning

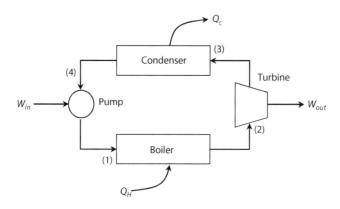

Figure 10.7 Basic components of a steam powered system.

may be used for maintaining the health of workers at safe tolerance limits. Industrial buildings have to be designed according to their intended use. For instance, the manufacture of hygroscopic materials (paper, textiles, foods, etc.) will require relatively tight controls of relative humidity. On the other hand, the storage of furs will demand relatively low temperatures, while the ambient temperature in a facility that manufactures refractories might be acceptable [7].

Basically, an air-conditioning system consists of a fan unit which forces a mixture of fresh outdoor air and room air through a series of devices which increases or decreases its temperature, and increases or decreases its water-vapor content or humidity. In general, air conditioning equipment can be classified into two broad types: central (sometimes called field erected) and unitary (or packaged) [7].

As noted above, one of the main financial considerations when dealing with refrigeration and cryogenics is the cost of building and powering the unit. This is a costly element in the process, so it is important to efficiently transfer heat in the refrigeration and cryogenic processes. Since the cost of equipment can be expensive, various factors should be considered when choosing the applicable equipment.

Cryogenics. Cryogenics is not, in itself, s stand-alone field. It is the extension of many other fields of science that delve into the realm of the thermo-dynamic variable of temperature. When compared to room temperature, the properties of most substances change dramatically at extremely low temperatures. From a molecular perspective, the atoms in any substance at a lower temperature, while still vibrating, are compressed closer and closer together. Depending on the phase of the substance, various phenomena and changes to physical and chemical characteristics occur at these lower temperatures.

There are many accepted definitions of cryogenics. Some classify it simply as a "temperature range below −240 °F." Another more elaborate explanation defines it as: "the unusual and unexpected property variation appearing at low temperatures and which makes extrapolation from ambient to low temperature reliable" [7]. Webster's dictionary defines cryogenics as "the branch of physics that deals with the production and effects of very low temperatures" [11]. Cryogenics has also been referred to as: "all phenomena, process, techniques, or apparatus occurring or using temperatures below 120K" [7]. Combining all of these definitions, one of the authors of this book has provided the all-purpose definition: "cryogenics is the study of the production and effects of materials at low temperatures" [4].

There have been uses for cryogenic technologies as far back as the latter part of the 19th century. It became common knowledge in the 1840s that in order to store food at low temperatures for long periods of time, it needed to be frozen, a technology that is still utilized today.

At the beginning of the 20th century, the scientist Carl von Linde produced a double distillation column process that separated air into pure streams of its basic components of 78% nitrogen, 21% oxygen, and 1% argon. By 1912, it was discovered that minor modifications to the double distillation column process could separate many other gases from the input stream. Since there are trace amounts of neon, krypton, and xenon in air, the aforementioned distillation process, with minor modifications, was found capable of separating these gases into relatively pure streams. In the 1930s, the development of the sieve tray brought changed in cryogenic technology. (A sieve tray is a plate, utilized in distillation columns, with perforated holes about 5 – 6 mm in diameter, which enhances mass transfer [12].) These trays were highly popular in cryogenics due to their simplicity, versatility, capacity, and cost effectiveness.

When the "Space Race" hit the United States in the 1960s, cryogenic technologies were utilized to develop a process known as cryopumping, which is based on the freezing of gases on a cold surface. This process helped produce an ultrahigh vacuum here on Earth that would be similar to what was to be experienced in outer space. This led to many other discoveries, including rocket propulsion technologies which enabled astronauts to better prepare for their voyage(s) into space.

The *cryopreservation* process is based on the same principles as food storage (i.e., using extremely low temperatures to preserve a perishable item). Cryopreservation has become increasingly popular because of its appeal in preserving living cells. Whole cells and tissue can be preserved by this technique by stopping biological activity at extremely low temperatures. This preservation of organs by cryogenics has been a stepping stone for *cryosurgery* which relies on the cold temperatures to insure clean and precise incisions. More recently, *cryobiology* has been applied well below freezing temperatures to living organisms to observe how they "react." Most recently, incorporating electronic systems with cryo-technologies has provided valuable information on superconductivity. Extremely low temperatures have also led to systems that contain near-zero resistance throughout wires.

Liquefaction. Liquefaction is the process of converting a gaseous substance into a liquid. Depending on the liquefied material, various steps are

employed in an industrial process. Common to each is the use of the Joule
– Thomson effect [10, 14] (where the temperature changes as a fluid flows
through a valve), heat exchangers, and refrigerants to achieve the cryogenic temperatures. Generally, the methods of refrigeration and liquefaction include:

1. vaporization of a liquid
2. application of the Joule–Thomson effect (a throttling
   process)
3. expansion of a gas in a work producing engine

Liquid nitrogen is the best refrigerant for hydrogen and neon liquefaction
systems while liquid hydrogen is usually used for helium liquefaction.

The largest and most commonly used liquefaction process involves the
separation of air into nitrogen and oxygen. The process starts by taking air
compressed initially to 1500 psia through a four-stage compressor with
intercoolers [15]. The air is then compressed again to 2000 psia and cooled
down to about the freezing point of water in a precooler. The high-pressure
air is then further cooled by ammonia to about –70 °F. The air is then split
into two streams after this cooling stage. One steam leads to heat exchangers that cool most of the air by recycled cold gaseous nitrogen. This proceeds to an expansion valve which condenses most of the air, absorbing
heat in the process.

Hazards associated with the chemical properties of cryogenic fluids can
give rise to fires and explosions. In order for a fire or explosion to occur,
there must be a fuel and/or an oxidant, *and* an ignition source. Because
oxygen and air are prime candidates for cryogenic fluids, and are present in high concentrations, the chances of disasters occurring dramatically
increase, as oxygen will obviously act as the oxidizer. A source of fuel can
range from a noncompatible material to a flammable gas, or even a compatible material under extreme heat. An ignition source could be any electrical or mechanical spark or flame, any undesired thermodynamic event,
or even a chemical reaction [16–18].

Past experience has shown that cryogenic fluids can be used safely in
industrial environments as well as in typical laboratories provided all facilities are properly designed and maintained, and personnel handling these
fluids are adequately trained and supervised. There are many hazards associated with cryogenic fluids. However, the principal ones are those associated with the response of the human body and the surroundings to the
fluids and their vapors, and those associated with reactions between the

fluids and their surroundings. Potential hazards also exist in highly compressed gases because of their stored energy. In cryogenic systems, such high pressures are obtained by gas compression during liquefaction or refrigeration, by pumping of liquids to high pressure followed by evaporation, and by confinement to cryogenic liquids with subsequent evaporation. If this confined gas is suddenly released through a rupture or break in a line, a significant thrust may be experienced.

There are a few sources of personal hazard in the field of cryogenics. If the human body were to come in contact with a cryogenic fluid or a surface cooled by a cryogenic fluid, severe "cold burns" could result. Cold burns inflict damage similar to a regular burn, causing stinging sensations and accompanying pain. But, with cryogenics, the skin and/or tissue is essentially frozen, significantly damaging or destroying it. As with any typical burn or injury, the extent and "brutality" of a cold burn depends on the area and time of contact; medical assistance is strongly advised when one receives a burn of this type.

Protective insulated clothing should be worn during work with low temperature atmospheres to prevent "frost bite" when dealing with cryogenic liquids. Safety goggles (or in some cases, face shields), gloves and boots are integral parts of these uniforms. The objective of these precautions is to prevent any direct contact of the skin with the cryogenic fluid itself or with surfaces in contact with the cryogenic liquid. All areas in which cryogenic liquids are either stored or used should be clean and organized in a manner to prevent any avoidable accidents or fires and explosions that could result; this is especially true when working with systems using oxygen. Additional details are provided in Chapter 12.

Example 10.6 – Calculation of Heat Absorbed by an Evaporator.
A refrigerator's evaporator pressure is 0.2 MPa and has a working fluid that enters the unit as liquid with an enthalpy, h_{in}, of 230 kJ/kg. If the refrigerator's working fluid (see Table 10.3 for thermodynamics property data) exits the evaporator as dry and saturated vapor, calculate the heat absorbed by the evaporator, Q_C. See also Figure 10.6.

Table 10.3 Fluid data for Example 10.6.

P, MPa	T_{sat}, °C	h_f, kJ/kg	h_g, kJ/kg	h at 20 °C Superheated, kJ/kg
0.2	−10	190	390	410
0.6	20	230	410	430

SOLUTION: Determine the enthalpy of the fluid that exits from the evaporator, h_{out}. From the problem statement and data,

$$h_{out} = h_g \text{ at } 0.2\,\text{MPa}$$
$$= 390\,\text{kJ/kg}$$

Calculate the heat absorbed by the evaporator, Q_C, using the change in enthalpy across the evaporator.

$$Q_C = h_{out} - h_{in}; \ h_{in} = h_l$$
$$= 390 - 230$$
$$= 160\,\text{kJ/kg}$$

Example 10.7 – Heat Absorbed by a Refrigerator. The working fluid in a refrigerator enters a compressor at dry-saturated conditions at a pressure of 0.2 MPa and exits the compressor at 20°C superheated at a pressure of 0.6 MPa. Given the fluid data from Example 10.6, and given that the fluid leaves the condenser wet-saturated, what is the work input to the refrigerator?

SOLUTION: Employ the subscripts associated with Figure 10.6. Determine the fluid enthalpy on entering the compressor, h_1. From the problem statement and data,

$$h_1 = 390\,\text{kJ/kg}$$

Determine the fluid enthalpy on leaving the compressor, h_2. From the problem statement and data

$$h_2 = 430\,\text{kJ/kg (superheated)}$$

Finally, determine the fluid enthalpy on leaving the condenser, h_3. From the problem statement and data, and noting that the enthalpy changes across the expansion value is approximately zero [2, 10, 15],

$$h_4 = h_3 = 230\,\text{kJ/kg}$$

Calculate the heat rejected from the condenser, Q_H, using the change in enthalpy across the condenser:

$$Q_H = h_2 - h_3$$
$$= 430 - 230$$
$$= 200\,\text{kJ/kg}$$

Calculate the work input, W_{in}, using the change in enthalpy across the compressor:

$$W_{in} = h_2 - h_1$$
$$= 430 - 390$$
$$= 40 \, kJ/kg$$

Finally, one could also calculate the heat absorbed by the evaporator, Q_C, as in the previous example using the first law of thermodynamics [6]:

$$Q_C = Q_H - W_{in}$$
$$= 200 - 40$$
$$= 160 \, kJ/kg$$

Example 10.8 – Calculating a Refrigerator's COP Refer to Example 10.7. What is the refrigerator's COP?

SOLUTION: Determine the COP using equation (10.16) and values of Q_C and W_{in} from the previous example:

$$COP = \frac{Q_C}{W_{in}} \qquad (10.16)$$

$$COP = \frac{160}{40}$$
$$= 4.0$$

Comment: The COP for a refrigerator is defined in terms of the cooling load, Q_C; however, the COP for a heat pump is defined in terms of the heating load, Q_H.

10.3   Instrumentation and Controls

The operation of a process is dependent upon the control of the process variables [4–6]. These are defined as conditions in the process materials or apparatus which are subject to change. Because there may be several materials and several pertinent operating factors which may change in the simplest of processes, the maintenance of control over an entire process is an important aspect of process design. Many of the advances in processing technology during recent years have been due in part to the widespread use of automatic-control mechanisms. Naturally a comprehensive study of so broad a field of endeavor is beyond the scope of this text, and it is the intention here to introduce in a practical way, only the most elementary principles of process control.

Process Variables [7]. When a flow sheet is laid out for a process, the temperatures, pressures, and fluid-flow quantities are theoretically fixed in accordance with heat, pressure, and material balances. The translation of the flow sheet into an operable plant requires that special provision be made to assure the *relative* constancy of the various quantities and qualities. It is impossible to achieve absolute constancy in even the simplest of industrial operations and these do not include the multitude of complex operations which are normally encountered. Consider the simple case of a storage tank to which one pump continuously supplies liquid and an identical pump continuously removes liquid. Because of differences in suction and discharge, both pumps, acting independently, pump at different rates, and the liquid level in the storage tank cannot be expected to remain constant.

Similar factors influence nearly every other steady-state condition. Considering the utilities alone can include high and low-pressure steam, cooling water, electricity, compressed air, and fuel supply. When any single unit process in a plant is either shut down or started up, it may affect the supply of the utilities to the other unit processes. Furthermore, when the furnaces in a powerhouse are periodically re-fired, the temperature, pressure, and quantity of steam throughout the plant may show some variation. Similarly, a sudden change in the steam demand at some point in the plant for steaming out large vessels may cause a sufficient speed variation to affect the performance of turbine-driven pumps, compressors, and generators – including their discharge rates and pressures. Or, the temperature of cooling-tower water, varying with atmospheric conditions, might affect the total heat transfer at critical points in the process. Add to these the variations resulting from changes in the composition of the feed materials such as boiling points, heat capacities, or viscosities, and additional fluctuations may be anticipated in the pressure, temperature, and fluid flow of the process streams.

When using automatic controllers, the flow may be intentionally divided so that the bulk of the fluid flows through a continual by-pass and only a small amount through the controlling line. This reduces the cost of the controller by reducing its size. However, the use of a continual by-pass as an integral part of the flow-control system introduces several unfavorable factors in the maintenance of uniform control. Except where the flow lines are extremely large and when the control valve is expensive, it is undesirable to use a continual by-pass. Like temperature control, pressure and level control are often intimately associated with the control flow.

Instrumentation Control Symbols [19]. Most of the common arrangements for process equipment requires several interrelated instruments to assure reasonable control. To simplify the representation, the instrumentation flow-plan symbols employed below are those suggested in a survey questionnaire by a symbols subcommittee of the Instrument Society of America. The legend for a number of the common symbols is given in Figure 10.8. The first letters which appear as instrument designation are

$$F = \text{flow}$$
$$L = \text{level}$$
$$P = \text{pressure}$$
$$T = \text{temperature}$$

The second or third letters are

$$C = \text{controller}$$
$$G = \text{glass (as LG)}$$
$$I = \text{indicator}$$

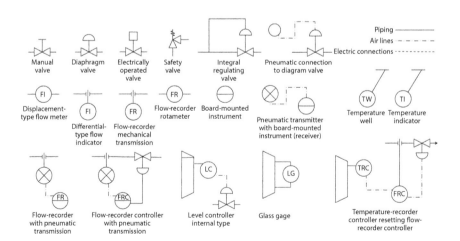

Figure 10.8 Typical instrument symbols. Instrument lines are usually lighter than piping or equipment.

R = recorder

S = safety

V = valve

W = well

It is often desirable to assemble all the instruments at a single point away from the locality at which the measurement and control actually occur. These are referred to as *board*-mounted instruments as differentiated from locally mounted instruments, and are designated by a horizontal line through the instrument. Often the instruments on a single control board are at some distance from the point of measurement. In such cases, the response is amplified by a pneumatic transmitter. This is particularly requisite for the control of low-pressure fluids at a centralized board.

In a typical instrumentation diagram, the air lines to the instruments and the by-passes around the controllers are usually omitted. Except where necessary for manual control, the indicating or recording features have also been omitted, although they are essential to the instrumentation of the process as a whole.

Whenever a control valve is installed directly in the inlet or outlet line of an exchanger, cooler, or heater, the fluid passing through the equipment must still retain sufficient available pressure so that, when operating, the control valve is capable of increasing or decreasing the flow of the line. Thus, an exchanger which at design flow utilizes nearly all of the available static pressure on a line has little pressure available for overcoming the effect of the control valve. An instrument is useless which operates under such conditions that the control valve is wide open, or unable to open further to have an impact on control.

Automatic controls are usually employed to measure, suppress, correct, and modify changes of the four principal types of process variation:

1. temperature
2. pressure
3. flow
4. level

There are, in addition, other potential controllable variables such as density, thermal conductivity, velocity, and composition.

While the traditional controller is satisfactory in many process applications, it does not perform well for processes with slow dynamics, time delays, frequent disturbances, or multivariable interactions. Several advanced control methods discussed below can be implemented via computer control, namely, feedback control, feedforward control, cascade control, time-delay compensation, selective and override control, adaptive control, fuzzy logic, and statistical process control. Several of these methods are discussed below.

Feedback Control [20, 21]. Feedback control is a very important aspect of process control. Its role is best described in terms of an example (see also Figure 10.9). Assume that one desires to maintain the temperature of a polymer reactor at 70 °C. Temperature is thus the controlled variable, and the desired temperature level 70 °C, is called the *set point*. In feedback control, the temperature is measured using a sensor (such as a thermocouple device). This information is then continuously relayed to a controller, and a device known as a *comparator* is used to compare the set point with the measured signal (or variable). The difference between the set point and the measured variable is defined as the *error*. With respect to the magnitude of error, the controller element in the feedback loop takes corrective action by adjusting the value of a process parameter, known as the *manipulated variable*.

The controller logic (how it handles the error) is an important process control criterion. Generally, feedback controllers are either proportional *P* (send signals to the final control element proportional to the error), proportional-integral *PI* (send signals to the final control element that is proportional to the magnitude of the error at any instant and to the sum of the error), and proportional-integral-derivative *PID* (send signals that are also based on the slope of the error).

In the example above, the manipulated variable may be cooling water flow through the reactor jacket. This adjustment or manipulation of the

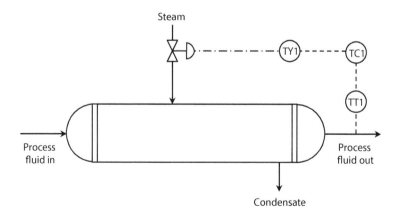

Figure 10.9 Controller action: feedback system.

flow rate is achieved by a *final control element*. In most chemical processes, the final control element is usually a pneumatic control valve. However, depending on the process parameter being controlled, the final control element could very well be a motor whose speed is regulated. Thus, the signal from the controller is sent to a final control valve that manipulates the manipulated variable in the process.

In addition to the controlled variable, other variables may disturb or affect the process. In the reactor example above, a change in the inlet temperature of the feed (or inlet) flow rate are considered to be *load* disturbances. A *servo* problem is one in which the response of the system to a change in set point is recorded, whereas a load or regulator problem is one in which the response of a system to a disturbance or load variable is measured.

Before selecting a controller, it is very important to determine its action. In the heat exchanger shown in Figure 10.9, steam is used to heat the process fluid. If the inlet temperature of the process fluid increases, this will result in an increase in the outlet temperature. Since the outlet temperature moves above the set point (or desired temperature), the controller must close the steam valve (the control valve is air-to-open or fail-closed). This is achieved by the controller sending a lower output (pneumatic or current) signal to the control valve, i.e., an increase in the input signal from the controller to the valve. The action of the controller is considered to be *reverse*. If the input signal to the controller and the output signal from it act in the same direction, the controller is *direct acting*. To determine the action of the controller, it is important to consider the process requirements for control, and the action of the final control element. The controller action is usually set by a switch on electronic and pneumatic controllers. On microprocessor-based controllers, the setting can be made by changing the sign of the scale factor in the software (which then changes the sign of the proportional gain of the controller).

The function of a feedback controller in a process control loop are twofold: (1) to compare the process signal from the transmitter (the controlled or measured variable) with the set point, and (2) to send a signal to the final control element with the sole purpose of maintaining the controlled variable at its set point. The most common feedback controllers are the aforementioned proportional controllers (P control), proportional-integral controllers (PI control), and proportional-integral-derivative (PID) controllers. Theodore provides extensive details on these functions [21].

The following rules of thumb may be used in the selection of controllers:

1. In process control applications, where offset is not a major problem, proportional controllers can be used. Also, if the

process itself exhibits an integrating action, then a simple proportional controller will suffice. In general, P-only controllers are used in liquid-level and gas pressure (surge tanks) control applications.

2. PI controllers are used in applications in which offset has to be completely eliminated. In applications where the response of the process is very fast (small time constants), a PI controller can be used. A typical application is flow control, where the sluggish response of the integral action does not hamper the overall performance of the feedback control loop.

3. Processes with large time constants usually require the addition of the derivative mode of action to the control system. Typical examples are multicapacity processes, or temperature and composition control. The addition of the derivative mode makes the control action more robust by stabilizing its response.

The selection of a controller is based on qualitative, rather than qualitative considerations. It is important to note that with the development of digital or PC-based control, the choice of a P, PI, or PID controller is based on the reasons presented above and usually not on economic considerations.

The selection and tuning processes are based on the choice of a performance criterion which may be simple (based on steady-state performance) of the controlled variable, or more complex (based on the dynamic performance criteria) of the controlled variable. Simple performance criteria may include control objectives such as keeping the maximum deviation from the set point as small as possible, or returning to the set point as soon as possible in the event of a disturbance. Simple performance criteria are therefore based on some feature of the feedback control behavior such as decay ratio, overshoot, or rise time. In general, steady state performance criteria call for zero error at steady-state. It should be clear that the error or offset can be zero if integral action is selected. It is not possible to design a controller response that is based on multiple criterion as they may conflict with one another. The most popular steady-state criterion is the decay ratio, and a one-quarter (1/4) decay ratio provides a good compromise between a rapid rise time and a fairly good settling time.

Feedforward Control [20, 21]. Feedforward control has several advantages. Unlike feedback control, a feedforward control measures the disturbance directly, and takes preemptive action before the disturbance can affect the process. A conventional feedback controller measures the controlled variable and takes corrective action *after* the disturbance has been felt by the system. Good feedforward control relies to a large extent on sound knowledge of the process, which is the major drawback. Finally, the stability of a feedforward/feedback system is determined by the roots of the characteristic equation of the feedback loop (feedforward control does not affect the stability of the system). Vasudevan provides additional developmental material and illustrative examples [20].

Alarms are used in process plants to inform operators that plant conditions are outside of the normal operating range. *Trips* are devices which sense operation outside of normal range and automatically shut off or turn on some device. Alarms are used when the time constants involved are large enough for an operator to make adjustments to the process and return it to normal conditions. Trips are used when process time constants are very short and immediate action may be needed to prevent a disaster. As might be expected, complex plants may have several different priority levels of alarms and trips.

An everyday example of an alarm would be a high-temperature warning light on the dash board of a car. If it goes on, it indicates that there is a problem and the driver should do something about it. If a trip were installed here, it would shut off the motor when a certain high temperature point was reached. Household gas water heaters have trips on the fuel supply linked to a temperature element which sense whether the flame has gone out.

Adaptive and Fuzzy Logic Control. Process control problems inevitably require online tuning of the controller constants to achieve a satisfactory degree of control. If the process operating conditions or the environment changes significantly, the controller may have to be retuned. If these changes occur quite frequently, then adaptive control techniques should be considered. An adaptive control system is one in which the controller parameters are adjusted automatically to compensate for changing process conditions.

The subject of adaptive control is one of current interest. New algorithms continue to be developed, but these need to be field-tested before industrial acceptance can be expected. An adaptive controller is inherently nonlinear and therefore more complicated than the conventional PID controller.

The application of fuzzy logic to process control requires the concept of fuzzy rules and fuzzy inference. A fuzzy rule, also known as a fuzzy IF-THEN statement, has the form:

$$\text{If } x \text{ then } y$$

if input1 = high

and input2 = low

then output = medium

Three functions are required to perform logical inferencing with fuzzy rules. The fuzzy AND is the product of a rules input membership values, generating a weight for the rule's output. The fuzzy OR is a normalized sum of the weights assigned to each rule that contributes to a particular decision. The third function used is defuzzification, which generates a crisp final output. In one approach, the crisp output is the weighted average of the peak element values.

With a single feedback control architecture, information that is readily available to the algorithm includes the error signal, difference between the process variable and the set-point variable, change in error from pervious cycles to the correct cycle, changes to the set point variable, change of the manipulated variable from cycle to cycle, and changes in the process variable from past to present. In addition, multiple combinations of the system response data are available. As long as the irregularity lies in the dimension wherein fuzzy decisions are being based or associated, the result should be enhanced performance. This enhanced performance should be demonstrated in both the transient and steady-state response. If the system tends to have changing dynamic characteristics or exhibits nonlinearities, fuzzy logic control should offer a better alternate to using a constant PID setting. Most fuzzy logic software begins building its information base during the autotune function. In fact, the majority of the information used in the early stages of system start-up comes from the autotune solutions.

In addition to single-loop process controllers, products that have benefited from the implementation of fuzzy logic are camcorders, elevators, antilock braking systems, and televisions with automatic color, brightness, and sound control. Sometimes fuzzy logic controllers are combined with pattern recognition software such as artificial neural networks.

Expert Systems [2]. An expert system is a computer program that uses an expert's knowledge in a particular domain to solve a narrowly focused, complex problem. An offline system uses information entered manually and produces results in visual form to guide the user in solving the problem at hand An online system uses information taken directly from process measurements to perform tasks automatically or instruct or alert operating personnel at the plant.

Each expert system has a rule base created by the expert to respond as the expert would to sets of input information. Expert systems used for plant diagnostics and management usually have an open rule base, which can be changed and augmented as more experience accumulates and more tasks are automated. The system begins as an empty shell with an assortment of functions such as equation solving, logic, and simulation, as well as input and display tools to allow an expert to construct a proprietary rule base. The "expert" in this case would be the person or persons having the deepest knowledge about the process, its problems, its symptoms, and remedies. Converting these inputs to meaningful outputs is the principal task in constructing a rule base. First-principles models (deep knowledge) produce the most accurate results, although heuristics are always required to establish limits. Often modeling tools such as artificial neural nets are used to develop relationships among the process variables.

A number of process control vendors offer comprehensive, object-oriented software environments for building and deploying expert systems. Advantages of such software include transforming complex real-time data to useful information through knowledge-based reasoning and analysis, monitoring for potential problems before they adversely impact operations, diagnosing root causes of time-critical problems to speed up resolution, and recommending or taking corrective actions to help ensure successful recovery.

10.4   Batch and Unsteady-State Processes

The relationships of the preceding chapters in Part II have primarily applied to the steady-state in which the heat flow and source-temperature were constant with time. *Unsteady-state* processes are those in which the heat flow, the temperature, or both vary with time at a fixed point. Batch heat-transfer processes are typical unsteady-state processes in which discontinuous heat changes occur with specific quantities of material as when heating a given quantity of liquid in a tank or when a cold furnace is started up, or hot furnace shut down. Still other common problems involve the rates at which heat is conducted through a material while the temperature of the heat source varies. The daily periodic variations of the heat of the sun on various objects or the quenching of steel in an oil bath are examples of the latter. Other equipment, based on the characteristics of the unsteady state, are the regenerative furnaces used in the steel industry, the batch reactor, drying equipment, and equipment in processes employing fixed and moving-bed catalysts.

In batch processes, the time requirement for heat transfer to liquids, can usually be modified by increasing the circulation of the batch fluid, the heat transfer medium, or both. The reasons for using a batch rather than a continuous heat-transfer operation are dictated by numerous factors. Some of the common reasons are (1) the liquid being processed is not continuously available, (2) the heating or cooling medium is not continuously available, (3) the requirements of reaction time or treating time necessitates holdup, (4) the economics of intermittently processing a large batch justifies the accumulation of a small continuous stream, (5) cleaning or regeneration is a significant part of the total operating period, and (6) the simplified operation of most batch processes is advantageous.

In order to treat the most common applications of batch and unsteady-state heat transfer systematically, it is preferable to divide processes between liquid (fluid) heating or cooling and solid heating or cooling. The commonest examples, as listed by Kern [19], are outlined below.

1. Heating and cooling liquids
   a. Liquid batches
   b. Batch distillation
   c. Batch reactors
2. Heating and cooling solids
   a. Constant medium temperature
   b. Periodically varying temperature
   c. Regenerators
   d. Granular material in beds
   e. Granular material in fluidized beds

The presentation to follow will primarily key on the heating and cooling of liquid batches. Batch distillation, heating and cooling solids, and drying operations are also briefly reviewed.

Heating and Cooling Liquid Batches. It is not always possible to distinguish between the presence or absence of agitation in a liquid batch, although the two premises lead to different requirements for the accomplishment of a batch temperature change in a given period of time. When a mechanical agitator is installed in a tank or vessel, there is no need to question that the tank fluid is agitated. When there is no mechanical agitator, but the liquid is continuously recirculated, the conclusion that the batch is agitated is one of discretion. Where the heating element is an external exchanger, it is safer to assume agitation. However, heat of moving effects [10] can generally be safely neglected.

In the derivation of batch equations given below by Kern [19], the temperature, T, refers to the hot batch liquid *or* heating medium while t refers to the cold batch liquid or cooling medium. The following cases are treated by Kern [19].

1. *Heating and cooling agitated batches, countercurrent*
   a. Coil-in-tank or jacketed vessel, isothermal medium
   b. Coil-in-tank or jacketed vessel, nonisothermal medium
   c. External exchanger, isothermal medium
   d. External exchanger, nonisothermal medium
   e. External exchanger, liquid continuously added to tank, isothermal medium
   f. External exchanger, liquid continuously added to tank, nonisothermal medium
2. *Heating and cooling agitated batches, parallel flow-counter-current*
   a. External 1–2 exchanger
   b. External 1–2 exchanger, liquid continuously added to tank
   c. External 2–4 exchanger
   d. External 2–4 exchanger, liquid continuously added to tank
3. *Heating and cooling batches without agitation*
   a. External counterflow exchanger, isothermal medium
   b. External counterflow exchanger, nonisothermal medium
   c. External 1–2 exchanger
   d. External 2–4 exchanger

Kern [19] provides describing equations for the above 14 cases. Perry [2] also provides solutions for many similar cases. For example, Kern [19] provides the following describing equation for a coil-in-tank or jacketed vessel, isothermal heating medium for a vessel containing M lb of liquid with heat capacity c and initial temperature t_1 heated by a condensing medium of the temperature T_1. The batch temperature t_2 at any time θ is given by

$$\ln\left(\frac{T_1 - t_1}{T_1 - t_2}\right) = \frac{UA\theta}{Mc} \tag{10.20}$$

Batch Distillation [19]. In a batch distillation, a still pot is charged with a batch of liquid, and heat is supplied by a coil or a natural or forced circulation reboiler. In some high-temperature installations, the still pot may be directly fired. Batch distillation is usually employed when there is an insufficiency of charge stock to warrant continuous operation and the assemblies are often relatively small. In batch distillation [12] the composition and temperature of the residual liquid in the still constantly vary, and usually the same applies to the condensate except when the still has been charged with or forms a constant-boiling mixture. During this process, it is possible to obtain an initial overhead fraction which is purer than that obtainable with the same reflux by continuous distillation. This is particularly true when the overhead product is sold in different grades with a premium for purity. It is also possible, by constantly varying the reflux ratio, to obtain a nearly-uniform overhead composition although its quantity constantly decreases. The latter option is usually too costly to be general.

The composition change during the batch distillation of a binary mixture is given by the Rayleigh equation [12].

$$\ln\left(\frac{L_1}{L_2}\right) = \int_{x_2}^{x_1} \frac{1}{y-x}\,dx \tag{10.21}$$

where,

L_1 = moles liquid charge to still
L_2 = moles residue after distillation
x_1 = mole fraction of volatile component in charge liquid
x_2 = mole fraction of volatile component in residue
y = mole fraction of vapor in equilibrium with x

The temperature, T, of the charge must be obtained from a boiling-point curve if the mixture is not ideal and does not follow Raoul's [10] and Henry's [10] laws.

The Rayleigh equation does not contain any term with a unit of time. The time allotted to the distillation is therefore independent of any feed quantity. If a batch represents the intermittent accumulation of several hours' flow of charge stock, the rate of distillation must be such that the still pot will be vacated and ready to receive the next charge. If the distillation occurs infrequently, the rate of distillation can be determined economically from the optimum relationship between fixed and operating charges.

A common practice for obtaining the heat load for the reboiler and condenser without resorting to a distillation curve is to take the total heat load and divide it by the time allotted to the distillation. This will give a fictitious average hourly heat load, Q, which is greater than that at the cut-off but usually less than the initial heat load. The value of Q so obtained is combined with the value of U and ΔT occurring at both the start and the cutoff temperatures, and the larger calculated surface plus some allowance for the error is used. If the heating element is also used to preheat the charge, it is entirely likely that the preheat rate may be limiting. The time for preheating can be obtained from the batch heating equation in the preceding subsection.

Heating and Cooling Solids [19]. Since the appearance of Fourier's [22] early work the conduction of heat in solids has attracted the interest and attention of many leading mathematicians and physicists. It is possible to present only some of the simplest and most representative cases here and to suggest the overall nature of the study.

In the treatment of unsteady-state conduction the simplest types of problems are those in which the surface of the solid suddenly attains a new temperature which is maintained constant. This can happen only when the film coefficient from the surface to some isothermal heat-transfer medium is infinite, and although there are not many practical applications of this type, it is an important steppingstone to the solution of numerous problems. Ordinarily, heating or cooling involves a finite film coefficient or else a contact resistance develops between the medium and the surface so that the surface never attains the temperature of the medium. Moreover, the temperature of the surface changes continuously as the solid is heated even though the temperature of the medium remained constant. It is also possible that the temperature of the medium itself varies, but this class of problem was treated separately by Kern [19]. Kern treated also those with finite film coefficients or contact resistances as well as those with infinite coefficients. The following applications are considered by Kern [19].

1. *Sudden change of the surface temperature (infinite coefficient)*
    a. Wall of infinite thickness heated on one side
    b. Wall of finite thickness heated on one side
    c. Wall of finite thickness heated on both sides (slab)
    d. Square bar, cube, cylinder of infinite length, cylinder with length equal to its diameter, sphere
2. *Change due to media having contact resistances*
    a. Wall of finite thickness
    b. Cylinder of infinite length, sphere, semi-infinite solid
    c. Newman's method for common and composite shapes
    d. Graphical determination of the time-temperature distribution

The reader should note that several of the above cases received treatment in Part I, Chapter 2. As noted at that time, the prediction of the unsteady-state temperature distribution in solids can be accomplished by using the conduction equations; these energy balance equations can usually be easily solved to calculate the spatial and time variation of the temperature within the solid. The classic work of Carslaw and Jaeger [23] provide a near infinite number of applications with solutions.

Batch Reactors [24]. Batch reactors are commonly used in experimental studies. Their industrial applications are somewhat limited. They are used for gas phase, e.g., combustion reactions since small quantities (mass) of product are produced with even a very large-sized reactor. It is used for liquid phase reactions when small quantities of reactants are to be processed. It finds its major application in the pharmaceutical industry. As a rule, batch reactors are less expensive to purchase but more expensive to operate than either continuous stirred tank or tubular flow reactors.

A batch reactor is a solid vessel or container. It may be open or closed. Reactants are usually added to the reactor simultaneously. The contents are then mixed (if necessary) to ensure no spatial variations in the concentration of species present. The reaction then proceeds. The concentration of reactants and products changes with time; thus, this is a *transient* or *unsteady state* operation. The reaction is terminated when the desired chemical change has been achieved. The contents are then discharged and sent elsewhere, usually for further processing.

The extent of a chemical reaction and/or the amount of product produced can be affected by the relative quantities of reactants introduced to the reactor. For two reactants, each is usually introduced through separate feed lines normally located at or near the top of the reactor. Both are usually fed simultaneously over a short period of time. Mixing is accomplished with the aid of a turning/spinning impeller. The reaction is assumed to begin after both reactants are in the reactor. No spatial variations in concentration, temperature, etc., are generally assumed.

The describing equation for chemical reaction mass transfer is obtained by applying the conservation law for either mass or moles on a time-rate basis to the contents of a batch reactor. It is best to work with moles rather than mass since the rate of reaction, r_A, is most conveniently described in terms of molar concentrations. The describing equation for species A in a batch reactor take the form:

$$t = \frac{N_{A_0}}{V} \int_0^X \left(-\frac{1}{r_A}\right) dX \qquad (10.22)$$

$$= C_{A_0} \int\limits_{0}^{X} \left(-\frac{1}{r_A} \right) dX \tag{10.23}$$

Theodore [24] provides a more detailed presentation that is accompanied with numerous illustrative examples.

Drying [12]. In many unit operations, it is necessary to perform calculations involving the properties of mixtures of air and water vapor. Such calculations often require knowledge of the amount of water vapor carried by air under various conditions, the thermal properties of such mixtures, and the changes in enthalpy content and moisture content as air containing some moisture is brought into contact with water or wet solids and other similar processes. Some mass transfer operations in the chemical process industry involve simultaneous heat and mass transfer.

Drying involves the *removal* of relatively small amounts of water from solids. In many applications such as in corn processing, drying equipment follows an evaporation step to provide an ultra-high solids content product stream. Drying, in either a batch or continuous process, removes liquid as a vapor by passing warm gas (usually air) over, or indirectly heating, the solid phase.

The drying process is carried out in one of the three basic dryer types The first is a continuous tunnel dryer. In a continuous dryer, supporting trays with wet solids are move through an enclosed system while warm air blows over the trays. Similar in concept to the continuous tunnel dryer, rotary dryers consist of an inclined rotating hollow cylinder. The wet solids are fed on one side and hot air is usually passed countercurrently over the wet solids. The dried solids then pass out the opposite side of the dryer unit. The final type of dryer is a spray dryer. In spray dryers, a liquid or slurry is sprayed through a nozzle, and fine droplets are dried by a hot gas, passed either concurrently countercurrently, past the falling droplets. This unit has found wide application in air pollution control [21, 25].

10.5 Operation, Maintenance, and Inspection (OM&I)

Operation, maintenance, and inspection (OM&I) issues generally apply to equipment, processes, and plants [26, 27]. However, the presentation to follow will address OM&I from a heat exchanger perspective. These issues obviously vary with the type of heat exchanger under consideration. For the purposes of this section, the material will primarily address condensers since most of the heat exchangers in industrial use can be employed for condensation operations. However, this material can be applied to virtually all heat exchangers which have been used in process operations for decades: shell-and-tube, double-pipe, finned, air-cooled, flat-plate, spiralplate, barometric jet, spray, and other heat exchangers. Because of the generic nature of the material to follow, the reader should note that this can also be applied to most other equipment.

Section contents include installation procedures; clearing provisions; foundations; leveling; piping considerations; operation; start-up; shutdown; cleaning; maintenance and inspection; testing; and, improving operation and performance. The section concludes with a presentation on equipment purchasing guidelines. Note that the bulk of the material for this chapter has been drawn from the original work of Connery [26] and subsequent publication based on his work [27].

Installation Procedures. The preparation of a condenser or heat exchanger for installation begins on delivery of the unit from the manufacturer. Condensers are shipped domestically using skids for complete units, and boxes or crates for bare tube bundles. Units are normally removed from trucks using a crane or forklift. Lifting devices should be attached to lugs provided for that purpose (i.e., for lifting of the complete unit as opposed to individual parts), or used with slings wrapped around the main shell. The supports are acceptable lugs for lifting, provided that the complete sets of supports are used together; any nozzles should not be used for attachment of lifting cables. On delivery of the unit, the general condition should be noted to determine any damage sustained during transit and prior to receipt. Any dents or cracks should be reported to the manufacturer prior to attempting to install the unit. Flanged connections are usually blanked with suitable pipe plugs. These closures are necessary to avoid entry of debris into the unit during shipping and handling, and should remain in place until actual piping connections are made.

Clearing Provisions. Sufficient clearance is required for at least inspection of the unit or in-place maintenance. The inspection of heat exchangers requires minimal clearances or the following: access to inspection parts (if provided), removal of channel or bonnet covers, inspection of tube sheets, and tube-to-tube sheet joints. If the removal of tubes or tube bundles is anticipated, provision should be provided in the equipment layout.

Foundations. Heat exchangers must be supported on structures of sufficient rigidity and strength to avoid imposing excessive strains due to settling. Horizontal units with saddle-type shell supports are normally supplied with slotted holes in one support to allow for expansion. Foundation bolts in these supports should be loose enough to allow movement.

Leveling. Heat exchangers should be carefully leveled and squared to ensure proper drainage, venting, and alignment with attached piping. On occasions, these units are purposely angled to facilitate venting and drainage, and the alignment with piping then becomes the prime concern.

Piping Considerations. The following guidelines for piping are necessary to avoid excessive strains, mechanical vibrations, and access for regular inspection:

1. Sufficient support devices are required to prevent the weight of piping and fitting that may be imposed on the unit.
2. Piping should have sufficient expansion joints or bends to minimize expansion stresses arising with temperature excursions.
3. Surge drums or sufficient length of piping to the condenser should be provided to minimize pulsations and potential mechanical vibrations.
4. Valves and bypasses should be provided to permit inspection or maintenance in order to isolate the condenser during periods other than complete system shutdown (outage).
5. Plugged drains and vents are normally provided and located at low and high points of shell-tube sides not otherwise drained or vented. These connections are functional during start-up, operation, and shutdown, and should be piped up to either continuous or periodic use and never left unplugged.
6. Instrument and controls connections should be provided either on condenser nozzles (if applicable) or in the piping close to the condenser. Pressure and temperature indicators should be installed to validate the initial performance of the unit as well as to demonstrate the need for inspection or maintenance.

Operation. The maximum allowable working pressures and temperatures are normally indicated on the heat exchanger's nameplate. These excursion values should *not* be exceeded. Special precautions should be taken if any individual part of the unit is designed for a maximum temperature lower than the unit as a whole (the most common example is some copper alloy tubing with a maximum allowable temperature lower than the actual inlet gas temperature). This is necessary to compensate for the low strength levels of some brasses or other copper alloys at elevated temperatures. In addition, maintaining an adequate flow of the cooling medium may be required at all times.

Equipment such as condensers are designed for a particular fluid throughout. Generally, a reasonable overload can be tolerated without causing damage. If the equipment is operated at excessive flow rates, erosion or destructive vibrations could result. Erosion could occur at normally acceptable flow rates if other conditions, such as entrained liquids or particulates in a gas stream or abrasive solids in a liquid stream, are present. Evidence of erosion should be investigated to determine the cause. Vibration can be propagated by problems other than flow overloads, e.g., improper design, fluid misdistribution, or corrosion-erosion of internal flow-directing devices such as baffles. Considerable study and research have been conducted to develop a reliable vibration analysis procedure to predict or correct damaging vibration.

Start-up. Most equipment, and exchangers in particular, should be warmed up slowly and uniformly; the higher the temperature ranges, the slower the warm-up should be. This is generally accomplished by introducing the coolant or heated fluid and increasing the flow rate to the design level and gradually adding the other stream. For fixed-tube-sheet units with different shell-and-tube material, expansion of shell and tubes should be considered. The respective areas should be vented to ensure complete distribution. It is recommended that gasketed joints be inspected after continuous full-flowing operation has been established. Handling, temperature fluctuations, and yielding of gaskets or bolting may necessitate retightening of the bolting.

Shutdown. Exchangers are usually cooled down by shutting off the vapor stream first and then the cooling stream. Again, fixed-tube-sheet units require consideration of differential expansion of the shell and tubes. Condensers containing flammable, corrosive, or high-freezing-point fluids should be thoroughly drained for any prolonged outages.

Maintenance and Inspection. Recommended maintenance of all equipment, including exchangers, requires regular inspection to ensure the mechanical integrity of the unit and a level of performance consistent with the original design criteria. A brief general inspection should be performed on a regular basis while the unit is operating. Vibratory disturbances, leaking gasketed joints, excessive pressure drop, decreased heat thermal efficiency indicated by higher gas outlet temperatures or lower outlet coolant temperature, lower condensate rates, and intermixing of fluids, are all signs that a thorough inspection and maintenance procedure are required.

Complete inspection requires a shutdown of the unit for access to internals plus pressure testing and cleaning. Scheduling can be determined only from experience and general inspections. For exchangers, tube internals and exteriors, where accessible, should be visually inspected for fouling, corrosion, and/or damage. The nature of any metal deterioration should be investigated to properly determine the anticipated life of the equipment or possible corrective action. Potential causes of deterioration include general corrosion, integranular corrosion, stress cracking, galvanic corrosion, impingement, erosion attack, and the lack of a formal maintenance-and-inspection program.

Cleaning. Fouling of exchangers occurs because of the deposition of foreign material on the interior or exterior of the tubes. Evidence of fouling during operation is increased pressure drop and a general decrease in performance. Fouling can be so severe that the tubes become completely plugged, resulting in thermal stresses, leading to mechanical damage of the equipment.

The nature of the deposited fouling determines the method of cleaning that should be employed. Soft deposits can be removed by steam, hot water, various chemical solvents, or brushing. Cooling water is sometimes treated with four parts of chlorine per million parts water to prevent algae growth and the consequent reduction in the overall heat transfer coefficient of the exchanger. Past experience usually determines the method to be used. Chemical cleaning should be performed by contractors and workers specialized in the field who will consider the deposit to be removed and the materials of construction. If the cleaning method involves elevated temperatures, consideration should be given to thermal stresses induced in the tubes, e.g., steaming out individual tubes can loosen the tube-to-tube sheet joints.

Mechanical methods of cleaning are useful for both soft and hard deposits There are numerous tools for cleaning tube interiors: brushes, scrapers, and various rotating cutter-type devices. The exchanger manufacturer or suppliers of tube tools should be consulted in the selection of the correct tool for a particular deposit. When cutting or scraping deposits, care should naturally be exercised to avoid damaging the tubes.

Cleaning of tube exteriors is generally performed using chemicals, steam, or other suitable fluids. Mechanical cleaning is performed but requires that the tubes be exposed, as in a typical air-cooled condenser, or capable of being exposed, as in a removable bundle shell-and-tube condenser. The layout pattern of the tubes must provide sufficient intersecting empty lanes between the tubes, e.g., as in a square pitch. Mechanical cleaning of tube bundles, if necessary and as noted, requires the utmost care to avoid damaging the tubes (or fins, if present).

Testing. Proper maintenance requires testing of the unit to check the integrity of the following: tubes, tube-to-tube sheet joints, wells, and gasketed joints. The normal procedure consists in pressuring the shell with water or air at the nameplate-specified test pressure and viewing the shell welds and the face of the tube sheet for leaks in the tube sheet joints or tubes. Any water employed should be at or near ambient temperature to avoid false indications due to condensation. Pneumatic testing requires extra care because of the destructive nature of a rupture or explosion, or fire hazards when residual flammable material are present [28]. Condensers of the straight-tube floating head construction require a test gland to perform the test. Tube bundles without shells are tested by pressuring the tubes and viewing the length of the tubes and back face of the tube sheets.

Corrective action for leaking tube-to-tube sheet joint requires expanding the tube end with a suitable roller-type tube expander. Good practice calls for an approximate 8% reduction in wall thickness after metal-to-metal contact between the tube and the tube hole. Tube expansion should not extend beyond 1/8 inch of the inner tube sheet face to avoid cutting the tube.

Defective tubes can be either *replaced* or *plugged*. Replacing tubes requires special tools and equipment. The user should contact the manufacturer or a qualified repair contractor. Plugging of tubes, although a temporary solution, is acceptable provided that the percentage of the total number of tubes per tube pass to be plugged is not excessive (see also Part II, Chapter 7). One of the authors has addressed this problem for several types of equipment from a risk perspective [28]. The type of plug to be used is a tapered one-piece or two-piece metal plug suitable for the tube material and inside diameter. Care should be exercised in seating plugs to avoid damaging the tube sheets.

Improving Operation and Performance. Within constraints of the existing system, *improving operation and performance* generally refers to maintaining or improving operation of original (or consistent) performance. Several factors previously mentioned are critical to the design and performance of a condenser: operating pressure, amount of noncondensable gases in the vapor stream, coolant temperature and flow rate, fouling resistance, and mechanical soundness. Any pressure drop in the vapor line upstream of the condenser should be minimized. Deaerators or similar devices should be operational where necessary to remove gases in solution with liquids. Proper and regular venting of equipment and leak-proof gasketed joints in vacuum systems are all necessary to prevent gas binding and alteration of any condensation equilibrium. Coolant flow rates and temperatures should be checked regularly to ensure that they are in accordance with the original design and performance criteria. The importance of this can be illustrated simply by comparing the winter and summer performance of a condenser using cooling tower or river water. Decreased performance due to fouling will generally be exhibited by a gradual decrease in thermal efficiency and should be corrected as soon as possible. Mechanical malfunctions can also be gradual, but will eventually lead to a near-total or total inability for the unit to perform.

Fouling and mechanical soundness can be controlled only by regular and complete maintenance. In some cases, fouling is much worse than originally predicted and requires frequent cleaning regardless of the precautions taken in the original design. These cases require special action to alleviate the problems associated with fouling.

Most condenser manufacturers will provide designs for alternate conditions as a guide to estimating the cost of improving efficiency via other coolant flow rates and temperatures as well as alternate configurations (i.e., vertical, horizontal, shell side, or tube side).

Equipment Purchasing Guidelines [21, 27]. Purchasing guidelines obviously vary with the equipment in question. This section provides information as it applies to a heat exchanger. However, the recommendations can be adapted or modified for other equipment.

Prior to the purchase of a heat exchanger (for example), experience has shown that the following points should be emphasized:

1. Refrain from purchasing any heat exchanger without reviewing *certified independent test data* on its performance under a similar application. Request that the manufacturer provide performance information and design specifications.
2. In the event that sufficient performance data are unavailable, request that the equipment supplier provide a small pilot model for evaluation under existing conditions.
3. Prepare a good set of specifications. Include a strong performance guarantee from the manufacturer to ensure that the heat exchanger will meet all design criteria and specific process conditions.
4. Closely review the overall process, other equipment, and economic fundamentals.
5. Make a careful energy balance study.

6. Refrain from purchasing any heat exchanger until firm installation cost estimates have been added to the total cost. Escalating installation costs are the rule rather than the exception.

7. Give operation and maintenance costs high priority on the list of exchanger selection factors.

8. Refrain from purchasing any heat exchanger until a solid commitment for assistance from the vendor(s) is obtained. Make every effort to ensure that the exchanger is compatible with the (plant) process.

9. The specification should include written assurance of prompt technical assistance from the supplier. This, together with a completely understandable operating manual (with parts list, full schematics, *consistent units and notations*, etc.) is essential and is too often forgotten in the rush to get the heat exchanger operating.

10. Schedules can be critical. In such cases, delivery guarantees should be obtained from the manufacturers and penalties identified.

11. The heat exchanger should be of fail-safe design with built-in indicators to show when performance is deteriorating.

12. Perhaps most importantly, withhold 10 to 15% of the purchase price until satisfactory operation is clearly demonstrated.

These 12 points have, in a practical sense, become the "bible" to those involved with and/or responsible for the purchase of equipment.

The usual design procurement, installation, and/or start-up problems can be further compounded by any one or a combination of the following:

1. Unfamiliarity of chemical engineers with heat exchangers
2. New suppliers, frequently with unproven heat exchanger equipment
3. Lack of industry standards
4. Compliance schedules that are too tight
5. Vague specifications
6. Weak guarantees
7. Unreliable delivery schedules
8. Process reliability problems

These eight problems represent the standard that the purchaser should also be aware of.

10.6   Economics and Finance

An understanding of the economics and finances involved in applied engineering is important in making decisions at both the engineering and management levels [25–27, 29–31]. Every engineer should be able to execute an economic evaluation of a proposed project. If the project is not profitable, it should obviously not be pursued, and the earlier such a project can be identified, the fewer are the resources that will be wasted.

Before the cost of a unit (e.g., a heat exchanger) or process or facility can be evaluated, the factors contributing to the cost must be recognized. There are two major contributing factors: capital cost and operation costs; these are discussed in the next two sections. Once the total cost of the unit or process has been estimated, the chemical engineer must determine whether the project will be profitable. This involves converting all cost contributions to an annualized basis, a method that is discussed in later subsections. If more than one project proposal is under study, this method provides a basis for comparing alternate proposals and for choosing the best proposal. Project optimization is the subject of a later section, where a brief description of a perturbation analysis is presented.

Detailed cost estimates are beyond the scope of this text. Such procedures are capable of producing accuracies in the neighborhood of ±5 percent; however, such estimates generally require many months of engineering work. This section is designed to give the reader a basis for a *preliminary cost analysis* only, with an expected accuracy of approximately ±20 percent.

The following topics are discussed in this section; capital costs, operating costs, project evaluation, perturbation studies in optimization, and principles of accounting. Note that materials presented in this chapter was adapted primarily from Shen *et al.* [29], Theodore and Neuser [30], Theodore [31], and Theodore and Moy [3].

Capital Costs. *Equipment cost* is a function of many variables, one of the most significant of which is capacity. Other important variables vary with the cost of equipment or process. Preliminary estimates are often based on simple cost-capacity relationships that are valid when the other variables are confined to narrow ranges of values; these relationships can be represented by an approximate linear (on log-log coordinates) cost equipment relationship of the form:

$$C = \alpha Q^{\beta} \qquad\qquad (10.24)$$

where,

> C = cost
> Q = some measure of equipment or process capacity
> α, β = empirical "constants" that depend mainly on equipment type

It should be emphasized that this procedure is suitable for rough estimation only; actual estimates from vendors are preferable. Only major pieces of equipment are included in this analysis; smaller peripheral equipment such as pumps and compressors may not be included. Similar methods for estimating costs are available in the literature. If greater accuracy is needed, however, actual quotes from vendors — as noted — should be used.

Again, the equipment cost estimation model just described is useful for a very preliminary estimation. If more accurate values are needed and the old price data are available, the use of an indexing method [21] may be preferable, although a bit more time consuming. The method consists of adjusting the earlier cost data to present values using factors that correct for inflation.

A number of such indices were available in the past; some of the most commonly used were:

1. Chemical Engineering Fabricated Equipment Cost Index (FECI) [32] past values of which are listed in Table 10.4
2. Chemical Engineering Plant Cost Index [32] listed in Table 10.5
3. Marshall and Swift (M&S) Equipment Cost Index [33] listed in Table 10.6

Other indices for construction, labor, buildings, engineering, and so on are also available in the literature [32–35]. Generally, it is not wise to use past cost data older than 5 to 10 years as in Tables 10.4 to 10.6, even when using the cost indices. Within that time span, the technologies used in developing indices may have changed drastically. Using the indices could cause the estimates to greatly exceed the actual costs. Such an error might lead to the choice of an alternative proposal other than the least costly one. Note that the three indices listed in Tables 10.4 to 10.6 range over the years only to the end of the twentieth century. The base years for Tables 10.4 and 10.5 are 1957–1959 with an index of 100. The Marshall and Swift index uses 1926 as a base year with a cost index of 100. Only the M&S index has survived the passage of time.

The usual technique for determining the *capital costs* (i.e., total capital costs, which include equipment design, purchase, and installation) for a facility is based on the *factored method* of establishing direct and indirect installation costs as a function of the known equipment costs. This is basically a *modified Lang method*, in which cost factors are applied to known equipment costs [35]. This is a three-step method as detailed before.

1. Obtain directly from vendors (or, if less accuracy is acceptable, from one of the estimation techniques discussed above) the purchase price of the primary and auxiliary heat exchanger equipment. The total base price, designated by X, which should include instrumentation, control, taxes, and freight costs, serves as the basis for estimating the direct and indirect installation costs. The installation costs are obtained by multiplying X by the cost factors, which can be adjusted to more closely model the proposed calculation by using adjustment factors that may be available in order to into account for the complexity and sensitivity of the system [34, 35].

Table 10.4 Fabricated equipment cost index.

1999	434.1
1998	435.6
1997	430.4
1996	425.5
1995	425.4
1994	401.6
1993	391.2
1957–1959	100

Source: Ref [32]

Table 10.5 Plant cost index.

1999	390.6
1998	389.5
1997	385.5
1996	381.7
1995	381.1
1994	368.1
1993	359.2
1992	358.2
1991	361.3
1990	357.6
1989	355.4
1957–1959	100

Source: Ref [32]

Table 10.6 Marshall & Swift equipment cost index.

1999	1068.3
1998	1061.9
1997	1056.8
1996	1039.1
1995	1027.5
1994	993.4
1993	964.2
1992	943.1
1991	930.6
1990	915.1
1926	100

Source: Ref [33]

2. Estimate the *direct installation cost* by summing all the cost factors involved in the direct installation costs, which can include piping, insulation, foundation, and supports. The sum of these factors is designated as the *direct installation cost factor* (DCF). The direct installation costs are then the product of the DCF and X.

3. Once the direct and indirect installation costs (ICF) have been calculated, the *total capital cost* (TCC) may be evaluated as

$$TCC = X + (DCF)(X) + (ICF)(X) \qquad (10.25)$$

This TCC cost is then converted to *annualized* capital costs with the use of the *capital recovery factor* (CRF), which is described later. The *annualized capital cost* (ACC) is the product of the CRF and the TCC essentially represents the total installed equipment cost distributed over the lifetime of the facility.

Perry's handbook offers the following information [2]. Basic costs of shell-and-tube heat exchangers made in the United States of carbon steel construction in 1958 are provided. Cost data for shell-and-tube exchangers from 15 sources were correlated and found to be consistent when scaled by the Marshall and Swift index referenced above. These costs should be updated by use of the aforementioned Marshall and Swift Index. Note that during periods of high and low demand for heat exchangers the prices in the marketplace may vary significantly from those determined by this method.

In addition, small heat exchangers and exchangers bought in small quantities are likely to be more costly than indicated. Standard heat exchangers (which are in some instances off-the-shelf items) are available in sizes ranging from 1.9 to 37 m² (20 to 400 ft²) at costs lower than for custom-built units. Steel costs are approximately one-half, admiralty tube-side costs are two-thirds, and stainless costs are three-fourths of those for equivalent custom-build exchangers [2].

Operating Costs. Operating costs can vary from site to site, plant to plant, and equipment to equipment, since these costs, in part, reflect local conditions, e.g., staffing practices, labor, and utility costs. Operating costs, like capital costs, may be separated into two categories: direct and indirect costs. *Direct costs* cover material and labor and are directly involved in operating the facility. These can include labor, materials, maintenance, labor, maintenance suppliers, replacement parts, waste (e.g., residue) disposal fees, utilities, and laboratory costs. *Indirect costs* are operating costs associated with but not directly involved in operating the unit or facility in question; costs such as overhead (e.g., building-land leasing and office supplies), administrative fees, local property taxes, and insurance fees fall into this category [34, 35].

Project Evaluation. Although this subsection primarily deals with a plant or process, it may be applied to heat exchanger equipment or other economic issues. In comparing alternate processes or different options of a particular process from an economic perspective, the total capital cost should be converted to an annual basis by distributing it over the projected lifetime of the facility. The sum of both the *annualized capital costs* (ACCs) and the annual operating costs (AOCs) is known as the *total annualized cost* (TAC) of the facility. The economic merit of the proposed facility, process or scheme can be examined once the total annual cost is available. Alternate facilities or options may also be compared. A small flaw in this procedure is the assumption that the operating costs remain constant through the lifetime of the facility. However, since the analysis is geared to comparing different alternatives, the changes with time should be somewhat uniform among the various alternatives, resulting in little loss of accuracy.

The aforementioned conversion of the total capital cost to an annualized basis involves an economic parameter known as the *capital recovery factor* (CRF), an approach routinely employed by one of the authors in the past. These factors can be found in any standard economics text [35] or can be calculated directly from

$$\text{CRF} = \frac{i(1+i)^n}{(1+i)^{n-1} - 1} \tag{10.26}$$

where,

    n = projected lifetime of the system, years
    i = annual interest rate (expressed as a fraction)

The CRF calculated from Equation (10.26) is a positive, *fractional* number. The ACC is computed by multiplying the TCC by the CRF. The annualized capital cost reflects the cost associated with recovering the initial capital outlay over the depreciable life of the system [21].

Investment and operating costs can be accounted for in other ways, such as the popular *present-worth* analysis [30]. However, the capital recovery method is preferred because of its simplicity and versatility. This is especially true when comparing systems having different depreciable lives. There are usually other considerations in such decisions besides the economics, but if all the other factors are equal, the proposal with the lowest total annualized cost should be the most attractive.

If an industrial system is under consideration for construction, the total annualized cost should be sufficient to determine whether the proposal is economically attractive in comparison to other proposals. If, however, a commercial process is being considered, the profitability of the proposed operation becomes an additional factor. One difficulty in this analysis is estimating the revenue generated from the facility because both technology and costs can change from year to year. Another factor affecting the revenue generated is the availability and raw materials to be handled by the facility. If the revenue that will be generated from the facility can be reasonably estimated, a rate of return can be calculated. This method of analysis is known as the *discounted cash flow method using an end-of-year convention*, where the cash flows are assumed to be generated at the end of the year, rather than throughout the year (the latter obviously is the real case) [21, 30]. An expanded explanation of this method can be found in any engineering economics text. The data required for this analysis are the TCC, the annual after-tax cash flow (A), and the working capital (WC).

Perturbation Studies in Optimization [36, 37]. Once a particular process heat exchanger or project has been selected, it is common practice to optimize the process from a capital cost and operation-and-maintenance standpoint. There are many optimizing procedures available, most of them too detailed for meaningful application to most studies. These sophisticated optimization techniques, some of which are routinely used in the design of conventional chemical and petrochemical plants, invariably involve computer calculations. Use of these techniques in many chemical engineering applications is not warranted, however.

One simple optimization procedure that is recommended is the perturbation study. This involves a systematic change (or perturbation) of variables, one by one, in an attempt to locate the optimum design from a cost-and-operation perspective. To be practical, this often means that the chemical engineer must limit the number of unknowns by assigning constant values to those process variables that are known beforehand to play an insignificant role. Reasonable guesses and simple or shortcut mathematical methods can further simplify the procedure. Much information can be gathered from this type of study since it usually identifies those variables that significantly impact on the overall performance of the process and also helps identify the major contributors to the annualized cost.

Cost-Benefit Analysis [21, 30]. An ability to estimate the cost of a project or system is important for many reasons. From a business standpoint, it is essential to have sufficient knowledge of the cost so that adequate funds can be secured to carry out the project. Where several options exist, it is often useful to determine the costs of each candidate system to determine which approaches are affordable and which are most cost-effective. It may also become apparent that, rather than spending funds on a particular project or process, a more cost-effective approach may involve changing the nature of the process.

Cost-benefit analysis (CBA) is a tool which provides a technique for economic evaluation of policy and/or projects. It has been used primarily as an instrument for evaluating programs and projects. The technique provides quantitative information to analysts on the costs and benefits of various choices which they might make.

The basic premise of CBA is rather simple and easy to accept: benefits and costs of a program should be weighted prior to deciding on a particular choice. However, performing CBA can be a demanding technique which requires a great deal of information; quantification of the potential benefits and costs of a program is not an insignificant exercise. Information must first be identified and structured so as to allow comparison in common measures of value and time. The value of benefits and costs are generally measured in monetary terms, and the time value of money must also be considered. Establishing monetary values by time allows a cost or a benefit for one year to be compared with that cost or benefit at some other time.

A common difficulty in measuring both costs and benefits of a program is the fact that the analysis is forward looking and thus requires an estimate of what a particular strategy *will* cost, which is more difficult than quantifying what a program actually *does* cost. All costs must be considered for each alternative program. Measuring benefits can also be a difficult process. Many benefits can also be difficult to process. Benefits that are generated from a project are also often difficult, if not impossible, to identify and quantify.

Cost-benefit analysis provides an economic tool to assist chemical engineer decision-makers in evaluating alternative programs or policy decisions for their economic feasibility. It establishes a systematic approach for decision making which attempts to comprehensively address all consequences of deciding on a particular alternative. However, there are many limitations to the method, which should be recognized. As noted, the estimation of costs and benefits is sometimes difficult. All benefits and costs must be identified individually, with appropriate monetary values assigned and time periods determined. Establishing a monetary value for intangible or other benefits is, as noted above, uncertain at best.

A *sensitivity analysis* will involve adjusting inputs into the model to evaluate the effects on the ultimate outcome. For example, a range of discount rates may be applied to the analysis to evaluate the sensitivity of the model to a change in the discount rate. Programs that are more sensitive to changes may be scrutinized further before taking action point. CBA may be internally or externally driven. A facility or plant faced with major expenditures for updating equipment would naturally look for the solution that provides the greatest benefit for the least expense. While this would typically be the lowest-cost solution, other factors — including environmental and safety issues [28] — may lead to consideration of higher-cost options.

There is much more to selecting a system or equipment or unit than simply picking one with the lowest purchase cost. Although the initial costs must certainly be considered relative to the project budget, long-term annual costs are typically used in comparing the costs for the purpose of selecting the most suitable strategy. In addition to the initial purchase and installation costs, those long-term annual costs are greatly influenced by equipment life expectancy, utility and environmental (if applicable) costs, and maintenance and labor costs.

Principle of Accounting [21, 30]. *Accounting* is the science of recording business transactions in a systematic manner. Financial statements are both the basis for and the result of management decisions; practicing chemical engineers are rarely involved. Such statements can tell a manager or a chemical engineer a great deal about a company, provided that one can interpret the information correctly.

Since a fair allocation of costs requires considerable technical knowledge of operations in the chemical process industry, a close liaison between the senior process engineer (usually a chemical engineer) and the accountants in a company is desirable. Indeed, the success of a company depends on a combination of financial, technical, and managerial skills.

Accounting has also been defined as the language of business. The different departments of management use it to communicate within a broad context of financial and cost terms. Chemical engineers who do not bother to learn the language of accountancy deny themselves the most important means available for communicating with top management. They might be regarded by upper management as lacking business knowhow. Some chemical engineers have only themselves to blame for their low status within the company hierarchy since they seem determined to hide themselves from business realities behind the screen of their specialized technical expertise. However, an increasing number of chemical engineers are becoming involved in decisions that are related to business.

Chemical engineers involved in feasibility studies and detailed process evaluations are dependent on financial information from the aforementioned company accountants, especially information regarding the way the company intends to allocate its overhead costs. It is vital that the chemical engineer correctly interpret such information or data and be able, if necessary to make the accountant understand the effect of the chosen method of allocation.

The method of allocating overheads can seriously affect the assigned costs of a project and hence the apparent cash flow for that project. Since these cash flows are often used to assess profitability by such methods as present net worth (PNW) (see earlier section), unfair allocation of overhead costs can result in a wrong choice between alternative projects.

In addition to understanding the principles of accounting and obtaining a working knowledge of its practical techniques, chemical engineers should be aware of possible inaccuracies of accounting information in the same wat that they allow for errors in any technical information data. Theodore [21] provides additional details on the principles of accounting.

Example 10.9 – Comparing Two Insulation Thicknesses. It is desired to determine whether to use 1-inch-thick or 2-inch-thick insulation for a steam pipe. Flynn & Theodore Consultants have determined that the following economic data needs to be provided in order to perform a meaningful analysis.

1. Cost of heat loss w/o insulation
2. Cost of each insulation and the corresponding heat loss
3. Interest rate

4. Insulation lifetime
5. Depreciation
6. Salvage value

Comment on the role of the above economic analysis.

Solution: This is obviously an open-ended problem. The approach to the solution is left as an exercise for the reader.

Example 10.10 – Comparing Two Types of Exchangers. Plans are underway to purchase a heat exchanger for a facility in Dumpsville in the state of Egabrag. The company is still undecided as to whether to purchase a double-pipe (DP) or tube-and-bundle (TB) exchanger for the site. The DP unit is less expensive to purchase and operate than a comparable TB system. However, projected energy recovery income from the TB unit is higher as it can handle a larger quantity stream. Based on economic and financial data provided below in Table 10.7, select the exchanger that will yield the higher annual profit.

Calculations should be based on interest rates in the 2–18% range and process lifetime of 8–15 years for the both incinerators.

Solution: The solution for the case of $i = 0.12$ (12%) and $n = 12$ follows.

1. Calculate the capital recovery factor, CRF.

$$CRF = \frac{i}{1-(1+i)^{-n}}$$
$$= \frac{0.12}{1-(1+0.12)^{-12}}$$
$$= 0.1614$$

Table 10.7 Financial data of the two types of exchangers.

Cost/credits	Double pipe	Tube-and-bundle
Capital ($)	2,625,000	2,975,000
Installation ($)	1,575,000	1,700,000
Operation ($/yr)	400,000	550,000
Maintenance ($/yr)	650,000	775,000
Income ($/yr)	2,000,000	2,500,000

2. Determine the annual capital and installation costs for the DP unit.

$$(\text{Costs}) = (\text{Capital} + \text{Installation})(\text{CRF})$$
$$= (2,625,000 + 1,575,000)(0.1614)$$
$$= \$677,800 / \text{yr}$$

3. Determine the annual capital and installation costs for the TB unit.

$$(\text{Costs}) = (\text{Capital} + \text{Installation})(\text{CRF})$$
$$= (2,975,000 + 1,700,000)(0.1614)$$
$$= \$754,450 / \text{yr}$$

4. Complete the following (see Table 10.8) which provides a comparison of the costs and credits for both incinerators.
5. Calculate the profit for each unit on an annualized basis. The profit is the difference between the total annual cost and the income credit.

$$DP(\text{profit}) = 2,000,000 - 1,805,000 = \$195,000 / \text{yr}$$

$$TB(\text{profit}) = 2,500,000 - 2,003,000 = \$497,000 / \text{yr}$$

The tube-and-bundle exchanger yields a far greater profit than the double pipe exchanger for the case of $i = 0.12$ (12%) and $n = 12$. Similar calculations

Table 10.8 Cost comparison for Example 10.10.

	DP	TB
Total installed ($/yr)	755,000	678,000
Operation ($/yr)	400,000	550,000
Maintenance ($/yr)	650,000	775,000
Total annual cost ($/yr)	1,805,000	2,003,000
Income credit ($/yr)	2,000,000	2,500,000

should be performed for a host of different combination of values of i and n to assist in the selection process.

Example 10.11 – Insulation Selection. Ricci and Theodore (R&T) Consultants have been assigned the job of selecting insulation for all the plant piping at a local power plant. Included in the plant piping ate 8000 ft of 1 inch schedule 40 steel (1% C) pipe carrying steam at 240 °F. It is estimated that the heat transfer coefficient for condensing steam on the inside of the pipe is 2000 Btu/hr·ft². °F. The air temperature outside of the pipe can drop to 20 °F, and with wind motion the outside heat transfer coefficient can be as high as 100 Btu/hr·ft· °F. R&T have decided to use a fiberglass insulation having a thermal conductivity of 0.01 Btu/hr·ft· °F. It is available in 6 ft lengths in the four thicknesses listed below:

3/8 inch thick @ $1.51 per 6 ft length

1/2 inch thick @ $3.54 per 6 ft length

3/4 inch thick @ $5.54 per 6 ft length

1 inch thick @ $8.36 per 6 ft length

Calculate the energy saved per dollar of insulation investment in going from ½-in. to ¾-in. thick insulation. Express the results in units of Btu/hr per dollar

SOLUTION: Employ cylindrical coordinates. Calculate the outside log-mean diameter of the pipe. For the 1-inch pipe schedule 40:

$$D_i = 1.049\,in = 0.0874\,ft$$
$$D_o = 1.315\,in = 0.1096\,ft$$

The log-mean diameter of the pipe is therefore

$$\bar{D} = \frac{D_o - D_i}{\ln(D_o / D_i)}$$
$$= \frac{0.1096 - 0.0874}{\ln(0.1096 / 0.0874)}$$
$$= 0.09808\,ft$$

Provide an expression for the insulation outside diameter in terms of Δx_I, the insulation thickness in inches:

$$D_I = \frac{1.315 + 2(\Delta x_I)}{12 \text{ in/ft}}$$

Thus,

Thickness, in	D_I, ft
1/2	0.193
3/4	0.235

Obtain the relationship between the insulation thickness and the insulation log-mean diameter of the pipe:

$$\bar{D}_I = \frac{D_I - 0.1096}{\ln\left(D_I / 0.1096\right)}$$

Therefore,

Thickness, in	$\bar{D}_I$, ft
1/2	0.147
3/4	0.164

The pipe wall resistance and pipe thickness are:

$$R_w = \frac{\Delta x_w}{k \pi \bar{D}_w L}$$

$$\Delta x_w = \frac{0.1096 - 0.0874}{2} = 0.0111 \text{ft}$$

For steel:

$$k = 24.8 \, \text{Btu/hr} \cdot \text{ft} \cdot {}^\circ\text{F}$$

Therefore, for an 800 ft pipe, the wall resistance is:

$$R_w = \frac{0.0111 \text{ft}}{\left(24.8 \text{Btu/hr} \cdot \text{ft} \cdot {}^\circ\text{F}\right)\left(\pi\right)\left(0.09808 \, \text{ft}\right)\left(8000 \, \text{ft}\right)}$$
$$= 1.816 \times 10^{-7} \, \text{hr} \cdot {}^\circ\text{F/Btu}$$

The inside steam convection resistance is:

$$R_i = \frac{1}{h_i \pi D_i L}$$

$$= \frac{1}{\left(2000 \text{Btu/hr} \cdot \text{ft}^2 \cdot {}^\circ\text{F}\right)\left(\pi\right)\left(0.0874 \text{ft}\right)\left(8000 \text{ft}\right)}$$
$$= 2.28 \times 10^{7} \, \text{hr} \cdot {}^\circ\text{F/Btu}$$

Express the insulation resistance in terms of Δx_I:

$$R_I = \frac{\Delta x_I}{k_I \pi \overline{D}_I L}$$

Thickness, in	R_I, hr · °F/Btu
1/2	1.125×10^{-3}
3/4	1.514×10^{-3}

Also express the outside air convection resistance in terms of the thickness, Δx_I:

$$R_o = \frac{1}{h_I \pi D_I L}$$

Thickness, in	R_o, hr · °F/Btu

1/2	2.062×10^{-6}
3/4	1.696×10^{-6}

The total resistance, R, is,

$$R = R_i + R_w + R_l + R_o$$

The total resistance for each thickness is therefore:

Thickness, in	R, hr$\cdot$°F/Btu
1/2	1.128×10^{-3}
3/4	1.156×10^{-3}

The overall outside heat transfer coefficient, U_o, is given by,

$$U_o = \frac{1}{R\pi D_I L}$$

Therefore,

Thickness, in	U_o,Btu/hr$\cdot$ft$^2\cdot$°F
1/2	0.183
3/4	0.112

The inside overall heat transfer coefficient, U_i, is,

$$U_i = \frac{1}{R\pi D_i L}$$

Therefore,

Thickness, in	U_i, Btu/hr · ft² · °F
1/2	0.404
3/4	0.300

The energy loss is,

$$Q = \frac{\Delta T}{R} = UA\Delta T$$

with,

$$A_i = \pi D_i L = \pi (0.0874)(8000) = 2195 \, \text{ft}^2$$

$$\Delta T = 240 - 20 = 220°F$$

Therefore,

Thickness, in	Q, Btu/hr
1/2	1.951×10^5
3/4	1.451×10^5

Finally, calculate the energy saved per dollar of insulation investment going from ½ in. to ¾ in.

$$\frac{\text{Energy}}{\$} = \frac{\Delta Q}{(8000)(\Delta \text{Unit cost per 6 ft})}$$

Substituting:

$$\frac{\text{Energy}}{\$} = \frac{\left(1.951 \times 10^5\right) - \left(1.451 \times 10^5\right)}{(8000)\left[(\$5.54 - \$3.54)/6\right]} = 18.7 \, \text{Btu/hr} \cdot \$$$

Note: It is suggested that the reader perform additional calculations for the other insulation thicknesses — particularly that between the ¾ in. and 1 in. insulation — since either of these pipes may not be the choice for the most cost-effective system.

Example 10.12 – Describing Two Cost Methods. The economic insulation thickness in the previous example may be determined by various methods. Two of these are the minimum-total-cost and the incremental-cost method (or marginal cost method). Describe differences between the two methods.

SOLUTION: The minimum-total-cost method involves the actual calculations of lost energy and insulation costs for each insulation thickness. The thickness producing the lowest total cost is the optimal economic solution. The optimum thickness is determined to be the point where the last dollar invested in insulation results in exactly $1 in energy-cost savings. The incremental-cost method provides a simplified and direct solution for the least-cost-thickness. The total-cost method does *not* in general provide a satisfactory means for making most insulation investment decisions, since an economic return on investment is required by investors and the method does not properly consider this factor.

References

1. Webster's New World Dictionary, Second College Edition, Prentice-Hall, Upper Saddle River, NJ, 1971.
2. D. Green and R. Perry (editors), *Perry's Chemical Engineers' Handbook*, 8th edition, McGraw-Hill, New York City, NY, 2008.
3. L. Theodore and E. Moy, *Hazardous Waste Incineration*, A Theodore Tutorial, Theodore Tutorials, East Williston, NY, originally published by the USEPA/APTI, RTP, NC, 1996.
4. J. Santoleri, J. Reynolds, and L. Theodore, *Introduction to Hazardous Waste Incineration*, 2nd edition, (adapted from), John Wiley & Sons, Hoboken, NJ, 2000.
5. L. Theodore and J. Reynolds, *Heat Transfer*, A Theodore Tutorial, Theodore Tutorials, East Williston, NY originally published by the USEPA/APTI, RTP, NC, 1992.
6. L. Theodore, *Heat Transfer Applications for the Practicing Engineer*, John Wiley & Sons, Hoboken, NJ, 2013.
7. D. Green and R. Perry (editors), *Perry's Chemical Engineers' Handbook*, 7th edition, McGraw-Hill, New York City, NY, 1998.
8. C. Mockler, term project submitted to L. Theodore (adopted from), Manhattan College, Riverdale, NY, 2008.

9. R. Kirk and D. Othmer, *Encyclopedia of Chemical Technology*, 4th edition, Vol. 7, "Cryogenics," p.659, John Wiley & Sons, Hoboken, NJ, 2001.

10. L. Theodore, F. Ricci, and T. VanVliet, *Thermodynamics for the Practicing Engineer*, John Wiley & Sons, Hoboken, NJ, 2009.

11. Merriam-Webster Online Dictionary, 2008, Merriam-Webster Online. http://www.merriam-webster.com/dictionary/cryogenics

12. L. Theodore and F. Ricci, *Mass Transfer Operations for the Practicing Engineer*, John Wiley & Sons, Hoboken, NJ, 2010.

13. J. Brock, C. Frisbie, and A. Jersey, "Cryogenics," term project submitted to L. Theodore, Manhattan College, Riverdale, NY, 2009.

14. J. Smith, H. Van Ness, and M. Abbott, *Introduction to Chemical Engineering Thermodynamics*, 7th edition, McGraw-Hill, New York City, NY, 2005.

15. P. Abulencia and L. Theodore, *Fluid Flow for the Practicing Chemical Engineer*, John Wiley & Sons, Hoboken, NJ, 2009.

16. M. Zabetkis, *Safety with Cryogenic Fluids*, Plenum, New York City, NY, 1967.

17. J. Reynolds, J. Jeris, and L. Theodore, *Handbook of Chemical and Environmental Engineering Calculations*, John Wiley & Sons, Hoboken, NJ, 2002.

18. L. Theodore, *Nanotechnology: Basic Calculations for Engineers and Scientists*, John Wiley & Sons, Hoboken, NJ, 2006.

19. D. Kern, "Process Heat Transfer," McGraw-Hill, New York City, NY, 1950.

20. P. T. Vasudevan, *Process Dynamics and Control*. A Theodore Tutorial, Theodore Tutorials, East Williston, NY, originally published by USEPA/ APTI, RTP, NC, 1996.

21. L. Theodore, *Chemical Engineering: The Essential Reference*, McGraw-Hill, New York City, NY, 2014.

22. J. Fourier, *Theorie Analytique de la chaleur*, other information unknown, 1822.

23. H. Carslow and J. Jaeger, *Conduction of Heat in Solids*, 2nd edition, Oxford University Press, London, 1959.

24. L. Theodore, *Chemical Reactor Analysis and Applications for the Practicing Engineer*, John Wiley & Sons, Hoboken, NJ, 2012.

25. L. Theodore, *Air Pollution Control Equipment Calculations*, John Wiley & Sons, Hoboken, NJ. 2010.

26. W. Connery et al., "Energy and the Environment," *Proceedings of the 3rd National Conference*, AIChE, NY, 1992.

27. W. Connery, "Condensers," in L. Theodore and A. J. Buonicore, eds., *Air Pollution Control Equipment*, Chap. 6, A Theodore Tutorials, East Williston, NY, originally published by Prentice Hall, Upper Saddle River, NJ, 1992.

28. L. Theodore and R. Dupont, *Environmental Health and Hazard Risk Assessment: Principles and Calculations*, CRC Press/Taylor & Francis Group, Boca Raton, Fla., 2012.

29. T. Shen, Y. Choi, and L. Theodore, EPA Manual *Hazardous Waste Incineration*, (USEPA Manual) USEPA, RTP, NC, 1985.

30. L. Theodore and K. Neuser, *Engineering Economics and Finance*, A Theodore Tutorial, East Williston, NY, originally published by USEPA/APTI, RTP, NC, 1996.

31. L. Theodore, personal notes, East Williston, NY, 1990.

32. Economic Indicators, "Chemical Engineering Plant Cost Index," *Chem. Eng.*, New York City, NY, 2000.

33. Economic Indicators, "Marshall and Swift Equipment Cost Index," *Chem. Eng.*, New York City, NY, 1996.

34. R. Neveril, *Capital and Operating Costs and Selected Air Pollution Control Systems*, Card, Inc., Niles, Ill., US EPA Report 450/5-80-002, Dec. 1978.

35. W. Vatavuk and R. Reveril, "Factors for Estimating Capital and Operating Costs," *Chem. Eng.*, New York City, NY, Nov. 3, 1980.

36. C. Prochaska and L. Theodore, *Introduction to Mathematical Methods for Environmental Engineers and Scientists*, Scrivener-Wiley, Beverly, MA, 2018.

37. L. Theodore and K. Behan, *Introduction to Optimization for Environmental and Chemical Engineers*, CRC Press/Taylor & Francis Group, Boca Raton, FL, 2018.

11

Entropy Consideration and Analysis

Contributing author: Anet Kashoa

Introduction

As noted in Chapter 1, the law of conservation of energy is defined by many as the first law of thermodynamics. The second law of thermodynamics is referred to as the "limiting law." Unlike the first law, the second law places "restrictions" on energy relationships. For instance, it dictates that heat cannot spontaneously flow from a cold to a warmer body. Historically, the study of the second law was developed by individuals such as Carnot, Clausius and Kelvin in the middle of the nineteenth century. This development was made purely on a macroscopic scale and is referred to as the "classical approach" to the second law.

The first law of thermodynamics may also be viewed as a conservation law concerned with energy transformations. Regardless of the types of energy involved in processes—thermal, mechanical, electrical, elastic, magnetic, etc.—the change in the energy of a steady-state system is equal to the difference between energy input and energy output. The first law also allows free convertibility from one from of energy to another, as long as the overall quantity is conserved. Thus, this law places no restriction on the conversion of work into heat, or on its counterpart—the conversion of heat into work. However, the second law is another matter and its impact on the design of heat exchangers can be significant.

Environmental concerns involving conservation of energy issues gained increasing prominence during and immediately after the OPEC oil embargo of 1973. In addition, global population growth has led to an increasing demand for energy. Although the use of energy has resulted in great benefits, the environmental and human health impact of this energy use has become a concern. One of the keys to reducing and/or eliminating this problem will be achieved through what has come to be referred to as "meaningful" energy conservation in heat exchangers design [1–4].

The design methodology presented in Chapters 6, 7, and 8 can be extended to include second law analysis. This delves into entropy considerations, an aspect that has received little attention in the literature. This chapter attempts to correct this "oversight" by introducing, defining, and applying the concept of *quality energy*. Theodore, Ricci, and VanVliet [3] provided additional details.

This chapter contains the following four topics:

11.1 Qualitative Review of the Second Law
11.2 Describing Equations
11.3 The Heat Exchanger Dilemma
11.4 Application to a Heat Exchanger Network

11.1   Qualitative Review of The Second Law

The brief discussion of energy conversion above leads to an important second law consideration—energy has "quality" as well as quantity [2, 3]. Because work is 100% convertible to heat, whereas the reverse situation is not true, work is a more valuable form of energy than heat. Although it is not as obvious, it can also be shown through second law arguments that heat also has "quality" in terms of the temperature at which it is discharged from a system. The higher the temperature, the greater the possible energy transformation into work. Thus, thermal energy stored at high temperatures is generally more useful to society than that available at lower temperatures. While there is an immense quantity of energy stored in the oceans, or deep under the earth's surface, for example, its present availability to society for performing useful tasks is quite low. This implies, as noted above, that thermal energy loses some of its "quality" or is degraded when it is transferred by means of heat transfer from one temperature to a lower one. Other forms of energy degradation include energy transformations due to frictional effects and electrical resistances. Such effects are highly undesirable if the use of energy for practical purposes is to be maximized [1–3].

The second law provides some means of measuring the aforementioned energy degradation through a thermodynamic term referred to as entropy. It is normally designated as S with units of energy per absolute temperature (e.g., Btu/°R or cal/K). Furthermore, entropy calculations can provide quantitative information on both the *quality* of energy and energy degradation [2, 3].

In line with the discussion regarding the *quality* of energy, individuals at home and in the workplace are often instructed to "conserve energy." However, this comment, if taken literally, is a misnomer since energy is automatically conserved by the provisions of the first law. In reality, the comment "conserve energy" addresses only the concern associated with the *quality* of energy. If the light in a room is not turned off, energy is degraded although energy is still conserved; i.e., the electrical energy is converted to internal energy (which heats up the room). Note, however, that this energy transformation will produce a token rise in temperature of the room from which little, if any, *quality* energy can be recovered and used again (for lighting or other useful purposes) [1].

There are a number of other phenomena that cannot be explained by the law of conservation of energy. It is the previously mentioned second law of thermodynamics that provides an understanding and analysis of their diverse effects. However, among these considerations, it is the second law that can allow the measuring of the aforementioned *quality* of energy, including its effect on the design and performance of heat exchangers.

11.2 Describing Equations

Key equations on entropy calculations and heat exchanger design receive treatment in this section [2, 3]. If ΔS_{syst} and ΔS_{surr} represent the entropy change of a system and surroundings, respectively, it can be shown [1–3] that for a closed system, the second law dictates that the total entropy change of the universe, ΔS_{total}, is given by:

$$\Delta S_{total} = \Delta S_{syst} + \Delta S_{surr} \geq 0 \tag{11.1}$$

In effect, the second law requires that for any real processes, the total (or overall) entropy change is positive; the only exception is if the process is reversible (the driving force for heat transfer is at all times zero) and then:

$$\left(\Delta S_{total}\right)_{rev} = 0 \tag{11.2}$$

Thus, no real process can occur for which that total entropy change is zero or negative. The fundamental facts relative to the entropy concept are that the entropy change of a system may be positive (+), negative (−), or zero; the entropy change of the surroundings during this process may likewise be positive, negative, or zero. However, their sum, by necessity, must be greater than or equal to zero.

To illustrate the concept of "quality" energy [2], consider the insulated space pictured in Figure 11.1(a)-(b). In state (a), the system contains air and steam that are separated; in state (b), the partition between the two sub-systems is opened and both components are mixed. The overall system is insulated (i.e., $Q = 0$), closed to mass flow, and without the input of mechanical work (i.e., $W = 0$), so that one can conclude from the first law $\left(Q + W = \Delta U\right)$ that:

$$\Delta U = 0 \tag{11.3}$$

and,

$$U_a = U_b \tag{11.4}$$

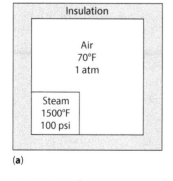
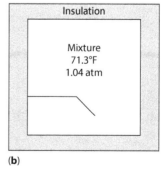

(a)   (b)

Figure 11.1 Entropy analysis.

where,

$$U = \text{internal energy}$$

Although the energy levels are the same in both systems, one notes that in state (a) the system has the capability of doing useful work (because of the high-temperature high-pressure steam) while in state (b) the system does not. If an entropy analysis is performed (see later example) on both states (a) and (b), one would deduce that (as noted earlier):

$$S_a < S_b \qquad (11.5)$$

In effect, the entropy level has increased for the system and the energy within the system (although conserved) has lost its ability to do useful work due to the irreversible nature of the mixing process. It is in this manner that the concept of entropy can be used to determine a system's ability to either do useful work or lose its ability to do useful work. Thus, the second law leads to the conclusion that the greater the irreversibility of a process, the greater the entropy increase and the greater the amount of energy that becomes unavailable for doing useful work.

Consider now the entropy change of gases. The entropy change of an ideal gas undergoing a change of state from pressure P_1 to P_2 at a constant temperature T is given by [2],

$$\Delta S_T = R \ln\left(\frac{P_1}{P_2}\right) \qquad (11.6)$$

where R is the ideal gas law constant in consistent units. The entropy change of an ideal gas (as well as a liquid) undergoing a change of state from T_1 to T_2 at a constant pressure is given by [2],

$$\Delta S_P = C_P \ln\left(\frac{T_1}{T_2}\right) \qquad (11.7)$$

where c_P is the heat capacity at constant pressure in consistent units. Correspondingly, the entropy change for an ideal gas undergoing a change from (P_1, T_1) to (P_2, T_2) is [2],

$$\Delta S = R \ln\left(\frac{P_1}{P_2}\right) + C_P \ln\left(\frac{T_1}{T_2}\right) \qquad (11.8)$$

Heat exchanger design equations are again reviewed in light of the development above. If Q represents the rate of heat transfer between a hot and cold fluid flowing in a heat exchanger, application of the conservation on law for energy gives,

$$Q_H = \dot{m}_H c_{P,H} \left(T_{H,i} - T_{H,o} \right) \tag{11.9}$$

and,

$$Q_C = \dot{m}_C c_{P,C} \left(T_{C,o} - T_{C,i} \right) \tag{11.10}$$

where the superscripts H and C refer to the hot and cold fluids, respectively; i and o refer to the fluid inlet and outlet temperature, respectively; $\dot{m}$ represents the mass flow rate, and c_p is once again the heat capacity at constant pressure (assumed constant). In addition, if there is no heat lost from the heat exchanger to the surroundings.

$$Q_H = Q_C \tag{11.11}$$

As noted in the previous chapters, the following equation (The Heat Transfer Equation) relates Q to the temperature difference driving force and area between the hot and cold fluids:

$$Q = UA\Delta T_{LM} \tag{11.12}$$

This may be viewed once again as the heat exchanger design equation. As noted, the terms U, A, and ΔT_{LM} defined earlier represent the overall heat transfer coefficient (a function of the resistance to heat transfer), the area of heat transfer, and the log mean temperature difference (LMTD) driving force, respectively. For some of the exchangers reviewed in Part II, the latter term is given by,

$$\Delta T_{LM} = \text{LMTD} = \frac{\Delta T_2 - \Delta T_1}{\ln\left(\Delta T_2 / \Delta T_1 \right)} \tag{11.13}$$

where ΔT_2 and ΔT_1 represent the temperature difference between the hot and cold fluid at each end of the exchanger, respectively. If $\Delta T_1 = \Delta T_2 = \Delta T$, then $\Delta T_{LM} = \Delta T$. Finally, for purposes of the analysis to follow, Equation (11.12) is rearranged in the form,

$$\frac{Q}{U\Delta T_{LM}} = A \tag{11.14}$$

11.3   The Heat Exchanger Dilemma

One of the areas where quality energy calculations can be quite useful is in the design and specification of process (operating) conditions for heat exchangers [2–4]. The quantity of heat recovered in an exchanger is not alone in influencing size and cost. As the temperature difference driving force (LMTD) in the exchanger approaches zero, the "quality" heat recovered increases. However, there are practical considerations on how small the driving force can be.

Most heat exchangers are designed with the requirement specification that the temperature difference between the hot and cold fluid be at all times positive and be at least 20 °F. This temperature difference is referred to by some as the *approach temperature*. However, as it will be demonstrated in the examples to follow, the corresponding entropy change is also related to this difference, with large temperature difference driving forces resulting in large irreversibilites and the associated accompanying large entropy changes.

The individual designing a heat exchanger is faced with two choices. He/she may decide to design with a large LMTD that results in both a more compact (smaller area) exchanger (see Equation 11.14) and a large entropy increase (and hence a larger loss of "quality" energy). Alternately, a design with a small driving force results in both a larger heat exchanger and a smaller entropy change (hence a smaller loss of *quality* energy).

Regarding the amount of cooling medium for a given heat transfer duty, the design engineer thus has the option of circulating a large quantity with a small temperature increase or a small quantity with a large temperature increase. The temperature change (or difference) of the coolant temperature affects the LMTD. If a large coolant quantity is used, the LMTD is larger and less heat transfer area A is required. Although this will reduce the original investment and fixed charges (capital and operating costs were discussed earlier), the amount of *quality* energy recovered will also be smaller, owing to the greater quantity of coolant employed. It is therefore apparent that an optimum must exit between the two choices: too much coolant, smaller surface, and the recovery of less *quality* energy *or* too little coolant, larger surface, and the recovery of more quality energy. In the limit, as the temperature difference approaches zero $(\text{LMTD} \rightarrow 0)$ the area requirement, A, approaches infinity $(A \rightarrow \infty)$, the entropy change approaches zero $(\Delta S \rightarrow 0)$ and the aforementioned recovered "quality" energy increases to a maximum. Clearly, cost must be minimized, but just as clearly, the *quality* energy recovered must be considered in the analysis. This dilemma is addressed below.

Consider the modes of operation for the three heat exchangers shown in Figure 11.2. Note that for the purpose of analysis, $m_C = m_H = 1.0$ lb and $c_p = 1.0 \text{ Btu} / \text{lb} \cdot °\text{F}$.

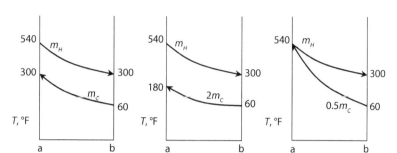

Figure 11.2 Heat exchanger operation.

For operation (a):

$$\Delta T_{LM,A} = 540 - 300 = 300 - 60$$
$$= 240°F$$

The entropy change for the hot fluid is,

$$\Delta S_H = m_H c_P \ln\left(\frac{T_2}{T_1}\right)$$
$$= (1)(1)\ln\left(\frac{300 + 460}{540 + 460}\right)$$
$$= \ln\left(\frac{760}{1000}\right)$$
$$= -0.2744 \text{ Btu / °R}$$

and,

$$\Delta S_C = m_C c_P \ln\left(\frac{t_2}{t_1}\right)$$
$$= (1)(1)\ln\left(\frac{300 + 460}{60 + 460}\right)$$
$$= \ln\left(\frac{760}{520}\right)$$
$$= 0.3795 \text{ Btu / °R}$$

The total entropy increase is therefore,

$$\Delta S_{T,A} = -0.2744 + 0.3795$$
$$= 0.1054 \text{ Btu} / {}^\circ R$$

For operation (b):

$$\Delta T_{LM,B} = \frac{360 - 240}{\ln(360 / 240)}$$
$$= 296{}^\circ F$$

The entropy can be calculated in a manner similar to that of operation (a):

$$\Delta S_H = -0.2744 \text{ Btu} / {}^\circ R$$

and,

$$\Delta S_C = (2)(1)\ln\left(\frac{180 + 460}{60 + 460}\right)$$
$$= 2\ln\left(\frac{640}{520}\right)$$
$$= 0.4153 \text{ Btu} / {}^\circ R$$

The total entropy increase for operation (b) is therefore,

$$\Delta S_{T,B} = -0.2744 + 0.4153$$
$$= 0.1409 \text{ Btu} / {}^\circ R$$

For operation (c):

$$\Delta T_{LM} = \frac{240}{\infty} = 0{}^\circ F$$

The entropy change for the hot fluid is again,

$$\Delta S_H = -0.2744 \text{ Btu} / {}^\circ R$$

and,

$$\Delta S_C = (0.5)(1)\ln\left(\frac{540+460}{60+460}\right)$$

$$= 0.5\ln\left(\frac{1000}{520}\right)$$

$$= 0.3270 \text{ Btu} / {}^\circ R$$

The total entropy change for (c) is therefore,

$$\Delta S_{T,C} = -0.2744 + 0.3270$$

$$= 0.0526 \text{ Btu} / {}^\circ R$$

A summary of the results for operation A, B, and C plus the heat exchanger area requirement (A) and quality energy (QE) analysis is provided in Table 11.1. Once concludes that as the ΔT_{LM} (or LMTD) increases, the area requirement decreases (see earlier equation (11.14)); however, the QE available correspondingly decreases. Alternatively, if ΔT_{LM} decreases, both A and QE increase.

Consider now the operation of heat exchangers A and B, as provided in Figure 11.3. For Case I, one notes, using the same analysis as above, that,

$$\text{LMTD}_A = \text{LMTD}_B$$

$$A_A = A_B; \quad A_A = A_B = A, \quad \text{and} \quad A_T = 2A$$

$$\Delta S_A = \Delta S_B; \quad \Delta S_T = 2\Delta S_A = 2\Delta S_B$$

End result: Two m_C streams ($m_{C,A}$ and $m_{C,B}$) are heated to 300 °F where $m_{C,A} = m_{C,B} = m_C$. Two m_H streams are cooled to 300 °F where $m_{H,A} = m_{H,B} = m_H$.

Table 11.1 Heat exchanger-entropy analysis results.

	ΔT_{LM}, °F	ΔS_T, Btu / °R	A	QE
Operation A	240	0.1054	Moderate	Moderate
Operation B	296	0.1409	Lower	Lower
Operation C	0	0.0526	∞	Higher

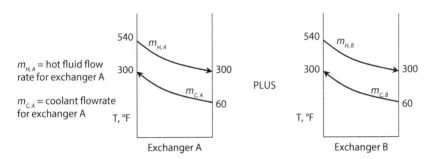

Figure 11.3 Heat exchanger comparison; Case I.

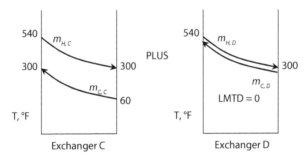

Figure 11.4 Heat exchanger comparison; Case II.

Consider Case II described by Figure 11.4. Here the coolant discharge from exchanger C serves as the inlet coolant to exchanger D. For Case II, one notes that,

$$\text{LMTD}_C > \text{LMTD}_D; \quad \text{LMTD}_D = 0$$
$$A_C < A_D; \quad A_D = \infty \quad \text{and} \quad A_T = \infty$$
$$\Delta S_A = \Delta S_C > \Delta S_D; \quad \Delta S_D = 0 \quad \text{and} \quad \Delta S_T = 2\Delta S_A = 2\Delta S_B$$

Also,

$$\text{LMTD}_A = \text{LMTD}_B = \text{LMTD}_C$$
$$A_A = A_B = A_C$$

End result: One $m_C \left(m_{C,C} \right)$ results at 540 °F.

Case I/Case II comparison: One m_C stream results at 540 °F (II) vs. two m_C at 300 °F (I)

$$A_T = 2A(\text{I})\,\text{vs.}\infty(\text{II})$$
$$\Delta S(\text{II}) < \Delta S(\text{I})$$

Thus, for Case II, stream $m_{C,D}$ *can* heat, for example, another fluid to 520 °F while stream $m_{C,A}$ and/or $m_{C,B}$ for Case I *cannot*.

Consider Case III (see Figure 11.5). Here, twice the coolant is employed in each exchanger. For Case III, one notes that:

1. LMTD has increased.
2. A has decreased.
3. ΔS has increased.

Also,

$$\text{LMTD}_F = \text{LMTD}_E > \text{LMTD}_C = \text{LMTD}_A = \text{LMTD}_B > \text{LMTD}_D$$
$$A_F = A_E < A_A = A_B = A_C < A_D$$
$$\Delta S_F = \Delta S_E > \Delta S_A = \Delta S_B = \Delta S_C > \Delta S_D$$

End result: Four m_C streams result at 180 °F, A has decreased and ΔS has increased. The above analysis is now extended to the four examples that follow.

Example 11.1 – Comparing Entropy Changes for Two Exchangers: Case I. Refer to Case I, Figure 11.3, above. Calculate the entropy change of the two exchangers, A and B. For the purpose of analysis, arbitrarily assume $c_p = 1.0\,\text{Btu}/\text{lb}\cdot°\text{F}$ and $m_H = m_C = 1.0$ lb.

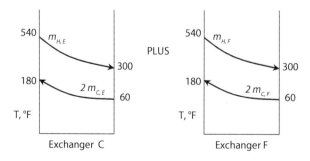

Figure 11.5 Heat exchanger comparison; Case III.

SOLUTION: For the hot fluid in Exchanger A,

$$\Delta S_H = mc_P \ln\left(\frac{T_2}{T_1}\right)$$

$$= (1)(1)\ln\left(\frac{300+460}{540+460}\right)$$

$$= \ln\left(\frac{760}{1000}\right)$$

$$= -0.2744 \text{ Btu}/^\circ\text{F}$$

For the cold fluid,

$$\Delta S_C = mc_P \ln\left(\frac{t_2}{t_1}\right)$$

$$= (1)(1)\ln\left(\frac{300+460}{60+460}\right)$$

$$= \ln\left(\frac{760}{520}\right)$$

$$= 0.3795 \text{ Btu}/^\circ\text{F}$$

Therefore, for Exchanger A

$$\Delta S_A = \Delta S_H + \Delta S_C$$

$$= -0.2744 + 0.3795$$

$$-0.1051 \text{ Btu}/^\circ\text{F}$$

Since A and B are identical exchangers.

$$\Delta S_A = \Delta S_B$$

and,

$$\Delta S_{Total,I} = (2)(0.1051)$$

$$= 0.2102 \text{ Btu}/^\circ\text{F}$$

As expected, there is a positive (+) entropy change.

Example 11.2 – Comparing Entropy Changes for Two Exchangers: Case II. Refer to Case II, Figure 11.4. Calculate the entropy change of exchangers C and D.

SOLUTION: Consider exchanger C first.

$$\Delta S_C = \Delta S_{H,C} + \Delta S_{C,C}$$

This is given by the result in the previous example, i.e.,

$$\Delta S_C = 0.1051 \, \text{Btu} \, / \, ^\circ F$$

Consider exchanger D. Since the temperature difference driving force is zero, the operation is reversible. Therefore,

$$\Delta S_D = 0$$

The total entropy change is then,

$$\Delta S_{\text{Total,II}} = \Delta S_C + \Delta S_D$$
$$= 0.1051 + 0$$
$$= 0.1051 \, \text{Btu} \, / \, ^\circ F$$

Example 11.3 – Comparing Entropy Changes for Two Exchangers: Case III. Refer to Case III, Figure 11.5. Calculate the entropy change in exchangers E and F.

SOLUTION: Consider exchanger E,

$$\Delta S_E = \Delta S_{H,E} + \Delta S_{C,E}$$
$$= -0.2744 + (2)(1)\ln\left(\frac{180 + 460}{60 + 460}\right)$$
$$= -0.2744 + 0.4153$$
$$= 0.1409 \, \text{Btu} \, / \, ^\circ F$$

Since E and F are identical exchangers,

$$\Delta S_E = \Delta S_F$$

and,

$$\Delta S_{\text{Total,III}} = \Delta S_E + \Delta S_F$$
$$= (2)(0.1409)$$
$$= 0.2818 \text{ Btu} / °F$$

Example 11.4 – Entropy Change Comparison. Comment on the results of Examples 11.1, 11.2, and 11.3.

SOLUTION: The calculated results of the three cases paint a clear picture. As LMTD decreases, the area cost requirement (per ft^2 of heat exchanger area) increases. In addition, the entropy change decreases, and the *quality* energy increases. Thus, from a conservation of *quality* energy perspective, the second law *should* be considered in heat exchanger design applications. Hence, both the cost of the exchanger and the economic factors associated with the *quality* of the recovered energy must be considered in the analysis. Note that pressure drop, materials of construction, etc., are not included in this analysis.

11.4 Application to a Heat Exchanger Network

In many chemical and petrochemical plants there are cold streams that must be heated and hot streams that must be cooled [5, 6]. Rather than use steam to do all the heating and cooling water to do all the cooling, it is often advantageous to have some of the hot streams heat the cold ones and some of the cold streams to cool the hot ones.

Highly interconnected networks of exchangers can save a great deal of energy in a chemical plant. The more interconnected they are, however, the harder the plant is to control, start-up and shut-down. Often auxiliary heat sources and cooling sources must be included in the plant design in order to ensure that the plant can operate smoothly.

Consider the following application [5]. A plant has three streams to be heated and three streams to be cooling. Cooling water (90 °F supply, 155 °F return) and steam (saturated at 250 psia) are available. Information on the three streams to be heated is provided in Table 11.2, while the three streams to be cooled are provided in Table 11.3. The sensible heating duties for all streams are first calculated. The results are shown in Table 11.4. The total heating and cooling duties can now be computed and compared.

Table 11.2 Streams to be heated.

Stream	Flowrate, lb/hr	C_p, Btu/lb·°F	T_{in}, °F	T_{out}, °F
1	50,000	0.65	70	300
2	60,000	0.58	120	310
3	80,000	0.78	90	250

Table 11.3 Streams to be cooled.

Stream	Flowrate, lb/hr	C_p, Btu/lb·°F	T_{in}, °F	T_{out}, °F
1	60,000	0.70	420	120
2	40,000	0.52	300	100
3	35,000	0.60	240	90

Table 11.4 Duty requirements.

Stream	Duty, Btu/hr
1	7,475,000
2	6,612,000
3	9,984,000
4	12,600,000
5	4,160,000
6	3,150,000

Heating: $7,475,000 + 6,612,000 + 9,984,000 = 24,071,000 \, \text{Btu} / \text{hr}$
Cooling: $12,600,000 + 4,160,000 + 3,150,000 = 19,910,000 \, \text{Btu} / \text{hr}$
Heating $-$ Cooling $= 24,071,000 - 19,910,000 = 4,161,000 \, \text{Btu} / \text{hr}$
At a minimum, 4,431,000 Btu/hr will have to be supplied by steam or another hot medium.

Example 11.5 – Network Design. Design a system of heat exchangers to accommodate the process described in the previous example. In effect, devise a network of heat exchangers that will make full use of heating and cooling streams against each other, using utilities only if necessary.

SOLUTION: Kauffman's [5] solution is provided in Figure 11.6, which represents a system of heat exchangers that will transfer heat from the hot streams to the cold ones in the amounts desired. It is important to note that this is but one of *many* possible schemes. The optimum system would require a trial-and-error procedure that would examine a host of different schemes. Obviously, the economics would also come into play.

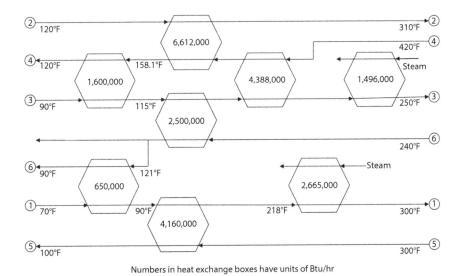

Numbers in heat exchange boxes have units of Btu/hr

Figure 11.6 Flow diagram of a network system of heat exchangers.

As noted in the previous example, in many chemical and petrochemical plants there are cold streams that must be heated and hot streams that must be cooled. The problem of optimum heat exchanger networks to accomplish this has been extensively studied and is available in the literature. This example provides one simple illustration.

Highly interconnected networks of exchangers can save a great deal of *quality* energy in a chemical plant, and as noted above, an economic analysis should be included. Theodore [7] extended the above example to include two other considerations.

1. *Outline* how to solve the above problem if the heat exchanger cost is a function of the heat exchanger area. The network should be such that the total cost of the resulting network is *minimized*. In effect, the economics involved are to be included with the outline to the solution.

2. *Outline* how to solve the two previous problems so that the network's entropy is decreased and the network's costs are minimized. Is there a unique solution? Justify your answer to this question. If there is not a unique solution, *outline* how a practicing chemical engineer could arrive at a *reasonable* solution to this heat exchanger network problem. As noted above, the more interconnected they are, however, the harder the plant is to control, start-up, and shut down.

Often auxiliary heat sources and cooling sources must be included in the plant design in order to ensure that the plant can operate smoothly.

Finally, there are many possible combinations of interconnected heat exchangers that will work. Detailed cost analyses would be needed to determine which one is best. For each exchanger in the network, one needs to make sure that the duties on each side are equal and that there are no *temperature crossovers*. These occur when the stream being cooled is colder than the lowest temperature of the stream being heated at some point in the exchanger. This is (of course) forbidden by the second law of thermo-dynamics [2, 3].

References

1. M.K. Theodore and L. Theodore, *Introduction to Environmental Management*, CRC Press/Taylor & Francis Group, Boca Raton, FL, 2009.
2. L. Theodore and J. Reynolds, *Thermodynamics*, A Theodore Tutorial, Theodore Tutorials, East Williston, NY, originally published by the USEPA/APTI, RTP, NC, 1994.
3. L. Theodore, F. Ricci, and T. VanVliet, *Thermodynamics for the Practicing Engineer*, John Wiley & Sons, Hoboken, NY, 2009.
4. L. Theodore, *Heat Transfer Applications for the Practicing Engineer*, John Wiley & Sons, Hoboken, NY, 2011.
5. D. Kauffman, *Process Synthesis and Design*, A Theodore Tutorial, Theodore Tutorials, East Williston, NY, originally published by the USEPA/APTI, RTP, NC, 1992.
6. I. Farag and J. Reynolds, *Heat Transfer*, A Theodore Tutorial, Theodore Tutorials, originally published by the USEPA/APTI, RTP, NC, 1994.
7. P. Abulencia and L. Theodore, *Open-Ended Problems: A Future Chemical Engineering Education Approach*, Scrivener-Wiley, Beverly, MA, 2015.

12

Health and Safety Concerns

Introduction

The role of health and safety in the chemical process industries has increased dramatically in the last 45 years. Concerns associated with heat exchangers have correspondingly increased but not nearly at the same rate. Nonetheless, the practicing engineer involved with heat transfer processes and equipment should possess a working understanding of health and safety effects that can arise in dealing with heat exchangers. This chapter attempts to introduce the reader to this topic by presenting a generic overview of health and safety that is complimented with illustrative examples — some of which are directly related to heat exchangers.

This last chapter not only discusses the dangers posed by hazardous substances but also examines the general subject of health, safety, and accident management. The laws and legislation passed to protect workers, the public, and the environment from the effects of these chemicals and accidents are reviewed. The chapter also discusses the regulations and emergency planning. In effect, the chapter addresses topics that one would classify as health, safety, and accident management. The bulk of the material has been adapted from Theodore *et al.* [1], Theodore and Theodore [2], Theodore [3], Flynn and Theodore [4], and Theodore *et al.* [5].

Two general types of potential chemical health, safety, and accident exposures or concerns exist. These are classified as,

1. *Chronic.* Continuous exposure occurs over long periods of time, generally several months to years. Concentrations of inhaled contaminants are usually relatively low. Direct skin contact by immersion, by splash, or by contaminated air involving contact with substances exhibiting low dermal activity.

2. *Acute.* Exposures occur for relatively short periods of time, generally seconds or minutes to 1 to 2 days. The concentration of contaminants is usually high relative to their protection criteria. In addition to inhalation, airborne substances might directly contact the skin, or liquids and sludge may be splashed on the skin or into the eyes leading to toxic effects.

These two types of exposures will be revisited in Section 12.5.

In general, acute exposures to chemicals in air are more typical in transportation accidents, explosions, and fires, or releases at chemical manufacturing or storage facilities. High concentrations of contaminants in air rarely persist for long periods of time. Acute skin exposure may occur when workers come in close contact with the substances in order to control a release e.g., while patching a tank car, offloading a corrosive material, uprighting a drum, or containing and treating a spilled material.

Chronic exposures, on the other hand, are usually associated with longer-term removal and remedial operations. Contaminated soil and debris from emergency operations may be involved in the round-the-clock discharge of pollutants to the atmosphere, soil and groundwater. They may be polluted, or temporary impoundment systems may contain diluted chemicals. Abandoned waste sites typically represent chronic exposure problems. As cleanup activities start at these sites, personnel engaged in certain operations such as sampling, handling containers, or bulking compatible liquids, face an increased risk of acute exposure. These exposures stem from splashes of liquids or from the release of vapors, gases, or particulates that might be generated.

In any specific incident, the hazardous and/or toxic properties of the materials may be the only problem that represents a potential risk. For example, if a tank car containing liquefied natural gas is involved in an accident, but remains intact, the risk from fire and explosion is low. In other incidents, the risks to response personnel are high, e.g., when toxic or flammable vapors are released from a ruptured tube in a heat exchanger. The continued concern for health and safety of response personnel requires that the risks, both real and potential, at an accident be assessed, and appropriate measures instituted to reduce or eliminate the threat to response personnel.

From a health perspective, specific chemicals and chemical groups affect different parts of the body. One chemical, such as one with either a very low or a very high pH, may affect the skin, whereas another, such as carbon tetrachloride, might attack the liver. Some chemicals will affect more than one organ or system. When this occur, the organ or system under attack is referred to as the *target organ*. The damage done to a target organ can differ in severity depending on chemical composition, length of exposure, and the concentration of the chemical.

Another health effect is involved when two different chemicals enter the body simultaneously. The result can be intensified or compounded. A *synergistic effect* results when one substance intensifies the damage done by the other. Synergism complicates almost any exposure due to a lack of toxicological information. For just one chemical, it may typically take a toxicological research facility approximately 2 years of studies to generate valid data. The data produced in that 2 year time frame applies only to the effect of that one chemical acting alone. With the addition of another chemical, the original chemical may have a totally different effect on the body. This fact results in a great many unknowns when dealing with toxic substances, and therefore increases risks due to a lack of dependable information.

The previous paragraphs primarily discussed problems associated with exposure to chemicals. In the chemical industry, there is also a high risk of accidents due to the nature of the processes and the materials used. Although precautions are taken to ensure that all processes run smoothly, there is always (unfortunately) room for error, and accidents will occur. This is especially true for highly technical complicated operations, as well as processes (including heat exchangers) operating under extreme conditions such as high temperatures and pressures.

In general, accidents are caused by one or more of the following factors (see also the beginning of Section 12.6):

1. Equipment breakdown
2. Human error
3. Terrorism
4. Fire exposure and explosions
5. Control system failure(s)

6. Natural causes
7. Utilities and ancillary system outages
8. Faulty siting and plant layout

These causes are usually at the root of most industrial accidents. Although there is no way to guarantee that these problems will not arise with heat exchangers, steps can be taken to minimize the number, as well as the severity, of incidents.

For the readers, it should be noted that health risk addresses risk that arise from health and health-related problems. Chemicals are generally the culprit. Both the effect on and exposure to a receptor (in this case, generally a human) ultimately determine the risk to the individual for the health problem of concern. The risk can be described in either qualitative or quantitative terms, and there are various terms that may be used, e.g., 10 individuals will become sick, or 1×10^{-6} (one in a million) will die, or something as simple as "it is a major problem."

The second category of environmental risk is hazard risk. This class of risk is employed to describe risks associated with hazards or hazard-related problems, e.g., accidents, negative events, and catastrophes. Unlike most health problems, these usually occur over a short period of time, e.g., seconds or minutes. Both the probability and the consequence associated with the accident/event ultimately determine the hazard risk. Once again, the risk can be described in either qualitative or quantitative terms, and there are various terms that may be used.

Once a risk has been calculated, one needs to gauge the estimated consequences (or opportunities if examining financial/economic scenarios) and evaluate and prioritize options for risk management or mitigation. These potentially strategic evaluations are usually fraught with uncertainties at numerous levels. Thus, the risk assessment process is normally followed by option analyses; these options can be based on decision-making procedures that are beyond the scope of this book. However, it is fair to say that there may be a full range of outcomes and consequences to various scenarios. It should also be noted that risk assessment is a dynamic process that can very definitely be a function of time. Some of the material in this paragraph is addressed in this chapter, but some only receive superficial treatment.

Environmental risk is one type of risk. Although this chapter primarily addresses this class of risk, there are others. Consider for example, financial risk that can arise in the chemical process industry. The cost of the capital of a proposed new venture, e.g., purchasing a heat exchanger, is primarily dependent on three factors: (1) the proportion of equity to debt financing, (2) the method of financing involved, and (3) the risk inherent in the proposed project. As with environmental risk, uncertainties abound in these economic analyses. Discounted cash flow rates of return or net present values for these projects can rarely be predicted with absolute certainty because of a host of factors [3]. This topic is also beyond the scope of this chapter.

Since emergency planning and response to both health and hazard problems are two topics directly related to environmental risk, they are briefly covered in this chapter. Although these two subjects are primarily concerned with hazards, they can — and often do — play an important role in any comprehensive environmental risk analysis.

The material to follow — all six sections — attempts to examine the aforementioned topic areas in a clear and comprehensive manner. Essay material is complemented with numerous illustrative examples, many of which have been drawn from real-world experiences and applications. Topics covered in this chapter include:

12.1   Definitions
12.2   Legislation
12.3   Material Safety Data Sheets (MSDSs)
12.4   Health Risk versus Hazard Risk
12.5   Health Risk Assessment
12.6   Hazard Risk Assessment

12.1   Definitions

Material in this section was drawn directly from the work of Theodore, Reynolds, and Morris [6].

Acute (risk). Risks associated with short periods of time. For health risk, it usually represents short exposures to high concentrations of a hazardous agent.

Atmospheric dispersion. The mixing of a gas or vapor (usually from a discharge point) with air in the lower atmosphere. The mixing is the result of convective motion and turbulent eddies.

Atmospheric stability. A measure of the degree of atmospheric turbulence, often defined in terms of the vertical temperature gradient in the lower atmosphere.

Auto-ignition temperature (AIT). The lowest temperature at which a flammable gas in air will ignite without an ignition source.

Average rate of death (ROD). The average number of fatalities that can be expected per unit time (usually on an annual basis) from all possible risks and/or incidents.

C (ceiling). The term used to describe the maximum allowable exposure concentration of a hazardous agent related to industrial exposures with hazardous vapors.

Cancer. A tumor formed by mutated cells.

Carcinogen. A cancer-causing chemical.

Catastrophe. A major loss in terms of death, injuries, and damage.

Cause–consequence. A method for determining the possible consequences or outcomes arising from a logical combination of input events or conditions that determine a cause.

CAS. Chemical Abstract Service; CAS numbers are used to identify chemicals and mixtures of chemicals.

Chronic (risk). Risks associated with long-term chemical exposure duration, usually at low concentrations.

Conditional probability. The probability of occurrence of an event given that a precursor event has occurred.

Confidence interval. A range of values of a variable with a specific probability that the true value of the variable lies within this range. The conventional confidence interval probability is the 95% confidence interval, defining the range of a variable in which its true value falls with 95% confidence.

Confidence limits. The upper and lower range of values of a variable defining its specific confidence interval.

Consequences. A measure of the expected effects of an incident outcome or cause.

Continuous release. Emissions that are of an extended or continuous duration.

Deflagration. The chemical reaction of a substance in which the reaction front advances into the unreacted substance present at less than sonic velocity.

Delphi method. A polling of experts that involves the following:

1. Select a group of experts (usually three or more).
2. Solicit, in isolation, their independent estimates of the value of a particular parameter and their reason for the choice.
3. Provide all initial analysis results to all experts and allow them to then revise their initial values.
4. Use the average of the final estimates as the best estimate of the parameter. Use the standard deviation of the estimates as a measure of uncertainty.

The procedure is iterative, with feedback between iterations. One of the authors modestly refers to it as the Theodore method [2, 3, 6].

Dermal. Applied to the skin.

Detonation. A release of energy caused by a rapid chemical reaction of a substance in which the reaction front advances into the unreacted substance present at greater than sonic velocity.

Dose. The amount of a substance available for interaction with metabolic processes or biologically significant receptors after crossing the outer boundary of an organism. The *potential* dose is the amount ingested, inhaled, or applied to the skin. The *applied* dose is the amount of a substance presented to an absorption barrier and available for absorption (although not necessarily having yet crossed the outer boundary of the organism). The *absorbed* dose is the amount crossing a specific absorption barrier (e.g., the exchange boundaries of skin, lung, and digestive tract) through uptake processes. *Internal dose* is a more general term denoting the amount absorbed without respect to specific absorption barriers or exchange boundaries. The amount of the chemical available for interaction by any particular organ or cell is termed the *deliverable dose* for that organ or cell.

Dose rate. This represents the dose per unit time, for example, in mg/day, also referred to as *dosage*. Dose rates are often expressed on a per-unit bodyweight basis, yielding units such as mg/kg-day. They are often expressed as averages over some time period, e.g., a lifetime.

Dose-response curve. A graphical representation of the quantitative relationship between the administered, applied, or internal dose of a chemical or agent and a specific biological response to that chemical or agent.

Episodic release. A massive release of limited or short duration, usually associated with an accident.

Equipment reliability. The probability that, when operating under stated conditions, the equipment will perform its intended purpose for a specified period of time.

Event. An occurrence associated with an incident either as the cause or a contributing cause of the incident, or as a response to an initiating event.

Event sequence. A specific sequence of events composed of initiating events and intermediate events that may lead to a problem or an incident.

Event tree analysis (ETA). A graphical logic model that identifies and attempts to quantify possible outcomes following an initiating event.

Explosion. A release of energy that causes a pressure discontinuity or blast wave.

Exposure period. The duration of an exposure.

External event. A natural or man-made event; often an accident.

Failure frequency. The frequency (relative to time) of failure.

Failure mode. A symptom, condition, or manner in which a failure occurs.

Failure probability. The probability that failure will occur, usually in a given time interval.

Failure rate. The number of failures divided by the total elapsed time during which these failures occur.

Fatal accident rate (FAR). The estimated number of fatalities per 10^8 exposure hours (roughly 1000 employee working lifetimes).

Fault tree. A method for representing the logical combinations of events that lead to a particular outcome (top event).

Fault tree analysis (FTA). A logic model that identifies and attempts to quantify possible causes of an event.

Federal Register. A daily government publication of laws and regulations promulgated by the U.S. Federal Government.

Flammability limits. The range in which a gaseous compound *in air* will explode or burst into flames if ignited.

Frequency. Number of occurrences of an event per unit time.

Gaussian model. A plume dispersion model based on mixing and turbulence in the lower atmosphere.

Half-life. The time required for a chemical concentration or quantity to decrease by half its current value.

Hazard (problem). An event associated with an accident which has the potential for causing damage to people, property, or the environment.

Hazard and operability study (HAZOP). A technique to identify process hazards and potential operating problems using a series of guide words that key on process deviations.

Hazard risk assessment (HZRA). A technique associated with quantifying the risk of a hazard employing probability and consequence information.

Health (problem). An environmental problem normally associated with and arising from the continuous emission of a chemical into the environment.

Health risk assessment (HRA). A technique associated with quantifying the risk of a health problem employing toxicology and exposure information.

Human error. Actions by engineers, operators, managers, etc., that may contribute to or result in accidents.

Human error probability. The ratio between the actual number of human errors and the number of opportunities for human error.

Human factors. Factors attempting to match human capacities and limitations.

Human reliability. A measure of human errors.

Immediately dangerous to life and health (IDLH). The concentration representing the maximum level of a pollutant from which an individual could escape within 30 minutes without impairing symptoms or irreversible health effects.

Incident. An event.

Individual risk. The risk to an individual.

Ingestion. The intake of a chemical through the mouth.

Initiating event. The first event in an event sequence.

Instantaneous release. Emissions that occur over a very short duration.

Intermediate event. An event that propagates or mitigates the initiating event during an event sequence.

Isopleth. A concentration plot at specific locations, usually downwind from a release source.

Lethal concentration (LC). The concentration of a chemical that will kill a test animal, usually based on 1–4 h exposure duration.

Lethal concentration 50 (LC$_{50}$). The concentration of a chemical that will kill 50% of test animals, usually based on 1–4 h exposure duration.

Lethal dose (LD). The quantity of a chemical that will kill a test animal, usually normalized to a unit of body weight.

Lethal dose 50 (LD$_{50}$). The quantity of a chemical that will kill 50% of test animals, usually normalized to a unit of body weight.

LEL/LFL. The lower explosive/flammability limit of a chemical in air that will produce an explosion or flame if ignited.

Level of concern (LOC). The concentration of a chemical above which there may be adverse human health effects.

Likelihood. A measure of the expected probability or frequency of occurrence of an event.

Limit of detection (LOD). The minimum concentration of a substance being measured that, in a given matrix and with a specified method, has a 99% probability of being identified, qualitatively or quantitatively measured, and reported to be greater than zero.

Lowest observed effect level (LOEL). In dose-response experiments, the lowest exposure level at which there are statistically biologically significant increases in the frequency or severity of any effect between the exposed population and its appropriate control group.

Malignant. A cancerous tumor.

Maximally exposed individual / maximum exposed individual (MEI). The single individual with the highest exposure in a given population. This term has historically been defined in various ways, including as defined here, and is also synonymous with a worse case or bounding estimate.

Mutagen. A chemical capable of changing a living cell.

Permissible exposure limit (PEL). Expressed as a time-weighted average (TWA), the PEL is the concentration of a substance to which most workers can be exposed without adverse effects, averaged over a normal 8-hour workday or a 40-hour workweek. Also, the permissible exposure limit of a chemical in air, established by the Occupational Safety and Health Administration (OSHA).

Personal protection equipment (PPE). Material/equipment worn to protect a worker from exposure to hazardous agents.

Precision. The degree of "exactness" of repeated measurements.

ppm. The parts per million of a chemical in air — almost always on a volume basis; often designated as ppmv as opposed to ppmm (mass basis).

ppb. The parts per billion of a chemical in air — almost always on a volume basis; often designated as ppbv as opposed to ppbm (mass basis).

Maximum individual risk. The highest individual risk in an exposed population.

No observed adverse effect level (NOAEL). In dose–response experiments, an exposure level at which there are no statistically biologically significant increases in the frequency or severity of adverse effects between the exposed population and its appropriate control; some effects may be produced at this level, but they are not considered to be adverse, nor precursors to specific adverse effects. In an experiment with more than one NOAEL, the regulatory focus is primarily on the highest one, leading to the common usage of the term NOAEL to mean the highest exposure level without adverse effect.

No observed effect level (NOEL). In dose–response experiments, an exposure level at which there are no statistically biologically significant increases in the frequency or severity of any effect between the exposed population and its appropriate control group.

Probability. An expression for the likelihood of occurrence of an event or an event sequence, usually over an interval of time.

Protective system. Systems, such as pressure vessel or heat exchanger relief valves, that function to prevent or mitigate the occurrence of an accident or incident.

Recommended exposure limit (REL). NIOSH-recommended exposure limit for an 8 or 10-hour TWA exposure and/or ceiling.

Reference dose (RfD). The EPA's preferred toxicity value for evaluating noncarcinogenic effects resulting from exposures at Superfund sites.

Risk. A measure of economic loss or human injury in terms of both the incident likelihood and the magnitude of the loss or injury.

Risk analysis. The engineering evaluation of incident consequences, frequencies, and risk assessment results.

Risk assessment. The process by which risk estimates are made.

Risk contour. Lines on a risk graph that connect points of equal risk.

Risk estimation. Combining the estimated consequences and likelihood of a risk.

Risk management. The application of management policies, procedures, and practices in analyzing, assessing, and controlling risk.

Risk perception. The perception of risk that is a function of age, race, sex, personal history and background, familiarity with the potential risk, dread factors, perceived benefits of the risk causing action, marital status, residence, etc.

Societal risk. A measure of risk to a group of individuals.

Source term. The estimation of the release of a hazardous agent from a source.

Time of failure. The time period associated with the inability to perform a duty or intended function.

TLV. The threshold limit value (established by the American Council of Government Industrial Hygienists (ACGIH)). The concentration of a chemical in air that produces no adverse effects.

TLV-C. The ceiling exposure limit representing the maximum concentration of a chemical in air that should never be exceeded.

TLV-STEL. The short-term exposure limit (maximum concentration in air) for a continuous 15-minute averaged exposure duration.

TLV-TWA. The allowable time weighted average concentration of a chemical in air for an 8-hour workday/40-hour workweek that produces no adverse effect.

Top event. The accident, event, or incident at the "top" of a fault tree that is traced downward to more basic failures using logic gates to determine their causes.

Toxic dose. The combination of concentration and exposure period for a toxic agent to produce a specific harmful effect.

UEL/UFL. The upper explosive/flammability limit of a chemical in air that will produce an explosion or flame if ignited.

Uncertainty. A measure, often quantitative, of the degree of doubt or lack of certainty associated with an estimate.

The reader should also note that there are other risks — in addition to environmental ones — that the practicing engineer and applied scientist must be proficient in understanding. Perhaps the most important of these is financial risk. And, although this chapter is primarily concerned with the aforementioned environmental health risk and hazard risk, the authors would be negligent if this topic of financial risk were not at least qualitatively mentioned.

Example 12.1 – Annual vs. Lifetime Risk Comparison. Compare annual versus lifetime risks.

Solution: A time frame must be included with a risk estimate for the numbers to be meaningful. For both health and hazard risks, annual or lifetime risks are commonly used. Direct evidence is usually expressed annually because the information is often collected and summarized annually. However, predictive information is commonly expressed as a lifetime probability, e.g., when expressing cancer risk or a terrorist-related risk.

Example 12.2 – Interpreting Cancer Risk. Describe what a cancer risk number of 10^{-6} probability means.

Solution: A cancer risk number usually represents a lifetime probability of developing cancer risk. A risk of 10^{-6} indicates an individual has a 1 in 1,000,000 chance of developing cancer throughout a lifetime (assumed to be 70 years). One generally can also assume an upper 95% confidence limit on the maximum likelihood estimate. Since the predicted risk is an upper bound, the actual risk is unlikely to be higher but may be much lower than the predicted risk.

Example 12.3 – Dose Units. Comment on the units associated with the various dose terms.

Solution: Doses generally are expressed in terms of the quantity administered per unit body weight, quantity per skin surface area, or quantity per unit volume of the respired air. In addition, doses are also expressed over the duration of time that the dose was administered. Dose amounts are generally expressed as milligrams (one thousandth of one gram) per kilogram body weight (mg/kg), in some cases grams per kilogram (g/kg), micrograms (one millionth of a gram) per kilogram (μg/kg), or nanograms (one billionth of a gram) per kilogram (ng/kg) are used. Volume measurements of dose can be converted to weight units by appropriate calculations. Densities can be obtained from standard reference texts. Where densities are not available, liquids are assumed to have a density of 1 g/mL. All body weights are converted to kilograms (kg) for uniformity. Concentrations of a gaseous substance in air are generally listed as parts of vapor or gas per million *by volume* (ppm$_v$). Concentrations of liquid or solid substances are usually expressed as parts per million *by weight* (ppm$_w$) *or mass* (ppm$_m$). Other units include any mass per unit volume combination of units.

Example 12.4 – Concentration Unit. Convert concentration units to parts per million by volume (ppm$_v$) from mg/m^3 at a standard temperature and pressure of 0 °C and 1.0 atm.

SOLUTION: Set x as the concentration in mg/m^3 at 0 °C and 1.0 atm. Apply the appropriate conversion factors:

$$\text{ppm}_v = x \left(\frac{\text{mg}}{\text{m}^3} \right) \left(\frac{\text{m}^3}{\text{L}} \right) \left(\frac{22.4\,\text{L}}{\text{gmol}} \right) \left(\frac{\text{g}}{10^3\text{mg}} \right) \left(\frac{\text{gmol}}{\text{MW}(\text{g})} \right)$$

where, MW = molecular weight, g/gmol. The above expression results in the following,

$$\text{ppm}_v = x \left[\frac{22.4}{(10^6)(\text{MW})} \right] \tag{12.1}$$

One notes that the conversion is *not* possible if the molecular weight (MW) of the gas is unknown or not specified.

12.2   Legislation

The concern for health and safety is reflected in the legislation summarized in the literature [6–8]. Although the Clean Air Act does not cover emergency planning and response in a clear and comprehensive manner, certain elements of the act are particularly significant. These include implementation plans and national emission standards for hazardous air pollutants. The Clean Water Act as well as other legislation pertaining to water pollution provides emergency planning and response that is more developed than it is for air. The Resource Conservation and Recovery Act (RCRA) and the Comprehensive Environmental Response, Compensation, and Liability Act (CERCLA) are two important pieces of legislation that are concerned with preventing releases, and with the requirements for the cleanup of hazardous and toxic sites. RCRA and CERCLA also contain specific sections that address emergency planning and response. The Superfund Amendments and Reauthorization Act (SARA) is another important piece of legislation. SARA deals with the cleanup of hazardous waste sites as well as emergency planning and response. Title III, which is the heart of SARA, establishes requirements for emergency planning and "community right to know" for federal, state, and local governments, as well as industry. Title III is a major stepping-stone in the protection of the environment, but its principal thrust is to facilitate planning in the event of a catastrophe. The Occupational Safety and Health Act (OSHAct) was enacted by Congress in 1970 and established the Occupational Safety and Health Administration (OSHA) which addressed safety in the workplace. Both EPA and OSHA are mandated to reduce the exposure of hazardous substances over land, sea, and air. The OSHAct is limited to conditions that exist in the workplace where its jurisdiction covers both safety and health. Frequently, two agencies may regulate the same substance but in a different manner if they are overlapping environmental organizations.

Developed under the Clean Air Act's (CAA's) Section 112(r), the Risk Management Program (RMP) rule (40 CFR Part 68) is designed to reduce the risk of accidental releases of acutely toxic, flammable, and explosive substances. A list of the regulated substances (138 chemicals) along with their threshold quantities is provided in the Code of Federal Regulations at 40 CFR 68.130.

A brief overview of OSHA and RMP is provided in the next two subsections.

Occupational Safety and Health Act (OSHA) [6–8]. Congress intended that OSHA be enforced through specific standards in an effort to achieve a safe and healthy working environment. A "general duty clause" was added to attempt to cover those obvious situations for which no specific standard existed. The OSHA standards are an extensive compilation of regulations, some that apply to all employers (such as eye and face protection) and some that apply to workers who are engaged in a specific type of work (such as welding or crane operation). Employers are obligated to familiarize themselves with the standards and comply with them at all times.

Health issues, most importantly contaminants in the workplace, have become OSHA's primary concern. Health problems are complex and difficult to define. Because of this, OSHA has been slow to implement health standards. To be complete, each standard requires medical surveillance, record keeping, monitoring and physical reviews. On the other side of the ledger, safety hazards are aspects of the work environment that are expected to cause death or serious physical harm immediately or before the imminence of such dangers can be eliminated.

Probably one of the most important safety and health standards ever adopted is the OSHA hazard communication standard, more properly known as the "right to know" laws. The hazard communication standard

requires employers to communicate information to the employees on hazardous chemicals that exist within the workplace. The program requires
employers to craft a written hazard communication program, keep *material safety data sheets* (MSDSs) — see next section for more details — for all
hazardous chemicals at the workplace and provide employees with training on those hazardous chemicals, and assure that proper warning labels
are in place.

As noted in the previous paragraph, each company must develop a
health and safety program for its workers. For example, OSHA has regulations governing employee health and safety at hazardous waste operations
and during emergency responses to hazardous substance releases. These
regulations (29 CFR 1910.120) contain general requirements for,

1. Safety and health programs
2. Training and educational programs
3. Work practices along with personal protective equipment
4. Site characterization and analysis
5. Site control and evacuation
6. Engineering controls
7. Exposure monitoring and medical surveillance
8. Materials handling and decontamination
9. Emergency procedures
10. Illumination
11. Sanitation

The EPA Standard Operating Safety Guides supplement these regulations.
However, OSHA's regulations must be used for specific legal requirements
for an industry. Other OSHA regulations pertain to employees working
with hazardous materials or at hazardous waste sites. These, as well as state
and local regulations, must also be considered when developing worker
health and safety programs [9].

The OSHA Hazard Communication Standard was first promulgated
on November 25, 1983, and can be found in 29 CFR Part 1910.120. The
standard was developed to inform workers who are exposed to hazardous chemicals of the risk associated with specific chemicals. The purpose
of the standard is to ensure that the hazards of all chemicals produced or
imported are evaluated, and that information concerning chemical hazards is conveyed to *both* employers and employees.

Information on chemical hazards must be dispatched from the manufacturers to employers via the aforementioned *material safety data sheets*
(MSDSs) and container labels. These data must then be communicated

Table 12.1 RMP Approach.

Program	Description
1	Facilities submit RMP, complete registration of processes, analyze worst-case release scenario, complete 5-year accident history, coordinate with local emergency planning and response agencies; and, certify that the source's worst-case release would not reach the nearest public receptors.
2	Facilities submit RMP, complete registration of processes, develop and implement a management system; conduct a hazard risk assessment; implement certain prevention steps; develop and implement an emergency response program; and, submit data on prevention program elements.
3	Facilities submit RMP, complete registration of processes, develop and implement a management system; conduct a hazard risk assessment; implement prevention requirements; develop and implement an emergency response program; and, provide data on prevention program elements.

to employees by means of comprehensive hazard communications programs, which include training programs as well as the MSDSs and container labels.

USEPA's Risk Management Program (RMP). In the RMP rule, EPA requires a *Risk Management Plan* that summarizes how a facility is to comply with EPA's RMP requirements. It details methods and results of hazard assessment, accident prevention, and emergency response programs instituted at the facility. The hazard assessment shows the area surrounding the facility and the population potentially affected by accidental releases. EPA requirements include a three-tiered approach for affected facilities. A facility is affected if a process unit manufactures, processes, uses, stores, or otherwise handles any of the listed chemicals at or above the threshold quantities. The RMP approach is summarized in Table 12.1.

12.3   Material Safety Data Sheets

The MSDS have, over the past 35+ years, become the major media for transmitting health and hazard information concerning chemicals. MSDSs have also played a key role in health and hazard communication programs [6–8]. MSDS preparers and users have discovered that there are many aspects to their preparation that can affect their utility. Since MSDSs have multiple audiences, their preparation has become a complicated process. The preparation and use of MSDS have thus become an evolving and ongoing process. On the global scene, there is increasing pressure to try to develop an international standard for MSDS, so the same document can be used in diffcrent international markets.

The MSDS is a detailed document prepared by the manufacturer or importer of a chemical that describes its formulation, composition, precautions for proper use and handling as well as the potential physical and health problems it might pose to users. The MSDS details routes of exposure, emergency and first aid procedures, and control measures in case of an uncontrolled release. Information on an MSDS aids in the selection of safe products and helps prepare employers and employees to respond effectively to daily exposure situations as well as to emergency releases. It is also a source of information for identifying chemical hazards (accidents), a topic treated later in this chapter.

In line with the Occupational Safety and Health Administration (OSHA) requirements, employers must maintain a complete and accurate MSDS for each hazardous chemical that is used in their facility. They are entitled to obtain this information automatically upon purchase of the material. When new and significant information becomes available concerning a product's hazards or ways to protect against the hazards, chemical manufacturers, importers, or distributors *must* add it to their MSDS within 3 months and provide this updated information to their customers with the next shipment of the chemical. If there are multiple suppliers of the same chemical, there is no need to retain multiple MSDSs for that chemical.

While MSDSs are not required to be physically attached to a shipment, they must accompany or precede the shipment. When the manufacturer/supplier fails to send an MSDS with a shipment labelled as hazardous, the employer must obtain one from the chemical manufacturer, importer, or distributor as soon as possible. Similarly, if the MSDS is incomplete or unclear, the employer should contact the manufacturer or importer to seek clarification or obtain missing information. When an employer is unable to obtain an MSDS from a supplier or manufacturer, he or she should submit a written complaint, with complete background information, to the nearest OSHA area office. OSHA will then call and send a certified letter to the supplier or manufacturer to obtain the needed information. If the supplier or manufacturer still fails to respond within a reasonable time, OSHA will inspect the supplier or manufacturer and take appropriate enforcement action against them.

It is important to note that OSHA specifies the information to be included on an MSDS but does not prescribe the precise format for an MSDS.

A non-mandatory MSDS form that meets the Hazard Communication Standard requirements can be used as is or expanded as needed. The MSDS must include at least the information presented in Table 12.2. In reviewing this material, the reader should understand the effect and importance of each subsection in helping to identify a chemical hazard, particularly the section on health hazards.

Table 12.2 MSDS Information.

Chemical identity	• The chemical and common names must be provided for single chemical substances. • An identity on the MSDS must be cross-referenced to the identity found on the label.
Hazardous ingredients	• For a hazardous chemical mixture that has been tested as a whole to determine its hazards, the chemical and common names of the ingredients that are associated with the hazards and the common name of the mixture must be listed. • If the chemical is a mixture that has not been tested as a whole, the chemical and common names of all ingredients determined to be health hazards and compromising 1% or greater of the composition must be listed. • Chemical and common names of carcinogens must be listed if they are present in the mixture at levels of 0.1% or greater. • All components of a mixture that have been determined to pose a physical hazard must be listed. • Chemical and common names of all ingredients determined to be health hazards and comprising less than 1% (0.1% for carcinogens) of the mixture must also be listed if they can still exceed an established OSHA Permissible Exposure Limit (PEL) or the ACGIH Threshold Limit Value (TLV) or present a health risk to exposed employees in these concentrations.
Physical and chemical characteristics	• The physical and chemical characteristics of the hazardous substance must be listed. These include items such as boiling and freezing points, density, vapor pressure, specific gravity, solubility,

(Continued)

Table 12.2 Cont.

	volatility, and the product's general appearance and odor. These characteristics provide important information for designing safe and healthy work practices.
Fire and explosion hazard data	• The compound's potential for fire and explosion must be described. Also, the fire hazards of the chemical and the conditions under which it could ignite or explode must be identified. Recommended extinguishing agents and firefighting methods must be described.
Reactivity data	• This section of the MSDS presents information about other chemicals and substances with which the chemical is incompatible or with which it reacts. Information on any hazardous decomposition products, such as carbon monoxide, must be included.
Health problems	• The acute and chronic health problem of the chemical, together with signs and symptoms of exposure, must be listed. In addition, any medical conditions that are aggravated by exposure to the compound must be included. The specific types of chemical health hazards defined in the standard include carcinogens, corrosives, toxins, irritants, sensitizers, mutagens, teratogens, and effect on target organs (i.e., liver, kidney, nervous system, blood, lungs, mucous membranes, reproductive system, skin, eyes, etc.). • The route of entry section describes the primary pathway by which the chemical enters the body. There are three principal routes of entry: inhalation, skin, and ingestion. • This section of the MSDS supplies the PEL, the TLV, and other exposure levels used or recommended by the chemical manufacture. • If OSHA, the National Toxicology Program (NTP), or the International Agency for Research on Cancer (IARC) list the compound as a carcinogen (cancer causing agent), it must be indicated as such on the MSDS.

(Continued)

Table 12.2 Cont.

Precautions for safe handling and use	• The standard requires the preparer to describe the precautions for safe handling and use. These include recommended industrial hygiene practices, precautions to be taken during repair and maintenance of equipment, and procedures for cleaning up spills and leaks. Some manufacturers also use this section to include useful information not specifically required by the standard, such as EPA waste disposal methods plus state and local requirements.
Control measures	• The standard requires the preparer of the MSDS to list any generally applicable control measures. These include engineering controls, safe handling procedures, and personal protective equipment. Information is often included on the use of goggles, gloves, body suits, respirators, and face shields.
Employer's responsibilities	• Employers must ensure that each employee has a basic knowledge of how to find information on an MSDS and how to properly make use of that information. Employers also must ensure the following: 1. Complete and accurate MSDSs are made available during each work shift to employees when they are in their work areas. 2. Information is provided for each hazardous chemical.

Example 12.5 – Qualitatively Describing MSDS. Qualitatively describe in "layman language" the health and safety information that is provided on an MSDS.

SOLUTION: An MSDS serves as a reference source for information on a hazardous substance. The MSDS identifies the substance, the producer or seller of the substance, and the location of the producer or seller; explains why the substance is hazardous and how a person can be exposed to the substance; identifies conditions that increase the hazard; explains safe handling procedures; identifies proper protective clothing or devices to be used when working with the substance; explains the steps that should be taken if a person is exposed to the substance; and, explains the steps that should be taken if there is a spill or emergency situation.

Example 12.6 – Limitations of MSDS. Discuss some of the limitations of MSDS use.

SOLUTION: The information contained in the MSDS is highly variable and is dependent on the supplier's knowledge and expertise. Work is in progress to improve the quality of information in MSDSs and (as noted earlier) to standardize the MSDS format. The American National Standards Institute (ANSI) has developed a MSDS format.

Understanding the terminology used in MSDSs can be a problem for some employees. Additional training and use of supplemental information can help in dealing with this problem.

12.4   Health Risk versus Hazard Risk

People face all kinds of risks every day, some voluntarily and others involuntarily [6–8]. Therefore, risk plays a very important role in today's world. Studies on cancer caused a turning point in the world of risk because it opened the eyes of risk engineers, scientists, and health professionals to the world of risk assessments.

Both health risk assessments (HRAs) and hazard risk assessment (HZRA) are ultimately concerned with characterizing risk. As noted earlier, the description of risk is closely related to probability. For example, a probability of unity, i.e., 1.0, indicates that a health/hazard problem *will* occur; alternately, a probability of zero, i.e., 0.0, indicates that a health/hazard problem definitely will *not* occur.

Unfortunately, the word "risk" has come to mean different things to different people. Although defined earlier, here are two additional definitions. Webster defines risk as "… the chance of injury, danger or loss … to expose to the chance of injury, damage, or loss." Stander and Theodore [8] have defined it as "a combination of uncertainty and change." To compound this problem, there are *two* types of risk that environmental professionals are concerned with: *health risk* and *hazard risk*. However, these two classes of risk have been used interchangeably by practitioners, researchers, and regulators. Because of this confusion, one of the main objectives of this chapter is to both define and clarify the differences between these two risks.

Irrespective of the category of the applicable risks for a system, the total risk, R, is given by the summation of the risk R from all n events/scenarios, i.e.,

$$R = \sum_{i=1}^{n} R_i \qquad (12.2)$$

In addition, the magnitude of each risk can be described relative to the total risk.

Regarding human health risk, concern arises because chemicals can possibly elude natural defense mechanisms upon entering the human body. Exposure to chemicals can lead to various pathways of entry into the human body. As noted earlier, these include inhalation, skin absorption (absorption), and ingestion (digestion system). It is fair to say that the dominant route of human exposure to hazardous chemicals is via inhalation. Note also that two types of potential exposures exist relative to the concentration and duration of the exposure: chronic and acute; both were described in the introduction to this chapter. Alternatively, hazard risk, which is classified in the *acute* category, is described as (1) a ratio of hazards (e.g., an explosion) to failures of safeguards; (2) a triplet combination of event, probability, and consequences; or even (3) a measure of economic loss or human injury in terms of both an incident likelihood and the magnitude of the loss or injury.

Health risk and its assessments are addressed, in part, under the Clean Air Act (CAA) Section 112(d) and (f), where "EPA must promulgate (along with methods of calculating) residual risk standards for the source category as necessary to provide an ample margin of safety to protect public health." Regarding hazard risk, the CAA's Section 112(r), the *Risk Management Program* (RMP) rule (40 CFR Part 68), is designed to reduce the risk of accidental releases of acutely toxic, flammable, and explosive substances [8]. Both of these risk terms are addressed in this chapter and the next two sections.

12.5   Health Risk Assessment

No study of health and safety would be complete without introducing the general subjects of health risk assessment and hazard risk assessment [6–8]. The concluding two sections of this chapter attempts to satisfy this need. Although much of the material of this and the next/last section is generic in content, the authors have attempted — via some of the illustrative examples — to relate the two topics to heat transfer processes and equipment.

For the purposes of this and the last section, a few definitions of common terms will again be defined. *Health risk* is the probability that individuals or the environment will suffer adverse health consequences as a result of an exposure to a substance. The amount of risk is determined by a combination of the concentration the person or the environment is exposed to, the rate of intake (or dose) of the substance, and the toxicity of the substance. *Risk assessment* is the procedure used to attempt to quantify or estimate this risk. *Risk-based decision making* also distinguishes between the terms *point of exposure* and the *point of compliance*. The *point of exposure* is the point at which the environment or the individual comes into contact with the chemical release. An individual may be exposed by methods such as inhalation of vapors, as well as physical contact with the substance. The *point of compliance* is a point in between the point of release of the chemical (i.e., the source), and the point of exposure. The point of compliance is selected to provide a safety buffer for effected individuals and/or environments.

Since 1970, the field of risk assessment has received widespread attention within the engineering, scientific, and regulatory committees. It has also attracted the attention of the public. Properly conducted risk assessments have received fairly broad acceptance because they put into perspective the aforementioned terms *toxic, health problems,* and *risk. Toxicity* is an inherent property of all substances. It states that all chemical and physical agents can produce adverse health effects at some dose or under specific exposure conditions. In contrast, exposure to a chemical that has the capacity to produce a particular type of adverse effect represents a health problem. As noted, risk is the probability or likelihood that an adverse outcome will occur in a person or a group that is exposed to a particular concentration or dose of the hazardous agent. Therefore, health risk is generally a function of both exposure and dose. Consequently, *health risk assessment* is defined as the process or procedure used to estimate the likelihood that humans or ecological systems will be adversely affected by a chemical or physical agent under a specific set of conditions [9,10].

The term *risk assessment* has been used to describe or predict not only the likelihood of an adverse response to a chemical or physical agent but also or of any unwanted event. This latter subject is treated in more detail in the next section. These include risks such as explosions or injuries in the workspace; natural catastrophes; injury or death due to various voluntary activities such as skiing, ski diving, flying or bungee jumping; diseases; death due to natural causes; and, many others [11].

Risk assessment and *risk management* are two different processes, but they are closely related. Risk assessment and risk management provide a framework not only for setting regulatory priorities but also for making decisions that cut across different environmental areas. *Risk management* generally refers to a decision-making process that involves such considerations as risk assessment, technology feasibility, economic information about costs and benefits, regulatory requirements, and public concerns. Therefore, risk assessment supports risk management in that the choices on whether and to what extent to control future exposure to the suspected hazards may be determined [12]. Regarding both risk assessment and risk management, this section will address this subject primarily from a health perspective; as noted, the next section will examine risk assessment from a safety and accident perspective.

Health risk assessments provide an orderly, explicit way to deal with scientific issues in evaluating whether a health problem exists and what the magnitude of the problem may be. This evaluation typically involves large uncertainties because the available scientific data are limited and the mechanisms for adverse health impacts or environmental damage are only imperfectly understood.

When examining risk, how does one decide how safe is "safe" or how clean is "clean"? To begin with, one has to look at both sides of the risk equation, i.e., both the toxicity of a pollutant and the extent of public exposure. Information is required for both the current and the potential exposure, considering all possible exposure pathways. It should be remembered that there are always uncertainties in conducting a comprehensive health risk assessment and these assumptions must be included in the analysis. In addition to human health risks, one also needs to look at potential ecological or other environmental effects.

In recent years, several guidelines and handbooks have been produced to help explain approaches for doing health risk assessments. As discussed by a special National Academy of Sciences committee convened in 1983, most human or environmental health problems can be evaluated by dissecting the analysis into four parts: health problem identification, dose-response assessment or toxicity assessment, exposure assessment, and risk characterization (see Figure 12.1). For some perceived health problems the risk assessment might stop with the first step, identification, if no adverse effect is identified, or if an agency elects to take regulatory action without further analysis [13]. Regarding health problem identification, a problem is defined as a toxic agent or a set of conditions that has the potential to cause adverse effects to human health or the environment. The identification process involves an evaluation of various forms of information in order to identify the different health problems. Dose-response or toxicity assessment is required in an overall assessment; responses/effects can vary widely since all chemicals and contaminants vary in their capacity to cause adverse effects. This step frequently requires that assumptions be made to relate experimental data from animals and humans. Exposure assessment is the determination of the magnitude, frequency, duration, and routes of exposure of human populations and ecosystems. Finally, in risk characterization, toxicology, and exposure data/information are combined to obtain qualitative or quantitative expressions of risk.

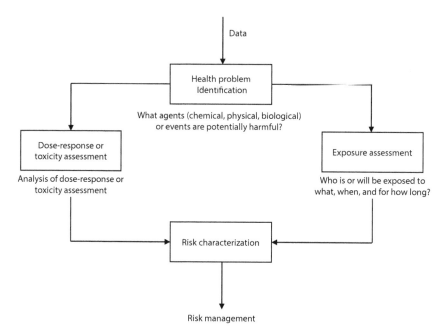

Figure 12.1 The health risk evaluation process [6–9, 13].

Risk assessment also involves the integration of the information and analysis associated with the above four steps to provide a complete characterization of the nature and magnitude of risk and the degree of confidence associated with this characterization. A critical component of the assessment is a full elucidation of the aforementioned uncertainties associated with each of the major steps. All of the essential problems of toxicology are encompassed under this broad concept of risk assessment. Risk assessment takes into account all of the available dose-response data. It should treat uncertainty not by the application of arbitrary safety factors, but by stating them in quantitatively and qualitatively explicit terms, so that they are not hidden from decision-makers. Risk assessment, defined in this broad way, forces an assessor to confront all the scientific uncertainties and to set forth in explicit terms the means used in specific cases to deal with these uncertainties [10]. In effect, risk characterization is the process of estimating the incidence of a health effect under the various conditions of human or animal exposure described in the exposure assessment. As noted above, it is performed by combining the exposure and dose-response assessments; the summary effects of the uncertainties in the preceding steps should also be described in this step.

Summarizing, both risk assessment and risk management, the examples in this section primarily addresses this subject from a health perspective; the next section will primarily address this subject from a safety and accident perspective.

The reader should once again note that two general types of potential risks exist. These were classified earlier in the Introduction to this chapter as:

1. *Acute.* Exposures that occur for relatively short periods of time, generally from minutes to one or two days. Concentrations of (toxic) air contaminants are usually high relative to their protection criteria. In addition to inhalation, airborne substances might directly contact the skin, or liquids and sludges may be splashed on the skin or into the eyes, leading to adverse health effects. This subject area falls, in a general sense, in the domain of hazard risk assessment (HZRA), a topic treated in the next section.

2. *Chronic.* Continuous exposure usually occurs over long periods of time, generally several months to years. Concentrations of inhaled (toxic) contaminants are usually relatively low. This subject falls in the general domain of health risk assessment (HRA) and it is this subject that is addressed in this section. Thus, in contrast to the acute (short-term) exposures that predominate in hazard risk assessment, chronic (long-term) exposures are the major concern in health risk assessments.

Finally, there are two major types of risk: maximum individual risk and population risk. Maximum individual risk is defined exactly as it implies, i.e., the maximum risk to an individual person. This person is considered to have a 70-year lifetime exposure to a process or chemical. Population risk is basically the risk to a population. It is expressed as a certain number of deaths per thousand *or* per million people. These risks are often based on very conservative assumptions — due to numerous uncertainties — that may yield too high a risk.

Example 12.7 – Toxicology Definition. Describe toxicology in non-technical terms.

SOLUTION: As noted in this chapter, toxicology may be viewed as the "science of poisons," where the term poison refers to a chemical that is capable of injuring or killing an organism. Alternately, one may describe it as the "study of harmful effects of chemicals." Still others refer to it as the "study of a chemical's ability to cause damage to the body."

Example 12.8 – Toxicology vs. Epidemiology. Briefly describe the difference between toxicology and epidemiology.

SOLUTION: Toxicology deals with the adverse effects of chemical substances on living things. Most of the information on toxic effects of chem icals in humans provided by toxicology is derived from animal studies. Epidemiology analyzes the relationship between a chemical and the disease in an exposed population. Information on toxic effects of chemicals in humans provided by epidemiology is derived from direct human evidence.

Example 12.9 – Dose-Response and Toxicology Details. Provide a more detailed presentation on dose-response and/or toxicity.

SOLUTION: Dose-response assessment is the process of characterizing the relation between the dose of an agent administered or received and the incidence of an adverse health effect in exposed populations, and estimating the incidence of the effect as a function of exposure to the agent. This process considers such important factors as intensity of exposure, age pattern of exposure, and other possible variables that might affect responses such as sex, lifestyle, and other modifying factors. A dose-response assessment usually requires extrapolation from high to low doses and extrapolation from animals to humans, or even one laboratory animal species to a wildlife species. A dose-response assessment should describe and justify the methods of extrapolation used to predict incidence, and it should characterize the statistical and biological uncertainties in these methods. As noted, when possible, the uncertainties should be described numerically rather than qualitatively.

Toxicologists tend to focus their attention primarily on extrapolations from cancer bioassays. However, there is also a need to evaluate the risks of lower doses to see how they affect the various organs and systems in the body. Many scientific papers focus on the use of a safety factor or uncertainty factor approach since all adverse effects other than cancer and muta-tion-based developmental effects are believed to have a threshold — a dose below which no adverse effect should occur. Several researchers have discussed various approaches to setting acceptable daily intakes or exposure limits for developmental and reproductive toxicants. It is thought that an acceptable limit of exposure could be determined using cancer models, but today they are considered inappropriate because of thresholds [1,13].

Dangers are not necessarily defined by the presence of a particular chemical, but rather by the amount of that substance one is exposed to, also known as the aforementioned dose. A dose is usually expressed in milli-grams of chemical received per kilogram of body weight per day. For toxic substances other than carcinogens, a threshold dose must be exceeded before a health effect will occur, and for many substances, there is a dosage below which there is no harm; i.e., a health effect will occur or at least will be detected at the threshold. For carcinogens, it is assumed that there is *no* threshold, and, therefore, any substance that produces cancer is assumed to produce cancer at any concentration. It is vital to establish the link to cancer and to determine if that risk is acceptable. Analyses of cancer risks are much more complex than those for non-cancer risks.

For a variety of reasons, it is difficult to precisely evaluate toxic responses caused by acute exposures to hazardous materials. First, humans experience a wide range of acute adverse health effects including irritation, narcosis, asphyxiation, sensitization, blindness, organ system damage, and death. In addition, the severity of many of these effects varies with intensity and duration of exposure. Second, there is a high degree of variation in response among individuals in a typical population. Third, for the overwhelming majority of substances encountered in industry, there is insufficient data on toxic responses of humans to permit an accurate or precise assessment of the substance's health potential. Fourth, many releases involve multiple components. There are presently no rules on how these types of releases should be evaluated. Fifth, there are no toxicology testing protocols that exist for studying episodic releases on animals. In general, this has been a neglected area of toxicology research. However, there are many useful measures available to employ as benchmarks for predicting the likelihood that a release event will result in serious injury or death.

Not all contaminants or chemicals are equal in their capacity to cause adverse effects. Thus, some clean-up standards or action levels are based in part on the compounds' toxicological properties. Toxicity data employed are derived largely from animal experiments in which the animals (primarily mice and rats) are exposed to increasingly higher concentrations or doses. As described above, responses or effects can vary widely from no observable effect to temporary and reversible effects, to permanent injury to organs, to chronic functional impairment, and ultimately death.

Example 12.10 – Dose-Response Relationships. As noted above, a dose–response relationship provides a mathematical formula or graph for estimating a person's risk of illness at each exposure level for human toxins. To estimate a dose–response relationship, measurements of health effects are needed for at least one dose level of the toxic agent compared to an unexposed group. However, there is one important difference between the dose–response curve commonly used for estimating the risk of cancer and the ones used for estimating the risk of all other illnesses: the existence of a threshold dose, i.e., the aforementioned highest dose at which there is no risk of illness for the non-carcinogenic risk. Because a single cancerous cell may be sufficient to cause a clinical case of cancer, EPA's and many of the dose–response models for cancer of others assume that the threshold dose level for cancer is *zero*. In other words, people's risk of cancer is present even at very low doses. However, the increased cancer risk at very low doses is likely to also be very low.

Draw a straight-line model showing the level of cancer risk increasing at a constant rate as the dose level increases. The model should illustrate increasing risk of cancer for the toxic agent. (It is accepted by scientists and engineers that the human body is capable of adjusting to varying amounts of cell damage without showing signs of illness. Therefore, the EPA has developed models for noncancer illnesses that include a threshold dose level that is *greater* than zero; this means that at low doses, there may be no risk of adverse health effects. For noncancer health effects, such as permanent liver or kidney damage, temporary skin rashes, or asthma attacks, information from human or animal studies is used to estimate the threshold dose levels.)

SOLUTION: Figure 12.2 shows the linear cancer dose–response relationship plotted from data on a dose of 100 µg/d. This dose caused an extra chance of cancer of about 1 in 100 in the study animals that received the dose. The straight-line model developed here indicates that the level of cancer risk increases at a constant rate as the dose level increases, and this rate of increasing cancer risk is known as the aforementioned slope factor for the toxic agent.

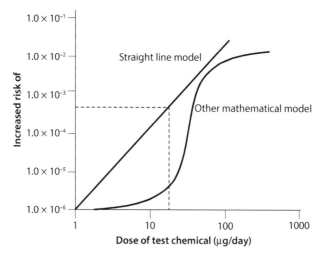

Figure 12.2 Cancer dose-response curve highlighting the straight line dose response model.

Example 12.11 – Straight-Line Model and Uncertainty Factor. With reference to non-cancer health effects develop a straight-line model to show the EPA's methodology in which the EPA adjusts the observed threshold from animal studies downward to the human threshold by dividing by uncertainty factors (UFs) that range from 1 to 10,000.

SOLUTION: Figure 12.3 illustrates the *noncancer* dose–response curve, which was drawn after converting uncertainties from animal to human data. Since individuals vary in their susceptibility to the harmful effects of toxic agents, EPA adjusts the observed threshold dose downward by dividing by UFs that range from 1 to 10,000. This new adjusted value is known as the human threshold at which EPA expects no appreciable risk of harmful health effects for most of the general population. This example shows the application of a UF of approximately 2 to animal threshold dose results.

Example 12.12 – Heat Exchanger Health Risk. A heat exchanger is located in a relatively large laboratory with a volume of 1100 m³ at 22 °C and 1 atm. The exchanger can leak as much as 0.75 gmol of a hydrocarbon (HC) from the flowing liquid into the room if the exchanger ruptures. A hydrocarbon mole fraction in the air greater than 4250 parts per billion (ppb) constitutes a health and safety hazard.

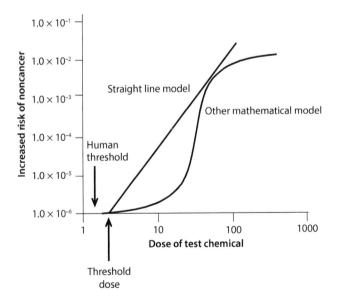

Figure 12.3 Noncancer dose-response curve highlighting the straight line dose-response model and the threshold dose.

Suppose the heat exchanger fails and the maximum amount of the HC is released instantaneously. Assume the air flow in the room is sufficient to cause the room to behave as continuously stirred tank reactor (CSTR) [14]; i.e., the air composition is spatially uniform. Calculate the ppb of hydrocarbon in the room. Is there a health risk? From a treatment point-of-view, what can be done to decrease the environmental hazard or to improve the safe operation of the exchanger?

SOLUTION: Calculate the total number of gmols of air in the room, n_{air}. Assuming that air is an ideal gas, 1 gmol of air occupies 22.4 liters (0.0224 m³) at standard temperature and pressure (273 K, 1 atm). Since the room temperature is not at 273 K,

$$n_{air} = \left(1100\,\text{m}^3\right)\left(\frac{1\,\text{gmol}}{0.0224\,\text{STP m}^3}\right)\left(\frac{273\,\text{K}}{295\,\text{K}}\right)$$

$$= 45,445\,\text{gmol}$$

Note: STP m³ indicates the volume (in m³) that the gas would have at a specified standard temperature and pressure.

The mole fraction of hydrocarbon in the room, x_{HC}, is,

$$x_{HC} = \frac{0.75\,\text{gmol HC}}{45,445\,\text{gmol air} + 0.75\,\text{gmol HC}} = 16.5\,\text{ppm} = 16,500\,\text{ppb}$$

Since 16,500 ppb > 4250 ppb, the health problem presents a significant health risk. To implement safety measures, the potential rupture area housing the exchanger should be vented directly into a hood or a duct to capture any leakage in the event of a rupture. Another alternative is to input liquid substitution, a source reduction measure [15], i.e., where an input substitution is the replacement of the fluid flowing in the heat exchanger.

Example 12.13 – Heat Exchanger Leak. An explosive emission from a heat exchanger leak were ignited by an unknown ignition source in a 10 ft × 10 ft × 10 ft space. The room ventilator was shut off and the fire was fought with a 10 lb CO_2 fire extinguisher and the extinguisher was completely emptied in putting out the fire.

The level of CO_2 that serves as an immediate danger to life and health (IDLH), set by the National Institute for Occupational Safety and Health (NIOSH), is 50,000 ppm [13]. At that level, vomiting, dizziness, disorientation, and breathing difficulties occur after a 30 minute exposure; at a 100,000 ppm_v, death can occur after a few minutes, even if the oxygen in the atmosphere would otherwise support life.

Assume that the gas mixture in the room is uniformly mixed, that the temperature in the room is 30 °C (warmed by the fire above the normal room temperature of 20 °C), and that the ambient pressure is 1 atm.

Calculate the concentration of CO_2 in the area after the fire extinguisher is emptied. Does it exceed the IDLH value?

SOLUTION:

First calculate the number of moles of CO_2 discharged by the fire extinguisher:

$$y = \text{Moles of } CO_2 = \frac{(101 \text{ lb } CO_2)(454 \text{ g / lb})}{44 \text{ g / gmol } CO_2} = 103 \text{ gmol } CO_2$$

Calculate the volume of the room:

$$\text{Room volume} = (10 \text{ ft})(10 \text{ ft})(10 \text{ ft})\left(\frac{0.0283 \text{ m}^3}{\text{ft}^3}\right)$$

$$= 28.3 \text{ m}^3 = 28,300 \text{ } L$$

Next, calculate the total number of moles of gas in the room by applying the ideal gas law:

$$\text{Moles of gas} = \frac{PV}{RT} = \frac{(1 \text{ atm})(28,300 \text{ L})}{(0.08206 \text{ atm} \cdot L / \text{gmol} \cdot K)(303 \text{ K})}$$

$$= 1138 \text{ gmol of gas}$$

Calculate the concentration, or mole fraction, of CO_2 in the room:

$$\text{Mole fraction} = \frac{\text{gmol } CO_2}{\text{gmol gas}} = \frac{103 \text{ gmol } CO_2}{1138 \text{ gmol of gas}} = 0.0905$$

Convert this fraction to a percent and compare to the IDLH and lethal levels:

$$\% CO_2 = (\text{Mole fraction})(100) = (0.0905)(100) = 9.05\%$$

The IDLH level is 5.0% and the lethal level is 10.0%. Therefore, the level in the room of 9.05% does exceed the IDLH level for CO_2. It is also dangerously close to the lethal level. The engineer extinguishing the fire is in great danger and should take appropriate safety measures immediately.

Example 12.14 – Health Problems from Tube Leaks. Provide an example of how a tube leak in a heat exchanger can result in a health problem for those individuals in the vicinity of the heat exchanger.

SOLUTION: Depending on the chemicals present, exposure to either the coolant or the hot fluid in an exchanger *may* create a health problem. However, there is also the possibility that when the two fluids mix — due to a tube leak or a failed tube — the resultant mixture might be reactive, toxic, or carcinogenic. A subsequent emission to the atmosphere could create a major health problem, or if explosive, a hazard problem (as discussed in the next section).

12.6 Hazard Risk Assessment

The previous section discussed problems associated with exposure to chemicals. In the chemical industry, there is also a high risk of accidents due to the nature of the processes and the materials used. Although precautions are taken to ensure that all processes run smoothly, there is always (unfortunately) room for error, and accidents will occur. This is especially true for highly technical complicated operations, as well as processes (including heat exchangers) operating under extreme conditions such as high temperatures and pressures; some heat exchangers fit within this latter category.

In general, accidents are caused by one or more of the following factors (see also Introduction to this chapter):

1. Equipment breakdown
2. Human error
3. Terrorism
4. Fire exposure and explosions
5. Control system failure(s)
6. Natural causes
7. Utilities and ancillary system outages
8. Faulty siting and plant layout

These causes are usually at the root of most industrial accidents. Although there is no way to guarantee that these problems will not arise with heat exchangers, steps can be taken to minimize the number, as well as the severity, of incidents.

Hazard risk evaluation serves a dual purpose. It estimates the probability that an accident will occur and also assesses the severity of the consequences of an accident. Consequences may include damage to the surrounding environment, financial loss, or injury to life. This section is primarily concerned with the methods used to identify hazards and the causes and consequences of accidents. (Issues dealing with health risks were discussed in the previous section.) Risk assessment of accidents provides an effect way to help ensure that a mishap does not occur or reduces the likelihood of an accident. The result of a hazard risk assessment can also allow concerned parties to take precautions to help prevent an accident before it happens.

The first thing a practicing engineer needs to understand is exactly what an accident is. An *accident* is defined as an unexpected event that has undesirable consequences. The causes of accidents have to be identified in order to help prevent accidents from occurring. Any situation or characteristic of a system, plant, or process that has the potential to cause damage to life, property, or the environment is considered a hazard. A hazard can also be defined as any characteristic that has the potential to cause an accident. The severity of a hazard plays a large part in the potential amount of damage a hazard can cause if it occurs. Risk is the probability that human injury, damage to property, damage to the environment, or financial loss will occur.

An *acceptable risk* is a risk whose probability is unlikely to occur during the lifetime of the problem or plant or process equipment. An acceptable risk can also be defined as an accident that has a high probability of occurring, but with negligible consequences. Risks can be ranked qualitatively in categories of high, medium, and low. Risk can also be ranked quantitatively as the annual number of fatalities per million affected individuals. This is normally denoted as a number times one millionth, e.g., 3 $\times 10^{-6}$; this representation indicates that on average, three individuals will die every year for every million individuals. A quantitative approach that has become popular in industry is the *fatal accident rate* (FAR) concept. This determines or estimates the number of fatalities over the lifetimes of 1000 workers. The *lifetime* of a worker is defined as 10^5 hours, which is based on a 40-hour work week for 50 years. A reasonable FAR for a chemical plant is 3.0, with 4.0 usually taken as a maximum. A FAR of 3.0 means that there are three deaths for every 1000 workers over a 50-year period [1,13]. Interestingly the FAR for an individual at home is approximately 3.0.

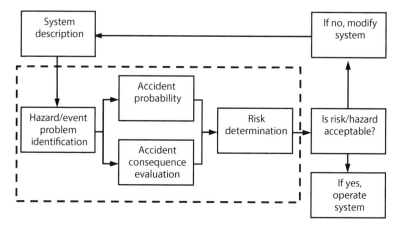

Figure 12.4 HZRA flowchart for a chemical process.

There are several steps in evaluating the risk of an accident (see also Figure 12.4). If the system in question is a chemical plant, the following guidelines apply.

1. A brief description of the equipment and chemicals used in the plant is needed.
2. Any hazard in the system has to be identified. Hazards that may occur in a chemical plant include:
   a. Corrosion
   b. Explosions
   c. Fires
   d. Rupture of a pressurized vessel
   e. Runaway reactions
   f. Slippage
   g. Unexpected leaks

3. The event or series of events that will initiate an accident must be identified. An event could be a failure to follow correct safety procedures, to improperly repair equipment, or the failure of a safety mechanism.
4. The probability that the accident will occur has to be determined. For example, if a heat exchanger has a 10-year life, what is the probability that the temperature in the exchanger will exceed the specified temperature range? The probability can be ranked qualitatively from low to high. A low probability means that it is unlikely for the event to occur in the

life of the exchanger. A high probability means that the event will probably occur during the life of the exchanger.

5. The severity of the consequences of the accident must be determined.
6. If the probability of the accident and the severity of its consequences are low, then the risk is usually deemed acceptable and the plant should be allowed to operate. If the probability of occurrence is too high or the damage to the surroundings it too great, then the risk is usually unacceptable and the system needs to be modified to minimize these effects.

The heart of the hazard risk assessment algorithm provided is enclosed in the dashed line box in Figure 12.4. This algorithm allows for re-evaluation of the process if the risk is deemed unacceptable (the process is repeated starting with step 1 or 2) [1,13].

The reader should note that hazard risk assessment plus the accompanying calculations receives an extensive treatment by Theodore and Dupont [13].

Finally, the extent of the need for emergency planning is often significant, and continues to expand as new regulations on safety are introduced. Planning for an industrial emergency must begin before a plant is constructed. A new plant will have to pass all safety measures and OSHA standards. This is emphasized by Armenante, who states that "The first line of defense against industrial accidents begins at the design stage. It should be obvious that it is much easier to prevent an accident rather than to try to rectify the situation once an accident has occurred" [15].

Successful emergency planning begins with a thorough understanding of the event or potential disaster being planned for. The impacts on public health and the environment must also be estimated. Some of the types of emergencies that should be included in the plan are [13]:

1. Natural disasters such as earthquakes, tornadoes, hurricanes, and floods
2. Explosions and fires
3. Acts of terrorism
4. Hazardous chemical leaks
5. Power or utility failures
6. Radiation accidents
7. Transportation accidents

The affected area or emergency zone must be studied in depth in order to estimate the impact on the public or the environment. A hazardous gas leak, fire, or explosion may cause a toxic cloud to spread over a great distance as it did in Bhopal, India [13].

Example 12.15 – System Describing Uncertainty. Describe some of the uncertainties that can arise in preparing a description of a system.
Solution: Uncertainties can include:

1. Process descriptions or drawings that are incorrect or out of date
2. Procedures that do not represent actual operation
3. Weather data from the nearest available site that may be inappropriate for the system location under study
4. Site area maps and population data that may be incorrect or out of date

Example 12.16 – Potential Consequences from a Distillation Reboiler Leak. What are potential consequences if a pinhole leak develops in a tube in the reboiler of a distillation column?
Solution:
See also Example 12.14. Potential consequences of a pinhole leak in this piece of process equipment are as follows:

1. Changes in pressure
2. Changes in temperature
3. Chemical reaction, with accompanying overpressure, over-temperature, and formation of other phases
4. Leakage of toxics/flammables to an undesirable location
5. Corrosion, embrittlement, or similar effect

Example 12.17 – Causes of a Storage Tank Leak [16]. A toxic liquid is employed as a coolant in a heat exchanger. Consider the release of this fluid from a storage tank. List and discuss possible causes for the release.
Solution: Some possible causes for a toxic release from a storage tank are:

1. Rupture in a storage tank
2. Fire in a tank farm
3. Explosion of a storage tank
4. Collapse of a tank due to earthquake
5. Rupture in a main line
6. Leak from a line or from a tank

Example 12.18 – Equipment Failure Rate [17]. Discuss the three major factors that often influence equipment (such as heat exchangers) failure rates.

SOLUTION: The variation of equipment failure rate with time-in-service is usually represented by three regions:

1. At initial startup, the rate of equipment failure is high due to factors such as improper installation or problems as a result of defective equipment.
2. The rate of failure declines when the equipment is under normal operation. At this point, failures are chance occurrences.
3. The rate of failure increases as the equipment ages. This can be termed as wear-out failure.

This failure rate with time will be discussed in Example 12.20; its graphical representation is known as the "bathtub curve" (or Weibull distribution to statisticians) because of its shape [18,19].

Example 12.19 – Defective Boiler Tubes [19]. Two boiler tubes are drawn in succession from a lot of 100 tubes, of which 10 tubes are defective. What is the probability that both tubes are defective if (a) the first is replaced before the second is drawn and (b) the first is not replaced before the second is drawn?

SOLUTION: The probability of event A, $P(A)$, can be interpreted as a theoretical relative frequency; i.e., a number about which the relative frequency of event A tends to cluster as n, the number of times a random equipment (on the event) is performed, increases indefinitely. This is the objective interpretation of probability. Probability can also be interpreted subjectively as a measure of the degree of belief, on a scale from 0 to 1, that the event A occurs. This interpretation is frequently used in ordinary conversation. For example, if someone says, "The probability I (the author) will go to the racetrack today is 90%," then 90% is a measure of the person's belief that he or she will go to the racetrack. This interpretation is also used when, in the absence of concrete data needed to estimate an unknown probability on the basis of observed relative frequency, the personal opinion of an expert is sought to provide the estimate.

The conditional probability of event B given A is denoted by $P(B|A)$ and defined as follows:

$$P(B|A) = \frac{P(AB)}{P(A)} \qquad (12.3)$$

where $P(AB)$ is the probability that events A and B will occur. $P(B|A)$ can be interpreted as the proportions of A occurrences that also include the occurrence of B.

First determine the probability that the first tube is defective, $P(A)$. Since 10 out of 100 are defective:

$$P(A) = \frac{10}{100}$$

Determine the probability that the second tube is defective if the first is replaced, $P(B)$. Since the first tube is replaced, the probability for the defective tube is the same:

$$P(B) = \frac{10}{100}$$

Determine the probability that the two tubes are defective if the first is replaced, i.e., $P(AB)$:

$$P(AB) = P(A)P(B)$$
$$= \left(\frac{10}{100}\right)\left(\frac{10}{100}\right)$$
$$= \frac{1}{100}$$

Determine the probability that the second tube is defective if the first tube is *not* replaced, i.e., $P(B|A)$. Since the remaining lot contained 99 tubes:

$$P(B|A) = \frac{10-1}{100-1} = \frac{9}{99}$$

Finally, determine the probability that both tubes are defective if the first tube is not replaced, that is, $P'(AB)$:

$$P'(AB) = P(A)P(B|A)$$
$$= \left(\frac{10}{100}\right)\left(\frac{9}{99}\right)$$
$$= \frac{1}{110}$$

Conditional probability can be used to formulate a definition for the independence of two events A and B. Event B is defined to be independent of event A only if $P(B|A) = P(B)$. Similarly, event A is defined to be independent of event B if and only if $P(A|B) = P(A)$. From the definition of conditional probability, one can deduce the logically equivalent definition of the independence of event A and event B if and only if

$$P(AB) = P(A)P(B).$$

Example 12.20 – Coolant Sprinkler Spray Failure [18, 19]. A coolant sprinkler system in a reactor has 20 independent spray components each of which fails with a probability of 0.1. The coolant heat transfer system is considered to "fail" only if four or more of the sprays fail. What is the probability that the sprinkler system fails?

SOLUTION: Several probability distributions figure prominently in reliability calculations. The binomial distribution is one of them. Consider n independent performances of a random experiment with mutually exclusive outcomes which can be classified as "success" or "failure". The words "success" and "failure" are to be regarded as labels for two mutually exclusive categories of outcomes of the random experiment. They do not necessarily have the ordinary connotation of success or failure. Assume that P, the probability of success on any performances of the random experiment, is constant. Let $q = 1 - P$ be the probability of failure. The probability distribution of x, the number of successes in n performances of the random experiment is the binomial distribution with probability distribution function specified by [18],

$$f(x) = \frac{P^x q^{n-x} n!}{x!(n-x)!}; \; x = 0,1,2,\ldots,n \tag{12.4}$$

where $f(x)$ is the probability of x successes in n performances and n is the number of independent performances of a random experiment. The binomial distribution can therefore be used to calculate the reliability of a redundant system. A redundant system consisting of n identical components is a system which fails only if more than r components fail. Typical examples include single-usage equipment such as missile engines, short-life batteries, and flash bulbs which are required to operate for one time period and are not reused.

Assume that the n components are independent with respect to failure, and that the reliability of each is $1 - P$. One may associate "success" with the failure of a component. Then x, the number of failures, has a binomial pdf and the reliability of the random system is [18],

$$P(x \leq r) = \sum_{x=0}^{r} \frac{P^x q^{n-x} n!}{x!(n-x)!}$$ (12.5)

Consider now the coolant sprinkler spray system in the problem statement. Let x denote the number of components which fail. Identify the value of n, P, and q from the problem statement.

$$n = 20$$
$$P = 0.1$$
$$q = 0.9$$

Calculate the probability that the sprinkler system fails (i.e., $P(X \geq 4)$) by using the binomial distribution provided by Equation (12.4).

$$P(X \geq 4) = \sum_{x=4}^{20} \frac{P^x q^{n-x} n!}{x!(n-x)!}$$

Note that calculation can be simplified by the fact that $P(X \geq 4)$. $= 1 - P(X \leq 3)$ Therefore,

$$P(X \geq 4) = 1 - P(X \leq 3)$$
$$= 1 - \sum_{x=0}^{3} \frac{(0.1)^x (0.9)^{20-x} 20!}{x!(20-x)!}$$
$$= 0.13 = 13\%$$

Example 12.21 – Time of Heat Exchanger Failure [18, 19]. Assume the time to failure (in hours), t, of a tube in a heat exchanger has a Weibull distribution with $\alpha = 1.3 \times 10^{-3}$ and $\beta = 0.77$. Find the probability that a tube in a heat exchanger will fail in 1000 hours.

SOLUTION: Frequently, and as discussed earlier, the failure rate of equipment exhibits three states: a break-in stage with a declining failure rate, a useful life stage characterized by a fairly constant failure rate, and a wear out period characterized by an increasing failure rate. A failure rate curve exhibiting these three phases is called a *bathtub curve*. The Weibull distribution provides a mathematical model of all three states of the bathtub curve. The probability distribution function is given by [18, 19]:

$$f(t) = \alpha\beta t^{(\beta-1)} \cdot \exp\left(-\int_0^t \alpha\beta t^{(\beta-1)} dt\right) \tag{12.6}$$

$$= \alpha\beta t^{(\beta-1)} e^{-\alpha t^\beta} \; ; t > 0, \alpha > 0, \beta > 0$$

where α and β are constants.

The probability distribution function (pdf) for the heat exchanger tube is then defined as,

$$f(t) = \alpha\beta t^{(\beta-1)} e^{-\alpha t^\beta} = \left(1.3 \times 10^{-3}\right)\left(0.77\right)\left(t^{0.77-1}\right)e^{-\left(1.3\times10^{-3}\right)t^{0.77}} \; ; t > 0$$

The probability that a heat exchanger will fail within 1000 hours (i.e., $(P(t) < 1000)$) can now be calculated as,

$$\left(P(t) < 1000\right) = \int_0^{1000} f(t) dt = 1 - e^{-\left(1.3\times10^{-3}\right)1000^{0.77}}$$

$$= 0.23 = 23\%$$

Example 12.22 – The HAZOP Procedure [1, 13]. Discuss the HAZOP (Hazard and Operability Procedure) as it applies to a chemical plant.

SOLUTION: Specific details regarding this procedure are available in the literature.(13) The overall HAZOP method, however, is summarized in the following steps:

1. Define objective(s).
2. Define plant limits.
3. Appoint and train a team.
4. Obtain (generate) complete preparative work.
5. Conduct examination meetings in order to:
   a. select a manageable portion of the process.
   b. review the flow sheet and operating instructions.
   c. agree on how the process is intended to operate.
   d. state and record the intention.
   e. search for possible ways to deviate from the intention, utilizing the HAZOP "guide" words [1, 13].

    f. determine possible causes for the deviation.
    g. determine possible consequences of the deviation.
    h. recommend action(s) to be taken.
6. Issue meeting reports.
7. Follow up on recommendations.

After any serious hazards have been identified with a HAZOP study or some other type of qualitative approach, a quantitative examination should be performed. Hazard quantification or hazard analysis (HAZAN) involves the estimation of the expected frequencies or probabilities of events with adverse or potentially adverse consequences. It logically ties together historical occurrences, experience, and imagination. To analyze the sequence of events that lead to an accident or failure, event and fault trees are used to represent the possible failure sequences [18].

References

1. L. Theodore, J. Reynolds, and K. Morris, *Accident and Emergency Management*, A Theodore Tutorial, Theodore Tutorials, East Williston, NY, originally published by USEPA/APTI, RTP, NC, 1992.
2. M.K. Theodore and L. Theodore, *Introduction to Environmental Management*, CRC Press/Taylor & Francis Group, Boca Raton, FL, 2010.
3. L. Theodore, *Chemical Engineering: The Essential Reference*, McGraw-Hill, New York City, NY, 2014.
4. A.M. Flynn and L. Theodore, *Accident and Emergency Management in the Chemical Process Industries*, CRC Press/Taylor & Francis Group, Boca Raton, FL, originally published by Marcel Dekker, New York City, NY, 2004.
5. L. Theodore, M. Hyland, Y. McGuinn, E. Schoen, and F. Taylor, *Principles of Accident and Emergency Management*, USEPA Manual, Air Pollution Training Institute, RTP, NC, 1988.
6. L. Theodore, J. Reynolds, and K. Morris, *Concise Dictionary of Environmental Terms (adapted from)*, Gordon and Breach Science Publishers, Amsterdam, The Netherlands, 1997.
7. G. Burke, B. Singh, and L. Theodore, *Handbook of Environmental Management and Technology*, 2nd edition, John Wiley & Sons, Hoboken, NJ, 2001.
8. L. Stander and L. Theodore, *Environmental Regulatory Calculations Handbook*, John Wiley & Sons, Hoboken, NJ, 2008.
9. EPA Office of Emergency and Remedial Division, *Standard Operating Safety Guides*, Washington D.C., July 1988.
10. P. Armenante, *Contingency Planning for Industrial Emergencies*, Van Nostrand Reinhold, New York City, NY 1991.

11. D. Paustenbach, *The Risk Assessment of Environmental and Human Health Hazards: A Textbook of Case Studies*, John Wiley & Sons, Hoboken, NJ, 1989.

12. G. Holmes, B. Singh, and L. Theodore, *Handbook of Environmental Management & Technology*, John Wiley & Sons, Hoboken, NJ, 1993.

13. L. Theodore and R. Dupont, *Environmental Health Risk and Hazard Risk Assessment: Principles and Calculations*, CRC Press/Taylor & Francis Group, Boca Raton, FL, 2012.

14. L. Theodore, *Chemical Reactor Analysis and Applications for the Practicing Engineer*, John Wiley & Sons, Hoboken, NJ, 2012.

15. R. Dupont, K. Ganeson, and L. Theodore, *Pollution Prevention: Sustainability, Industrial Ecology, and Green Science and Engineering*, 2nd edition, CRC Press/Taylor & Francis Group, Boca Raton, FL, 2000.

16. K. Ganeson, L. Theodore, and J. Reynolds, *Air Toxics — Problems and Solutions*, Gordon and Breach Publishers, Amsterdam, The Netherlands, 1996.

17. R. Dupont, T. Baxter, and L. Theodore, *Environmental Management: Problems and Solutions*, CRC Press/Taylor & Francis Group, Boca Raton, FL, 1998.

18. F. Taylor and L. Theodore, *Probability and Statistics*, A Theodore Tutorial, East Williston, NY, originally published by the USEPA/APTI, RTP, NC, 1999.

19. S. Shaefer and L. Theodore, *Probability and Statistics Applications in Environmental Science*, CRC Press/Taylor & Francis Group, Boca Raton, FL, 2007.

Appendix

Appendix Table of Contents

AT.1 Conversion Constants   652
AT.2 Thermodynamic Properties of Steam/ Steam Tables   656
AT.3 Properties of Water (Saturated Liquid)   665
AT.4 Properties of Air at 1 atm   667
AT.5 Properties of Selected Liquids at 1 atm and 20 °C (68 °F)   668
AT.6 Properties of Selected Gases at 1 atm and 20 °C (68 °F)   670
AT.7 Dimensions, Capacities, and Weights of Standard Steel Pipes   672
AT.8 Dimensions of Heat Exchanger Tubes   674
AT.9 Tube-Sheet Layouts (Tube Counts) on a Square Pitch   676
AT.10 Tube Sheet Layouts (Tube Counts) on a Triangular Pitch   678
AT.11 Approximate Design Overall Heat Transfer
      Coefficients (Btu/hr·ft2·°F)   681
AT.12 Approximate Design Fouling Coefficient Factors
      (hr·ft2·°F/Btu)   682

Figures

AF.1 Fanning Friction Factor (f) vs. Reynolds Number (Re) Plot   686
AF.2 Psychometric Chart: Low Temperatures:
      Barometric Pressure, 29.92 in. Hg.   687
AF.3 Psychometric Chart: High Temperatures:
      Barometric Pressure, 29.92 in. Hg.   688

Table AT.1 Conversion Constants.

To convert from	To	Multiply by
Length		
m	cm	10^7
m	mm	10^3
m	microns (µm)	10^6
m	angstroms (Å)	10^{10}
m	in	39.37
m	ft	3.281
m	mi	6.214×10^{-4}
ft	in	12
ft	m	0.3048
ft	cm	30.48
ft	mi	1.894×10^{-4}
Mass		
kg	g	1000
kg	lb	2.205
kg	oz	35.24
kg	ton	2.268×10^{-4}
kg	grains	1.543×10^4
lb	oz	16
lb	ton	5×10^{-4}
lb	g	453.6
lb	kg	0.4536
lb	grains	7000
Time		
s	min	0.01667
s	h	2.78×10^{-4}
s	day	1.157×10^{-7}
s	week	1.653×10^{-6}
s	yr	3.171×10^{-8}
Force		
N	kg·m/s^2	1
N	dynes	10^5
N	g·cm/s^2	10^5

(*Continued*)

Table AT.1 Cont.

To convert from	To	Multiply by
N	lb_f	0.2248
N	$lb \cdot ft/s^2$	7.233
lb_f	N	4.448
lb_f	dynes	4.448×10^5
lb_f	$g \cdot cm/s^2$	4.448×10^5
lb_f	$lb \cdot ft/s^2$	32.17
Pressure		
atm	N/m^2 (Pa)	1.013×10^5
atm	kPa	101.3
atm	bars	1.013
atm	$dynes/cm^2$	1.013×10^6
atm	lb_f/in^2 (psi)	14.696
atm	mm Hg at 0 °C (torr)	760
atm	in Hg at 0 °C	29.92
atm	ft H_2O at 4 °C	33.9
atm	in H_2O at 4 °C	406.8
psi	atm	6.80×10^{-2}
psi	mm Hg at 0 °C (torr)	51.71
psi	in H_2O at 4 °C	27.70
in H_2O at 4 °C	atm	2.458×10^{-3}
in H_2O at 4 °C	psi	0.0361
in H_2O at 4 °C	mm Hg at 0 °C (torr)	1.868
Volume		
m^3	L	1000
m^3	cm^3 (cc, mL)	10^6
m^3	ft^3	35.31
m^3	gal (U.S.)	264.2
m^3	qt	1057
ft^3	in^3	1728
ft^3	gal (U.S.)	7.48
ft^3	m^3	0.02832
ft^3	L	28.32

(Continued)

Table AT.1 Cont.

To convert from	To	Multiply by
Energy		
J	N·m	1
J	erg	10^7
J	dyne·cm	10^7
J	kW·h	2.778×10^{-7}
J	cal	0.2390
J	ft·lb$_f$	0.7376
J	Btu	9.486×10^{-4}
cal	J	4.186
cal	Btu	3.974×10^{-3}
cal	ft·lb$_f$	3.088
Btu	ft·lb$_f$	778
Btu	hp·h	3.929×10^{-4}
Btu	cal	252
Btu	kW·h	2.93×10^{-4}
ft·lb$_f$	cal	0.3239
ft·lb$_f$	J	1.356
ft·lb$_f$	Btu	1.285×10^{-3}
Power		
W	J/s	1
W	cal/s	0.2390
W	ft·lb/s	0.7376
W	kW	10^{-3}
kW	Btu/s	0.949
kW	hp	1.341
hp	ft·lb/s	550
hp	kW	0.7457
hp	cal/s	178.2
hp	Btu/s	0.707
Viscosity		
P (poise)	g/cm·s	1
P	cP (centipoise)	100
P	kg/m·h	360

(*Continued*)

Table AT.1 Cont.

To convert from	To	Multiply by
P	lb/ft·s	6.72×10^{-2}
P	lb/ft·h	241.9
P	lb/m·s	5.6×10^{-3}
lb/ft·s	P	14.88
lb/ft·s	g/cm·s	14.88
lb/ft·s	kg/m·h	5.357×10^{3}
lb/ft·s	lb/ft·h	3600
Heat Capacity		
cal/g·°C	Btu/lb·°F	1
cal/g·°C	kcal/kg·°C	1
cal/g·°C	cal/gmol·°C	Molecular weight
cal/gmol·°C	Btu/lbmol·°F	1
J/g·°C	Btu/lb·°F	0.2389
Btu/lb·°F	cal/g·°C	1
Btu/lb·°F	J/g·°C	4.186
Btu/lb·°F	Btu/lbmol·°F	Molecular weight

Table AT.2 Thermodynamic Properties of Steam (Steam Tables).
Saturated Steam*

Temperature,	Absolute pressure,	Specific volume, ft³/lb			Enthalpy, Btu/lb			Entropy, Btu/lb·°R		
		Saturated liquid,	Evaporation difference,	Saturated vapor,	Saturated liquid,	Evaporation difference,	Saturated vapor,	Saturated liquid,	Evaporation difference,	Saturated vapor,
T, °F	P, lb_f/in^2	v_1	v_{vap}	v_g	h_1	h_{vap}	h_g	s_1	s_{vap}	s_g
32	0.08854	0.01602	3306	3306	0.00	1075.8	1075.8	0.0000	2.1877	2.1877
35	0.09995	0.01602	2947	2947	3.02	1074.1	1077.1	0.0061	2.1709	2.1770
40	0.12170	0.01602	2444	2444	8.05	1071.3	1079.3	0.0162	2.1435	2.1597
45	0.14752	0.01602	2036.4	2036.4	13.06	1068.4	1081.5	0.0262	2.1167	2.1429
50	0.17811	0.01603	1703.2	1703.2	18.07	1065.6	1083.7	0.0361	2.0903	2.1264
60	0.2563	0.01604	1206.6	1206.7	28.06	1059.9	1088.0	0.0555	2.0393	2.0948
70	0.3631	0.01606	867.8	867.9	38.04	1054.3	1092.3	0.0745	1.9902	2.0647
80	0.5069	0.01608	633.1	633.1	48.02	1048.6	1096.6	0.0932	1.9428	2.0360
90	0.6982	0.01610	468.0	468.0	57.99	1042.9	1100.9	0.1115	1.8972	2.0087
100	0.9492	0.01613	350.3	350.4	67.97	1037.2	1105.2	0.1295	1.8531	1.9826
110	1.2748	0.01617	265.3	265.4	77.94	1031.6	1109.5	0.1471	1.8106	1.9577
120	1.6924	0.01620	203.25	203.27	87.92	1025.8	1113.7	0.1645	1.7694	1.9339
130	2.2225	0.01625	157.32	157.34	97.90	1020.0	1117.9	0.1816	1.7296	1.9112
140	2.8886	0.01629	122.99	123.01	107.89	1014.1	1122.0	0.1984	1.6910	1.8894
150	3.718	0.01634	97.06	97.07	117.89	1008.2	1126.1	0.2149	1.6537	1.8685

(*Continued*)

Table AT.2 Cont. (Saturated Steam)

Temperature, T, °F	Absolute pressure, P, lb_f/in^2	Specific volume, ft³/lb			Enthalpy, Btu/lb			Entropy, Btu/lb·°R		
		Saturated liquid, v_l	Evaporation difference, v_{vap}	Saturated vapor, v_g	Saturated liquid, h_l	Evaporation difference, h_{vap}	Saturated vapor, h_g	Saturated liquid, s_l	Evaporation difference, s_{vap}	Saturated vapor, s_g
160	4.741	0.01639	77.27	77.29	127.89	1002.3	1130.2	0.2311	1.6174	1.8485
170	5.992	0.01645	62.04	62.06	137.90	996.3	1134.2	0.2472	1.5822	1.8293
180	7.510	0.01651	50.21	50.23	147.92	990.2	1138.1	0.2630	1.5480	1.8109
190	9.339	0.01657	40.94	40.96	157.95	984.1	1142.0	0.2785	1.5147	1.7932
200	11.526	0.01663	33.62	33.64	167.99	977.9	1145.9	0.2938	1.4824	1.7762
210	14.123	0.01670	27.80	27.82	178.05	971.6	1149.7	0.3090	1.4508	1.7598
212	14.696	0.01672	26.78	26.80	180.07	970.3	1150.4	0.3120	1.4446	1.7566
220	17.186	0.01677	23.13	23.15	188.13	965.2	1153.4	0.3239	1.4201	1.7440
230	20.780	0.01684	19.365	19.382	198.23	958.8	1157.0	0.3387	1.3901	1.7288
240	24.969	0.01692	16.306	16.323	208.34	952.2	1160.5	0.3531	1.3609	1.7140
250	29.825	0.01700	13.804	13.821	218.48	945.5	1164.0	0.3675	1.3323	1.6998
260	35.429	0.01709	11.746	11.763	228.64	938.7	1167.3	0.3817	1.3043	1.6860
270	41.858	0.01717	10.044	10.061	238.84	931.8	1170.6	0.3958	1.2769	1.6727
280	49.203	0.01726	8.628	8.645	249.06	924.7	1173.8	0.4096	1.2501	1.6597
290	57.556	0.01735	7.444	7.461	259.31	917.5	1176.8	0.4234	1.2238	1.6472
300	67.013	0.01745	6.449	6.466	269.59	910.1	1179.7	0.4369	1.1980	1.6350
310	77.68	0.01755	5.609	5.626	279.92	902.6	1182.5	0.4504	1.1727	1.6231
320	89.66	0.01765	4.896	4.914	290.28	894.9	1185.2	0.4637	1.1478	1.6115

(Continued)

Table AT.2 Cont. (Saturated Steam)

Temperature, T, °F	Absolute pressure, P, lb_f/in^2	Specific volume, ft³/lb			Enthalpy, Btu/lb			Entropy, Btu/lb·°R		
		Saturated liquid, v_l	Evaporation difference, v_{vap}	Saturated vapor, v_g	Saturated liquid, h_l	Evaporation difference, h_{vap}	Saturated vapor, h_g	Saturated liquid, s_l	Evaporation difference, s_{vap}	Saturated vapor, s_g
330	103.06	0.01776	4.289	4.307	300.68	887.0	1187.7	0.4769	1.1233	1.6002
340	118.01	0.01787	3.770	3.788	311.13	879.0	1190.1	0.4900	1.0992	1.5891
350	134.63	0.01799	3.324	3.342	321.63	870.7	1192.3	0.5029	1.0754	1.5783
360	153.04	0.01811	2.939	2.957	332.18	862.2	1194.4	0.5158	1.0519	1.5677
370	173.37	0.01823	2.606	2.625	342.79	853.5	1196.3	0.5286	1.0287	1.5573
380	195.77	0.01836	2.317	2.335	353.45	844.6	1198.1	0.5413	1.0059	1.5471
390	220.37	0.01850	2.0651	2.0836	364.17	835.4	1199.6	0.5539	0.9832	1.5371
400	247.31	0.01864	1.8447	1.8633	374.97	826.0	1201.0	0.5664	0.9608	1.5272
410	276.75	0.01878	1.6512	1.6700	385.83	816.3	1202.1	0.5788	0.9386	1.5174
420	308.83	0.01894	1.4811	1.5000	396.77	806.3	1203.1	0.5912	0.9166	1.5078
430	343.72	0.01910	1.3308	1.3499	407.79	796.0	1203.8	0.6035	0.8947	1.4982
440	381.59	0.01926	1.1979	1.2171	418.90	785.4	1204.3	0.6158	0.8730	1.4887
450	422.6	0.0194	1.0799	1.0993	430.1	774.5	1204.6	0.6280	0.8513	1.4793
460	466.9	0.0196	0.9748	0.9944	441.4	763.2	1204.6	0.6402	0.8298	1.4700
470	514.7	0.0198	0.8811	0.9009	452.8	751.5	1204.3	0.6523	0.8083	1.4606
480	566.1	0.0200	0.7972	0.8172	464.4	739.4	1203.7	0.6645	0.7868	1.4513

(Continued)

Table AT.2 Cont. (Saturated Steam)

Temperature,	Absolute pressure,	Specific volume, ft³/lb			Enthalpy, Btu/lb			Entropy, Btu/lb·°R		
		Saturated liquid,	Evaporation difference,	Saturated vapor,	Saturated liquid,	Evaporation difference,	Saturated vapor,	Saturated liquid,	Evaporation difference,	Saturated vapor,
T, °F	P, lb$_f$/in²	v_l	v_{vap}	v_g	h_l	h_{vap}	h_g	s_l	s_{vap}	s_g
490	621.4	0.0202	0.7221	0.7423	476.0	726.8	1202.8	0.6766	0.7653	1.4419
500	680.8	0.0204	0.6545	0.6749	487.8	713.9	1201.7	0.6887	0.7438	1.4325
520	812.4	0.0209	0.5385	0.5594	511.9	686.4	1198.2	0.7130	0.7006	1.4136
540	962.5	0.0215	0.4434	0.4649	536.6	656.6	1193.2	0.7374	0.6568	1.3942
560	1133.1	0.0221	0.3647	0.3868	562.2	624.2	1186.4	0.7621	0.6121	1.3742
580	1325.8	0.0228	0.2989	0.3217	588.9	588.4	1177.3	0.7872	0.5659	1.3532
600	1542.9	0.0236	0.2432	0.2668	617.0	548.5	1165.5	0.8131	0.5176	1.3307
620	1786.6	0.0247	0.1955	0.2201	646.7	503.6	1150.3	0.8398	0.4664	1.3062
640	2059.7	0.0260	0.1538	0.1798	678.6	452.0	1130.5	0.8679	0.4110	1.2789
660	2365.4	0.0278	0.1165	0.1442	714.2	390.2	1104.4	0.8987	0.3485	1.2472
680	2708.1	0.0305	0.0810	0.1115	757.3	309.9	1067.2	0.9351	0.2719	1.2071
700	3093.7	0.0369	0.0392	0.0761	823.3	172.1	995.4	0.9905	0.1484	1.1389

*Adapted from *Thermodynamic Properties of Steam*, by Joseph H. Keenan and Fredrick G. Keyes. Copyright 1936, by Joseph H. Keenan and Frederick G. Keyes, Published by John Wiley & Sons, Inc., Hoboken, NJ.

Table AT.2 Thermodynamic Properties of Steam (Steam Tables).

Superheated Steam*

Absolute pressure lb$_f$/in^2 (saturated temperature)		Temperature, °F												
		200	220	300	350	400	450	500	550	600	700	800	900	1000
1	v	392.6	404.5	452.3	482.2	512.0	541.8	571.6	601.4	631.2	690.8	750.4	809.9	869.5
	h	1150.4	1159.5	1195.8	1218.7	1241.7	1264.9	1288.3	1312.0	1335.7	1383.8	1432.8	1482.7	1533.5
(101.74)	s	2.0512	2.0647	2.1153	2.1444	2.1720	2.1983	2.2233	2.2468	2.2702	2.3137	2.3542	2.3923	2.4283
5	v	78.16	80.59	90.25	96.26	102.26	108.24	114.22	120.19	126.16	138.10	150.03	161.95	173.87
	h	1148.8	1158.1	1195.0	1218.1	1241.2	1264.5	1288.0	1311.7	1335.4	1383.6	1432.7	1482.6	1533.4
(162.24)	s	1.8718	1.8857	1.9370	1.9664	1.9942	2.0205	2.0456	2.0692	2.0927	2.1361	2.1767	2.2148	2.2509
10	v	38.85	40.09	45.00	48.03	51.04	54.05	57.05	60.04	63.03	69.01	74.98	80.95	86.92
	h	1146.6	1156.2	1193.9	1217.2	1240.6	1264.0	1287.5	1311.3	1335.1	1383.4	1432.5	1482.4	1533.2
(193.21)	s	1.7927	1.8071	1.8595	1.8892	1.9172	1.9436	1.9689	1.9924	2.0160	2.0596	2.1002	2.1383	2.1744
14.696	v		27.15	30.53	32.62	34.68	36.73	38.78	40.82	42.86	46.94	51.00	55.07	59.13
	h		1154.4	1192.8	1216.4	1239.9	1263.5	1287.1	1310.9	1334.8	1383.2	1432.3	1482.3	1533.1
(212.00)	s		1.7624	1.8160	1.8460	1.8743	1.9008	1.9261	1.9498	1.9734	2.0170	2.0576	2.0958	2.1319
20	v			22.36	23.91	25.43	26.95	28.46	29.97	31.47	34.47	37.46	40.45	43.44
	h			1191.6	1215.6	1239.2	1262.9	1286.6	1310.5	1334.4	1382.9	1432.1	1482.1	1533.0
(227.96)	s			1.7808	1.8112	1.8396	1.8664	1.8918	1.9160	1.9392	1.9829	2.0235	2.0618	2.0978

(Continued)

Table AT.2 Cont. (Superheated Steam)

Absolute pressure lb$_f$/in^2 (saturated temperature)		Temperature, °F												
		200	220	300	350	400	450	500	550	600	700	800	900	1000
40 (267.25)	v			11.040	11.843	12.628	13.401	14.168	14.93	15.688	17.198	18.702	20.20	21.70
	h			1186.8	1211.9	1236.5	1260.7	1284.8	1308.9	1333.1	1381.9	1431.3	1481.4	1532.4
	s			1.6994	1.7314	1.7608	1.7881	1.8140	1.8384	1.8619	1.9058	1.9467	1.9850	2.0214
60 (292.71)	v			7.259	7.818	8.357	8.884	9.403	9.916	10.427	11.441	12.449	13.452	14.454
	h			1181.6	1208.2	1233.6	1258.5	1283.0	1307.4	1331.8	1380.9	1430.5	1480.8	1531.9
	s			1.6492	1.6830	1.7135	1.7416	1.7678	1.7926	1.8162	1.8605	1.9015	1.9400	1.9762
80 (312.03)	v				5.803	6.220	6.624	7.020	7.410	7.797	8.562	9.322	10.077	10.830
	h				1204.3	1230.7	1256.1	1281.1	1305.8	1330.5	1379.9	1429.7	1480.1	1531.3
	s				1.6475	1.6791	1.7078	1.7346	1.7598	1.7836	1.8281	1.8694	1.9079	1.9442
100 (327.81)	v				4.592	4.937	5.268	5.589	5.905	6.218	6.835	7.446	8.052	8.656
	h				1200.1	1227.6	1253.7	1279.1	1304.2	1329.1	1378.9	1428.9	1479.5	1530.8
	s				1.6188	1.6518	1.6813	1.7085	1.7339	1.7581	1.8029	1.8443	1.8829	1.9193
120 (341.25)	v				3.783	4.081	4.363	4.636	4.902	5.165	5.683	6.195	6.702	7.207
	h				1195.7	1224.4	1251.3	1277.2	1302.5	1327.7	1377.8	1428.1	1478.8	1530.2
	s				1.5944	1.6287	1.6591	1.6869	1.7127	1.7370	1.7822	1.8237	1.8625	1.8990
140	v					3.468	3.715	3.954	4.186	4.413	4.861	5.301	5.738	6.172
	h					1221.1	1248.7	1275.2	1300.9	1326.4	1376.8	1427.3	1478.2	1529.7

(Continued)

Table AT.2   Cont. (Superheated Steam)

Absolute pressure lb_f/in² (saturated temperature)		Temperature, °F												
		200	220	300	350	400	450	500	550	600	700	800	900	1000
(353.02)	s					1.6087	1.6399	1.6683	1.6945	1.7190	1.7645	1.8063	1.8451	1.8817
	v					3.008	3.230	3.443	3.648	3.849	4.244	4.631	5.015	5.396
160	h					1217.6	1246.1	1273.1	1299.3	1325.0	1375.7	1426.4	1477.5	1529.1
(363.53)	s					1.5908	1.6230	1.6519	1.6785	1.7033	1.7491	1.7911	1.8301	1.8667
	v					2.649	2.852	3.044	3.229	3.411	3.764	4.110	4.452	4.792
180	h					1214.0	1243.5	1271.0	1297.6	1323.5	1374.7	1425.6	1476.8	1528.6
(373.06)	s					1.5745	1.6077	1.6373	1.6642	1.6894	1.7355	1.7776	1.8167	1.8534
	v					2.361	2.549	2.726	2.895	3.060	3.380	3.693	4.002	4.309
200	h					1210.3	1240.7	1268.9	1295.8	1322.1	1373.6	1424.8	1476.2	1528.0
(381.79)	s					1.5594	1.5937	1.6240	1.6513	1.6767	1.7232	1.7655	1.8048	1.8415
	v					2.125	2.301	2.465	2.621	2.772	3.066	3.352	3.634	3.913
220	h					1206.5	1237.9	1266.7	1294.1	1320.7	1372.6	1424.0	1475.5	1527.5
(389.86)	s					1.5453	1.5805	1.6117	1.6395	1.6652	1.7120	1.7545	1.7939	1.8308
	v					1.9276	2.094	2.247	2.393	2.533	2.804	3.068	3.327	3.584
240	h					1202.5	1234.9	1264.5	1292.4	1319.2	1371.5	1423.2	1474.8	1526.9

(Continued)

Table AT.2 Cont. (Superheated Steam)

Absolute pressure lb_f/in² (saturated temperature)		Temperature, °F 200	220	300	350	400	450	500	550	600	700	800	900	1000
(397.37)	s					1.5319	1.5686	1.6003	1.6286	1.6546	1.7017	1.7444	1.7839	1.8209
	v						1.9183	2.063	2.199	2.330	2.582	2.827	3.067	3.305
260	h						1232.0	1262.3	1290.5	1317.7	1370.4	1422.3	1474.2	1526.3
(404.42)	s						1.5573	1.5897	1.6184	1.6447	1.6922	1.7352	1.7748	1.8118
	v						1.7674	1.9047	2.033	2.156	2.392	2.621	2.845	3.066
280	h						1228.9	1260.0	1288.7	1316.2	1369.4	1421.5	1473.5	1525.8
(411.05)	s						1.5464	1.5796	1.6087	1.6354	1.6834	1.7265	1.7662	1.8033
	v						1.6364	1.7675	1.8891	2.005	2.227	2.442	2.652	2.859
300	h						1225.8	1257.6	1286.8	1314.7	1368.3	1420.6	1472.8	1525.2
(417.33)	s						1.5360	1.5701	1.5998	1.6268	1.6751	1.7184	1.7582	1.7954
	v						1.3734	1.4923	1.6010	1.7036	1.8980	2.084	2.266	2.445
350	h						1217.7	1251.5	1282.1	1310.9	1365.5	1418.5	1471.1	1523.8
(431.72)	s						1.5119	1.5481	1.5792	1.6070	1.6563	1.7002	1.7403	1.7777
	v						1.1744	1.2851	1.3843	1.4770	1.6508	1.8161	1.9767	2.134
400	h						1208.8	1245.1	1277.2	1306.9	1362.7	1416.4	1469.4	1522.4
(444.59)	s						1.4892	1.5281	1.5607	1.5894	1.6398	1.6842	1.7247	1.7623

Table AT.2 Thermodynamic Properties of Steam (Steam Tables).

Saturated Steam - Ice

Tempera-ture,	Absolute pressure,	Specific volume, ft³/lb		Enthalpy, Btu/lb			Entropy, Btu/lb·°R		
		Saturated liquid,	Saturated vapor,	Saturated liquid,	Evaporation difference,	Saturated vapor,	Saturated liquid,	Evaporation difference,	Saturated vapor,
T, °F	P, lb_f/in^2	v_i	$v_g \times 10^{-3}$	h_i	h_{sub}	h_g	s_i	s_{sub}	s_g
32	0.0885	0.01747	3.306	−143.35	1219.1	1075.8	−0.2916	2.4793	2.1877
30	0.0808	0.01747	3.609	−144.35	1219.3	1074.9	−0.2936	2.4897	2.1961
20	0.0505	0.01745	5.658	−149.31	1219.9	1070.6	−0.3038	2.5425	2.2387
10	0.0309	0.01744	9.05	−154.17	1220.4	1066.2	−0.3141	2.5977	2.2836
0	0.0185	0.01742	14.77	−158.93	1220.7	1061.8	−0.3241	2.6546	2.3305
−10	0.0108	0.01741	24.67	−163.59	1221.0	1057.4	−0.3346	2.7143	2.3797
−20	0.0062	0.01739	42.2	−168.16	1221.2	1053.0	−0.3448	2.7764	2.4316
−30	0.0035	0.01738	74.1	−172.63	1221.2	1048.6	−0.3551	2.8411	2.4860

Table AT.3 Properties of Water (Saturated Liquid).

Temperature,		Heat capacity, c_p,	Density, ρ,	Viscosity, μ,	Thermal conductivity, k,	Prandtl number, Pr	Free convection coefficient, $\dfrac{g\beta\rho^2 c_p}{\mu k}$
°F	°C	kJ/kg·K	kg/m³	kg/m·s	W/m·°C		1/m³·°C
32	0	4.225	999.8	1.79×10^{-3}	0.556	13.25	1.91×10^{9}
40	4.44	4.208	999.8	1.55×10^{-3}	0.575	11.35	1.91×10^{9}
50	10	4.195	999.2	1.31×10^{-3}	0.585	9.40	6.34×10^{9}
60	15.56	4.186	998.6	1.12×10^{-3}	0.595	7.88	1.08×10^{10}
70	21.11	4.179	997.4	9.8×10^{-4}	0.604	6.78	1.46×10^{10}
80	26.67	4.179	995.8	8.6×10^{-4}	0.614	5.85	1.91×10^{10}
90	32.22	4.174	994.9	7.65×10^{-4}	0.623	5.12	2.48×10^{10}
100	37.78	4.174	993.0	6.82×10^{-4}	0.630	4.53	3.3×10^{10}
110	43.33	4.174	990.6	6.16×10^{-4}	0.637	4.04	4.19×10^{10}
120	48.89	4.174	988.8	5.62×10^{-4}	0.644	3.64	4.89×10^{10}
130	54.44	4.179	985.7	5.13×10^{-4}	0.649	3.30	5.66×10^{10}
140	60	4.179	983.3	4.71×10^{-4}	0.654	3.01	6.48×10^{10}
150	65.55	4.183	980.3	4.3×10^{-4}	0.659	2.73	7.62×10^{10}
160	71.11	4.186	977.3	4.01×10^{-4}	0.665	2.53	8.84×10^{10}
170	76.67	4.191	973.7	3.72×10^{-4}	0.668	2.33	9.85×10^{10}

(Continued)

Table AT.3 Cont.

Temperature, °F	°C	Heat capacity, c_p, kJ/kg·K	Density, ρ, kg/m³	Viscosity, μ, kg/m·s	Thermal conductivity, k, W/m·°C	Prandtl number, Pr	Free convection coefficient, $\dfrac{g\beta\rho^2 c_p}{\mu k}$ 1/m³·°C
180	82.22	4.195	970.2	3.47×10^{-4}	0.673	2.16	1.09×10^{11}
190	87.78	4.199	966.7	3.27×10^{-4}	0.675	2.03	
200	93.33	4.204	963.2	3.06×10^{-4}	0.678	1.90	
220	104.4	4.216	955.1	2.67×10^{-4}	0.684	1.66	
240	115.6	4.229	946.7	2.44×10^{-4}	0.685	1.51	
260	126.7	4.250	937.2	2.19×10^{-4}	0.685	1.36	
280	137.8	4.271	928.1	1.98×10^{-4}	0.685	1.24	
300	148.9	4.296	918.0	1.86×10^{-4}	0.684	1.17	
350	176.7	4.371	890.4	1.57×10^{-4}	0.677	1.02	
400	204.4	4.467	859.4	1.36×10^{-4}	0.655	1.00	
450	232.2	4.585	825.7	1.20×10^{-4}	0.646	0.85	
500	260	4.731	785.2	1.07×10^{-4}	0.616	0.83	
550	287.7	5.024	735.5	9.51×10^{-5}			
600	315.6	5.703	678.7	8.68×10^{-5}			

Note: $\mathrm{Gr}_x \mathrm{Pr} = \text{Rayleigh number}$, $\mathrm{Ra}_x = \left(\dfrac{g\beta\rho^2 c_p}{\mu k} \right) L^3 \Delta T$.

Table AT.4 Properties of Air at 1 atm.

Temperature, °C	Density, ρ, kg/m³	Dynamic viscosity, λ, kg/m·s ($\times 10^5$)	Kinematic viscosity, ν, m²/s ($\times 10^5$)	Heat capacity, c_p, J/kg·K	Thermal conductivity, k, W/m·K ($\times 10^2$)	Thermal expansion coefficient, β, K (10^3)	Prandtl number, Pr
-40	1.52	1.51	0.98		2.0		
-20	1.40	1.61	1.15	1004.8	2.21		
0	1.29	1.71	1.32	1004.8	2.42	3.65	0.715
10	1.248	1.76	1.41	1004.8	2.49	3.53	0.713
20	2.205	1.81	1.50	1004.8			
30	1.165	1.86	1.60	1004.8			
40	1.128	1.90	1.68	1004.8	2.7		
50	1.09	1.95	1.79	1007.0	2.8		
60	1.060	2.00	1.87	1009.0			
80	1.000	2.09	2.09	1009.0			
100	0.946	2.17	2.30	1009.0	3.12		
150	0.835	2.38	2.85	1017.0	3.53		
200	0.746	2.57	3.45	1025.8	3.88		0.686
250	0.675	2.75	4.07	1034.1	4.24		0.680
300	0.616	2.93	4.76				
400	0.525	3.25	6.19				
500	0.457	3.55	7.77		5.73		0.709

Example: At 50 °C, the air properties are: density = 1.09 kg/m³ (0.00211 slug/ft³ = 0.679 lb/ft³), dynamic or absolute viscosity = 0.00001195 kg/m·s (4.073 × 10⁻⁷ slug/ft·s = 1.31 × 10⁻⁵ lb/ft·s), thermal conductivity, k = 0.028 W/m·K, coefficient of thermal expansion, β = 1/T = 1/(273 + 50) = 0.0031 K⁻¹. The Prandtl number, Pr = $c_p \mu / k$ = 0.7.

Table AT.5 Properties of Selected Liquids at 1 atm and 20 °C (68 °F).

Liquid	Density, ρ, kg/m³	Dynamic viscosity, λ, kg/m·s ($\times 10^4$)	Kinematic viscosity, ν, m²/s ($\times 10^6$)	Surface tension, σ, N/m ($\times 10^2$)	Vapor pressure, P', kPa	Sound velocity, c, m/s
Acetone	785	3.16	0.403	2.31	27.6	1174
Ammonia	608	2.20	0.362	2.13	910.0	
Benzene	881	6.51	0.739	2.88	10.1	1298
Carbon disulfide	1272					
Carbon tetrachloride	1590	9.67	0.608	2.70	1.20	924
Castor oil	970	9000	927.8			1474
Crude oil	856	72	8.4	3.0		
Engine oil (unused)	888	7994	900.2			
Ethanol (or ethyl alcohol)	789	11	1.4	2.28	5.7	1144
Ethylene glycol	1117	214	19.16	3.27		1644
Freon-12	1330	2.63	0.198	1.58		
Fuel oil, heavy	908	1324	145.9			
Fuel oil, medium	854	32.7	3.82			
Gasoline	680	2.92	0.429	2.16	55.1	
Glycerin	1260	14,900	1183	6.33	0.14	1909
Kerosene	804	1.92	0.239	2.8	3.11	1320
Mercury	13,550	15.6	0.115	48.4	1.1×10^{-6}	1450

(Continued)

Table AT.5 Cont.

Liquid	Density, ρ, kg/m^3	Dynamic viscosity, λ, kg/m·s ($\times 10^4$)	Kinematic viscosity, ν, m^2/s ($\times 10^6$)	Surface tension, σ, N/m ($\times 10^2$)	Vapor pressure, P', kPa	Sound velocity, c, m/s
Methanol	791	5.98	0.756	2.25	13.4	1103
Milk (skimmed)	1041	14	1.34			
Milk (whole)	1030	21.2	2.06			
Olive oil	919	840	91.4			
Pentane	624					
Soybean oil	919	400	43.5			
SAE 10 oil	917	1040	113.4	3.6		
SAE 30 oil	917	2900	316.2	3.5		
Seawater	1025	10.7	1.04	7.28	2.34	1535
Turpentine	862	14.9	1.73			
Water	998	10.0	1.06	7.28	2.34	1498

Example: At 20 °C, the properties of liquid methanol are: density = 791 kg/m^3 (or SG = 0.791), dynamic or absolute viscosity = 0.000598 kg/m·s (or 0.598 cP), kinematic viscosity = 0.756 $\times 10^6$ m^2/s (0.756 cP = 8.14 $\times 10^{-6}$ ft^2/s), surface tension = 0.0225 N/m (0.00154 lb$_f$/ft), vapor (or saturation) pressure = 13,400 Pa (1.943 psi).

Table AT.6 Properties of Selected Gases at 1 atm and 20 °C (68 °F).

Gas	Molecular weight, MW	Density, ρ, kg/m^3	Viscosity Dynamic, μ, kg/m·s ($\times 10^5$)	Kinematic, v, m^2/s ($\times 10^6$)	Ratio of heat capacities, k	T_{crit}, K	P_{crit}, atm
Acetylene	26	1.09	0.97	8.3	1.30	309.5	61.6
Air (dry)	28.96	1.20	1.80	15.0	1.40	133	37
Ammonia	17.03	0.74	1.01	13.6	1.31	405	111.3
Argon	39.944	1.66	2.24	13.5	1.67		
Butane	58.1	2.49			1.11	425.2	37.5
Carbon dioxide	44.01	1.83	1.48	8.09	1.30	304	72.9
Carbon monoxide	28.01	1.16	1.82	15.7	1.40	133	34.5
Chlorine	70.91	2.95	1.03	3.49	1.34	417	76.1
Ethane	30.07	1.25	0.85	6.8	1.19	305	48.2
Ethylene	28	1.17	0.97	8.3	1.22	283.1	50.5
Helium	4.003	0.166	1.97	118.7	1.66	5.26	2.26
Hydrogen	2.016	0.0838	0.905	108.0	1.41	33	12.8
Hydrogen chloride	36.5	1.53	1.34	8.76	1.41	324.6	81.5
Hydrogen sulfide	34.1	1.43	1.24	8.67	1.30	373.6	88.9
Methane	16.04	0.667	1.34	20.1	1.32	190	45.8
Methyl chloride	50.5	2.15			1.20	416.1	65.8
Natural gas	19.5	0.804			1.27		

(Continued)

Table AT.6 Cont.

Gas	Molecular weight, MW	Density, ρ, kg/m³	Viscosity Dynamic, μ, ($\times 10^5$) kg/m·s	Viscosity Kinematic, ν, ($\times 10^6$) m²/s	Ratio of heat capacities, k	T_{crit}, K	P_{crit}, atm
Nitrogen	28.02	1.16	1.76	15.2	1.40	126	33.5
Nitrogen oxide (NO)	30.01	1.23	1.90	15.4	1.40	179	65.0
Nitrous oxide (N₂O)	44.02	1.83	1.45	7.92	1.31	309	71.7
Oxygen	32.0	1.36	2.00	14.7	1.10	154	49.7
Propane	44.1	1.88			1.15	369.9	42.0
Sulfur dioxide	64	2.66	1.38	5.2	1.29	430	77.8
Water vapor	18.02	0.749	1.02	13.6	1.33	647	218.3

Example: At 20 °C, the properties of argon gas are: molecular weight = 39.944, density = 1.66 kg/m³ (0.00322 slug/ft³ = 0.104 lb/ft³), dynamic or absolute viscosity = 0.0000224 kg/m·s (0.0224 cP = 4.68 × 10⁻⁷ slug/ft·s), kinematic viscosity = 13.5 × 10⁻⁶ m²/s (13.5 cSt = 1.45 × 10⁻⁴ ft²/s = 0.523 ft²/h), heat capacity ratio = 1.67.

Table AT.7 Dimensions, Capacities, and Weights of Standard Steel Pipes.

Nominal pipe size, in	Schedule number	Outside diameter (OD), in	Wall thick-ness, in	Inside diameter (ID), in	Cross-sectional area of metal, in^2	Inside sectional area, ft^2	Pipe weight, lb/ft
1/8	40	0.405	0.068	0.269	0.072	0.00040	0.24
	80		0.095	0.215	0.093	0.00025	0.31
1/4	40	0.540	0.088	0.364	0.125	0.00072	0.42
	80		0.119	0.302	0.157	0.00050	0.54
3/8	40	0.675	0.091	0.493	0.167	0.00133	0.57
	80		0.126	0.423	0.217	0.00098	0.74
1/2	40	0.840	0.109	0.622	0.250	0.00211	0.85
	80		0.147	0.546	0.320	0.00163	1.09
3/4	40	1.050	0.113	0.824	0.333	0.00371	1.13
	80		0.154	0.742	0.433	0.00300	1.47
1	40	1.315	0.133	1.049	0.494	0.00600	1.68
	80		0.179	0.957	0.639	0.00499	2.17
1 1/4	40	1.660	0.140	1.380	0.668	0.01040	2.27
	80		0.191	1.278	0.881	0.00891	3.00
1 1/2	40	1.900	0.145	1.610	0.800	0.01414	2.72
	80		0.200	1.500	1.069	0.01225	3.63
2	40	2.375	0.154	2.067	1.075	0.02330	3.65
	80		0.218	1.939	1.477	0.02050	5.02

(*Continued*)

Table AT.7 Cont.

Nominal pipe size, in	Schedule number	Outside diameter (OD), in	Wall thickness, in	Inside diameter (ID), in	Cross-sectional area of metal, in²	Inside sectional area, ft²	Pipe weight, lb/ft
2 1/2	40	2.875	0.203	2.469	1.704	0.03322	5.79
	80		0.276	2.323	2.254	0.02942	7.66
3	40	3.500	0.216	3.068	2.228	0.05130	7.58
	80		0.300	2.900	3.016	0.04587	10.25
3 1/2	40	4.000	0.226	3.548	2.680	0.06870	9.11
	80		0.318	3.364	3.678	0.06170	12.51
4	40	4.500	0.237	4.026	3.17	0.08840	10.79
	80		0.337	3.826	4.41	0.07986	14.98
5	40	5.563	0.258	5.047	4.30	0.1390	14.62
	80		0.375	4.813	6.11	0.1263	20.78
6	40	6.625	0.280	6.065	5.58	0.2006	18.97
	80		0.432	5.761	8.40	0.1810	28.57
8	40	8.625	0.322	7.981	8.396	0.3474	28.55
	80		0.500	7.625	12.76	0.3171	43.39
10	40	10.75	0.365	10.020	11.91	0.5475	40.48
	80		0.594	9.562	18.95	0.4987	64.40
12	40	12.75	0.406	11.938	15.74	0.7773	53.86
	80		0.688	11.374	26.07	0.7056	88.57

Table AT.8 Dimensions of Heat Exchanger Tubes.

Tube OD, in	BWG gauge	Thickness, in	Tube inside diameter (ID), in	Flow area, in²	Surface area, per foot of length, ft²/ft	
					External	Internal
1/4	22	0.028	0.194	0.0295	0.0655	0.0508
1/4	24	0.022	0.206	0.0333	0.0655	0.0539
1/2	18	0.049	0.402	0.1269	0.1309	0.1052
1/2	20	0.035	0.430	0.1452	0.1309	0.1126
1/2	22	0.028	0.444	0.1548	0.1309	0.1162
3/4	10	0.134	0.482	0.1825	0.1963	0.1262
3/4	14	0.083	0.584	0.2679	0.1963	0.1529
3/4	16	0.065	0.620	0.3019	0.1963	0.1623
3/4	18	0.049	0.652	0.3339	0.1963	0.1707
1	8	0.165	0.670	0.3526	0.2618	0.1754
1	14	0.083	0.834	0.5463	0.2618	0.2183
1	16	0.065	0.870	0.5945	0.2618	0.2278
1	18	0.049	0.902	0.6390	0.2618	0.2361
1 1/4	8	0.165	0.920	0.6648	0.3272	0.2409
1 1/4	14	0.083	1.084	0.9229	0.3272	0.2838
1 1/4	16	0.065	1.120	0.9852	0.3272	0.2932
1 1/4	18	0.049	1.152	1.042	0.3272	0.3016

(Continued)

Table AT.8 Cont.

Tube OD, in	BWG gauge	Thickness, in	Tube inside diameter (ID), in	Flow area, in²	Surface area, per foot of length, ft²/ft	
					External	Internal
2	11	0.120	1.760	2.433	0.5236	0.4608
2	12	0.109	1.782	2.494	0.5236	0.4665
2	13	0.095	1.810	2.573	0.5236	0.4739
2	14	0.083	1.834	2.642	0.5236	0.4801

(1 in = 25.4 mm; 1 in² = 645.16 mm²; 1 ft = 0.3048 m; 1 ft² = 0.0929 m²)

Table AT.9 Tube-Sheet Layouts (Tube Counts) on Square Pitch.

¾ in. OD tubes on 1-in. square pitch					
Shell ID, in.	1-P	2-P	4-P	6-P	8-P
8	32	26	20	20	
10	52	52	40	36	
12	81	76	68	68	60
13¼	97	90	82	76	70
15¼	137	124	116	108	108
17¼	177	166	158	150	142
19¼	224	220	204	192	188
21¼	277	270	246	240	234
23¼	341	324	308	302	292
25	413	394	370	356	346
27	481	460	432	420	408
29	553	526	480	468	456
31	657	640	600	580	560
33	749	718	688	676	648
35	845	824	780	766	748
37	934	914	886	866	838
39	1049	1024	982	968	948

1 in. OD tubes on 1¼-in. square pitch					
Shell ID, in.	1-P	2-P	4-P	6-P	8-P
8	21	16	14		
10	32	32	26	24	
12	48	45	40	38	36
13¼	61	56	52	48	44
15¼	81	76	68	68	64
17¼	112	112	96	90	82
19¼	138	132	128	122	116
21¼	177	166	158	152	148
23¼	213	208	192	184	184
25	260	252	238	226	222
27	300	288	278	268	260
29	341	326	300	294	286
31	406	398	380	368	358
33	465	460	432	420	414
35	522	518	488	484	472
37	596	574	562	544	532
39	665	644	624	612	600

(Continued)

Table AT.9 Cont.

¾ in. OD tubes on 1-in. square pitch

1¼ in. OD tubes on 1 9/16-in. square pitch

Shell ID, in.	1-P	2-P	4-P	6-P	8-P
10	16	12	10		
12	30	24	22	16	16
13¼	32	30	30	22	22
15¼	44	40	37	35	31
17¼	56	53	51	48	44
19¼	78	73	71	64	56
21¼	96	90	86	82	78
23¼	127	112	106	102	96
25	140	135	127	123	115
27	166	160	151	146	140
29	193	188	178	174	166
31	226	220	209	202	193
33	258	252	244	238	226
35	293	287	275	268	258
37	334	322	311	304	293
39	370	362	348	342	336

1 in. OD tubes on 1¼-in. square pitch

1½ in. OD tubes on 1 7/8-in. square pitch

Shell ID, in.	1-P	2-P	4-P	6-P	8-P
12	16	16	12	12	
13¼	22	22	16	16	
15¼	29	29	25	24	22
17¼	39	39	34	32	29
19¼	50	48	45	43	39
21¼	62	60	57	54	50
23¼	78	74	70	66	62
25	94	90	86	84	78
27	112	108	102	98	94
29	131	127	120	116	112
31	151	146	141	138	131
33	176	170	164	160	151
35	202	196	188	182	176
37	224	220	217	210	202
39	252	246	237	230	224

Table AT.10 Tube-Sheet Layouts (Tube Counts) on Triangular Pitch.

¾ in. OD tubes on 15/16-in. triangular pitch

Shell ID, in.	1-P	2-P	4-P	6-P	8-P
8	36	32	26	24	18
10	62	56	47	42	36
12	109	98	86	82	78
13¼	127	114	96	90	86
15¼	170	160	140	136	128
17¼	239	224	194	188	178
19¼	301	282	252	244	234
21¼	361	342	314	306	290
23¼	442	420	386	378	364
25	532	506	468	446	434
27	637	602	550	536	524
29	721	692	640	620	594
31	847	822	766	722	720
33	974	938	878	852	825
35	1102	1068	1004	988	958
37	1240	1200	1144	1104	1072
39	1377	1330	1258	1248	1212

¾ in. OD tubes on 1-in. triangular pitch

Shell ID, in.	1-P	2-P	4-P	6-P	8-P
8	37	30	24	24	
10	61	52	40	35	
12	92	82	76	74	70
13¼	109	106	86	82	74
15¼	151	138	122	118	110
17¼	203	196	178	172	166
19¼	262	250	226	216	210
21¼	316	302	278	272	260
23¼	384	376	352	342	328
25	470	452	422	394	382
27	559	534	488	474	464
29	630	604	556	538	508
31	745	728	678	666	640
33	856	830	774	760	732
35	970	938	882	864	848
37	1074	1044	1012	986	870
39	1206	1176	1128	1100	1078

(*Continued*)

Table AT.10 Cont.

¾ in. OD tubes on 15/16-in. triangular pitch

1 in. OD tubes on 1¼-in. triangular pitch

Shell ID, in.	1-P	2-P	4-P	6-P	8-P
8	21	16	16	14	
10	32	32	26	24	
12	55	52	48	46	44
13 ¼	68	66	58	54	50
15 ¼	91	86	80	74	72
17 ¼	131	118	106	104	94
19 ¼	163	152	140	136	128
21 ¼	199	188	170	164	160
23 ¼	241	232	212	212	202
25	294	282	256	252	242
27	349	334	302	296	286
29	397	376	338	334	316
31	472	454	430	424	400
33	538	522	486	470	454
35	608	592	562	546	532
37	674	664	632	614	598

¾ in. OD tubes on 1-in. triangular pitch

1¼ in. OD tubes on 1 9/16-in. triangular pitch

Shell ID, in.	1-P	2-P	4-P	6-P	8-P
10	20	18	14		
12	32	30	26	22	20
13 ¼	38	36	32	28	26
15 ¼	54	51	45	42	38
17 ¼	69	66	62	58	54
19 ¼	95	91	86	78	69
21 ¼	117	112	105	101	95
23 ¼	140	136	130	123	117
25	170	164	155	150	140
27	202	196	185	179	170
29	235	228	217	212	202
31	275	270	255	245	235
33	315	305	297	288	275
35	357	348	335	327	315
37	407	390	380	374	357

(Continued)

Table AT.10 Cont.

¾ in. OD tubes on 15/16 -in. triangular pitch

Shell ID, in.	1-P	2-P	4-P	6-P	8-P
39	766	736	700	688	672

1½ in. OD tubes on 1 7/8 -in. triangular pitch

Shell ID, in.	1-P	2-P	4-P	6-P	8-P
12	18	14	14	12	12
13 ¼	27	22	18	16	14
15 ¼	36	34	32	30	27
17 ¼	48	44	42	38	36
19 ¼	61	58	55	51	48
21 ¼	76	72	70	66	61
23 ¼	95	91	86	80	76
25	115	110	105	98	95
27	136	131	125	118	115
29	160	154	147	141	136
31	184	177	172	165	160
33	215	206	200	190	184
35	246	238	230	220	215
37	275	268	260	252	246
39	307	299	290	284	275

¾ in. OD tubes on 1-in. triangular pitch

Shell ID, in.	1-P	2-P	4-P	6-P	8-P
39	449	436	425	419	407

Table AT.11 Approximate Design Overall Heat Transfer Coefficients (Btu/hr·ft²·°F).

Values include total dirt factors of 0.0003 and allowable pressure drops of 5 to 10 psi on the controlling stream.

Coolers		
Hot fluid	**Cold fluid**	**Overall U_D**
Water	Water	250–500 §
Methanol	Water	250–500 §
Ammonia	Water	250–500 §
Aqueous solutions	Water	250–500 §
Light organics*	Water	75–100
Medium organics†	Water	50–125
Heavy organics‡	Water	5–75 ‖
Gases	Water	2–50 ¶
Water	Water	100–200
Light organics	Brine	40–100
Heaters		
Hot fluid	**Cold fluid**	**Overall U_D**
Steam	Water	200–700 §
Steam	Methanol	200–700 §
Steam	Ammonia	200–700 §
Steam	Aqueous solutions:	
Steam	Less than 2.0 cP	200–700
Steam	More than 2.0 cP	100–500 §
Steam	Light organics	100–500
Steam	Medium organics	50–100
Steam	Heavy organics	6–60
Steam	Gases	5–50 ¶
Exchangers		
Hot fluid	**Cold fluid**	**Overall U_D**
Water	Water	250–500 §
Aqueous solutions	Aqueous solutions	250–500 §
Light organics	Light organics	40–75

(Continued)

Table AT.11 Cont.

Medium organics	Medium organics	20–60
Heavy organics	Heavy organics	10–40
Heavy organics	Light organics	30–60
Light organics	Heavy organics	10–40

* *Light organics* are fluids with viscosities of less than 0.6 centipoise and include benzene, toluene, acetone, ethanol, methyl ethyl ketone, gasoline, light kerosene, and naphtha.

† *Medium organics* have viscosities of 0.5 to 1.0 centipoise and include kerosene, straw oil, hot gas oil, hot absorber oil and some crudes.

‡ *Heavy organics* have viscosities above 1.0 centipoise and include cold gas oil, lube oils, fuel oils, reduced crude oils, tars, and asphalts.

§ Dirt factor 0.001.

‖ Pressure drop 20 to 30 psi.

¶ These rates are greatly influenced by the operating pressure.

Table AT.12 Approximate Design Fouling Coefficient Factors (hr·ft^2·°F/Btu).

Temperature of heating	Up to 240 °F		240 – 400 °F†	
Temperature of water	125 °F or less		Over 125 °F	
Water	Water velocity, fps		Water velocity, fps	
	3 ft and less	Over 3 ft	3 ft and less	Over 3 ft
Sea water	0.0005	0.0005	0.001	0.001
Brackish water	0.002	0.001	0.003	0.002
Cooling tower and artificial spray pond:				
Treated make-up	0.001	0.001	0.002	0.002
Untreated	0.003	0.003	0.005	0.004
City or well water (such as Great Lakes)	0.001	0.001	0.002	0.002
Great Lakes	0.001	0.001	0.002	0.002
River water:				
Minimum	0.002	0.001	0.003	0.022
Mississippi	0.003	0.002	0.004	0.003
Delaware, Schuylkill	0.003	0.002	0.004	0.003
East River and New York Bay	0.003	0.002	0.004	0.003
Chicago sanitary canal	0.008	0.006	0.010	0.008
Muddy or silty	0.003	0.002	0.004	0.003

(*Continued*)

Table AT.12 Cont.

Temperature of heating	Up to 240 °F		240 – 400 °F†	
Temperature of water	125 °F or less		Over 125 °F	
Water	Water velocity, fps		Water velocity, fps	
	3 ft and less	Over 3 ft	3 ft and less	Over 3 ft
Hard (over 15 grains/gal)	0.003	0.003	0.005	0.005
Engine jacket	0.001	0.001	0.001	0.001
Distilled	0.0005	0.0005	0.0005	0.0005
Treated boiler feedwater	0.001	0.0005	0.001	0.001
Boiler blowdown	0.002	0.002	0.002	0.002

† Ratings in the last two columns are based on a temperature of the heating medium of 240 to 400 °F. If the heating medium temperature is over 400 °F, and the cooling medium is known to scale these ratings should be modified accordingly.

Table AT.12 Cont.

Petroleum Fractions

Oils (industrial):	
Fuel oil	0.005
Clean recirculating oil	0.001
Machinery and transformer oils	0.001
Quenching oil	0.004
Vegetable oils	0.003
Gases, vapors (industrial):	
Coke-oven gas, manufactured gas	0.01
Diesel-engine exhaust gas	0.01
Organic vapors	0.0005
Steam (non-oil bearing)	0.0
Alcohol vapors	0.0
Steam, exhaust (oil bearing from reciprocating engines)	0.001
Refrigerating vapors (condensing from reciprocating compressors)	0.002
Air	0.002
Liquids (industrial):	
Organic	0.001
Refrigerating liquids, heating, cooling, or evaporating	0.001

(Continued)

Table AT.12 Cont.

Brine (cooling)	0.001
Atmospheric distillation units:	
Residual bottoms, less than 25°API	0.005
Distillate bottoms, 25°API or above	0.002
Atmospheric distillation units:	
Overhead untreated vapors	0.0013
Overhead treated vapors	0.003
Side-stream cuts	0.0013
Vacuum distillation units:	
Overhead vapors to oil:	
From bubble tower (partial condenser)	0.001
From flash pot (no appreciable reflux)	0.003
Overhead vapors in water-cooled condensers:	
From bubble tower (final condenser)	0.001
From flash pot	0.04
Side stream:	
To oil	0.001
To water	0.002
Residual bottoms, less than 20°API	0.005
Distillate bottoms, over 20°API	0.002
Natural gasoline stabilizer units:	
Feed	0.0005
Overhead vapors	0.0005
Product coolers and exchangers	0.0005
Product reboilers	0.001
H_2S Removal Units:	
For overhead vapors	0.001
Solution exchanger coolers	0.0016
Reboiler	0.0016
Cracking units:	
Gas oil feed:	
Under 500 °F	0.002
500 °F and over	0.003

(Continued)

Table AT.12 Cont.

Naphtha feed:	
Under 500 °F	0.002
500 °F and over	0.004
Separator vapors (vapors from separator, flash pot, and vaporizer)	0.006
Bubble-tower vapors	0.002
Residuum	0.010
Absorption units:	
Gas	0.002
Fat oil	0.002
Lean oil	0.002
Overhead vapors	0.001
Gasoline	0.0005
Debutanizer, Depropanizer, Depentanizer, and Alkylation Units:	
Feed	0.001
Overhead vapors	0.001
Product coolers	0.001
Product reboilers	0.002
Reactor feed	0.002
Lube treating units:	
Solvent oil mixed feed	0.002
Overhead vapors	0.001
Refined oil	0.001
Refined oil heat exchangers water cooled‡	0.003
Gums and tars:	
Oil-cooled and steam generators	0.005
Water-cooled	0.003
Solvent	0.001
Deasphaltizing units:	
Feed oil	0.002
Solvent	0.001
Asphalt and resin:	
Oil-cooled and steam generators	0.005
Water-cooled	0.003

(Continued)

Table AT.12 Cont.

Solvent vapors	0.001
Refined oil	0.001
Refined oil water cooled	0.003
Dewaxing units:	
Lube oil	0.001
Solvent	0.001
Oil wax mix heating	0.001
Oil wax mix cooling‡	0.003

‡ Precautions must be taken against deposition of wax.

Table AT.12 Cont.
Crude Oil Streams

	0 – 199 °F			200 – 299 °F			300 – 499 °F			500 °F and over		
	Velocity, fps											
	Under 2 ft	2 – 4 ft	4 ft and over	Under 2 ft	2 – 4 ft	4 ft and over	Under 2 ft	2 – 4 ft	4 ft and over	Under 2 ft	2 – 4 ft	4 ft and over
Dry	0.003	0.002	0.002	0.003	0.002	0.002	0.004	0.003	0.002	0.005	0.004	0.003
Salt §	0.003	0.002	0.002	0.005	0.004	0.004	0.006	0.005	0.004	0.007	0.006	0.005

§ Refers to a wet crude — any crude that has not been dehydrated

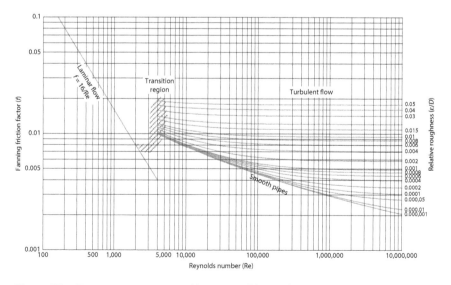

Figure AF.1 Fanning Friction Factor (f) vs. Reynolds Number (Re) Plot.

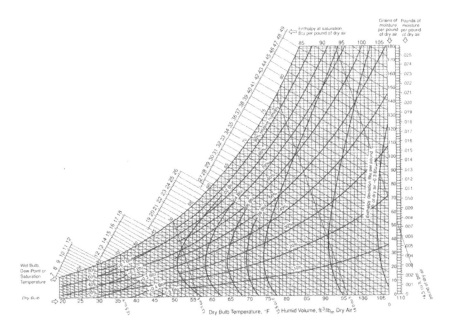

Figure AF.2 Psychometric Chart: Low Temperatures. Barometric pressure, 29.92 in. Hg.

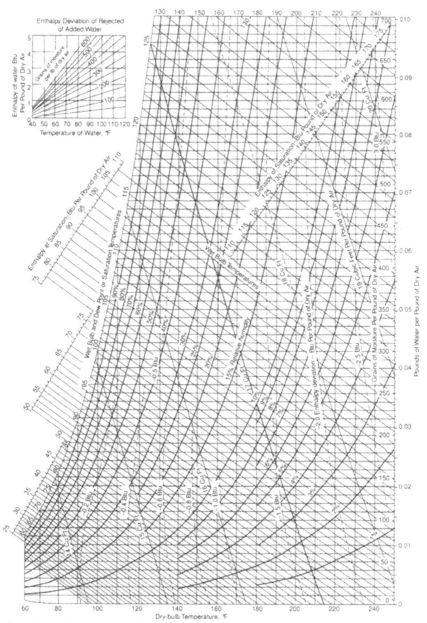

Figure AF.3 Psychometric Chart: High Temperatures. Barometric pressure, 29.92 in. Hg.

Index

Absorptivity, 138–140
Acceptable risk, 641
Acfm, 18
Acute, 608, 611, 626, 629, 633, 635
Adiabatic lapse rate, 122
Air cooling, 457
Allowable dirt factor, 204, 225
Allowable pressure drop, 252, 263, 268, 273, 277, 340, 365, 375, 390, 426, 683
Amagat's law, 20
Annualized capital cost, 572–574
Annular, 34, 185, 218, 221–224, 227, 238–239, 241, 243, 254–255, 385, 388, 395, 400, 411–412
Annular fins, 385, 395, 400
Annulus, 3–4, 168–171, 185, 187, 189, 202, 217–224, 226–227, 230–231, 243, 252–253, 262–263, 265, 267, 269, 271, 273, 277–278, 280, 285, 366, 394, 420, 422, 434
Approach temperature, 198, 255, 283, 595
Area requirements, 232, 237–238, 326, 483
Arithmetic mean, 99, 171–172, 207, 418

Baffle, 162, 215, 283, 290, 292, 295–297, 302–305, 316–318, 320, 322, 340, 344, 350, 352, 359, 366–367, 377–379, 426, 438, 564
Baffle spacing, 295, 305, 317–318, 320, 322, 340
Batch, 69, 515, 555–561
Batch distillation, 69, 556–558

Batch reactors, 69, 556, 560
Bathtub curve, 645, 648–649
Bell-delaware method, 322
Biot number, 9, 93, 101, 105, 393, 401, 405, 408, 411, 518
Bivariant, 442
Black body, 138, 140, 143–146, 148–149, 151, 154
Boilers, 153, 163, 290, 439, 455, 470–476, 480–484, 487, 493, 511
Boiling, 84–85, 164–165, 312, 334, 433, 439–441, 444, 448, 452–454, 461, 467–469, 484, 486, 547, 558, 625
Boundary conditions, 8, 56, 66, 69, 174, 416
Buoyancy, 109, 113–114, 121, 124–125, 441, 451, 475, 505

Caloric temperature, 207
Candela, 7
Capital costs, 211, 213, 569–570, 572–573
Capital recovery factor, 572–573, 578
Characteristic wavelengths of radiation, 130
Charles' law, 17–18
Chronic, 608, 612, 626, 629, 633, 635
Clean air act (caa), 629
Clean overall coefficient, uc 202, 271, 279, 343, 347, 355, 362, 370, 424, 431
Cleaning, 69, 205, 217, 221, 255, 285, 294–295, 299, 303–304, 465, 556, 562, 565, 567, 627

Clearance, 294, 317–318, 562

Clearing provisions, 562

Cocurrent, 4, 167–171, 179–183, 185, 202, 220, 223, 237–238, 248–249, 254, 261, 273, 313, 323, 379

Cogeneration/combined heat and power (chp), 439, 488

Comprehensive environmental response, compensation, and liability act (cercla), 620

Compressibility coefficient, 19

Concentric pipes, 3, 34, 167–169, 185, 187, 217

Condensation, 25, 84, 164–165, 172–173, 182, 184, 247, 310–311, 334, 364–365, 370, 438–452, 459, 480, 494, 498–501, 505, 509, 537, 562, 566–567

Condensers, 93, 163, 283, 290, 308, 389, 438–439, 441, 449–450, 499, 503, 505–506, 511, 562, 564, 566, 586, 686

Condensing boilers, 470, 474, 476

Conductance, 44, 46, 84

Conduction, 4, 13, 26–28, 43, 45–47, 49–51, 53, 55, 57, 59, 61, 63–65, 67–71, 73, 75, 77, 79–81, 83, 93, 105, 121, 129–131, 166, 184, 188, 215, 310, 391, 407, 453, 481, 500–501, 516–518, 528, 530, 536, 559–560, 586

Conductors, 44

Conduits, 29, 32–35, 59, 89

Conservation laws, 21–22

Controlling film coefficient, 194–195

Controlling resistance, 194–197, 351, 381, 434

Convection, 4, 26–28, 45, 59–60, 70, 79–87, 89–91, 93–97, 99–101, 103, 105, 107–121, 123, 125, 127, 129–131, 148–152, 184, 188, 215, 386, 391, 405, 411, 415, 433, 441, 451–453, 471–473, 481–482, 493–495, 501, 517–518, 529–530, 582, 667–668

Convective resistances, 82, 87

Conversion constants, 7, 653–654

Cooling towers, 351, 439, 456–458, 461, 498–500, 502–503, 505–506

Cooling-tower performance, 506

Corrected fin height, 395

Cost-benefit analysis, 575–576

Countercurrent, 167–168, 170–173, 179–183, 185, 200–201, 207, 220, 222–223, 237–238, 245–247, 249–250, 254–255, 264, 268, 273, 277, 282, 298, 313–315, 323, 325–329, 331–332, 372, 376–377, 379, 483, 504, 557

Critical number, 14

Cryogenics, 515, 533–535, 541–542, 544, 586

Dalton's law, 20

Darcy friction factor, 92, 94

Degrees of freedom, 441–442, 445

Density, 6, 10–11, 14, 18, 32, 79, 97, 100, 102, 104, 108–110, 113, 121–122, 124, 198, 226, 252, 311, 323, 365, 390, 406, 439, 442, 452, 475, 484, 504, 537, 549, 619, 625, 667–673

Dimensional analysis, 4, 8–9

Dimensionless groups, 8–9, 96–97, 100

Direct contact condensation, 447

Dirt factor, 202, 204, 213, 225, 266, 272, 280, 285–286, 340, 343–344, 347–348, 351, 356, 359, 363, 366, 370, 378, 426, 431, 434, 683–684

Dittus-boelter equation, 94, 101, 228, 242, 257, 260, 310, 429, 484

Dose-response, 613, 616–617, 631–632, 634–637

Double pipe, 3, 163, 167, 185, 202, 204, 213, 217–225, 227, 229, 231–239, 241, 243, 245, 247–249, 251–255,

257, 259–263, 265, 267–269, 271,
273–277, 279, 281, 283–287, 290,
307–308, 322, 339, 366, 376, 382,
384, 388–389, 419, 449, 578–579
Dropwise condensation, 334, 447–448
Dynamic viscosity, 11

Effect of noncondensables, 454
Efficiency of an exchanger, 350, 372
Emissivity, 28, 132, 139–151
Emissivity correction factor, 147
Enthalpy, 23–25, 165–166, 183, 307,
440–441, 444–445, 456, 459,
461–464, 466–470, 474, 477–479,
493, 495, 498, 501, 509, 537,
539–540, 544–546, 561,
658–661, 666
Entropy, 22, 25, 28, 38–39, 444, 477,
480, 538, 589–593, 595–603, 605,
658–661, 666
Epidemiology, 634
Equipment cost, 450, 569–572, 587
Equipment purchasing guidelines,
562, 567
Equivalent diameter, 222, 227,
230–231, 262, 265, 269, 278, 316,
318–320, 324, 366, 396, 421, 428
Error function, 76
Evaporative cooler, 456–461
Evaporators, 163, 439, 441, 450,
454–456, 461, 463, 465, 498
Exchangers using water, 350
Exchangers without baffles, 366
Exit temperatures, 177, 200, 218,
244, 328
Extended-surface shell-and-tube
exchangers, 389

F factor, 313, 315, 323, 326, 328, 331
Factored method, 571
Fanning friction factor, 31–32, 653, 688
Fatal accident rate, 614, 641
Feedback control, 475, 550, 552, 554
Feedforward control, 550, 552–553

Film coefficient, 82–85, 87–89, 93,
130, 188, 192, 194–195, 207, 221,
225–226, 230, 248, 258–259, 261,
284, 316–317, 334, 339, 351, 355,
364, 376, 389–390, 448, 484, 559
Film condensation, 310–311, 447
Fin effectiveness, 382, 398–399, 403,
410, 413–414
Fin efficiency, 385, 390–391, 393,
395–397, 399–402, 404–406,
408–409, 411–414, 417, 419,
423, 430
Fin performance coefficient, 398
Finned heat exchangers, 381, 383, 385,
387, 389, 391, 393, 395, 397, 399,
401, 403, 405, 407, 409, 411, 413,
415, 417, 419, 421, 423, 425, 427,
429, 431, 433, 435
Fins, 70, 162, 381–385, 387–390,
392–396, 398–400, 403–414,
419–422, 426, 431, 433–434, 566
Fittings, 33–34, 218, 220–221, 251
Fixed-tube-sheet 1–2 exchangers,
292, 298
Floating head, 299–300, 302, 321, 566
Flow of heat through a composite wall,
52
Flow of heat through a pipe wall,
46, 54
Flow of heat through a plane wall, 46
Forced convection, 28, 70, 80–82,
84–85, 89–90, 93–94, 96, 99–100,
108, 110–111, 113, 115, 131,
452, 495
Fouling coefficients, 202–204, 261
Fouling factor, 202–203, 205, 233, 268,
276, 284–285, 308–309, 351–352,
373, 377, 379, 419, 426, 434, 454
Foundations, 562
Fourier's law, 45, 63–64
Free convection, 28, 45, 70, 79–85, 87,
89, 91, 93, 95, 97, 99, 101, 103,
105, 107–111, 113–121, 123, 125,
127, 149–150, 452, 667–668

Friction factor, 30–32, 40, 90–92, 94, 96, 222, 230, 253–254, 321–322, 395, 653, 688

Furnaces, 70, 153, 389, 439, 470, 472–473, 476, 482–483, 521, 547, 555

Fuzzy logic control, 553–554

Geometric mean, 171–172

Gibbs phase rule, 442, 445

Graetz number, 9, 115–116, 227–228, 309–310

Grashof number, 9, 109, 114–115

Green chemistry, 36–37

Green engineering, 36–37, 41, 216

Hairpin, 218–219, 221, 253, 266, 268, 272, 277, 280, 285–286, 290, 306, 314, 388, 419, 425, 434

Hazard risk assessment, 41, 216, 511, 586, 611, 615, 623, 628–629, 633, 640–641, 643, 651

Health risk, 39, 610–611, 615, 619, 625, 628–633, 637–638, 641, 651

Heat capacity, 12–13, 24–26, 97, 100, 105, 110, 165–166, 172–173, 183–184, 198, 200, 236, 244, 307, 327–328, 358, 373, 406, 439, 467, 470, 484, 487, 493–494, 496, 537, 547, 557, 593–594, 657, 673

Heat exchanger, 3–4, 13, 15–16, 19–20, 22–23, 25, 28–29, 31, 33, 35, 39, 93, 103, 106, 131, 159, 161–164, 167, 169–171, 173, 179, 181, 183–185, 187–188, 200, 202, 205–207, 209–219, 221–227, 229, 231–233, 235–239, 241, 243–245, 247–249, 251, 253–257, 259, 261–263, 265–267, 269, 271–275, 277, 279, 281–283, 285, 287, 289–293, 295–297, 299, 301–309, 311, 313–317, 319–321, 323, 325–331, 333–337, 339–341, 343, 345, 347, 349, 351, 353, 355, 357, 359, 361, 363, 365, 367, 369, 371–375, 377, 379–381, 383–385, 387, 389, 391, 393, 395, 397, 399, 401, 403, 405, 407, 409, 411, 413, 415, 417–419, 421, 423, 425, 427, 429, 431, 433, 435, 437–439, 441, 443, 445, 447, 449, 451, 453, 455–457, 459–461, 463, 465, 467, 469, 471, 473–475, 477, 479, 481–483, 485, 487, 489, 491, 493, 495, 497, 499–501, 503, 505–507, 509, 511, 515–516, 543, 551, 561–563, 567–569, 571–574, 578, 590–591, 594–596, 598–600, 603–607, 609–610, 617, 637–638, 640, 642, 644–645, 648–649, 653, 676, 687

Heat pipes, 439, 508, 510

Heat transfer equation, 112, 161–163, 165, 167, 169, 171, 173, 175, 177, 179, 181, 183, 185, 187, 189, 191, 193, 195, 197, 199, 201, 203, 205, 207–209, 211, 213–215, 225–226, 232, 234, 246, 251, 256, 258, 263, 284, 313, 324, 335–336, 338, 374, 433, 484, 594

Heat-exchanger tubes, 293

Homogenous condensation, 447

Hydraulic radius, 34, 222, 230, 318–319

Ideal gas, 16–20, 102, 109, 114, 122, 166, 497, 593, 638–639

Ideal gas law, 16–20, 102, 122, 497, 593, 639

Improving operation and performance, 562, 566

Individual film, 83, 339, 351

Individual film coefficient, 83, 339, 351

Industrial ecology, 35–36, 39–41, 216, 651

Initial conditions, 8, 64, 74, 349, 499

Inside pipe, 85, 218, 222, 224, 242

Installation procedures, 562

Instrumentation control symbols, 548
Insulation, 44, 58–60, 73, 192–193,
    255, 454, 483, 515–521, 526–532,
    535–536, 572, 577–578, 580–582,
    584–585, 592
Insulators, 27, 44, 516, 531

Jakob number, 441

Kern's design methodology, 213, 215,
    218, 262–263, 282, 290, 324, 339,
    349, 382, 418
Kinematic viscosity, 11, 14, 100, 110,
    671, 673
Kinetic energy, 13, 15, 23, 133, 483
Knudsen and katz equation, 93

Laminar flow, 14, 30–31, 34, 89–90,
    100, 227, 265–267, 276, 309, 311,
    342, 448
Laminar flow through a circular tube,
    30
Lapse rate, 120–123
Latent enthalpy, 25, 441, 445, 467, 498
Law of conservation of energy, 21, 23,
    589, 591
Law of conservation of mass, 21, 23
Law of conservation of momentum, 22
Laws of thermodynamics, 4, 22, 28
Leidenfrost phenomenon, 453
Ligament, 294
Liquefaction, 542–544
Liquid jacob number, 9
Log mean temperature difference
    (lmtd), 162, 171, 179, 214, 232,
    283, 313, 325–326, 328–329,
    374, 594
Logarithmic mean, 55, 171–172
Longitudinal fins, 382–385, 388, 420
Maintenance, 162, 211, 348, 357, 449,
    459, 471, 481–482, 515–516, 540,
    546, 548, 561–568, 573–574, 576,
    578–579, 627
Manipulated variable, 550–551, 554

Material safety data sheets (msds), 611,
    622–623
Mean film temperature, 93, 99, 118, 448
Mechanical-draft tower, 503
Modified lang method, 571
Monovariant, 442
Moody friction factor, 31

Natural-circulation towers, 503–504
Network design, 604
Newton's law of cooling, 28, 82, 84–85,
    88, 529
No-slip boundary condition, 108
Nusselt number, 9, 89–92, 98–101,
    105, 109, 111–112, 114–115, 117,
    227–228, 309–310
Observed temperature, 205

Occupational safety and health act
    (osha), 621
Operating costs, 162, 211, 213, 569,
    573–574, 587, 595
Operation, 8, 22, 35, 37, 127, 130,
    151, 164, 206, 210–212, 218, 229,
    235, 255–256, 261, 284–286, 292,
    298, 312, 314, 320–321, 326, 334,
    340, 350, 364, 379, 390, 437–440,
    451, 456–460, 464–466, 474, 490,
    492, 494, 498–499, 506, 511–512,
    515–516, 533–534, 546–547, 553,
    555–556, 558, 560–566, 568–569,
    574–576, 578–579, 586, 595–598,
    602, 608–609, 621–622, 638, 640,
    644–645
Outer tube limit, 300
Overall fin efficiency, 400, 404, 414
Overall heat transfer coefficient, u,
    161–162, 183–184, 187–189,
    191–192, 195, 197–198, 204,
    207–208, 214, 232–233, 236–237,
    239, 284, 305, 308–309, 312, 328,
    374, 463, 465, 467–468, 482–484,
    565, 583, 594, 653, 683
Overall surface effectiveness, 400, 403

Partial pressure, 19–20, 443, 454, 507
Partial volume, 19–20
Peclet number, 9, 89, 95, 98–99, 101
Peg fins, 385
Perturbation studies in optimization, 569, 574
Petroleum-refinery furnaces, 470, 472
Piping consideration, 562–563
Plume rise, 120, 123–127
Point of compliance, 630
Point of exposure, 630
Pollution prevention, 35–37, 39–41, 212, 216, 458, 460, 511, 651
Potential energy, 16, 23, 31, 132
Prandtl number, 9, 89, 92, 98–100, 109, 114–115, 228, 260, 310, 441, 669
Preliminary cost analysis, 569
Pressure drop, 4, 22, 28, 30–31, 35, 122, 213, 218, 230–231, 239, 251–254, 256, 263, 266–268, 272–273, 277, 281, 283–286, 292, 294, 304–306, 316, 320–322, 324, 340, 343–344, 348–349, 351, 354, 356, 359, 361, 363, 365–366, 369, 371, 374–375, 378–379, 390, 395, 419, 425–426, 432–434, 456, 462, 487, 504, 564–566, 603, 683–684
Pressure drops in pipes and annuli, 251
Prime movers, 32–33, 491–492
Process heat transfer, 3–5, 7, 9, 11, 13, 15, 17, 19, 21, 23, 25, 27, 29, 31, 33, 35, 37, 39–41, 44, 46, 48, 50, 52, 54, 56, 58, 60, 62, 64, 66, 68, 70, 72, 74, 76–77, 80, 82, 84, 86, 88, 90, 92, 94, 96, 98, 100, 102, 104, 106, 108, 110, 112, 114, 116, 118, 120, 122, 124, 126–127, 130, 132, 134, 136, 138, 140, 142, 144, 146, 148, 150, 152, 154, 156–157, 162, 164, 166, 168, 170, 172, 174, 176, 178, 180, 182, 184, 186, 188, 190, 192, 194, 196, 198, 200, 202, 204, 206, 208, 210, 212, 214, 216, 218, 220, 222, 224, 226, 228, 230,
232, 234, 236, 238, 240, 242, 244, 246, 248, 250, 252, 254, 256, 258, 260, 262, 264, 266, 268, 270, 272, 274, 276, 278, 280, 282, 284, 286, 290, 292, 294, 296, 298, 300, 302, 304, 306, 308, 310, 312, 314, 316, 318, 320, 322, 324, 326, 328, 330, 332, 334, 336, 338, 340, 342, 344, 346, 348, 350, 352, 354, 356, 358, 360, 362, 364, 366, 368, 370, 372, 374, 376, 378–380, 382, 384, 386, 388, 390, 392, 394, 396, 398, 400, 402, 404, 406, 408, 410, 412, 414, 416, 418, 420, 422, 424, 426, 428, 430, 432, 434–435, 438, 440, 442, 444, 446, 448, 450, 452, 454, 456, 458, 460, 462, 464, 466, 468, 470, 472, 474, 476, 478, 480, 482, 484, 486, 488, 490, 492, 494, 496, 498, 500, 502, 504, 506, 508, 510, 512, 516, 518, 520, 522, 524, 526, 528, 532, 534, 536, 538, 540, 542, 544, 546, 548, 550, 552, 554, 556, 558, 560, 562, 564, 566, 568, 570, 572, 574, 576, 578, 580, 582, 584, 586, 590, 592, 594, 596, 598, 600, 602, 604, 606, 608, 610, 612, 614, 616, 618, 620, 622, 624, 626, 628, 630, 632, 634, 636, 638, 640, 642, 644, 646, 648, 650
Process temperatures, 169, 205, 340, 372
Process variables, 4, 14, 68, 213, 264, 283, 546–547, 555, 575
Profile area, 392–393, 402, 406, 408, 412
Project evaluation, 569, 573
Projected perimeter, 396
Proportional-integral, 550–551
Proportional-integral-derivative, 550–551
Psychometric chart, 443–444, 507, 689–690
Pull-through floating head, 299–300

Quality energy, 35, 39, 283, 590–591, 595, 598, 603, 605
Quenchers, 439, 492, 494

Radiant energy, 28, 129, 132–135, 138, 145, 495
Radiation, 4, 26, 28, 45, 59, 70, 80–81, 84, 116, 123, 129–135, 137–139, 141, 143–155, 157, 162, 184, 264, 391, 453, 471, 481, 493, 495, 501, 536, 643
Radiation between two large planes, 148
Radiation heat transfer, 148, 150–152
Radiation heat transfer coefficient, 148, 150–152
Rankine cycle, 477–479
Rayleigh number, 109–112, 115–117, 668
Real gas, 19
Receiver, 3, 26, 28, 39–40, 166, 510, 548
Reflectivity, 138–139, 518
Refractory, 50, 471, 515–516, 521, 528, 541
Refrigeration, 109, 441, 515, 529–530, 533–539, 541, 543–544
Relative roughness, 32, 92, 688
Removable-bundle exchanger, 299
Resistance, 11, 43–44, 46, 50, 57–60, 80–83, 86–87, 96, 184, 187–189, 191, 193–197, 202–204, 206, 225, 230, 233, 257, 260–263, 266, 271, 309, 328, 338, 349, 351, 355, 381–382, 386–387, 391, 401–402, 405–407, 409–411, 413–414, 434, 438, 446–448, 454, 461, 463–464, 475, 484, 498, 500, 502, 510, 516–521, 528, 530, 532–533, 542, 559, 566, 581–583, 591, 594
Resistances in series, 46, 50, 57, 195
Resource conservation and recovery act (rcra), 620
Reynolds number, 9, 14–15, 32, 89, 92–93, 96, 98, 100–102, 116, 226, 230–231, 238, 252, 265, 267, 270, 278, 309, 311, 318, 320, 342, 346, 354, 361, 365, 369, 421, 428, 653, 688
Risk management program (rmp), 621, 623, 629
Risk-based decision making, 630

Saturation curve, 444, 461
Scfm, 18, 165–166
Segmental baffle, 296, 305, 317, 366
Self-diffusion, 501
Sensible enthalpy, 23, 165, 456
Sensitivity analysis, 576
Shell, 163, 216–217, 220, 254, 262, 289–293, 295–307, 309–337, 339–341, 343–345, 347, 349–353, 355, 357–361, 363, 365–369, 371, 373–377, 379–380, 382, 384, 388–389, 419, 426–427, 429, 434, 449, 451, 471, 482, 495, 500, 532, 555, 562–567, 572, 678–682
Shell-and-tube heat exchanger, 163, 216, 289–291, 293, 295, 297, 299, 301, 303–307, 309, 311, 313–317, 319, 321, 323, 325–331, 333, 335–337, 339, 341, 343, 345, 347, 349, 351, 353, 355, 357, 359, 361, 363, 365, 367, 369, 371, 373, 375, 377, 379–380, 572
Shell-side equivalent diameter, 316, 318, 324
Shell side film coefficients, 316–317
Shell-side mass velocity, 316–318
Shell-side pressure drop, 316, 320, 349, 366, 375
Shutdown, 205, 562–565
Sieder-tate equation, 228, 310
Solution exchangers, 350, 357
Specific gravity, 10–11, 268, 320–321, 359, 365, 373, 625
Specific heat, 12, 358
Specular reflection, 139
Stanton number, 9, 89, 93, 99–100, 228

Start-up, 554, 562–564, 568, 603, 605

Stationary tube-sheet exchangers, 292, 295

Steady-state conduction, 46, 63–64

Steam-generating boilers, 470

Sublimation, 25, 440, 445

Superfund amendments and reauthorization act (sara), 620

Surface condensation, 446–447

Surface condenser, 449, 463

Surface tension, 11–12, 441, 452–453, 671

Sustainability, 35, 37–40, 212, 216, 460, 464, 480–481, 483, 491, 651

Synergistic effect, 609

Target organ, 609, 626

Temperature approach, 169–170, 179, 183, 198, 223, 324

Temperature difference driving force, 44, 58, 161–162, 169, 171–172, 177, 179, 214, 223, 232, 283, 308, 313, 323, 325–326, 418, 452, 483, 594–595, 602

Temperature gradient, 27, 48–49, 56, 63, 100, 109, 121–122, 391, 498, 611

Testing, 562, 565–566, 635

Thermal diffusivity, 8, 14, 48, 64, 100, 105, 110, 138

Thermal equilibrium, 13, 139–140

Toxicity, 617, 629–632, 634–635

Transient, 68–71, 73, 554, 560

Transmissivity, 138–139

Transverse fins, 383–385, 388–389, 396, 400

Trombone heat exchanger, 238–239

True temperature difference, 207, 323–324, 340–341, 365, 376, 379

Tube pitch, 292, 294, 304, 318–319

Tube-side pressure drop, 306, 316, 321, 340, 375

Tubular element, 292

Tubular exchanger manufacturers association (tema), 290

Turbulent flow, 16, 31–32, 34, 89–90, 92, 95, 99, 228–229, 231, 238, 252, 265–267, 276, 304, 310, 342

Turbulent flow in rough pipes, 92

Turbulent flow through a circular conduit, 31

Two phase flow, 32

Unit resistance, 80–81

Unsteady-state heat conduction, 43, 45, 47, 49, 51, 53, 55, 57, 59, 61, 63, 65, 67–71, 73, 75, 77

Valves, 33, 563, 617

Vapor pressure, 443, 445, 454, 459, 501, 507–508, 625

Variance, 441

View factors, 132, 153–156

Viscosity, 11, 14, 97, 100, 110, 207, 226–229, 264, 274–277, 285–286, 306, 309–311, 358, 362, 365, 373, 377–378, 439, 455, 484, 537, 547, 656, 667–673, 684

Viscosity correction, 274–276

Volumetric equivalent diameter, 396

Waste heat boilers, 439, 470, 480–481, 483–484, 487, 493, 511

Weber, 7

Wetted perimeter, 34, 222, 230–231, 319, 366, 371, 420, 428

Wilson's method, 254, 259, 261–262

Printed in the USA
CPSIA information can be obtained
at www.ICGtesting.com
LVHW012357020923
756111LV00001BA/2